DNA Methylation Handbook

DNA Methylation Handbook

Edited by **Billy Malcolm**

R CALLISTO REFERENCE

New York

Published by Callisto Reference,
106 Park Avenue, Suite 200,
New York, NY 10016, USA
www.callistoreference.com

DNA Methylation Handbook
Edited by Billy Malcolm

International Standard Book Number: 978-1-63239-148-3 (Hardback)

Printed in the United States of America.

Contents

Preface   VII

Part 1   **Epigenetics Technology and Bioinformatics**   1

Chapter 1   **DNA Methylation Profiling
from High-Throughput Sequencing Data**   3
Michael Hackenberg, Guillermo Barturen
and José L. Oliver

Chapter 2   **Modelling DNA Methylation Dynamics**   29
Karthika Raghavan and Heather J. Ruskin

Chapter 3   **Inheritance of DNA Methylation in Plant Genome**   55
Tomoko Takamiya, Saeko Hosobuchi,
Kaliyamoorthy Seetharam, Yasufumi Murakami
and Hisato Okuizumi

Chapter 4   **GC₃ Biology in Eukaryotes and Prokaryotes**   79
Eran Elhaik and Tatiana Tatarinova

Chapter 5   **MethylMeter®: A Quantitative,
Sensitive, and Bisulfite-Free Method
for Analysis of DNA Methylation**   93
David R. McCarthy, Philip D. Cotter, and Michelle M. Hanna

Part 2   **Human and Animal Health**   117

Chapter 6   **DNA Methylation and Trinucleotide Repeat Expansion Diseases**   119
Mark A. Pook

Chapter 7   **DNA Methylation in Mammalian and Non-Mammalian Organisms**   135
Michael Moffat, James P. Reddington,
Sari Pennings and Richard R. Meehan

Chapter 8   **Aberrant DNA Methylation of Imprinted Loci
in Male and Female Germ Cells of Infertile Couples**   167
Takahiro Arima, Hiroaki Okae, Hitoshi Hiura,
Naoko Miyauchi, Fumi Sato, Akiko Sato
and Chika Hayashi

Chapter 9   **Could Tissue-Specific Genes be Silenced in Cattle
Carrying the Rob(1;29) Robertsonian Translocation?**   177
Alicia Postiglioni, Rody Artigas, Andrés Iriarte,
Wanda Iriarte, Nicolás Grasso and Gonzalo Rincón

Chapter 10   **Epigenetic Defects Related Reproductive Technologies:
Large Offspring Syndrome (LOS)**   193
Makoto Nagai, Makiko Meguro-Horike
and Shin-ichi Horike

Part 3   **Methylation Changes and Cancer**   209

Chapter 11   **Investigating the Role DNA Methylations Plays
in Developing Hepatocellular Carcinoma Associated
with Tyrosinemia Type 1 Using the Comet Assay**   211
Johannes F. Wentzel and Pieter J. Pretorius

Chapter 12   **DNA Methylation and Histone Deacetylation:
Interplay and Combined Therapy in Cancer**   227
Yi Qiu, Daniel Shabashvili, Xuehui Li, Priya K. Gopalan,
Min Chen and Maria Zajac-Kaye

Chapter 13   **DNA Methylation in Acute Leukemia**   289
Kristen H. Taylor and Michael X. Wang

Chapter 14   **The Importance of Aberrant DNA Methylation in Cancer**   317
Koraljka Gall Trošelj, Renata
Novak Kujundžić and Ivana Grbeša

Chapter 15   **Effects of Dietary Nutrients on DNA Methylation and Imprinting**   345
Ali A. Alshatwi and Gowhar Shafi

Chapter 16   **Epigenetic Alteration of Receptor Tyrosine Kinases in Cancer**   359
Anica Dricu, Stefana Oana Purcaru, Raluca Budiu,
Roxana Ola, Daniela Elise Tache, Anda Vlad

Permissions

List of Contributors

Preface

This book aims to educate readers about the concepts of DNA methylation. Epigenetics is one of the most thrilling and speedily increasing areas of modern genetics with functions in many disciplines from medication to cultivation. The most frequent form of epigenetic alteration is DNA methylation, which plays a key role in primary developmental procedures such as embryogenesis and also in the response of organisms to a broad variety of ecological stimuli. Certainly epigenetics is being fast regarded as one of the main mechanisms used by animals and plants to alter their genome and its appearance to adapt to a diverse number of ecological factors. This book is a compilation of research into DNA methylation and highlights recent advances in methodology and information of underlying mechanisms of this most significant of genetic procedures.

All of the data presented henceforth, was collaborated in the wake of recent advancements in the field. The aim of this book is to present the diversified developments from across the globe in a comprehensible manner. The opinions expressed in each chapter belong solely to the contributing authors. Their interpretations of the topics are the integral part of this book, which I have carefully compiled for a better understanding of the readers.

At the end, I would like to thank all those who dedicated their time and efforts for the successful completion of this book. I also wish to convey my gratitude towards my friends and family who supported me at every step.

Editor

Part 1

Epigenetics Technology and Bioinformatics

DNA Methylation Profiling from High-Throughput Sequencing Data

Michael Hackenberg, Guillermo Barturen and José L. Oliver

Dpto. de Genética, Facultad de Ciencias,
Universidad de Granada,Granada,
Lab. de Bioinformática, Inst. de Biotecnología,
Centro de Investigación Biomédica, Granada,
Spain

1. Introduction

The methylation of a cytosine at the carbon 5 position (5meC) is a common epigenetic mark in eukaryotic cells that is normally found in CpG and CpHpG (H=A,C,T) sequence contexts. The inactivation of one of the X-chromosomes in female cells, the allele-specific expression of imprinted genes or the role in the embryonic development which is shown by the early lethality of Dnmt3a and Dnmt3b deficient mice are only some of the important functions that DNA methylation plays (Chen and Li, 2004; Dodge, et al., 2005; Karpf and Matsui, 2005; Okano, et al., 1999). The methylation of the gene promoter region is commonly associated with silenced transcription; however recently it was shown that the DNA methylation in the gene body of transcribed genes is increased in both animals and plants (Hellman and Chess, 2007; Jones, 1999; Zhang, et al., 2006). Furthermore, CHG methylation relates to the silencing of transposons in plants (Miura, et al., 2009) . Given all these facts, it is clear that the DNA methylation pattern along the genome sequence carries valuable biological information and is crucial for our understanding on gene expression and developmental control. Furthermore, it can reveal how aberrant epigenetic changes might lead to dysregulation of gene expression and to the development of diseases such as cancer (Costello, et al., 2000).

A broad panoply of techniques to detect DNA methylation has been developed. The DNA methylation is erased by PCR and not detected by hybridization as the methyl-group is located within the major groove and not at the hydrogen bonds. Therefore, virtually all techniques rely on a methylation dependent pretreatment of the DNA before hybridization, amplification or sequencing. The three main classes of pretreatments are: digestion by methyl-sensitive endonucleases, methyl-sensitive immunoprecipitation and bisulfite conversion (reviewed by (Laird, 2010)). An important impulse for epigenetic research was the adoption of DNA microarrays to methylation profiling (Estecio, et al., 2007). This technique was initially used together with a methyl-sensitive digestion of the DNA, however meanwhile also immunoprecipitation and bisulfite conversion variants do exist (Bibikova and Fan, 2009). Microarrays usually (there are arrays for individual CpGs which cover several thousand sites) allow to obtain information of the "mean" methylation values of a given region, however the methylation pattern cannot be revealed at a single base pair resolution. The generation of whole genome, single-base-pair resolution methylation maps became feasible just recently with the advent of Next-Generation Sequencing (NGS) or High-Throughput Sequencing (HTS) methods like those from Illumina, Roche 454 and Solid (applied biosystems) to mention just the 3 with the highest diffusion (Shendure and Ji, 2008). These techniques are frequently termed whole-genome shotgun bisulfite sequencing and have been applied already in several methylome projects (Bock, et al., 2011; Laurent, et al., 2010; Lister, et al., 2009). The denatured genomic DNA is treated with sodium bisulfite which leads to the deamination of cytosines, preserving however methylcytosines. After sequencing the treated DNA, the methylation state of the individual cytosines can be profiled directly from the aligned sequence reads: an unconverted cytosine indicates methylation while a thymine instead of a cytosine will reveal an unmethylation. In order to obtain sufficient coverage, the genome is (re)sequenced at a typical coverage of 15x which obviously implies a notable bioinformatics challenge. Limitations and main bias effects of genome-wide DNA methylation technologies, as well as the factors involved in getting an unbiased view of a methylome have been recently reviewed (Robinson, et al., 2010).

In this chapter we will review the common steps in the analysis of whole genome single-base-pair resolution methylation data including the pre-processing of the reads, the alignment and the read out of the methylation information of individual cytosines. We will specially focus on the possible error sources which need to be taken into account in order to generate high quality methylation maps. Several tools have been already developed to convert the sequencing data into knowledge about the methylation levels. We will review the most used tools discussing both technical aspects like user-friendliness and speed, but also biologically relevant questions as the quality control. For one of these tools, NGSmethPipe, we will give a step by step tutorial including installation and methylation profiling for different data types and species. We will conclude the chapter with a brief discussion of NGSmethDB, a database for the storage of single-base resolution methylation maps that can be used to further analyze the obtained methylation maps.

2. Analysis workflow

A general analysis workflow to convert the sequencing data into methylation maps can be divided into 3 parts: (i) pre-processing of the reads, (ii) alignment and (iii) the profiling of the methylation states from the alignments. In all three steps several error controls can be applied which are therefore discussed in a separate section below. Some of these steps are shared by virtually all of the so far published tools, others however are unique to a single or few applications. In this section we will give the theoretical background and in the next section we will compare the different implementations into the software tools.

2.1 Preprocessing of the reads

The pre-processing of the reads can be grouped into: (i) elimination or manipulation of low quality reads, and (ii) preparation of the reads for the alignment step. It is known that Illumina sequence reads loose quality towards the 3' end. In order to only use the high quality part of the read, Lister *et al* (Lister, et al., 2009) proposed to trim the read to before the first occurrence of a low quality base call (PHRED score <= 2). Another step which might increase the alignment accuracy is the removal of the adapter sequences. If the DNA fragment is shorter than the read length, parts of the 3' adapter will also be sequenced. The adapter sequence will however not align to the genome which might lead to missed or incorrect mapping of the read. Depending on the alignment algorithm the adapter removal is a mandatory step before mapping the reads to a reference genome. For example, some programs perform a seed alignment, i.e. not the whole read is aligned but only a given subsequence of the 5' part of the read: the seed. If the adapter does not extend into the seed, those reads can be aligned even without removing the adapter. However, there are also methods that align the whole read. Those algorithms will very likely fail to map reads that contain adapter sequences (see section 3 for more details).

Like mentioned before, and explained in detail within the next section, many methods are based on alignments using a 3-letter alphabet. Those programs need to manipulate the reads before the alignment step replacing all remaining cytosines (those that are not converted by the bisulfite) by a thymine (see Figure 1).

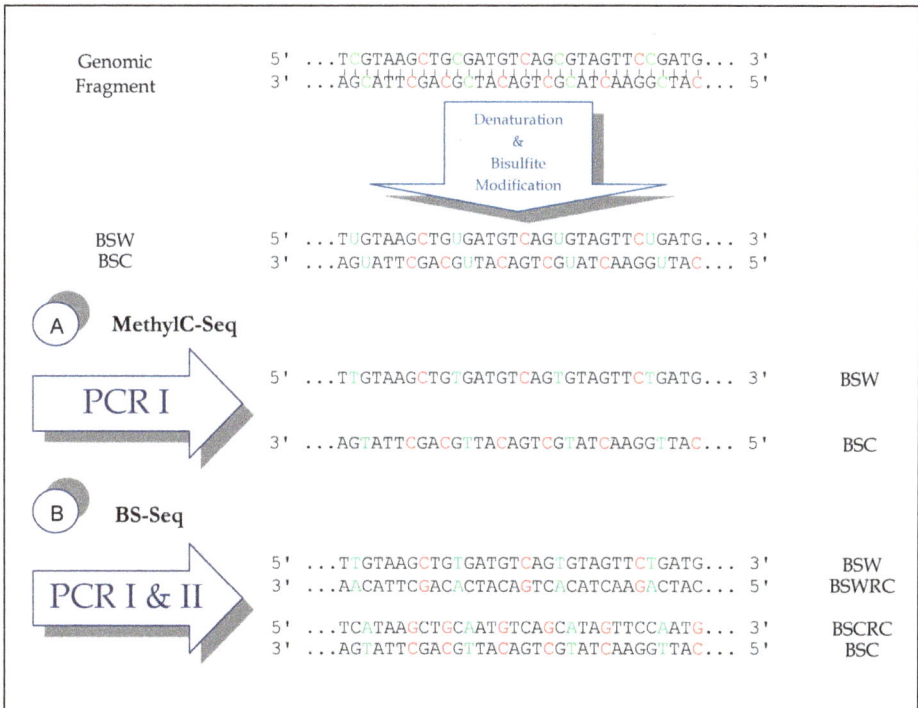

Fig. 1. Whole genome bisulfite sequencing: MethylC-Seq and BS-seq. After denaturing and bisulfite treatment, the genome DNA will lose the strand complementarity, as unmethylated cytosines are converted to uracils (green coloured cytosines). During the PCR, the uracils will be substituted by thymines. Here we show an illustration for the reads from MethyC-Seq (directional) and BS-Seq (non-directional). A) The MethylC-Seq protocol generates the library in a directional manner, resulting in either BSW (Bisulfite Watson) or BSC (Bisulfite Crick) reads. B) The BS-Seq protocol performs two consecutive PCRs which yields BSW and BSC reads, as well as their reverse complementary strands (BSWRC and BSCRC).

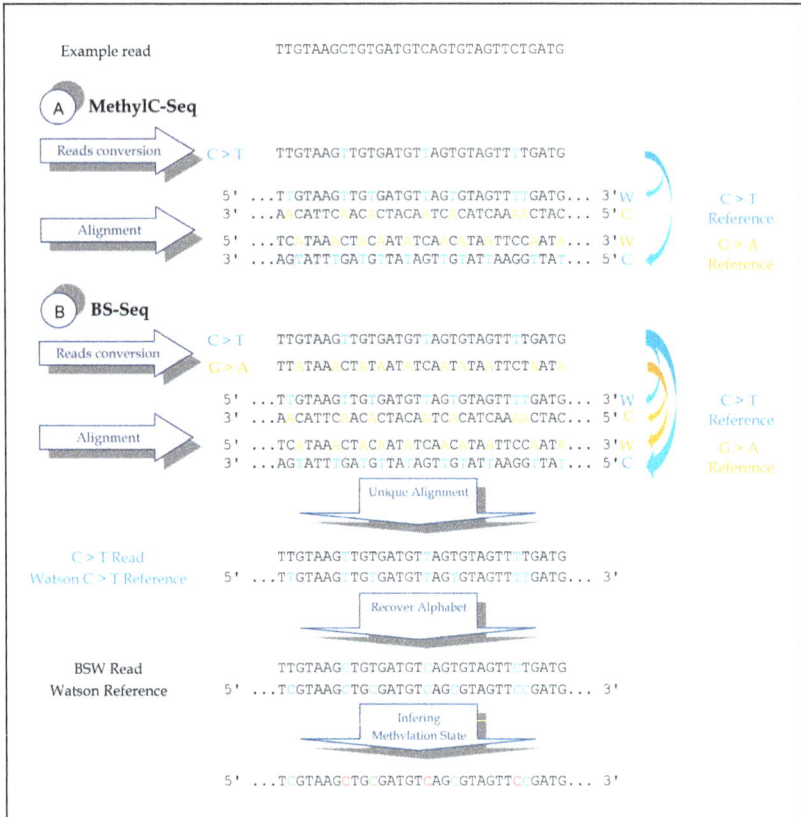

Fig. 2. Methylation profiling from bisulfite-treated reads. In order to retrieve the methylation information, the bisulfite treated reads must be first mapped to the reference genome. During the experimental protocol the unmethylated cytosines are converted to thymines. One way to deal with this reduced sequence complexity is to convert both, the reads and the reference sequence into a 3-letter alphabet. In the MethylC-Seq protocol (A), the reads come from the Watson (BSW) and Crick (BSC) bisulfite treated DNA fragments. Therefore, all cytosines of the input reads will be converted to thymines and will be tried to align to the C to T converted reference (blue arrows). For the BS-Seq protocol (B), the reads can map to the Watson and Crick C to T converted reference (reads BSW and BSC), as well as to the Watson and Crick G to A converted reference (reads BSWRC and BSCRC). Therefore, the reads must be converted into two different alphabets: C to T changed reads that will be mapped to the C to T converted reference (blue arrows) and G to A changed reads that will be mapped to the G to A converted reference (yellow arrows). If a best unique alignment exists for a read, both, the original read and reference sequences are recovered which allows directly to read out the methylation information: a C/T mismatch indicates an unmethylated cytosine (green colored), a cytosine in both the read and the reference indicates methylation (red colored). In case of reverse complement reads (G to A changed), a guanine in the reference and an adenine in the read indicates unmethylation and a guanine match between the read and reference sequence allows to infer methylation.

2.2 Alignment

After the treatment of the genomic DNA with sodium bisulfite, the sequence complexity is reduced as all cytosines with the exception of methylcytosines are converted into thymine. In mammal genomes methylation normally occurs exclusively in a CpG sequence context which only accounts for approx. 1.8% of all dinucleotides and therefore the majority of all cytosines is converted to T. This reduced complexity complicates the alignment process, i.e. the detection of the genome loci where the read originates from. A common philosophy in all developed programs and protocols is to just accept unique alignments, i.e. just one genome position exists to which the read aligns under a given set of parameters. In many cases it might occur that one and the same read has a unique alignment in a 4-letter presentation but exhibits mappings to several positions with the same quality in a 3-letter alphabet. In such cases, the information carried by the read is lost. The existence of sequencing errors complicates the correct read mapping further. An additional challenge in bisulfite read mapping is the increased search space. This is because the Watson and Crick strands of bisulfite treated DNA sequences are not complementary to each other as the bisulfite just acts on cytosines. As a consequence, both bisulfite Watson (BSW) and Crick (BSC) have their own reverse complementary strands, BSWRC and BSCRC. Frequently used experimental protocols are MethylC-Seq (Lister, et al., 2008), RRBS (Reduced Representation Bisulfite Sequencing, (Meissner, et al., 2005)) and BS-seq (Cokus, et al., 2008) (reviewed by ((Lister and Ecker, 2009))). Without going into the experimental details, the relevant difference is that in the BS-seq protocol two subsequent PCRs are performed which leads to the generation of four read types, BSW, BSWRC, BSC and BSCRC. On contrary, MethylC-Seq only generates BSW and BSC reads (see Figure 1). This difference has immediate consequences on the bioinformatics analysis (see figure 2).

After the bisulfite treatment, the reads cannot be aligned simply to the reference sequence as converted cytosines would lead to mismatches. In theory, it would be possible to increase the number of allowed mismatches in order to align the reads to the genome, however this would entail some serious drawbacks: (i) a high number of allowed mismatches will make the alignments less specific, i.e. a higher number of incorrect mappings will be the consequence leading to the incorrect inference of the methylation states, (ii) it would be virtually impossible to profile CpG dense regions (CpG islands) as those are frequently un-methylated presenting a high number of converted cytosines. Another possibility would be to generate for each read all combinations of possible T/C conversions which seems however computationally too demanding.

Given that we can rule out the increase of the allowed mismatches as a solution, two general approaches have been proposed to align bisulfite treated sequence reads. First, an alignment matrix can be used which gives the same weights to all matches and C/T mismatches (cytosine in the reference sequence and T in the read) and second, a three letter alphabet can be used to adapt the reference sequence to the reduced sequence complexity of bisulfite treated reads (see Figure 2 for an illustration). In theory, the first method could be slightly more accurate due to a higher sequence complexity, however it seems also clearly slower (see section 3.3). In the 3-letter alphabet approach, as a direct consequence of the lost of complementarity of bisulfite treated DNA, two different reference genomes must be prepared: 1) substituting all Cs by Ts and 2) substituting all Gs by As. Depending on the concrete experimental protocol, the reads need to be aligned only against the Watson strand from the C/T reference and the Crick strand from the G/A reference (MethylC-Seq) or against both strands of the two reference genomes.

2.3 Post-processing and output

Once the reads are aligned to the reference genome, the methylation information of the individual cytosines can be read out. In order to do so, both, the reads and the reference sequences need to be converted back to a 4-letter alphabet (see Figure 2). Methylcytosines are then indicated by C/C matches while unmethylated cytosines are given by a T/C mismatch in the alignment (also G/G and G/A in the case of BS-Seq). The methylation level of a given cytosine position in the genome is determined by the information of all reads that overlap this position. The methylation level of a cytosine is given simply by the number of methylcytosines divided by the total number of reads that map to the position. In this way, the methylation level is a number between 0 (completely unmethylated) and 1 (completely methylated). There are at least two reasons which can lead to intermediate methylation levels: (i) usually a cell population is used to extract DNA and intermediate values can indicate fluctuations at a given position between the individual cells and (ii) allele specific methylation is a well known phenomenon in imprinting. Bisulfite treatment together with sequencing has the advantage that the methylation level of each individual cytosine can be assessed. That means that, unlike in many other techniques not only the methylation levels of CpGs can be determined but also other sequence contexts like CHG or CHH. These sequence contexts will be particularly interesting in plants or embryonic stem cells (Lister, et al., 2009).

2.4 Quality control

There are several error sources which can compromise the quality of the methylation maps which ideally should be taken into account. This error sources include (i) wrong alignment of the reads, (ii) existence of Single Nucleotide Variants (SNV), (iii) sequencing errors, (iv) bisulfite failure.

2.4.1 Incorrect alignments

The reduced sequence complexity of bisulfite treated reads and the existence of sequencing errors can lead to wrong alignments, i.e. the read is aligned to a genome position where it does not originate from. This is particularity true for highly repetitive DNA sequences which are frequently CpG rich (like Alu retro-transposons) and methylated. When single end reads are used in the experimental assay, wrong mappings cannot be detected and the only way to control the number of incorrect alignments is the appropriate choice of the alignment parameters. The search for the best alignment is a highly parameterized task and each alignment algorithm has its own set of parameters depending whether seed alignment is possible, the base call probabilities are considered, etc. This is probably the main reason why no large scale comparison exists in order to fix the best parameters for the different algorithms. Apart from that, it is clear that the number of allowed mismatches and the minimum alignment length (length of the seed alignment in the methods were it applies) are crucial in the mapping accuracy. A higher number of allowed mismatches and a short alignment seed will permit to map more reads to the genome, however, the number of incorrect alignments will also increase (high sensitivity, low specificity). On the other hand, very strict parameters will impede the alignment of many "valid" reads and therefore many genome regions will be failed to be profiled. On contrary, the usage of paired-end or mate reads bears the big advantage that a considerable number of the wrong mappings can be detected and removed. In the paired-end technique, both ends of the DNA fragment are sequenced. Normally, the approximate fragment length distribution is known and therefore a narrow window on the genome can be established to which both reads must map. In this way, if the two mate reads are independently mapped to the genome and the best alignment is on different chromosomes or distanced far away on the same chromosome, these mappings can be eliminated as at least one of the two alignments will be incorrect.

2.4.2 Detection of SNVs

One frequently occurring type of variation are the Single Nucleotide Variants (SNVs) which are variations in just one nucleotide between the reference sequence and the sequenced genome. Many of these SNVs might be SNPs as their frequency is higher than 1% in the population. Nevertheless, we will call them SNVs as this is a more general and population genetics independent concept. Over two third of all SNPs are known to occur at the cytosine in a CpG sequence context (Tomso and Bell, 2003). These SNPs have usually two alleles, C and T. Given this high number of C/T SNPs, it can be supposed that the percentage of unknown C/T SNPs and C/T SNVs will be very similar. In the case of a C/T variation, normally the sequenced genome carries a thymine while in the reference genome a cytosine is annotated. If the presence of this sequence variant is unknown or ignored, the inference would be that the cytosine annotated in the reference genome is unmethylated. The correct conclusion however would have been that no cytosine exists in the genome, and therefore no methylation state can be detected. One possibility to take into account the existence of variation is to query a SNP database eliminating all positions with C/T alleles. The disadvantage is that also valuable information is lost when the sequenced genome carries the same allele as the reference genome. Another possibility is to detect the SNVs directly by means of the sequencing data which is possible for positions with a sufficiently high coverage. A C/T variation would manifest on the complementary DNA strand as an adenine, while bisulfite deamination would not affect the guanine on the complementary strand as the bisulfite is applied to denatured DNA (Weisenberger, et al., 2005). Currently, the detection of SNVs is exclusively implemented in the NGSmethPipe program while all other algorithms would interpret such C/T SNVs erroneously as unmethylated cytosines.

2.4.3 Bisulfite failure

Incomplete bisulfite conversion can be caused either by incomplete denaturing before applying the bisulfite treatment or reannealing during the bisulfite conversion. In any case, if the bisulfite has not acted the cytosines would remain unconverted within the read independently of its methylation state. If such reads are not detected, all cytosines would be inferred to be methylated. Lister et al. proposed to use non-CpG contexts to detect reads that are likely not bisulfite converted. This protocol proposes to discard those reads with more than 3 methylated cytosines in a non-CpG context. This measure should work fine in organisms or cell types were non-CpG methylation is virtually absent, however it might discard many valuable information when real non-CpG methylation exists like in embryonic stem cells.

2.4.4 Sequencing errors

Another source of incorrect profiling of the methylation state is the erroneously calling of a thymine instead of a cytosine. Such sequencing errors would be incorrectly interpreted as an unmethylated cytosine. Each base of a read has assigned a sequencing quality in form of a Phred Score which can be interpreted as the probability of the base to be incorrectly called. Therefore, in theory a probabilistic approach could be applied in order to control for the incorrect profiling due to sequencing errors (see section 3.6.2).

3. Bioinformatics tools

Several sofware applications and protocols have been developed so far. In this section we will discuss 9 tools: mrsFAST (Hach, et al., 2010), RMAP (Smith, et al., 2009), SOCS-B (Ondov, et al., 2010), BS-seeker (Chen, et al., 2010), BSMAP (Xi and Li, 2009), BRAT (Harris, et al., 2010), MethylCoder (Pedersen, et al., 2011) , Bismark (Krueger and Andrews, 2011) and NGSmethPipe (Barturen, et al., Submitted)). In table 1, the availability and some basic features are displayed.

Basically, we can distinguish two types of tools: (i) bisulfite alignment tools that perform the pre-processing and alignment but do not report methylation levels and (ii) full pipeline tools that perform all necessary steps from the pre-processing over the alignment to the methylation profiling and error control. We will concentrate the discussion mainly on the full pipeline tools as those will be the choice of many users with little or no bioinformatics background, mentioning the bisulfite aligners for advanced users. Apart from the available software packages, several protocols have been used and proposed (Bock, et al., 2010; Cokus, et al., 2008; Gu, et al., 2010; Harris, et al., 2010; Lister, et al., 2008; Lister, et al., 2009). We will mention those protocols whenever any of the implemented analysis steps have been proposed before in any of these works. Note finally, that several other programs like Methyl-Analyzer (Xin, et al., 2011) have been developed to generate methylation maps at a single-base-pair resolution for non-bisulfite data. Many of the analysis steps are shared by the tools; however, the concrete implementation might vary if other parameter sets are applied. Therefore, we will not discuss each tool within a separate section but analyze the differences directly within the section on the different analysis steps.

3.1 Implementation

The programming language is highly related to many features as the installation process and the speed. Most of the programs discussed here are implemented in C(++) and need to be compiled locally which might require the help of a system administrator in order to install the program. On the other hand, all tools that rely on an external alignment program are based on an interpreted scripting language like Perl (Bismark, NGSmethPipe) or Python (BS-seeker, MethylCoder, Methyl-Analyzer) which are normally installed on a standard Linux distribution. This implies a relatively easy installation or set-up process. Some of these tools rely on additional Perl or Python modules which can be easily installed from the command line. Therefore, the installation process is very similar for all of the Perl/Python based tools. Roughly, the set-up includes the following steps: (i) download the perl or python scripts to a local directory, (ii) "install" the alignment program (for example, Bowtie is delivered as binary files which just need to be downloaded), (iii) install additional

Program	Availability	Language	Sequence space / Color space	Multi-threads	Scope
mrsFAST	http://mrsfast.sourceforge.net/	C	yes/no	No	BS align
RMAP	http://www.cmb.usc.edu/peopl e/andrewds/rmap/	C++	yes/no	No	BS align
SOCS-B	http://solidsoftwaretools.com/gf /project/socs/	C++	no/yes	Yes	BS align
BS Seeker	http://pellegrini.mcdb.ucla.edu/ BS_Seeker/BS_Seeker.html	Python	yes/no	yes*	BS align
BSMAP	http://code.google.com/p/bsmap/	C++	yes/no	Yes	BS align
BRAT	http://compbio.cs.ucr.edu/brat/	C++	yes/no	No	Full
MethylCoder	https://github.com/brentp	Python/C	yes/yes	No	Full
Bismark	http://www.bioinformatics.bbsrc. ac.uk/projects/bismark/	Perl	yes/no	yes*	Full
NGSmethPi-pe	http://bioinfo2.ugr.es/NGSmeth Pipe	Perl	yes/no	Yes	Full

Table 1. The availability and basic features of the different software tools. The asterisk indicates those programs that have multi-threading in the alignment process through Bowtie, but all pre and post-processing steps are single-threaded. The column scope refers to whether the program reports the methylation levels of the individual cytosines or if only the bisulfite alignment is performed leaving the read out of the methylation information to the user.

modules if applies and (iv) prepare the reference genomes. For the last point, all programs provide scripts that take a multifasta file or a directory name as input yielding the two 3-letter reference genomes as output. Currently, all programs are tested only on Unix, Linux and/or Mac OSX platforms and no Windows support is available.

3.2 Input data and scope

The programs differ quite notably in the accepted input data and the number of implemented features (see table 1-3). Currently just one program, Methyl-Coder can be used for both, sequence space (Illumina, Roche 454) and color space input (SOLiD). With the exception of SOCS-B (SOLiD), all other tools presented here can only handle sequence space input. Another important difference is the availability to process the data from the different library preparation protocols (BS-seq, MethylC-Seq) and single/paired end reads. While all tools implement the MethylC-Seq protocol for single reads, mrsFAST, RMAP and Methyl-Coder do not implement BS-Seq. Methyl-Coder supplies a script that allows to convert BS-seq data into MethylC-Seq ("tagged_reads_prep.py"). Paired end support for directional reads is currently available by all tools with the exception of RMAP, SOCS-B and BS-Seeker while paired-end support for non-directional reads is only available in Bismark.

Program	Aligner	Method	Seed	Q	Single (non-dir)	PE (dir)	PE (non-dir)	MP	strand merge
mrsFAST	mrsFAST	3-letter	Yes	No	No	Yes	No	No	No
RMAP	RMAP	Matrix	Yes	Yes	No	No	No	No	No
SOCS-B	SOCS	Matrix	No	Yes	Yes	No	No	No	No
BS Seeker	Bowtie	3-letter	No	No	Yes	No	No	No	No
BSMAP	SOAP	4-letter	Yes	No	Yes	Yes	No	No	No
BRAT	BRAT	2-letter	Yes	No	Yes	Yes	No	Yes	No
Methyl-Coder	bowtie/ GSNAP	aligner dependent	Yes/ Yes	Yes/ No	No	Yes/ Yes	No	Yes	No
Bismark	Bowtie	3-letter	Yes	Yes	Yes	Yes	Yes	Yes	Yes
NGSmethPipe	Bowtie	3-letter	Yes	Yes	Yes	Yes	No	Yes	Yes

Table 2. Alignment method and output features. The 'Seed' column indicates whether a seed alignment methods is used, the Q column indicates whether the Phred Score base call quality values are used in the alignment process. The column PE (directional) and PE (non-directional) refer to the capability to perform paired end alignments for the two main sequencing protocols: MethylC-Seq (directional) and BS-Seq (non-directional). The columns 'MP' and 'strand merge' refer to the output options indicating whether the methylation profiling (MP) step is performed and if a strand merged output for palindromic sequences is possible. Note that, all programs that perform methylation profiling can report the methylation values for different sequence contexts (CpG, CHG, CHH).

3.3 Alignment

A huge number of short read alignment programs and methods have been developed so far (review by (Horner, et al., 2010)). Some of the tools discussed here implement its own alignment algorithm like mrsFast, BRAT or RMAP while others are based on an external aligner. Those that are based on an external aligner implement the conversion into a 3-letter alphabet as the mapping to a bisulfite treated 4-letter alphabet is only possible when manipulating internal parameters which are not accessible. In theory, every alignment method could be used to map the bisulfite treated reads to the 3-letter reference sequences. Currently, one of the most successful programs is Bowtie (Langmead, et al., 2009) which is used by Bismark, BS-seeker and NGSmethPipe. Methyl-Coder allows to chose between Bowtie and GNASP (Pedersen, et al., 2011). Although most full pipeline programs are based on Bowtie alignments, this does not at all imply that the programs produce identical or even similar results. There are basically two reasons: First, the alignment of short reads is a highly parameterized process and each program uses a different set of default parameters which in some cases cannot be changed and second, the programs differ greatly in the amount of implemented quality control features. The alignment parameters are crucial in order to control for the number of wrong alignments which is especially true for single end reads. Like mentioned before, relaxed parameters will lead to a high coverage (many cytosines can be profiled) however, a high number of incorrect alignments can also be expected leading to wrong methylation profiling. On the contrary, strict parameters might lead to a low coverage discarding a considerable amount of valuable information. Despite of this importance, the tools usually recommend some default parameters, however without basing them on a strict analysis. The first study trying to fix the best parameters for single end reads was carried out by (Barturen, et al., Submitted). These authors, in order to detect the best parameter set for the NGSmethPipe tool, used a golden standard generated by paired end reads to measure the alignment accuracy as a function of the seed length and number of mismatches.

Some alignment programs use the Phred Scores that are assigned to each base (see table 2) to improve the mapping accuracy. The reasoning is the following: mismatches of low quality bases are more likely due to sequencing errors while mismatches of high quality bases are more likely due to incorrect alignments. To take this fact into account, Bowtie calculates the sum of the Phred scores of all mismatches discarding an alignment if this sum (e value) is higher than a given threshold (70 by default, which corresponds to more than 2 high quality mismatches).

Finally NGSmethPipe performs a different 'best alignment' detection compared to other programs. The method is similar to the one used in the miRanalyzer tool (Hackenberg, et al., 2011) and functions as follows: (i) using Bowtie, a seed alignment (40 bp by default) with a given number of allowed mismatches in the seed (1 MM by default) is performed and the N best alignments (N depends on the number of possible read orientations, obtaining for each possible orientation up to 5 alignments) are obtained (--best --strata), (ii) the maximal N alignments are extended until the next mismatch occurs, (iii) if only one unique longest alignment exists (non-ambiguous mapping) after the extension step, this alignment is retained and extended as long as the 'global' number of allowed mismatches is not exceeded. Usually, just unique seed alignments are used for the methylation profiling and the seed alignment extension method proposed here has the advantage that it can disambiguate some read mappings leading to a higher number of mapped reads without compromising the quality.

3.4 Speed

Given that a resequencing experiment can easily produce up to 3.000 million sequence reads of length that currently vary between 36 - 100 nt, it is clear that the alignment speed is an important issue. Most of the CPU is consumed by the alignment process and by some specific output options like the strand joining which is however not implemented by all programs. A sound comparison of the speed performance for all 9 tools is currently impossible. The reason is that a comparison is only meaningful if it was performed on the same platform and CPU configuration which was not carried out for all these tools together. The Bismark authors for example compared their tool with the performance of BS Seeker on a set of reads containing approx. 15 million reads taken from SRR020138 (Lister, et al., 2009). The number of aligned reads and CPU time is 9,633,448 (64.2%)/42 min and 9,664,184 (64.4%)/29 min for Bismark and BS Seeker respectively. The BS Seeker authors (Chen, et al., 2010) compared the speed of their tool to RMAP showing that BS Seeker is over 10 times faster. Note that, the huge BS Seeker speed increase over RMAP is due to the faster Bowtie alignment algorithm while the slightly higher speed performance compared to Bismark (also based on Bowtie) is due to the different parameter set and choice of unique alignments. Finally, mrsFast was reported to be around twice as fast as Bowtie (Hach, et al., 2010). In summary, no final conclusion on the speed performance can be drawn as every author compared their tools to just one or very few other tools. It seems however, that the Bowtie based full pipeline programs and mrsFast might be the fastest tools currently available.

There are at least two other factors that influence the performance which are the availability of multi-threading (using more than one CPU) and memory issues. Right now, many of the programs discussed here do not support multi-threading. For example, mrsFast performs very well but it does run on only one CPU and it can be therefore easily be outperformed by programs based on Bowtie which supports multi-threading. Currently only NGSmethPipe and BSMAP are completely parallelized while Bismark for example uses the multi-threading capacity of Bowtie but all pre and post-processing steps are only single threaded. Finally, NGSmethPipe allows to adapt the memory usage to the resources of the user (see section 4).

3.5 Output

There are two important output features that a software tool should ideally implement: (i) the methylation levels for each cytosine in a given sequence context for each strand and (ii) an output for palindromic sequence contexts (like CpG) in which the information from both strands is merged together. For example, it has been observed that hemi-methylation (strand specific methylation of a palindromic sequence) is quite uncommon, and therefore the methylation levels from both strands can be merged together assigning the methylation level to the position of the C in the plus strand. Right now, only Bismark and NGSmethPipe implement the strand merge for palindromic sequence contexts. Furthermore, NGSmethPipe gives in the output the difference of the methylation levels between the two strands. In this way, hemi-methylation can be detected easily. Finally, NGSmethPipe is the only program that detects SNVs which are also reported in BED file format.

3.6 Quality control

The most disregarded aspect in the methylation profiling is the strict control of the different error sources. Most of the available programs do not implement quality controls like bisulfite failure check, removal of low quality reads or bases and the detection of sequence variation that can lead to a wrong inference of the methylation state. Currently, NGSmethPipe is the only algorithm that takes into account all these error sources.

In order to detect SNVs, the information from both strands needs to be considered. If NGSmethPipe detects a C/T mismatch after reconverting the alignments to a 4-letter alphabet, the information from the other strand is accessed: if the complementary base is an 'A' this means that the C/T mismatch is caused by an SNV, if on the other strand is a 'G', this means that in the genome we have an C:G pair and that the C/T mismatch is caused by the conversion of an unmethylated cytosine into a thymine. This is the theoretical background, however the experimental data will be influenced by fluctuations and noise, and therefore it is possible to find for one and the same position both, C/C and C/T on the forward strand and A and G on the reverse strand. Therefore, we need to parameterize this model reporting a SNV if a given threshold is surpassed.

NGSmethPipe first checks if at least one read exists that contains a C/T conversion for a given position. For these positions, the program calculates the "SNV-fraction" as the number of reads that have not got a G on the complementary strand (an A, C or T might indicate the existence of a SNV) divided by the total number of reads that map to the position on the complementary strand. We define a C/T conversion as caused by an SNV if the "SNV-fraction" is above a given threshold which is set by default to 0.7 in the tool.

Program	Trim reads	Trim tags	Remove adapter	Clonal reads	Bisulfite failure	Base call errors	SNVs
mrsFAST	no	No	No	No	No	No	No
RMAP	no	No	No	No	No	No	No
SOCS-B	yes	No	No	No	No	No	No
BS Seeker	no	Yes	Yes	No	Yes	No	No
BSMAP	yes	No	Yes	No	No	No	No
BRAT	yes	No	No	Yes	No	No	No
MethylCoder	no	No	No	No	No	No	No
Bismark	no	No	No	No	No	No	No
NGSmeth Pipe	yes	Yes	Yes	No	Yes	Yes	Yes

Table 3. Quality control features. The table indicates if the following quality control features are implemented into the programs: *trim reads*: the low quality end of the input reads is removed; *trim tags*: the tags in the BS-seq protocol are detected and removed, if the tag is not detected the read is trimmed on both ends by the tag size; *remove adapter*: detect the adapter sequence and remove it from the read; *clonal reads:* eliminate duplicated reads; *bisulfite failure:* detect those reads for those the bisulfite didn't acted; *Base call errors:* discard the information for low quality base calls; *SNVs:* detect single nucleotide variation and discard these positions.

3.6.1 SNV detection

Fig. 3. SNV detection. The figure shows an illustration of the SNV detection model and methylation profiling for reads from a MethylC-Seq protocol. Above the reference sequences the reads that map to the Watson strand are shown and below those that map to the Crick strand. Nucleotides colored in red mean potentially methylated positions (C/C matches); the green ones, potentially unmethylated positions (C/T mismatch and G's in the opposite strand) and the yellow one, mismatches to the reference sequence. If a position has both, yellow and green columns it is a good candidate to be detected as a SNVs. Table 4 shows the results inferred from this diagram.

Context	Start	Watson Level	Crick level	Merged Level	Watson SNV fraction	Crick SNV fraction	Result
CG (+/-)	2	0.1	0.2	0.15	0	0.9	rejected
CTT (-)	4	-	0	-	0	-	unMeth
CWG (+/-)	8	0.6	0.9	0.75	0	0	interMeth
CG (+/-)	11	0.2	0.1	0.15	0	0	unMeth
CAT (-)	13	-	0	0	1	-	rejected
CWG (+/-)	17	0.3	0.1	0.2	0.2	0	unMeth
CG (+/-)	20	1	0.9	0.95	0	0	meth
CTA (-)	22	-	0.1	-	0	-	unMeth
CCG (+)	27	0.2	-	-	-	0.2	unMeth
CG (+/-)	28	0.8	0.9	0.85	0.1	0	meth
CAT (-)	30	-	0	-	0.2	0	unMeth

Table 4. Inference from example in figure 3.As it has been explained in section 2.4.2, the existence of SNVs are an important error source in the profiling of methylation values. The table shows, both the methylation profiling and the detection of SNVs from the example in figure 3. Context, refers to the sequence context in the reference genome and in brackets their orientations are given. Start indicates to the coordinate of the first base in the Watson strand. Watson level and Crick level are the methylation levels for the two strands (fraction between methylcytosines and total number of reads). The merged level is the global fraction of methylcytosines for palindromic sequence contexts taking into account the cytosines of both strands. Watson SNV and Crick SNV are the fractions of mismatches (generally A/G mismatches) in the reverse complementary strand position. The last column shows the inferred methylation state (in red we mark the value on which the inference is based). No methylation level is assigned ('reject') if the 'SNV fraction' is above the threshold of 0.7, methylation ('meth') if the level is above 0.8, unmethylation ('unmeth')for levels below 0.2 and intermediate methylation between 0.2 and 0.8.

3.6.2 Sequencing errors

The contribution of the individual bases can be controlled as a function of their base call quality. Each base has assigned a Phred Score which varies between 0-93 in Sanger format and 0-64 in illumina 1.3+ format although in NGS sequencing no higher values than 60 in Sanger format and 40 in illumina 1.3+ format are normally achieved. The Phred Score has a very easy interpretation. For example, a score of 10 indicates a probability of 0.1 that the base call is wrong, while a Phred Score of 20 corresponds to a probability of 0.01 etc. The Phred Score threshold can be seen as an upper limit of wrongly inferred methylation states. For example, when setting Q>=20 (just accept bases with a probability less than 0.01 to be incorrectly called), less than 1% of all inferred methylation states are incorrect.

3.6.3 Bisulfite failures

NGSmethPipe implements the method proposed by Lister *et al.* to detect those reads where the bisulfite probably failed to act. While Lister et al discard all reads with more than 3 methylated cytosines in a non-CpG context, NGSmethPipe allows to set two different thresholds: (i) the absolute number of methylcytosines in a non-CpG context can be set as threshold and (ii) the fraction of methylcytosines and total number of non-CpG cytosines can be used with the advantage that the fraction is independent of the read length. As default parameter, a fraction of 0.9 is used.

3.6.4 Paired end incorrect alignment detection

Like mentioned before, by means of paired end reads the number of incorrect alignments can be lowered drastically. Some of the bisulfite aligners do not support paired end reads, however, all of the full pipeline programs discussed here do (see table 2).

4. Brief guideline to analyze the data with NGSmethPipe

NGSmethPipe is a program to analyze bisulfite sequencing data generated by means of BS-Seq and MethylC-Seq protocols. It is implemented in Perl and can be easily set-up as it needs no compilation. The modular structure allows both, running just one of the different sub-tasks or a full analysis by means of a meta-script. The main focus of NGSmethPipe is on the quality control of the generated methylation maps, implementing several quality related features that are currently only available in NGSmethPipe.

4.1 Main features

NGSmethPipe implements several exclusive features including both, aspects regarding the quality of the inferred methylation levels and technical issues like full multithreading (parallelization of the process) and memory scaling (the memory needs of the program can be adjusted to the resources of the user). The main properties of NGSmethPipe can be summarized as follows:

- The program implements three important quality filters: (i) putative bisulfite failures can be detected, (ii) the number of false inferences on the methylation state can be controlled by means of the Phred Scores (probability of a sequencing error), (iii) SNV (single nucleotide variants) can be detected and removed
- Usage of a "seed extension" method applied to the Bowtie alignments allowing to map a higher number of reads without compromising the mapping quality
- Extensive output options including all possible cytosine sequence contexts (CG, CHG and CHH; where H is A, T or C) and the possibility to join the information from both strands (detection of hemi-methylation)
- Complete statistics of the whole process, including aligned reads, discarded reads, discarded positions, chromosome data coverage, etc
- The memory and CPU needs can be adapted to the user's computer resources
- Single and paired end input is accepted
- Fastq input files are accepted in zip, gzip, bzip2 or uncompressed

4.2 Installation

Right now, NGSmethPipe is tested only for Linux platforms. In this section we will assume that the user has a Linux workstation with an installed Perl interpreter (by tying 'perl -v' in a terminal it can be checked easily if Perl is installed). The installation or set-up process can then be done in 3 simple steps:

1. Install two Perl modules that are needed. This can be done typing with super-user rights on the command line:
    *   perl MCPAN –e "install Bundle::Thread"
    *   perl -MCPAN -e "install IO::Uncompress::AnyUncompress"
2. Download the bowtie aligner http://bowtie-bio.sourceforge.net/index.shtml and extract the binary files to a folder (for example '/home/user/bowtie').
3. Download the NGSmethPipe files from http://bioinfo2.ugr.es/NGSmethPipe/ downloads/NGSmethPipe.tgz and uncompress them (tar xzvf NGSmethPipe.tgz). For the rest of this chapter we will assume that the directory is '/home/user/NGSmethPipe'. After uncompressing the tar file, the user should see 4 Perl scripts which are explained in the next section.

4.3 Structure of the program

NGSmethPipe is composed of 4 scripts which can be launched individually or as a pipeline by means of a meta-script. The scripts have the following functions:

*   'NGSmethPipeIndex.pl': Generation of two 3-letter reference sequences in bowtie format: one changing C by T in the Watson strand (C to T reference in figure 2) and one changing A by G in the Watson strand G/A Watson conversion (G to A reference in figure 2)
*   'NGSmethPipeAlign.pl': Pre-processing and alignment of the reads
*   'NGSmethPipeRatios.pl': Methylation profiling and quality control
*   NGSmethPipe.pl: This is the meta-script; it launches internally the other 3 scripts described above

4.4 Running NGSmethPipe

In this section, we will show how to prepare the reference sequences and how to perform the methylation profiling for different data sets.

4.4.1 Building the reference sequences

Before NGSmethPipe can be used, the reference genome must be downloaded in fasta format (for example from the UCSC Genome Bioinformatics site http://hgdownload.cse.ucsc.edu/goldenPath/hg19/chromosomes/) into a local directory (for example, 'home/user/sequences/hg19'). Note that in the current version, all chromosomes, contigs or scaffolds must be in single-fasta file format. No multi-fasta files are currently supported. The reference sequences can then be prepared using the following command:

perl NGSmethPipeIndex.pl seqDir=/home/user/sequences/hg19 bowtieDir=/home/user/bowtie

The only optional parameter is the number of threads (number of CPUs) to be used (p=4 by default). The output of the process consists of 12 bowtie index files and is also written into the input directory (seqDir). Bowtie generates for each reference 6 files. Here we obtain 12 files as we need two different 3-letter references: one changing C by T and one substituting G by A. Note that the meta script checks if converted reference sequences do exist and therefore, this step can be skipped when performing a full data analysis by means of the meta-script.

4.4.2 Download the example data

On the NGSmethPipe webpage (http://bioinfo2.ugr.es/NGSmethPipe/downloads/Examples.tgz), some examples can be downloaded to test the program. The test data contains both, the input reads and the reference sequences. After downloading (we will assume that the download directory is /home/user/Examples) and uncompressing with "tar xzvf NGSmethPipe.tgz", the user can see three folders: h1_exampleChr22 which contains sequencing data from the MethylC-Seq protocol for a region on the human chromosome 22, wt_shoots_example which contains reads from BS-Seq for *Arabidopsis thaliana* and wa09fibro_exampleChr21 which contains pair-end reads for a region in human chromosome 21.

4.4.3 Running the example data

In general, NGSmethPipe can be launched in two different ways: (i) by means of command line parameters or (ii) by using a configuration script. The parameter syntax is the same for both ways: parameter=<value>. An example for the configuration script can be seen within the three test data folders. To analyze MethylC-Seq data, only three mandatory parameters do exist, 'seqDir' (the folder with the reference genome single fasta files), 'inDir' (the input directory with the read files in (compressed) fastq format and 'bowtieDir' (the folder with the Bowtie binary files). The program can be launched giving the parameters on the command line:

*perl NGSmethPipe.pl seqDir=/home/user/Examples/h1_exampleChr22/seqDir/
inDir=/home/user/Examples/h1_exampleChr22/inDir/ bowtieDir=/home/user/bowtie/*

or by means of the configuration script:

perl NGSmethPipe.pl /home/user/Examples/h1_exampleChr22/NGSmethPipeConfigFile_h1.dat

If the first parameter on the command line is a file, NGSmethPipe will read and treat it automatically as a configuration file. These two commands will launch a full analysis with default parameters by consecutively launching the 3 scripts. Each of the scripts will write its own output files which are explained in detail in the section 4.4.5.

4.4.4 Important parameters

Multithreading: Two parameters, $'p'$ and $'maxChunk'$ allow to adapt the memory and CPU requirements of the process to the local resources. The number of CPUs can be controlled by $'p'$ (p=6 will generate 6 threads) and the memory needs can be fine-tuned by means of the $'maxChunk'$ parameter. By default, 10000 reads are processed by each thread which leads to a memory of approx. 2500Mb per thread for the human genome. The more reads are processed by a thread, the higher will be the addressed memory, but the speed will increase also as the hard disk access times decrease. It is important to adjust these parameters in order to exploit the available resources: number of processors and random access memory (the available RAM of the computer). For example, a high number of threads could increase the speed, however it will also increase the overall memory usage and together with a high value for 'maxChunk' it could exhaust the available random memory.

perl NGSmethPipe.pl seqDir=/home/user/Examples/h1_exampleChr22/seqDir/ inDir=/home/user/ Examples/h1_exampleChr22/inDir/ bowtieDir=/home/user/bowtie/ p=6 maxChunk=20000

BS-Seq protocol: For non-directional reads we need to provide the sequences of the forward and reverse tags which can be indicated by the *'fw'* and *'rc'* parameters respectively. Example reads can be found in the folder wtshoots_example (extracted from (Pellegrini, et al., 2010)).

perl NGSmethPipe.pl seqDir=/home/user/Examples/wtshoots_example/seqDir/ inDir=/home/user/reads/Examples/wtshoots_example/inDir/ bowtieDir=/home/user/bowtie/ fw=TCTGT rc=TCCAT

Pair-end protocol: In the case of pair-end reads the user has to provide the file suffix for each of the two mate files by means of 'm1' and 'm2'. NGSmethPipe will search within the 'inDir' directory for files with these suffixes running Bowtie in pair-end mode. Two important parameters implemented in Bowtie are the minimum and maximum insert sizes of the pairs, 'I' and 'X' respectively. Example reads can be found in the folder wa09finro_exampleChr21 (extracted from (Li, et al., 2010)).

perl NGSmethPipe.pl seqDir=/home/user/Examples/wa09fibro_exampleChr21/seqDir/ inDir=/home/user/reads/Examples/wa09fibro_exampleChr21/inDir/ bowtieDir=/home/user/bowtie/ m1=_1 m2=_2 I=0 X=500

Alignment Options: Like mentioned in section 3.3, the alignment parameters of NGSmethPipe have been chosen so that the percentage of incorrect alignments is lower than 1%. Nevertheless, if the user needs more coverage (accepting a higher percentage of false mappings and the corresponding consequence on the methylation levels), several alignment parameters can be manipulated: *'l'* sets the Bowtie seed length, 'n' sets the maximum number of mismatches within the seed region and *'m'* sets the maximum number of allowed mismatches in the whole alignment (the seed is extended until m+1 is reached).

perl NGSmethPipe.pl seqDir=/home/user/Examples/h1_exampleChr22/seqDir/ inDir=/home/user/Examples/h1_exampleChr22/inDir/ bowtieDir=/home/user/bowtie/ l=25 n=2 m=4

Quality control: Regarding the quality control, two parameters are of special interest: (i) the minimum Phred Score quality value of a "valid" base call (default:' minQ'=20) and (ii) the maximum number of nonCpG methylcytosines in order to detect bisulfite failure reads (default: "methNonCpGs'=0.9). The 'methNonCpGs' parameter can take both, values between 0 and 1 and integers. If a value between 0 and 1 is detected, it is interpreted as the fraction between methylated non-CpG contexts and total number of non-CpGs while integers are taken as 'number of methylated non-CpGs'. In the example below, the 'minQ' parameter is set to 40. This means that all positions with less than a Phred Score of 40 (probability of an erroneous base call less than 0.0001) are ignored. Setting the methNonCpGs to 3, the program will discard all reads with more than 3 methylated non-CpG contexts like proposed by (Lister, et al., 2009).

perl NGSmethPipe.pl seqDir=/home/user/Examples/h1_exampleChr22/seqDir/inDir=/home/user/ Examples/h1_exampleChr22/inDir/ bowtieDir=/home/user/bowtie/ minQ=40 methNonCpGs=3

Output parameters: Several output parameters do exist. If the user is only interested in a particular sequence context (CpG, CHG, CHH), it can be set with the 'pattern' parameter (for example pattern=CG). Next, the user can choose between reporting the methylation for each cytosine on both strands or merging the methylation levels of the two cytosines that belong to a palindromic sequence context (CpG, CWG). By default the strand merge is not performed, but it can be set with 'uniStrand=Y'. Finally, the output can also be reported in BED or WIG format (see http://genome.ucsc.edu/FAQ/FAQformat.html) for further analysis in a Genome Browser (see section 5). The example below, launches a full analysis for CpGs (pattern=CG), reporting the merged methylation levels (uniStrand=Y) and a BED file (bedOut=Y).

perl NGSmethPipe.pl seqDir=/home/user/Examples/h1_exampleChr22/seqDir/inDir=/home/user/ Examples/h1_exampleChr22/inDir/ bowtieDir=/home/user/bowtie/ pattern=CG uniStrand=Y bedOut=Y

4.4.5 Analyzing the output

When a full analysis is launched, each of the scripts will write its own output files.

NGSmethPipeIndex: Bowtie indexes will be stored in the fasta sequence directory, specified by 'seqDir'. The script will create two genomic indexes: in the first one, cytosines are converted to thymines and in the second one guanines to adenines. The indexes files have *.ebwt* extension, which is the output extension used by Bowtie.

NGSmethPipeAlign: The output of this script is stored into the 'outDir' folder, or into the reads directory by default. The alignments are reported in files with *.align* extension. The pair-end mode output will be the same as single-end mode, except a mate identifier at the end of the read id (/1 for the #1 mates and /2 for the #2 mates). The output file has 6 columns:

- ID: original identifier of the read
- Strand: the strand where the read maps (+ or -)
- Chromosome: chromosome where the read maps (chr1, chr2, chrX, etc…)
- Start position: start position of the read in the chromosome (0-based). The coordinate refers to the Watson-strand
- Read: The sequence of the read with its original alphabet, and with an asterisk (*) on the mismatch positions. In case of Crick-strand reads, the sequence returned is the reverse complement of the original sequence
- Quality line: Encoded Phred Quality Scores

Additionally, a log file is written giving the information about the used parameters, running time, and the number of processed and aligned reads.

NGSmethPipeRatios: This script extracts the methylation levels, either for each strand separated or merged together for palindromic sequence contexts. The output files are named after the analyzed pattern (methylation context): for example CG.output. The output file format depends on whether the uniStrand option is set or not.

- uniStrand=N
  - Chromosome: the chromosome
  - Start position: start position of the methylation context (1-based), in the positive strand
  - End position: end position (1-based)
  - Strand: the strand
  - Number of reads: number of reads covering the cytosine position
  - Methylation ratio: methylation level (number of methylcytosines divided by the number of reads mapped to the position)
- uniStrand=Y (by default)
  - Chromosome: the chromosome
  - Start position: start position of the methylation context (1-based), in the positive strand
  - End position: end position (1-based)
  - Number of reads: number of reads covering the context, on both strands
  - Methylation ratio: methylation level (number of methylcytosines divided by the number of reads mapped to the position)
  - Meth difference: the absolute difference between the methylation levels on each strand
  - Number of reads on the Watson strand
  - Methylation level on the Watson strand
  - Number of reads on the Crick strand
  - Methylation level on the Crick strand

The corresponding log file (RatiosCGStats.log) stores the number of reads discarded by the bisulfite check, the number of positions discarded by means of the Q value threshold and the coverage (% of positions covered) as a function of chromosome, strand and context.

5. Storage, browsing and data mining tools for methylation data: NGSmethDB

Powered by the emergence of whole-genome shotgun bisulfite sequencing techniques, many epigenetic projects are currently on the way (Bernstein, et al., 2010). In order to make this data available to all researchers several aims exist: (i) the data should be processed following the same protocol in order to make them comparable among each other, (ii) the results need to be stored in easily accessible databases including data browsing and download, (iii) the databases should implement basic data mining tools in order to compare different data sets retrieving only the relevant information. In order to attend these needs we developed NGSmethDB (Hackenberg, et al., 2011), a database for the storage, browsing and data mining of single-base-pair resolution methylation data which has the following main features:

- Based on the GBrowse (Stein, et al., 2002), the stored methylation data can be easily browsed and analyzed in a genomic context together with other annotations like RefSeq genes, CpG islands, Conserved elements, TFBSs, SNPs, etc
- The user can easily upload methylation data and analyze it together with publically available data in the database

- The raw data can be downloaded for all data sets for different read coverages (1, 5, 10)
- The implemented data mining tools allow for example to retrieve unmethylated cytosines or differentially methylated cytosines in a user defined set of tissues

5.1 Scope

At the end of August 2011, the database holds data for 3 species (human, mouse and *Arabidopis*) and 26 unique tissues. The data are available for 3 different levels of "read coverage": 1, 5 and 10. The read coverage is the number of reads that contribute to the methylation level of a cytosine. Finally, the DB hosts two different sequence contexts, CpG dinucleotids and CHG

5.2 Browsing

A web browser interface is set up by means of the GBrowse genome viewer that is connected through a MySQL backend to NGSmethDB. Features of the browser include the ability to scroll and zoom through arbitrary regions of a genome, to enter a region of the genome by searching for a landmark or performing a full text search of features, as well as the ability to enable and disable feature tracks and change their relative order and appearance. The user can also upload private annotations to view them in the context of the existing ones at the NGSmethDB web site. Apart from the methylation data, other functional annotations are available like CpG islands (Hackenberg, et al., 2006), RefSeq genes (Pruitt, et al., 2007) and Repetitive Elements (RepeatMasker track from UCSC) together with the chromosome sequences and the local G+C content. The methylation information of a given context is represented by the coordinate of the cytosine on the direct strand. To display the methylation values of the cytosines we use a color gradient from white (methylation value = 0, unmethylated in all reads) to red (methylation value = 1, methylated in all reads).

Directly with the NGSmethPipe BED output file (for example CG.bed), the user can upload his methylation results and compare them with other datasets in its genomic context. A brief example is shown in the Figures 4 and 5, for the pair-end reads example (wa09fibro_exampleChr21).

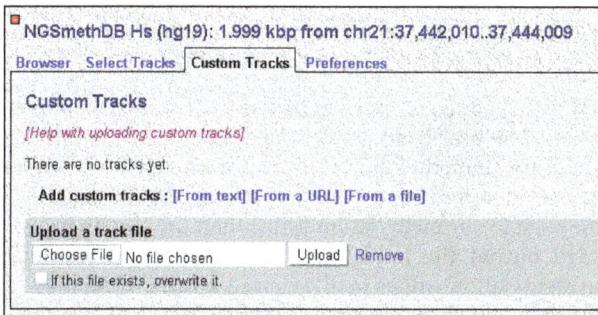

Fig. 4. Uploading BED files.The user can easily upload the bed files output from the NGSmethPipe to the NGSmethDB. In the 'Custom track' tab there are three options to upload tracks, selects the 'From a file' option and choose for example the CG.bed file from the 'outDir' or 'inDir' by default. For more information, see the help link.

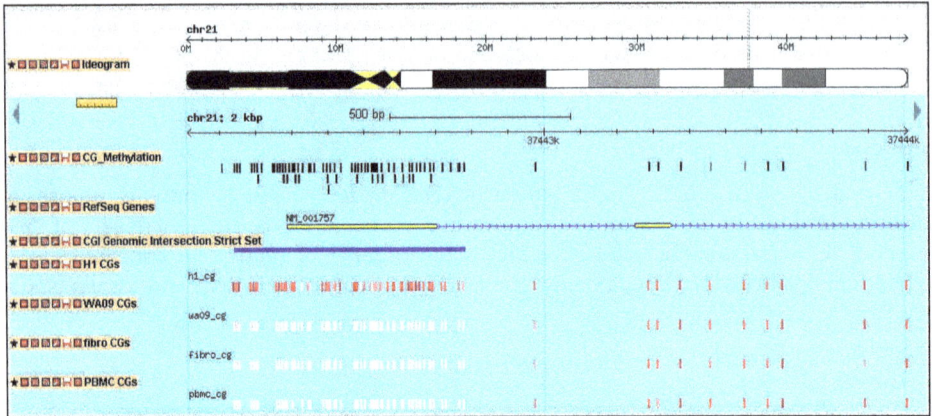

Fig. 5. Uploaded track example.The figure shows the pair-end wa09fibro example in grey scale: from black (methylated CpGs) to white (unmethylated CpGs). A region from the human chromosome 21 is depicted, where the CpG island overlapping the NM_001757 gene promoter is methylated in the uploaded track and H1 track, while it remains unmethylated in the rest of the displayed tissues. Note that the tracks from the database are shown in a different format (from red (methylated) to white (unmethylated)) and therefore the uploaded track can be easily distinguished. Once the track has been uploaded, the user can modify its features with the configure button at the left side of the track.

5.3 Data mining

Currently, the data can be accessed in 5 different ways. (i) Dump download, (ii) Retrieve unmethylated contexts, (iii) Retrieve differentially methylated contexts, (iv) Get methylation states of promoter regions and (v) Retrieve methylation data for chromosome region.

The 'Dump download' option shows first an overview of current database content, including a short description of the tissue, the genome coverage in %, a link to PubMed, and raw data files for #read ≥ 1, #read ≥ 5 and #read ≥ 10 coverage. The files show the chromosome, chromosome-start and chromosome-end coordinates, the sequence methylation context (either CpG or CWG), the number of reads and the cytosine methylation ratio.

The 'Retrieve unmethylated contexts' tool can be used to retrieve all unmethylated cytosines in a given set of tissues. The user has to select the sequence context (CG or CWG), the read coverage, the threshold for unmethylation (often a threshold of 0.2 is used, i.e. all cytosines with values ≤0.2 are considered to be unmethylated) and the tissues. The tool will detect all cytosine contexts showing lower methylation ratios than the chosen threshold in all selected tissues. The provided output file holds the chromosome, chromosome start- and end-coordinates and the methylation values in all selected tissues. Note that this tool can be also used to retrieve all CpGs which are present in every single analyzed tissue by setting the threshold to one. In doing so, cytosines with methylation data in all tissues will be reported regardless of its methylation state, i.e. cytosines that are not covered by at least the number of chosen coverage threshold (1, 5 or 10) in any of the analyzed tissues will not be reported in the output.

By means of the 'Retrieve differentially methylated contexts' tool all differentially methylated cytosine contexts can be determined in a given set of tissues. All parameters of the 'Retrieve unmethylated contexts' (see above) are available here, plus one additional parameter: the threshold for the methylation value which defines whether a cytosine is considered to be methylated (often a threshold of 0.8 is used, i.e. all cytosines with higher values than ≥0.8 are considered to be methylated). We define a cytosine as differentially methylated if it is unmethylated in at least one tissue and methylated in at least one other tissue. The tool reports those differentially methylated cytosine contexts that are either methylated or unmethylated in all analyzed tissues, i.e. those contexts that show intermediate methylation in only one tissue will not be reported.

Another tool allows depicting the methylation states of all cytosine contexts within the promoter region of RefSeq genes. The promoter region is defined from 1.5 kb upstream of the Transcription Start Site (TSS) to 500 bp downstream of the TSS. The output is displayed by default as an overview table that summarizes the fluctuation along the promoter as well as over the different tissues.

Finally, all methylation values for a selected set of tissues can be retrieved for a given chromosomal region, once the user provides the start and end chromosome coordinates.

6. Conclusions and outlook

The price for high-throughput DNA sequencing is dropping constantly and consequently the number of available whole genome shotgun bisulfite sequencing data is increasing at a very high rate. This implies a strong need for bioinformatics applications able to deal with this vast amount of data, converting them into high quality, single-base pair resolution methylation maps. In this chapter we review the most important bioinformatics tools that are currently available comparing them by several aspects including both, technical and biological issues. The quality control is still rather disregarded by many tools and currently NGSmethPipe is the program that implements the highest number of quality related features. This application is based on Bowtie with optimized alignment parameters and a seed extension method. It implements several quality control features like the detection of SNVs, the deletion of bisulfite failure reads, and the consideration of the base call qualities in the methylation profiling step. Furthermore, NGSmethPipe delivers output files that can be uploaded directly into the Genome Browser of the NGSmethDB database, thus allowing the user to analyze the custom data within the context of other tissues or genomic elements. Future directions will be to populate the NGSmethDB database by means of the NGSmethPipe application adding also other relevant data like histone methylation, expression or disease data.

7. Acknowledgment

This work was supported by the Ministry of Innovation and Science of the Spanish Government [BIO2010-20219 (M.H.), BIO2008-01353 (J.L.O.)]; 'Juan de la Cierva' grant (to M.H.) and Basque Country 'Programa de formación de investigadores' grant (to G.B.).

8. References

Barturen, G., *et al.* (Submitted) NGSmethPipe: A tool to generate high-quality methylation maps.

Bernstein, B.E., *et al.* (2010) The NIH Roadmap Epigenomics Mapping Consortium, *Nature biotechnology*, 28, 1045-1048.

Bibikova, M. and Fan, J.B. (2009) GoldenGate assay for DNA methylation profiling, *Methods in molecular biology (Clifton, N.J)*, 507, 149-163.

Bock, C., *et al.* (2011) Reference Maps of human ES and iPS cell variation enable high-throughput characterization of pluripotent cell lines, *Cell*, 144, 439-452.

Bock, C., *et al.* (2010) Quantitative comparison of genome-wide DNA methylation mapping technologies, *Nature biotechnology*, 28, 1106-1114.

Cokus, S.J., *et al.* (2008) Shotgun bisulphite sequencing of the Arabidopsis genome reveals DNA methylation patterning, *Nature*, 452, 215-219.

Costello, J.F., *et al.* (2000) Aberrant CpG-island methylation has non-random and tumour-type-specific patterns, *Nature genetics*, 24, 132-138.

Chen, P.Y., Cokus, S.J. and Pellegrini, M. (2010) BS Seeker: precise mapping for bisulfite sequencing, *BMC Bioinformatics*, 11, 203.

Chen, T. and Li, E. (2004) Structure and function of eukaryotic DNA methyltransferases, *Curr Top Dev Biol*, 60, 55-89.

Dodge, J.E., *et al.* (2005) Inactivation of Dnmt3b in mouse embryonic fibroblasts results in DNA hypomethylation, chromosomal instability, and spontaneous immortalization, *J Biol Chem*, 280, 17986-17991.

Estecio, M.R., *et al.* (2007) High-throughput methylation profiling by MCA coupled to CpG island microarray, *Genome Res*, 17, 1529-1536.

Gu, H., *et al.* (2010) Genome-scale DNA methylation mapping of clinical samples at single-nucleotide resolution, *Nature methods*, 7, 133-136.

Hackenberg, M., Barturen, G. and Oliver, J.L. (2011) NGSmethDB: a database for next-generation sequencing single-cytosine-resolution DNA methylation data, *Nucleic acids research*, 39, D75-79.

Hackenberg, M., *et al.* (2006) CpGcluster: A distance-based algorithm for CpG-island detection, *BMC Bioinformatics*, 7, 446.

Hackenberg, M., Rodriguez-Ezpeleta, N. and Aransay, A.M. (2011) miRanalyzer: an update on the detection and analysis of microRNAs in high-throughput sequencing experiments, *Nucleic acids research*, 39, W132-138.

Hach, F., *et al.* (2010) mrsFAST: a cache-oblivious algorithm for short-read mapping, *Nature methods*, 7, 576-577.

Harris, E.Y., *et al.* (2010) BRAT: bisulfite-treated reads analysis tool, *Bioinformatics (Oxford, England)*, 26, 572-573.

Harris, R.A., *et al.* (2010) Comparison of sequencing-based methods to profile DNA methylation and identification of monoallelic epigenetic modifications, *Nature biotechnology*, 28, 1097-1105.

Hellman, A. and Chess, A. (2007) Gene body-specific methylation on the active X chromosome, *Science*, 315, 1141-1143.

Horner, D.S., *et al.* (2010) Bioinformatics approaches for genomics and post genomics applications of next-generation sequencing, *Briefings in bioinformatics*, 11, 181-197.

Jones, P.A. (1999) The DNA methylation paradox, *Trends Genet*, 15, 34-37.

Karpf, A.R. and Matsui, S. (2005) Genetic disruption of cytosine DNA methyltransferase enzymes induces chromosomal instability in human cancer cells, *Cancer research*, 65, 8635-8639.

Krueger, F. and Andrews, S.R. (2011) Bismark: a flexible aligner and methylation caller for Bisulfite-Seq applications, *Bioinformatics (Oxford, England)*, 27, 1571-1572.

Laird, P.W. (2010) Principles and challenges of genomewide DNA methylation analysis, *Nat Rev Genet*, 11, 191-203.

Langmead, B., *et al.* (2009) Ultrafast and memory-efficient alignment of short DNA sequences to the human genome, *Genome biology*, 10, R25.

Laurent, L., *et al.* (2010) Dynamic changes in the human methylome during differentiation, *Genome Res*, 20, 320-331.

Li, Y., *et al.* (2010) The DNA methylome of human peripheral blood mononuclear cells, *PLoS Biol*, 8, e1000533.

Lister, R. and Ecker, J.R. (2009) Finding the fifth base: genome-wide sequencing of cytosine methylation, *Genome Res*, 19, 959-966.

Lister, R., *et al.* (2008) Highly integrated single-base resolution maps of the epigenome in Arabidopsis, *Cell*, 133, 523-536.

Lister, R., *et al.* (2009) Human DNA methylomes at base resolution show widespread epigenomic differences, *Nature*, 462, 315-322.

Meissner, A., *et al.* (2005) Reduced representation bisulfite sequencing for comparative high-resolution DNA methylation analysis, *Nucleic acids research*, 33, 5868-5877.

Miura, A., *et al.* (2009) An Arabidopsis jmjC domain protein protects transcribed genes from DNA methylation at CHG sites, *EMBO J*, 28, 1078-1086.

Okano, M., *et al.* (1999) DNA methyltransferases Dnmt3a and Dnmt3b are essential for de novo methylation and mammalian development, *Cell*, 99, 247-257.

Ondov, B.D., *et al.* (2010) An alignment algorithm for bisulfite sequencing using the Applied Biosystems SOLiD System, *Bioinformatics (Oxford, England)*, 26, 1901-1902.

Pedersen, B., *et al.* (2011) MethylCoder: software pipeline for bisulfite-treated sequences, *Bioinformatics (Oxford, England)*, 27, 2435-2436.

Pellegrini, M., *et al.* (2010) Conservation and divergence of methylation patterning in plants and animals, *Proceedings of the National Academy of Sciences of the United States of America*, 107, 8689-8694.

Pruitt, K.D., Tatusova, T. and Maglott, D.R. (2007) NCBI reference sequences (RefSeq): a curated non-redundant sequence database of genomes, transcripts and proteins, *Nucleic acids research*, 35, D61-65.

Robinson, M.D., *et al.* (2010) Protocol matters: which methylome are you actually studying?, *Epigenomics* 2, 587-598.

Shendure, J. and Ji, H. (2008) Next-generation DNA sequencing, *Nature biotechnology*, 26, 1135-1145.

Smith, A.D., *et al.* (2009) Updates to the RMAP short-read mapping software, *Bioinformatics (Oxford, England)*, 25, 2841-2842.

Stein, L.D., *et al.* (2002) The generic genome browser: a building block for a model organism system database, *Genome Res*, 12, 1599-1610.

Tomso, D.J. and Bell, D.A. (2003) Sequence context at human single nucleotide polymorphisms: overrepresentation of CpG dinucleotide at polymorphic sites and suppression of variation in CpG islands, *Journal of molecular biology*, 327, 303-308.

Weisenberger, D.J., *et al.* (2005) Analysis of repetitive element DNA methylation by MethyLight, *Nucleic acids research*, 33, 6823-6836.

Xi, Y. and Li, W. (2009) BSMAP: whole genome bisulfite sequence MAPping program, *BMC Bioinformatics*, 10, 232.

Xin, Y., Ge, Y. and Haghighi, F.G. (2011) Methyl-Analyzer--whole genome DNA methylation profiling, *Bioinformatics (Oxford, England)*, 27, 2296-2297.

Zhang, X., *et al.* (2006) Genome-wide high-resolution mapping and functional analysis of DNA methylation in arabidopsis, *Cell*, 126, 1189-1201.

2

Modelling DNA Methylation Dynamics

Karthika Raghavan and Heather J. Ruskin

Centre for Scientific Computing and Complex Systems Modeling (SCI SYM),
School of Computing, Dublin City University
Ireland

1. Introduction

"Epigenetics" as introduced by Conrad Waddington in 1946, is defined as a set of interactions between genes and the surrounding environment, which determines the phenotype or physical traits in an organism, (Murrell et al., 2005; Waddington, 1942). Initial research focused on genomic regions such as *heterochromatin* and *euchromatin* based on dense and relatively loose DNA packing, since these were known to contain inactive and active genes respectively, (Yasuhara et al., 2005). Subsequently, key roles of DNA methylation, Histone Modifications and other assistive proteins such as Methyl Binding Proteins (MBP) during gene expression and suppression were identified, (Baylin & Ohm, 2006; Jenuwein & Allis, 2001). An emergent and persistent view that every epigenetic event affects another, to strengthen or suppress gene expression has made this an active field of research. DNA methylation refers to the modification of DNA by addition of a methyl group to the cytosine base, and is the most stable, heritable and well conserved epigenetic change. It is introduced and maintained, (Riggs & Xiong, 2004; Ushijima et al., 2003) by an enzyme family called DNA Methyl Transferases (DNMT), (Doerfler et al., 1990). Methyl-Cytosine or "mC", often referred to as the fifth type of nucleotide plays an extremely important role in gene expression and other cellular activities. Although DM is defined a simple molecular modification, its effect, can range from altering the state of a single gene to controlling a whole section of chromosome in the human genome.

The human genome is largely made of complex sequences evolved over time due to replication, mutations and insertion of foreign DNA. Based on the nucleotide distribution and functional significance, the genome has been categorized into different block of sequences, namely genes or coding and non-coding regions. A special type of sequence located near genes, in relation to spread of DNA methylation and dinucleotide frequencies are the CpG islands[1]. These islands are mostly found near the promoters, (5'end), of genes and their methylation levels are closely monitored to investigate the spread of Cancer. Useful insight on epigenetic mechanisms may be found from analysing the DNA sequence patterns or the genotype of the organism, (Gertz et al., 2011; Glass et al., 2004; Segal & Widom, 2009). Since more than 90% of DM occurs in CG dinucleotides, (Raghavan et al., 2011), knowledge of the distribution and location of CG can be utilized to understand the biological

---

[1] DNA sequences are defined and classified as CpG islands if , (a) length of that DNA sequence >200 bp, (b) Total amount of Guanine and Cytosine nucleotides >50%, and, (c) the observed/expected ratio of CG dinucleotides for that given length of sequence, >60%, (Takai & Jones, 2002)

significance associated with determining the level of DM. A general overview of pattern analysis techniques is given and application of time series analyses in understanding "CG" dinucleotide occurrences in specific human sequences are discussed in detail in the following sections.

Histones are proteins that protect DNA from restriction enzymes and also act as bolsters in chromosome condensation, (Ito, 2007). A "Histone Core", made of nine types of histone proteins, is attached to DNA molecules whose length varies from 146bp to 148bp. In the histone core, a combination of modifications, within specific amino acids in each histone subtype leads to gene expression or inactivation, (Kouzarides, 2007). These modification patterns, unlike stable DNA methylation, are dynamic and activation of one change leads to successive modifications of other amino acids during cellular events, (Allis et al., 2007; Jung & Kim, 2009). Even though new findings with regard to the impact of several modifications have been recently reported, information is inconsistent and less precise with regard to how a network of histone modifications communicates and is influenced by DM. Despite this insufficiency, the interactions between histones and DNA methylation are known to be disrupted at some stage, during the onset of cancer, (Esteller, 2007). Hence, a novel stochastic model, based on Markov Chain, Monte Carlo class of algorithms, (MCMC), was recently developed to mimic the epigenetic system and predict the effects of dynamic histone modifications over DNA methylation and gene expression levels, (Raghavan et al., 2010), (Details are discussed in Background section).

In this chapter, the focus on modelling the feedback dynamics of DNA methylation is dealt with in four parts, consisting of: (1) DNA Methylation mechanisms, controlling factors – DNA sequence pattern analyses and Histone modifications and their association with disease initiation, (2) A background on the recent data explosion, multiple methods and modelling approaches developed so far to investigate DM mechanisms and associated factors, (3a) Description of methods to investigate CG distribution in human DNA sequences – Results obtained and their association with DM spread, (3b) Developments on a novel micromodel framework, (based on MCMC) used to investigate Histone modifications for different DM levels and, (4) Results obtained for DM and HM feedback influence. Finally, conclusions and future directions for continuing investigation are considered.

2. Background

DNA Methylation was initially addressed as one of the most primitive mechanisms that organisms utilize to (a) protect genomic DNA and initiate the host resistance mechanism towards foreign DNA insertion and subsequently, (b) control gene expression, (Doerfler & Böhm, 2006). From an evolutionary point of view as well, the catalytic domain in the structure of the methylation enzymes across all organisms has been preserved to perform methyl group addition. A major change however, in the level and functional utility of DNA methylation was noted in higher organisms such as eukaryotes, when DM mechanism evolved from protecting the genomic contents to controlling their level of gene expression. In humans, there are two ways by which DNA Methylation is established – (a)*De novo* methylation that establishes new DM patterns, (b) Maintenance methylation responsible for inheriting existing DM patterns. Within the family of methylating enzymes (DNMT), two types namely DNMT3a/b/L and DNMT1 establish DM patterns in these two ways, (Doerfler

& Böhm, 2006). The *De novo* methylation process carried out by DNMT3a/b/L, is responsible for methylating embryonic cells which are totally erased of any previous DM patterns and methylated based on the DNA sequence contents. These mechanisms are also responsible for establishing parental imprinting and X-chromosome inactivation that is set permanently within the organism enabling it to exhibit unique phenotypes from birth. On the other hand, DNMT1 distribution is dynamic across a cell during its lifetime. This enzyme type is highly biased towards hemi-methylated[2] DNA sequences, making it responsible for propagating methylation patterns after each cell cycle. DNMT1 is also known to interact with histone deacetylases enzyme and some methyl adding proteins, (e.g. HP1), to remove acetyl and add methyl groups in histones, (Allis et al., 2007; Turner, 2001).

Associated aberrations in DNA methylation

As elaborately discussed by Chahwan et al, "the significant role played by DM in epigenetic regulation is quite apparent when the cell is affected due to impaired methylation marks during establishment, maintenance or recognition". Such changes in the "methylation marks" are mainly attributed to the abnormal function of DNMT enzyme complex which leads to failure of DM mechanisms. This abnormality results in gene imprinting disorders and malignancy formation due to *hyper/hypo methylation* of specific sections in the chromosomes, (Chahwan et al., 2011). Among the most studied abnormalities recorded in connection to failure of DNMT enzyme complex, is Immunodeficiency–Centromere instability–Facial anomalies (ICF) syndrome. This is caused due to mutations associated with coding for DNMT3B enzymes leading to global hypomethylation of repeat regions located in the pericentromere of human chromosomes, (Ehrlich. et al., 2008). Prader-Willi syndrome, Angelman syndromes and specific type of cancers such as Wilm's tumour have also been associated with imprinting disorders characterized by growth abnormalities, (Chahwan et al., 2011). In these diseases, genetic mutations or altered DNA methylation cause improper imprinting patterns and lead to aberrant expression of the normally suppressed genes, (Chamberlain & Lalandea, 2010). Based on accumulative information in literature, (Chahwan et al., 2011), Cancer initiation is mainly attributed to the imbalanced connectivity between oncogenes and tumor suppressor genes. Hence a combination of genetic abnormalities such as mutations and aberrant DM spread trigger cancerous conditions leading to malignancies that spread across different systems in the human body, (Allis et al., 2007). For example, in Wilm's tumour, the loss of imprinting of *IGF2* gene is associated with spread cancer to lung, ovaries and colon area. In general the DNA methylation pattern when disrupted can lead to, (i) gene activation, promoting the over-expression of oncogenes, (b) chromosomal instability, due to demethylation and movement of retrotransposons and consequently acquire resistance to drugs, toxins or virus, (Chahwan et al., 2011). Apart from failure in the control exercised by DM, there are certain protein "Onco-modifications" recently categorized as definitive signatures during occurrence of malignancies. Some of the most frequently studied histone modifications, associated with DNA methylation and tumor progress are – acetylation of H3K18, H4K16 and H4K12, trimethylation of H3K4 and H4K20, acetylation/trimethylation of H3K9, trimethylation of H3K27, occurrence of histone variants and also other external proteins such as MBP, HP1 and Polycomb that play role in chromosome rearrangement, (Chi et al., 2010; Fullgrabe et al., 2011).

---

[2] DNA sequences which have one of its double strands methylated

The above considerations make a compelling case to model and understand the DNA methylation mechanisms. In the following subsections, analyses of DNA methylation frequency and influence of genotype or DNA sequence patterns in humans are discussed, followed by elaborations on the control by DNA methylation mechanisms over Histone modifications.

2.1 DNA sequences and patterns analysis – Dimension 1

The human genome, consisting of more than three billion base pairs, is very complex and efforts to comprehend its organization and contents are still ongoing, (Collins et al., 1998; Strachan & Read, 1999). The spread of DNA methylation in the genome is not randomly determined. Emerging evidence indicates that, although chromatin modeling factors, iRNA, histone modifications and even parental imprinting memory can influence methylation, the underlying genotype or DNA sequence has a stronger key role in enabling and propagating a spectrum of methylation patterns, (Doerfler & Böhm, 2006; Gertz et al., 2011). The nature of every biological cell is characterized by its preservation of the genetic and epigenetic contents also known as "dual inheritance" and in consequence it is of utmost importance to look at the underlying genetic pattern maps for further comprehension of the epigenetic phenomenon.

When it comes to studying the epigenome or methylation landscape in connection to the initiation of Cancer, the focus is on genes and their alleles, non coding regions, and also *CpG Islands*, (Takai & Jones, 2002). The islands are one of the main locations for studying DM patterns in association with cell adaptability to environmental stress, epigenetic control and disease onset, (Allis et al., 2007). Furthermore, repetitive sequences or "Retrotransposon" which mostly belong to the non-coding regions, contain highly methylated CG dinucleotides in the human genome. These regions are silenced and kept under control due to the fact that they can replicate quickly and place themselves in different locations within the genome. They are also the favoured loci of "foreign" DNA insertions, which tend to disturb the existing DNA methylation patterns, (Collins et al., 1998).

Information from literature indicates that a majority of DNA methylation occurs in nucleotides, specifically located in these repeat regions (non coding) and in CpG islands, (Raghavan et al., 2011). The CG dinucleotides are usually under-represented across the human genome as a whole but are densely located in certain repeat regions and islands which may be differentially methylated during cancer initiation, (Esteller, 2007). CG dinucleotides in these regions follow a specific pattern and thus are easy targets for enzyme recognition and consequently, for methylation. The indications are also that certain patterns of CG base pairs, that are accessible by the DNMTs enzyme complexes, appear near promoters and islands of non-expressed genes in the human genome. Emerging evidence from genome analyses for example, reveals that the *De novo* methylating enzymes such as DNMT3a/L, are biased toward CG dinucleotides, appearing after every 8-10bp near promoters of methylated genes, (Glass et al., 2004). Hence it is vital to perform a complete distribution or pattern analysis of nucleotides in human sequences, in particular of CG to understand how methylation is established and maintained based on the sequence patterns within the genome. Although there is no complete evidence about the nature of DNMT mechanisms in setting new methylation patterns, analysing the global periodicities or distributions of CG dinucleotides will help to reveal a part of the hidden picture.

2.1.1 Methods to analyse DNA patterns

Since the advent of DNA sequencing technologies, (França et al., 2002), deciphering the significance of sequence blocks has been an important focus for geneticists. Apart from encoding for proteins, the human genome is a reservoir of information that has inherent patterns, corresponding to chromosomal condensation and evidence of evolution through common patterns among organisms. Several pattern recognition/analysis techniques or time series analysis methods[3] have been explored starting from simple statistical measures to complicated transformation and decomposition methods such as the Discrete Wavelet Transformation(DWT). A well-known approach in sequence analysis is to calculate "Expected Frequency" based on the empirical probabilities of the occurrence of nucleotides. This method was proposed by Whittle, and further developed to apply on DNA sequences by Cowan, (Cowan, 1991; Whittle, 1955). In the latter, transition probabilities (for all 16 types of dinucleotides) in the form of a matrix were constructed from known DNA sequences, to predict patterns along a new sequence. This particular analysis was performed on specific sequences containing the same starting and ending nucleotides. Another tool developed to visualize sequences, was "GC-Profile" which was based on, calculating nucleotide frequencies from the total amount of G and C nucleotides, and use of quadratic equations to check for purine levels in small genomes, (Gao & Zhang, 2006).

A standard pattern analysis can be conducted using the Fourier Transformation (FT), which allows decomposition of the time/spatial components in the data and construction of a frequency map, (Morrison, 1994). Fields of application are wide in range with examples from – Physics (optics, acoustics and diffraction), Signal Processing and Communication Systems, Image Processing, Astronomy, and DNA sequence analysis, amongst others, (A'Hearn et al., 1974; Goodman, 2005; Salz & Weinstein, 1969). Early work using Fourier technique in DNA pattern recognition was carried out by Tiwari et al. In this method, small sequences from bacteria were first converted into four distinct sets of binary sequences, (each corresponding to location of a nucleotide), then analysed by applying Fourier. This was followed by a comparison between genes and non-coding, and identification of characteristic features/patterns such as 3bp periodicity in genes. This type of application gave rise to the phrase "Periodicity" of nucleotides i.e. count of appearance of specific patterns that appear in sequences. Subsequent research focused on these periodicities of small patterns (length upto 10 bp) in blocks of sequences. Thus the Fourier transformation was used to study frequency components of the sequences along a spatial axis where each nucleotide was represented by a directional vector. Periodicities in virus strains (SV40) were also studied to check for patterns of dinucleotides and their corresponding role in genome condensation, (Silverman & Linskera, 1986). The most prominent periodical pattern of 10-11bp, portrayed by pyridines (AA/TT/AT), which are involved in long range interactions of upto 147 bp and aid in nucleosome alignment, was confirmed through these attempts. Refinement of this method through introduction of new parameters included calculation of *autocorrelation*[4] for specific patterns from DNA sequences. More recently, further improvements have been employed and tested on example sequences, (Epps, 2009). Complete and significant analyses of patterns or

---

[3] Applied to study patterns along the spatial-varying data in DNA sequences.
[4] Autocorrelation of patterns is an extension for periodicity, i.e. appearance of a pattern after a lag or distance of "k" base pairs.

biological markers on sequences were identified by, (Herzel et al., 1999) and (Hosid et al., 2004) from *E.coli* genome. In the latter paper, authors discuss landmark periodicities in detail, along with supportive evidence of their biological significance inside the genome. This includes – 3bp spacing followed by all 16 dinucleotides in genes, 10-11bp spacing by pyridines, and some organism specific distributions. The corresponding power spectrum, that provide information on global periodicities, was calculated, (Hosid et al., 2004) using:

$$
f_p = \frac{\sqrt{\sum_{i=1}^{m} sin\left((2\pi * \frac{i}{p}) * (X - X')\right)^2 + \sum_{i=1}^{m} cos\left((2\pi * \frac{i}{p}) * (X - X')\right)^2}}{\left(2\pi * \sum_{i=1}^{m}(X - X')^2\right)} \tag{-1}
$$

f_p = Normalized wave function amplitude at period - p
X = Auto correlation profile of the dinucleotide
X' = Mean Auto Correlation
m = Maximum autocorrelation distance
p = Periodicity or in this case distance between identical patterns or nucleotides.

A Fourier analysis in our case involves calculating the auto correlation profile for desired dinucleotide/ nucleotide followed applying the formula shown above. More details on this approach and its application to study nucleotide distribution in genes, non-coding regions and CpG islands are discussed in the Methods section. The aim of this initiative was to understand the distribution of CG dinucleotides, similiar to the work of (Clay et al., 1995), and on different datasets containing genes, CpG islands and non-coding regions[5].

2.1.2 Note on Discrete Wavelet Transformation

An extension to the Fourier analysis, Discrete Wavelet Transformation, is the application of a set of orthonormal vectors in space to localize and study both frequency and time/spatial components for a given dataset, (Kaiser, 1994). The resulting coefficient matrix, a product of this family of vectors and input data helps to indicate regions of high and low frequencies along the spatial, (or sequential) axis based on an initial resolution factor, (e.g. Haar and Mortlet, (Kaiser, 1994)). Wavelets or specifically the method of DWT addressed here, have been quite extensively used to study financial markets, experimental data from Protein Mass Spectrometry and DNA sequence patterns amongst others, (Kwon et al., 2008). Although DWT is not quite often used as fourier, it has also been applied to visualise both frequency and location specific information of the DNA sequence patterns, (Tsonis et al., 1996; Zhao et al., 2001). Elaboration on this family of approaches, is not explicitly dealt in this chapter, hence more details on the method of Maximal Overlap Discrete Wavelet Transformation, (MODWT - extension to DWT), (Conlon et al., 2009), application to study patterns in DNA sequence and results thus obtained, are reported in (Raghavan et al., 2011).

So far we have discussed various methods and algorithms, used to detect nucleotide patterns in human DNA sequences and have considered in more detail the role of Fourier

---

[5] The non coding regions referred here in this analysis are the segments in-between exons/coding regions and are removed during translation or protein production phase

Transformation technique in investigating these patterns. In the next subsection, attempts to investigate the occurrence of histone modifications are reviewed. We describe ways to explore the relationship between these and DNA sequences. To test these approaches, we combine the results from Fourier analysis, or dinucleotide patterns with information on specific histone modification effects at fixed DNA methylation levels, using our recently developed, EpiGMP prediction tool.

2.2 Histone modifications – Dimension 2

Histones are closely linked to DNA molecules and play a vital part in encoding information from them. Over time, histone proteins have diversified from a few ancestors into five distinct types of subunits (2 copies of H2A, H2B, H3 and H4 each and a H1 subunit) in eukaryotes thus forming the octomeric structure of a nucleosome, (Allis et al., 2007). This nucleosome comprising of histone complex and 146 to148bp bp of DNA molecules on average, forms a "bead on string" structure. The histone octomer or core plays the most important role in condensing billions of DNA base pairs compactly within 23 pairs of chromosomes in the human genome. Covalent posttranslational histone modifications are mainly held responsible for chromatin architecture and propagation of many cellular events from simple gene expression to cell fate determination, differentiation, and, sometimes, disease onset. Thus, with more than one type of histone containing multiple types of modification (acetylation, methylation, phosphorylation, ubiquitination and sumoylation) in their tails present a potentially complex scenario, (Cedar & Bergman, 2009; Jenuwein & Allis, 2001; Kouzarides, 2007; Zheng & Hayes, 2003). DM and HM most often have a mutual feedback influence hence maintaining a strong dependency over one another. A very interesting fact about histone modifications is that though the exact mechanisms are unknown, they are memorized by the cells "post replication", especially those that aid in gene expression, methylation maintenance and chromosome structure stability. Among all the histone modifications, methylation (mono/di/tri) and acetylation have been most studied in regard to their influence over gene expression. These modifications are quite often noted to compete for the same type of residues and are also known to recruit antagonistic regulatory complexes such as trithorax and polycomb proteins, (Allis et al., 2007). For example, histone methylation was found to be important for DNA methylation maintenance at imprinted loci, which could lead to disorders such as the Prader-Willi syndrome, (Chahwan et al., 2011). Such individual experiments have helped unravel the connection step by step between levels of DM and specific histone modifications including special histone variants, (Barber et al., 2004; Ito, 2007; Meng et al., 2009; Sun et al., 2007; Taplick, 1998; Wyrick & Parra, 2008). Hence a complete picture of the molecular communications that control the cellular events is lacking. Consequently, attempts have been made to accumulate the *cross-talk information* from laboratory experiments and decipher the modification patterns in the human genome during different cellular events, (Bock et al., 2007; Yu et al., 2008).

2.2.1 Modeling DNA methylation and histone modification interactions

Epigenetics, as a field, is relatively new and models to study the associated phenomena are limited to date. The advent of favourable experimental techniques such as Protein Mass

Spectroscopy, (Sundararajan et al., 2006), ChIP-Seq and ChIP-on-Chip[6], (Collas, 2010), have led to new data and confirmed facts with regard to DNA-protein interactions and their role in cancer onset. Such experiments usually generate a large amount of data including measures such as direct count of modification detected along the genome after specific intervals of DNA sequences, (standard intervals are 200 or 400 base pairs for histone modifications detection). As discussed in detail, by Bock et al, extracting comprehensible epigenetic information is a three-stage process. First, the biochemical interactions are stored as genetic information in DNA libraries, followed by applying DNA experimental protocols such as tiling microarray, (special type of microarray experiment) along with ChIP-on-ChIP, and lastly applying computational algorithms to infer error free epigenetic information from these experiments. These algorithms are mainly quantitative and help to establish a pipeline for prediction of probable epigenetic events. An initial coarse attempt to define the epigenetic, genetic and environmental interdependencies paved the way for an in depth study of the molecular factors that trigger these effects, (Cowley & Atchley, 1992).

Among the many computational attempts to model and analyse epigenetic mechanisms some have successively identified correlated histone signatures during gene expression using data from ChIP-on-ChIP experiments and microarray based gene expression measurements, (Karlić et al., 2010; Yu et al., 2008). A Bayesian network model was constructed using the high-resolution maps from laboratory experiments to establish casual and combinatorial relationships among histone modifications and gene expression, (Yu et al., 2008). Quantitative measure of other proteins such as Polycomb, CTCF (insulating proteins) and Transcription factors were also included to build these models. Based on Bayesian networks, conditional probabilities and joint probability distribution measures of datasets were calculated and a finely clustered molecular modification network was obtained.

Repeated bootstrapping or random sampling verified the robustness of this Bayesian Network. For initial analysis, datasets containing information from ChIP-on-ChIP experiments ((Cuddapah et al., 2009) and (Boyer et al., 2006)) for histone protein modifications in human CD4+ (immunity), cells and gene expression measurements from microarray experiments (obtained from (Su et al., 2004)), were extracted for clustering (using k-means), followed by construction of the bayesian network.

Another quantitative model based on the same type of information such as data from ChIP-on-ChIP experiments, obtained from literature, (Cuddapah et al., 2009), was developed using Linear Regression (Karlić et al., 2010). In this case, a regression expression was used to build the model: ($N_{i,j}'=N_{i,j}+$constant), where, $N_{i,j}$ = count of j_{th} modification in i_{th} gene in template samples. This equation was modified by inclusion of more variables, to study multiple histone modifications, thus giving rise to more than one model type. Secondary information was also extracted and included in the model, namely, microarray expression data from literature, (Schones et al., 2008) and promoter blocks information from Unigene databases, (http://www.ncbi.nlm.nih.gov/unigene). Here, loci of new sets of ChIP-on-ChIP experimental results for histone modifications, were mapped on human genome using annotation track information obtained from University of California Santa Cruz genome browser, (http://genome.ucsc.edu). These multivariable models were

---

[6] Experiments conducted to check for protein-DNA interactions combining chromatin immuno precipitation and massively parallel DNA sequencing techniques or microarray (chip) experiments

applied on different sequence datasets which were based on Low CG or High CG dinucleotide concentration. The whole dataset thus obtained was divided into training and test sets namely – D1 and D2, where Pearson correlation coefficient values were used to confirm the accuracy of prediction(D1) over the test set, (D2). This model was also extended over different cells, (with initial trials being conducted on CD4+ human cells), for nine histone modifications and for confirmation on CD36+ and CD133+ human immune cells respectively.

Other model types based on Bayesian networks, have focused on developing tools to study DNA methylation and protein modifications, (Bock et al., 2007; Das et al., 2006; Jung & Kim, 2009; Su et al., 2010). Among those, two models by Jianzhang et al and Bock et al have mainly focused on identifying the function of CpG islands using information on Histone Modifications. These type of "reverse" models explain the feedback connectivity between the two epigenetic events (HM and DM). Bock's model was an important initiative in computational epigenetics, since a clear pipeline for analysis of epigenetic data was proposed. The training model used several inputs from the experimental datasets to identify *bonafide* CpG islands. Inputs included – CpG islands that qualified based on criteria defined, (Takai & Jones, 2002) and epigenetic datasets from experiments (such as lysine modifications in histones, transcription binding factors, MBP, and SP1 proteins). This work consisted of three main steps, the first of which involved identification of predictive parameters from the datasets, followed by cross validation and training of data using a linear support vector machine, and lastly comparison of CpG islands previously identified in chromosome 21. These elaborate measures took into account the level of histone modifications affecting the methylation status hence emphasizing on the strong connectivity between methylation levels and their corresponding epigenetic states. Similar to the model described, (Yu et al., 2008), another complementary attempt was made to construct regulatory patterns that appear in histone during high DNA methylation. A Bayesian network once again was used to predict a list of methylation modifications that leveraged the occurrence of DNA methylation (using the same datasets obtained from CD+4 cells in humans), (Jung & Kim, 2009). These independent and repeated attempts, on accumulation, helped to identify and confirm a definitive pattern and characteristic modifications that exist in epigenetic events in the human cells: for example, more acetylation modification appear during gene expression and more methylation modifications are preferred during gene suppression.

A major disadvantage in the development of these quantitative models was the restriction of obtaining results from a single source or studies performed to investigate a single disease onset. Such a scenario cannot account for the epigenetic events for all conditions due to absence of a general model framework that could definitively link different epigenetic events. This has ultimately indicated a need to develop a general predictive model that can report modifications occurring in genes associated with any type of cell or cancer (provided there is evidence on the role of genes in diseases). As a consequence, we recently developed a theoretical model based on cumulative information of the nature of epigenetic events and tested it on synthetic data, (Raghavan et al., 2010). The novelty of this micromodel lies in accounting for the dynamics in the epigenetic mechanisms based on a stored library of possible histone modifications as well as DM associated patterns in the DNA sequences. The model, which is based on MCMC algorithm, allows sampling of possible solutions of histone modifications, using probabilities of transition. Based on the accumulative knowledge on the nature of modifications as mentioned above, probabilistic cost functions are used to set the interdependencies between variables (HM and DM based patterns) in this model. This dependency, influences the random sampling and calculates the final output or rate of transcription (T) using exponential equations (T= $e^x * e^y * k$, "x" and "y" being histone modifications and DNA methylation respectively and "k" a constant value of transition probability – Figure 4). As a part of the validation, the initial probabilities of transition set have been assigned random values so as to investigate results, (Monte Carlo or boot strapping). Ultimately, our micromodel, in a simple and consistent manner can predict or forecast a possible network of molecular events that occur during specific cellular events such as gene expression and suppression.

3. Methods and modelling approaches

In this section, we discuss the current approaches and algorithms that were applied to study each epigenetic component influencing DNA methylation mechanisms. The use of Fourier Transformation to detect patterns in specific genes extracted from human genome databases is elaborated. This is followed by a detailed explanation of a stochastic algorithm recently developed, and its application on the gene datasets, to predict histone modifications corresponding to changes in DNA methylation levels.

3.1 Application of fourier transformation

The main aim is to use collateral data (or meta data) based on information from literature, (Yu et al., 2008) to refine our understanding of the complex epigenetic system. The focus here is to investigate the human genome for multiple patterns of specific dinucleotides (AA, TT, AT) and (CG - discussed here), that play a major role in epigenetics. As stated before, recurrent evidence, (Glass et al., 2004) suggests that distribution of specific dinucleotides control events like DNA methylation and chromatin remodeling. The methylating enzymes (DNMT) help to monitor the location and level of DNA methylation, in all types of cells based on these distributions. Hence among the available methods in time-series analyses, Fourier Transformation was chosen to study the frequency domain of specific components in spatially (or sequentially), varying DNA sequences.

Input data or DNA sequences obtained using Map viewer, NCBI database (www.ncbi.nlm.nih.gov) and UCSC genome browser (http://genome.ucsc.edu) were classified and tabulated into three sets namely - (i)19 Genes, (ii) non-coding regions near the genes and, (iii) All CpG islands in chromosome 21, for Fourier analysis. Details of specific genes, chosen due to their association with disease conditions, are given in Table 1.

Figure 1 shows how the CG patterns are screened for auto correlation, (associated with epigenetic mechanisms). Following screening, the amplitude of Fourier Wave Function for contributing periodicities was derived for the 19 genes, corresponding non coding regions and all CpG islands present in chromosome 21, (using equation 1).

3.2 Results on fourier methods

Fourier analysis of dinucleotide patterns in human DNA sequences, seeks to determine significant DM levels associated with these features. In particular, CG patterns are of interest, as this dinucleotide is known to be involved in DNA methylation. Figure 2 represents average

S.No.	Genes	Diseases associated with Genes
1.	PRSS7	Enterokinase Deficiency
2.	IFNGR2	Arthritis Lupus Erythematosus
3.	KCNE1	Jervell and Lange–Nielsen syndrome type 2 (JLNS2)
4.	MRAP	Glucocorticoid Deficiency type 2 (GCCD2)
5.	IFNAR2	Myeloid Leukemia, Hepatocellular Carcinoma, Behcet Syndrome, lung and bladder cancer
6.	SOD1	Amyotrophic Lateral Sclerosis type 1 (ALS1)
7.	KCNE2	Atrial fibrillation familial type 4 (ATFB4)
8.	ITGB2	Leukocyte Adhesion deficiency type I (LAD1)
9.	CBS	Atherosclerosis, Atherosclerosis, Coronary, Breast cancer and cystathionine beta-synthase deficiency
10.	FTCD	Glutamate Formiminotransferase Deficiency (GLUFORDE)
11.	PFKL	Mediterranean Myoclonus
12.	RUNX1	Asthma, Myeloblastic Leukemias
13.	COL6A1	Bethlem myopathy (BM)
14.	COL6A2	Bethlem myopathy (BM), Ullrich Congenital Muscular Dystrophy (UCMD), Autosomal Recessive Myosclerosis
15.	PCNT2	Microcephalic Osteodysplastic Primordial Dwarfism type 2 (MOPD2)
16.	CSTB	Neurodegenerative Disorder
17.	LIPI	Dyslipidemia
18.	TMPRSS3	Deafness and Nonsyndromic
19.	APP	Alzheimer's Disease, Dementia, Attention Deficit and Oppositional Defiant disorder

These gene sequences were used in Fourier Analyses.

Table 1. Dataset containing Genes and Diseases associated with them

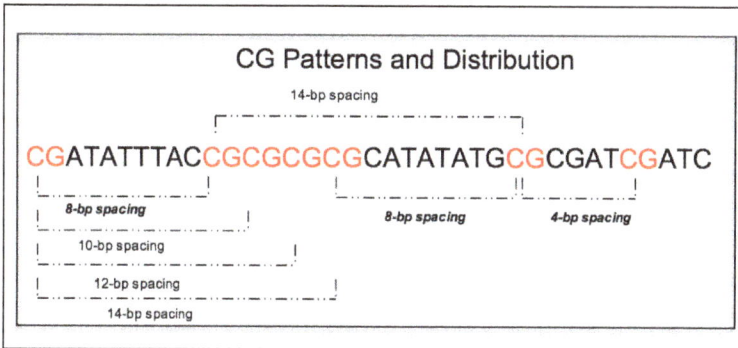

Fig. 1. Distribution of CG in Human DNA sequences.

amplitudes of the power spectrum for all values of CG periodicities possible. Genes/coding regions show an apparent peak at 3bp, which might be expected due to the codon bias in translating to amino acids, (Hosid et al., 2004). CpG islands, (throughout chromosome 21), also contribute to the peak at a periodicity of 3bp since these are present near the promoter

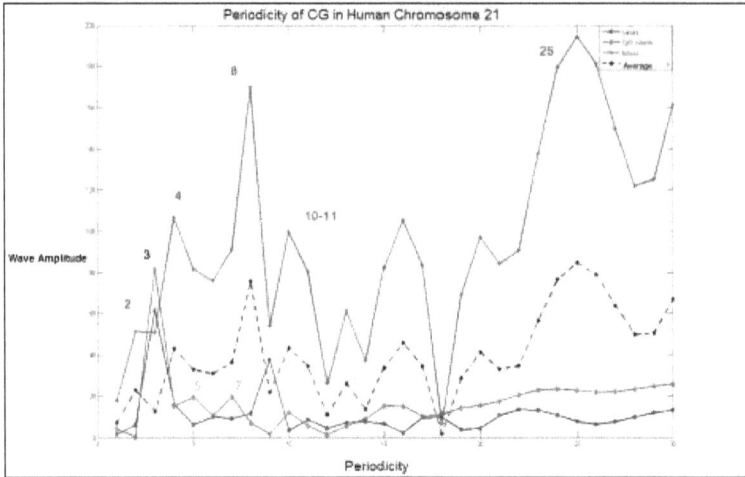

Fig. 2. Fourier analysis (Periodicity Vs Average Wave Amplitude) of global periodicities of CG dinucleotides in 19 Genes (blue line), non-coding near them (red line) and all CpG Islands (green line) in chromosome 21. The average of the 3 region levels is shown as a dotted line.

regions [7]. A 7bp spacing is also observed, probably due to repeats containing CG, in an island located near methylated regions, (Glass et al., 2007). The placement of CG after 3bp, in genes and even more densely clustered in CpG islands prevents the DNMT complex from naturally methylating those regions, (Glass et al., 2004). Hence spacing repeats of CG dinucleotides, can be used to confirm a CpG island, in addition to the dinucleotide based criteria in any input sequence, (Takai & Jones, 2002). One of the more prominent and interesting features can be noted in the non-coding regions, which display unexplored patterns (between 24 and 26bp). Research indicates that 8bp, and also 4bp intervals, (preferred by satellite/short repeats), (Glass et al., 2004), in this dinucleotide, attract DNA methylation complexes. In fact, genes that are silenced in germ cells by the *De novo* methylation mechanism, have these distributions near their promoters. Another peak, observed in Figure 2, between 10 to 11bp periodicity has been confirmed to support genomic structural condensation, (Glass et al., 2004). Other peaks, at periodicity of 15 and 20bp, are less persistent and are possibly due to noise in relation to dense repeat regions in chromosome 21.

The hitherto unexplored periodicity of an interval of length 24 to 26bp, in the non-coding region is less readily explained, but may be connected to DNA methylating mechanisms. A major clue, indicated in (Li. et al., 2010), is the appearance of several million repetitive 25-mers in the human genome. Although not uniform throughout the chromosome 21, this occurrence is known to be high, on average in the human genome. Furthermore in a recent paper, (Yin & Lin, 2007), the authors explain that piRNA or Piwi protein associated iRNA[8], which is significantly involved in cellular processes and propagation of *de novo* DNA methylation is usually of length 24 to 26 nucleotides, (Raghavan et al., 2011). This

---

[7] Promoters are blocks of DNA sequences that control expression for a set of Genes
[8] iRNA is an unusual type of single stranded RNA derived from DNA which help in blocking genomic information for protein production.

new evidence is only a part of the story of human DNA sequence analyses, especially with respect to differential gene expression, as controlled by epigenetics. The average plot as a test of confirmation, represented by dotted line in Figure 1, appears to retains the feature of major peaks at 8, 24, 25 and 26bp for all 22 chromosomes, which could be proposed as standard "marker patterns" of the human genome. Thus FT methods helped to identify possible CG distributions both previously reported and unexplored and to furnish supportive evidences on their corresponding biological significance. Following the initial data analysis, the sequences were investigated for possible histone modifications using our novel stochastic tool based on fixed initial DNA methylation levels.

3.3 Conceptualization of Epigenetic Micromodel – (EpiGMP)

The initial attempt to mimic the biological epigenetic structure is illustrated in reference, (Raghavan et al., 2010) which shows a simplified construction of our model. The status of epigenetic profile in the model is defined in terms of the corresponding DNA Methylation and associated Histone Modifications and model execution portrays the evolving interactions or interdependencies of the epigenetic elements. This section explains how histones were encoded and chosen for defined levels of DM. Information, (Kouzarides, 2007), on the number and type of amino acids for each histone type provides inputs to the model before the simulation. Table 2 gives the details of the number of amino acids, their positions, the

S.No.	H. Type	Amino acid No./String size	Amino Acid & Position	Modification	No. of States
1.	H1	zero	-	-	-
2.	H2A	Four	S1-R3-K5-K9	Ph-Met-Ace-Ace	16
3.	H2B	Ten	K5-S10-K11-K12 S14-K15-K16 K20-K23-K24	Ace/Met-Ph-Ace-Ace Ph-Ace-Ace Ace-Met-Ace	1536
4.	*H3	Six	R2-T3-K4 R8-K9-S10 T11-K14-R17 K18-T22-K23 R26-K27-S28 T32-K36-K37	Met-Ph-Met Met-Ace/Met-Ph Ph-Ace/Met-Met Ace/Met-Ph-Ace/Met Met-Ace/Met-Ph Ph-Ace/Met-Met	6300
5.	H4	Five	S1-R3-K5-K8-K12	Ph-Met-Ace-Ace-Ace/Met	48

Details of specific amino acids and their corresponding modifications in all histone types.
* - H3 has a special type of representation based on amino acid type and the corresponding modification. K - Lysine, S - Serine, T - Threonine, R - Arginine, Ace - Acetylation, Met - Methylation, Ph - Phosphorylation, citepThomas

Table 2. Amino Acid Positions and Modifications

corresponding modification types and the possible number of histone states generated, (Allis et al., 2007; Cedar & Bergman, 2009; Jenuwein & Allis, 2001; Kouzarides, 2007; Turner, 2001). These data are stored in the model as possible combinations of histone modifications that exist in the real epigenetic system. The modifications for each amino acid are assigned a value between 0 and 3 (acetyl -1, methyl -2 phosphate -3 and no modification - 0), which can generate libraries of strings with varying length based on histone type. These numerical strings represent histone modification state in a precise and encoded form. In the previous and current model versions, each string is considered as a node that can be visited during simulation based on a Markov chain - transition probability. A large number of strings exist for each histone type to be sampled due to the fact that each histone has many amino acid modifications, (Raghavan et al., 2010). For example, in case of H2A, a histone *state* or node whose string length is 4 here would be "3011". In this node, the Serine amino acid is phosphorylated and Lysine 5 and 9 are acetylated. A time-step or *Iteration* of the model

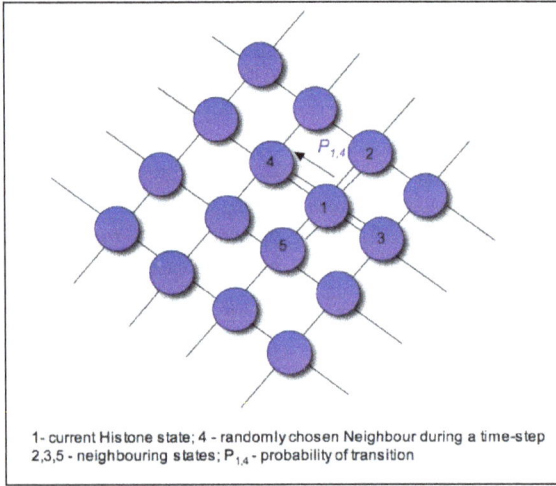

1- current Histone state; 4 - randomly chosen Neighbour during a time-step
2,3,5 - neighbouring states; $P_{1,4}$ - probability of transition

Fig. 3. The movement between active nodes or histone modifications in our model. Based on a random sampling, system shifts to node 4 from 1, based on an appropriate probability of transition. For example, if in case of H2A histone type, state 1 = "0000" and 4="3000", (Raghavan et al., 2010).

corresponds to moving between possible nodes, (i.e. if system chose to modify an amino acid) or remaining in the same node. Consequently, only one change or modification is made at each iteration when the model randomly samples between the possible histone states, based on probability of shift, (as shown in Figure 3). The potential shift to a "neighbouring state" from the current histone state is calculated during every iteration of the model. Computational graphs[9] or tables, of varying sizes based on the type of histone, are used in the system to store occurrence of dynamic modifications. These networks of graphs represent the level of modifications in all histone types and are used to calculate system outputs over several iterations. Our model can also handle multiple additions of the same modification in an amino acid (Mono/di/tri acetylation, methylation or phosphorylation, (Kouzarides, 2007)). Although this is invisible to the user, it is taken into account during calculation of global modification levels in each nucleosome. Hence for individual histone type, the modifications

---

[9] This is the application of graph theories which refers to use of appropriate data structures to store data whenever necessary.

are updated at each iteration, based on the influence of the DNA methylation values and output values of gene expression levels are calculated as depicted in Figure 4 and in reference, (Raghavan et al., 2010).

3.3.1 Epigenetic interdependency

A simple yet strong and well defined inter-dependency exists between histone evolution, transcription rate and level of DNA methylation inside each computational Block (or object, (Raghavan et al., 2010)). There are 3 main interactions in our model. The main dependency

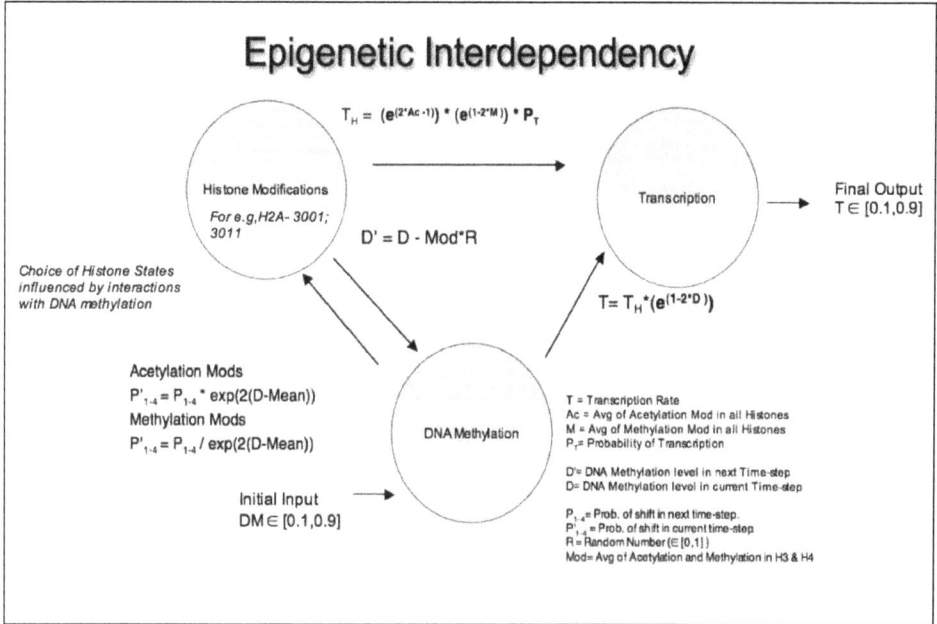

Fig. 4. Interactions between Epigenetic Elements in the Complex System. DM, associated with CG patterns in the DNA sequences and HM alter over each time step. *Transcription*, the output based on both parameters is calculated at regular intervals.

is mutually between Histone modifications and DNA methylation. Here the transition probability of histone states is altered by DNA methylation values, through use of exponential equations hence allowing the system to choose modifications preferentially. This crucial step is based on cumulative information extracted from laboratory experiments, which mention that specific patterns of modifications are explicitly preferred to other types during different levels of DNA methylation. Here, probabilities of shift, provide a window of control to introduce stress to the system so as to see how the output parameters fluctuate over several time-steps. The system is perturbed or subjected to stress through random initial probabilities for histone evolution, (or Monte Carlo based simulation) over different independent trials and subsequently system behaviour can be observed for changes in HM and DM based on their interactions.

Conversely, DM values are recalculated, *conditionally*, from average protein modification levels. This conditional step in DM calculation, has been implemented since literature states that DNA methylation levels are usually stable and less perturbed over several generations. The total output is expressed as "Transcription" which is calculated based on methylation levels in sequences and corresponding histone modifications. Details on the mathematical interdependency of the variables in the model are depicted clearly in Figure 4, (Raghavan et al., 2010). Results obtained from repeated simulation attempts are explained in the next section.

3.3.2 Simulation of combined model

The model consisting of DNA sequences and CG patterns together with histone states is executed to observe evolution of Histone modifications associated with DM in sequences similar to the real system. The steps given below explain the simulation process. The "Blocks" referred from here, are the computational representation of gene or island blocks of sequences within the EpiGMP model framework.

1. *Read and Store Inputs*
   (a) Histone Data -The possible combinations of Histone modifications as described in, (Raghavan et al., 2010) – states and transition probabilities.
   (b) DNA sequences with information on CG distribution throughout sequences are stored as well
   (c) User Selected Values are provided –
       i. Default Parameters: Maximum number of iterations(or time-steps), time-intervals and DNA methylation per a Block in a specific time-step.
       ii. Optional Parameters: preferred histone states in one or more blocks, set by the user (location during a time interval)

2. *Create Objects*
   (a) In one Block – Nucleosomes (number based on DNA sequence length) are created. Each nucleosome object, is assigned nine histone types (default) and 3 modification tables/graphs for each histone.

3. *Simulate*
   (a) Allow Markov Shifts among possible histone states for choice of solution.
   (b) For specific time-intervals, calculate DNA methylation if needed and output parameters: Transcription (based on interdependencies as in Figure 4).
   (c) Continue process till maximum number of iterations reached (for example 10,000 time steps).

4. *Store Outputs*
   (a) Results for the specified time interval, inside each Block –
       i. Transcription rate
       ii. DM value (assumed to be methylation of each CG dinucleotide)
       iii. Count of possible histone node visited per nucleosome

3.4 Model assumptions

As the major focus is on HM and DM progression, a few simplified assumptions were made to test the EpiGMP model reliability.

1. The model currently handles only three modifications i.e. Acetylation, Methylation and Phosphorylation as their biological role is known, (Kouzarides, 2007). More types of modifications can be included, given empirical or theoretical evidence on their significant contributions. (e.g. Role of Ubiquitination in H2B amino acids.)

2. One type of CG distribution, based on results from Fourier transformation method, i.e. CpG islands and gene blocks as shown in Table 1 are tested for prediction of possible histone modification under varying levels of DM.

3. H2A, H2B and H4 are encoded in a similiar fashion as explained above. However, H3 histone type has a large number of modifiable amino acids that can generate millions of possible histone states. Hence, to handle the large dataset, a special representation mode that could compress the possible histone states/nodes was developed. Methods to encode this histone type has been discussed in detail in, (Raghavan et al., 2010).

4. Independent simulation was carried out with three initial random transition probabilities. These values are generated by a system defined function (based on a pseudo random number generator - Mersenne Twister, which is robust, has a large range of period and a high order of dimensional equidistribution, (Matsumoto & Nishimura, 1998)). Hence the results obtained and discussed are the average of the three independent simulation trials.

This is a more advanced model in comparison to the one developed in (Raghavan et al., 2010), which considers both analysis of CG dinucleotide distributions and choice of histone modifications over the chosen sequences. The aim here was to observe histone evolution with DM associated sequence patterns in a manner similiar to real system and results thus obtained from this study are discussed in the next section.

4. Results and discussion

In order to investigate the system behaviour, 19 specific genes, and all CpG islands present in chromosome 21, were chosen. The datasets were preferred since they contain the maximum number of CG dinucleotides with 3bp intervals. These base pairs with specific distributions (usually associated with differentially expressed genes and promoters, (Allis et al., 2007)) were assigned DNA methylation values, based on equations shown in Figure 4. Outputs namely, Histone states, progress in transcription rate and DNA methylation, for the whole dataset were recorded every 1,000 time-steps (total number of time steps being 10,000). Although the system can trace and report evolution of all 4 types of histone, we discuss here only 2 types namely H4 and H2A. The following Figures 5 and 6 show the expected values of each histone node being chosen during several iterations over the 3 independent simulation trials.

The DNA methylation was set to a range of values, $\in [0.1, 1.0]$, for the 3 simulation runs (results not shown here). For initial values, (<0.2) of DM, the systems preferred least methylation modifications and inversely more acetylation changes. But for more sets of initial methylation values in the range $[0.3, 0.6]$, and those (>0.75), methylation was apparently chosen repeatedly among other histone modifications. This was due to evolution of DM values to a closed range of [0.95, 1.0] over a time period of ($> 10,000$) iterations. Hence to observe histone evolution we discuss in detail two sets of results observed under (i) Low DM (<0.15 or 15%), and (ii) High DM (>0.85 or 85%). These simulations demonstrate effective emulation of the biological process of transcription of genes (e.g. Onco-genes expression) for low DNA methylation levels and reverse case of high DNA methylation and gene suppression (e.g. silencing of tumor suppressor/control genes). Figure 5 contrasts the different modifications observed

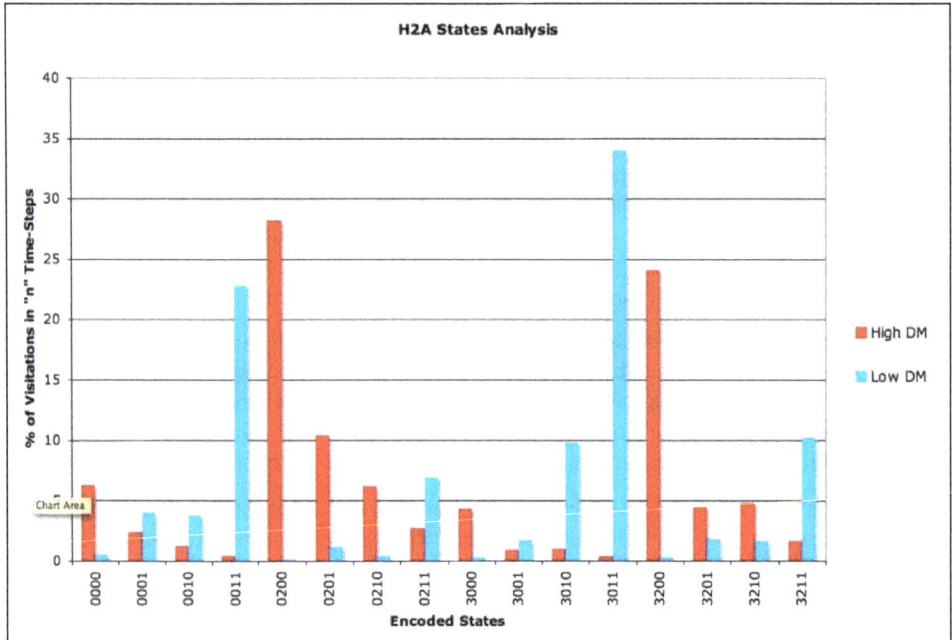

Fig. 5. A Comparison between the average (over 3 Simulation runs) preferences of H2A states for high (red) and low(blue) DNA Methylation Levels.

in H2A during high and low methylation conditions averaged over 3 simulation runs in all nucleosomes. During high methylation condition (DM level $> 85\%$), selective states such as the 5th and 13th were most preferred i.e Arginine was methylated in H2A most frequently. Evidence, (Eckert et al., 2008) indicates that specific cell types, do not contain this modification and hence develop into tumorous cells, (this is an explicit evidence of down regulation of methylation modification leading to tumor growth). Under lower DM conditions ($< 15\%$), the 4th and 12th states were most visited implying high priority to Lysine 5 and 9 modifications. Acetylation of Lysine 5 or (K5) is notably found more during gene expression while that of K9, is an unexplored modification, (Cuddapah et al., 2009; Wyrick & Parra, 2008). This hitherto unreported acetylation in H2A, could be a potential modification that supports gene expression. Figure 6 shows the preferences of H4 states for high and low DNA Methylation levels. Under low DM levels (initially set by the user), acetylated amino acids states, such as the 11th, 35th and 47th predominated i.e. states containing acetylated amino acids such as K5, K8 and K12 (see Table 2) were highly visited. Even when the probability assigned to the three preferred states was lowered for a test set, the system preferred the other two states

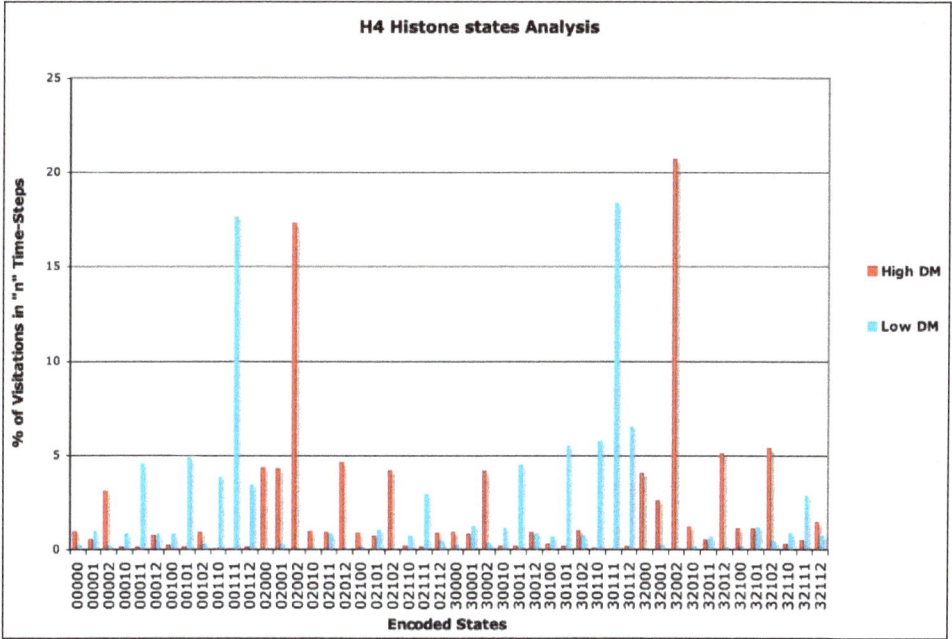

Fig. 6. A Comparison between the average (over 3 Simulation runs) preferences of H4 states for high (red) and low (blue) DNA Methylation Levels.

containing lysine acetylation. Such consistent results demonstrate the ability of our model to reproduce the presence of the modifications mentioned above, during transcription, (as reported, (Taplick, 1998; Zhang et al., 2007) in particular, during expression of oncogenes). For higher levels of DNA methylation (>0.85, Figure 6), the preference is more towards choosing methylated histone states leading to reduced transcription rate. During this high methylation condition, states such as the 15th, 39th and 45th i.e. methylation of K12 was predominantly high. Such strong evidence, (removal of acetylation and adding methylation to amino acids) of modification to a crucial lysine position in H4, is a potential indicator of transcription repression and initiation of DNA methylation. Similiar to the observation in H2B (as recorded in literature, (Zhang et al., 2003)), there is appearance of serine phosphorylation (states 39 and 35 in Figure 6) during both conditions of DM values, which show the importance of this specific modification during expression or otherwise. This suggests that the modification could be present from the time that the H4 histone complex was formed, (Barber et al., 2004) and aid in structural condensation.

Hence a stochastic model of this type can successfully simulate simple concepts to show the possible molecular modifications that appear during different genetic events. The DM fluctuation over specific time-intervals is associated with specific CG dinucleotides in the sequences. In this example, effect of DM and its influence on histone modifications have been effectively illustrated. Futhermore, the same model can be used to study other CG distributions such as 7bp spacing in CpG islands, which can be validated against information on disease associated genes.

5. Conclusion and future directions

In this chapter, the background to epigenetics, their association with diseases and the developments of computational methods and modelling approaches to understand the complexity in this field have been discussed. Significance of growth of experimental data in recent years, which enables detection of DNA methylation influence in disease onset has also been considered. Early attempts at computational methods and models dealing with (i) association of DNA sequences and DM, and (ii) Interdependencies between DM and HM have been explained in detail. Further, we propose approaches to analyse the two elements such as DNA sequence patterns and HM evolution and their influence over DNA methylation mechanisms. Finally, evaluation of success achieved through such computational attempts is illustrated briefly in our results section.

The application of Fourier techniques helped to understand how the sequence patterns appear within the genome and also postulate their control over DM. The results consist of a range of distributions, which are analysed in relation to possible biological significance. The broad spectrum thus obtained, can be attributed to the self-adapting and dynamic nature of the human genome exhibited through events such as self mutations (mC to T, (Doerfler & Böhm, 2006)) or reassignment of DNA methylation patterns across different cells. This ability of cells to dynamically adapt to environmental stimulus by introducing molecular modifications or positive mutations, (which changes nucleotide distributions), is also referred to as "Phenotypic Plasticity". Based on such analyses of the human DNA sequences, further investigations of dynamic histone protein modifications were predicted using novel stochastic modelling techniques.

The EpiGMP model, based on this stochastic approach, has reported histone modifications that were previously recorded and also unexplored modifications and compared them with data recorded through laboratory experiments. For example, the effect of H2A modifications such as Arginine methylation, are not as explicit and strong as H4 but their scattered presence in specific cells/cancer conditions indicates their contribution in the big picture. Hence, based on comparison with experimental and the model results, we conclude that histone modifications while not always consistent do have a role in controlling gene expression and chromosome condensation in human genome.

DNA methylation controls the direction of histone evolution, i.e. the states visited for high levels of DM are not visited for low levels and vice versa. This robust result, obtained for three simulation trials, is a good indication of the reliability of EpiGMP model. This consistency has helped to cluster and predict characteristic histone modifications under defined DNA methylation levels, thus efficiently emulating the real system to an accurate level. The idea behind designing a comprehensive model to mimic epigenetic mechanisms is to address and utilize all of the distributed data available in literature. A generic model, which can simulate conditions of any epigenetically associated disease and report results, is the ideal target. As mentioned in the background section, basic quantitative analyses have reinforced the presence of *apriori* patterns and hence this has given rise to a vital need to design a predictive model with a common framework that can be tested for most conditions. The main advantages of our approach lie in modelling (for all histone types simultaneously) cumulative information such as increased acetylation modifications which occur during gene expression and more methylation during suppression. A further advantage is the expandable layout, which can be developed to accommodate more data in future (incorporating more modifications and multiple sequence patterns).

5.1 Parallelization of EpiGMP model

Parallel computing is an approach, which carries out calculations simultaneously or in a parallel manner using many computational resources at the same time. It is extensively used when there is a high complexity of computation or the data are very large. In our case, the current model definitely requires parallelization, because the random algorithm has to compute outputs from a large sample space, for long iterations or time-steps and most importantly to study several molecular events at genome level. Simulation of the model when applied to objects of size of a chromosome (for more than 1 million time steps) would require heavy computational resources. As a consequence, a parallel and serial version of the model have been developed simultaneously, which is discussed in detail, (Raghavan & Ruskin., 2011; Raghavan et al., 2010).

The field of epigenetics is growing rapidly with important findings being reported on a regular basis. The complex epigenetic layer in humans also houses secondary events through which control is exercised within the cell. For example, chromatin dynamics, which rely on molecular interactions (DNA molecules and proteins such as polycomb), play a major role in long term silencing of genes. Our current work involves, applying this stochastic framework to real gene networks extracted from epigenetic databases such as StatEpigen, (http://statepigen.sci-sym.dcu.ie/) in order to predict cancer from simple molecular interactions. To improve realism further, future models must account for secondary effects such as chromatin remodeling, and also role of external proteins such as methyl binding proteins, transcription binding proteins, polycomb amongst others, (Allis et al., 2007) for cellular events. The final goal is to build integrated/hybrid models, combining agent-based and network approaches across several scales, which can be applied to precisely predict epigenetic events based on multiple factors. This "bottom-up" approach facilitates low-level information processing between different molecules so as to understand how the phenotype or physical appearance of an organism evolves at higher level especially under abnormal conditions.

The Fourier analysis on DNA sequences was performed using Matlab software and the source code is available on request. The serial version of EpiGMP model has been developed mainly using C++ language, while routines from OpenMP and MPI libraries were included for the parallel version.

6. Acknowledgements

We gratefully acknowledge financial support from Science Foundation Ireland, project 07/RFP/CMSF724, in the early stages of this work and, subsequently, Complexity-Net /IRCSET pilot award. We thank ICHEC, (Irish High End Computing Centre) for providing access to major computational facilities, required for background work.

7. References

A'Hearn, M. F., Ahern, F. J. & Zipoy, D. M. (1974). Polarization Fourier spectrometer for astronomy, *Applied Optics* 13(5): 1147–1157.

Allis, C. D., Jenuwein, T., Reinberg, D. & Caparros, M. L. (2007). *Epigenetics*, Cold Spring Harbor Press.

Barber, C. M., Turner, F. B., Wang, Y., Hagstrom, K., Taverna, S. D., Mollah, S., Ueberheide, B., Meyer, B. J., Hunt, D. F., Cheung, P. & Allis, C. D. (2004). The enhancement of histone H4 and H2A serine 1 phosphorylation during mitosis and s-phase is evolutionarily conserved, *Chromosoma* 112(7): 360–371.

Baylin, S. B. & Ohm, J. E. (2006). Epigenetic gene silencing in cancer – a mechanism for early oncogenic pathway addiction, *Nature Review Cancer* 6(2): 107–116.
URL: *http://dx.doi.org/10.1038/nrc1799*

Bock, C., Walter, J., Paulsen, M. & Lengauer, T. (2007). CpG island mapping by epigenome prediction, *PLoS Computational Biology* 3(6): e110.

Boyer, L. A., Plath, K., Zeitlinger, J., Brambrink, T., Medeiros., L. A., Lee, T. I., Levine, S. S., Wernig, M., Tajonar, A., Ray, M. K., Bell, G. W., Otte, A. P., Miguel Vidal, a. D. K. G., Young, R. A. & Jaenisch, R. (2006). Polycomb complexes repress developmental regulators in murine embryonic stem cells, *Nature* 441(7091): 349–353.

Cedar, H. & Bergman, Y. (2009). Linking DNA methylation and histone modification: Patterns and paradigms, *Nature Review Genetics* 10(5): 295–304.

Chahwan, R., Wontakal, S. N. & Roa, S. (2011). The multidimensional nature of epigenetic information and its role in disease, *Discovery Medicine* 11(58): 233–243.

Chamberlain, S. J. & Lalandea, M. (2010). Neurodevelopmental disorders involving genomic imprinting at human chromosome 15q11–q13, *Neurobiology of Disease* 39(1): 13–20.

Chi, P., Allis, C. D. & Wang, G. G. (2010). Covalent histone modifications – miswritten, misinterpreted and mis-erased in human cancers, *Nature Reviews Cancer* 10(7): 457–469.
URL: *http://dx.doi.org/10.1038/nrc2876*

Clay, O., Schaffner, W. & Matsuo, K. (1995). Periodicity of eight nucleotides in purine distribution around human genomic CpG dinucleotides, *Somatic Cell and Molecular Genetics* 21(2): 91–98.

Collas, P. (2010). The current state of chromatin immunoprecipitation, *Molecular Biotechnology* 45(1): 87–100.

Collins, F. S., Patrinos, A., Jordan, E., Chakravarti, A., Gesteland, R. & Walters, L. (1998). New Goals for the U.S. Human Genome Project: 1998–2003, *Science* 282(5389): 682–689.

Conlon, T., Ruskin, H. J. & Crane, M. (2009). Seizure characterization using frequency-dependent multivariate dynamics, *Computers in Biology and Medicine* 39(9): 760–767.

Cowan, R. (1991). Expected Frequencies of DNA Patterns using Whittle's Formula, *Journal of Applied Probability* 28(4): 886–892.

Cowley, D. E. & Atchley, W. R. (1992). Quantitative genetic models for development, epigenetic selection, and phenotypic evolution, *Evolution* 46(2): 495–518.

Cuddapah, S., Jothi, R., Schones, D. E., Roh, T., Cui, K. & Zhao, K. (2009). Global analysis of the insulator binding protein CTCF in chromatin barrier regions reveals demarcation of active and repressive domains, *Genome Research* 19(1): 24–32.

Das, R., Dimitrova, N., Xuan, Z., Rollins, R. A., Haghighi, F., Edwards, J. R., Ju, J., Bestor, T. H. & Zhang, M. Q. (2006). Computational prediction of methylation status in human genomic sequences, *Proceedings of the National Academy of Sciences* 103(28): 10713–10716.

Doerfler, W. & Böhm, P. (2006). *DNA Methylation: Basics Mechanisms*, first edn, Springer.

Doerfler, W., Toth, M., Kochaneka, S., Achtena, S., Freisem-Rabiena, U., Behn-Krappaa, A. & Orenda, G. (1990). Eukaryotic DNA methylation – facts and problems, *Febs Letter* 286(2): 329–333.

Eckert, D., Biermann, K., Nettersheim, D., Gillis, A., Steger, K., Jack, H., Muller, A., Looijenga, L. & Schorle, H. (2008). Expression of BLIMP1/PRMT5 and concurrent histone H2A/H4 arginine 3 dimethylation in fetal germ cells, CIS/IGCNU and germ cell tumors, *BMC Developmental Biology* 8: 106.

Ehrlich., M., Sanchez, C., Shao, C., Nishiyama, R., Kehrl, J., Kuick, R., Kubota, T. & Hanash, S. (2008). Icf, an immunodeficiency syndrome: DNA methyltransferase 3b involvement, chromosome anomalies, and gene dysregulation, *Autoimmunity* 41(4): 253–271.

Epps, J. (2009). A hybrid technique for the periodicity characterization of genomic sequence data, *EURASIP Journal on Bioinformatics and Systems Biology* 2009.

Esteller, M. (2007). Cancer epigenomics: DNA methylomes and histone-modification maps, *Nature Reviews Genetics* 8(4): 286–298.

França, L., Carrilho, E. & Kist., T. B. (2002). A review of DNA sequencing techniques., *Quarterly Reviews of Biophysics* 35(2): 169–200.

Fullgrabe, J., Kavanagh, E. & Joseph, B. (2011). Histone Onco – Modifications, *Oncogene* 30(31): 3391–3403.
URL: *http://dx.doi.org/10.1038/onc.2011.121*

Gao, F. & Zhang, C.-T. (2006). GC-Profile: a web-based tool for visualizing and analyzing the variation of GC content in genomic sequences, *Nucleic Acids Research* 34(2): 686–691.

Gertz, J., Varley, K. E., Reddy, T. E., Bowling, K. M. & Pauli, F. (2011). Analysis of DNA methylation in a three-generation family reveals widespread genetic influence on epigenetic regulation, *PLoS Genetics* 7(8): e1002228.

Glass, J. L., Fazzari, M. L., Ferguson-Smith, A. C. & Greally, J. M. (2004). CG di-nucleotide periodicities recognized by the dnmt-3a-dnmt-3l complex are distinctive at retro-elements and imprinted domains, *Mammalian Genome* 20(9-10): 633–643.

Glass, J. L., Thompson, R. F., Khulan, B., Figueroa, M. E., Olivier, E. N., Oakley, E. J., Zant, G. V., Bouhassira, E. E., Melnick, A., Golden, A., Fazzari, M. J. & Greally, J. M. (2007). CG dinucleotide clustering is a species-specific property of the genome, *Nucleic Acid Research* 35(20): 6798–6807.

Goodman, J. W. (2005). *Introduction to Fourier Optics*, third edn, Roberts and Company.

Herzel, H., Weiss, O. & Trifonov, E. N. (1999). 10-11 bp periodicities in complete genomes reflect protein structure and DNA folding., *Bioinformatics* 15(3): 187–193.

Hosid, S., Trifonov, E. N. & Bolshoy, A. (2004). Sequence periodicity of Escherichia coli is concentrated in intergenic regions, *BMC Molecular Biology* 5(1): 14.
URL: *http://www.biomedcentral.com/1471-2199/5/14*

Ito, T. (2007). Role of histone modification in chromatin dynamics, *Journal of Biochemistry* 141(5): 609–614.

Jenuwein, T. & Allis, C. D. (2001). Translating the histone code, *Science* 293(5532): 1074–1080.

Jung, I. & Kim, D. (2009). Regulatory patterns of histone modifications to control the DNA methylation status at CpG islands, *IBC* 1(4): 1–7.

Kaiser, G. (1994). *A Friendly Guide to Wavelets*, sixth edn, Birkhäuser.

Karlić, R., Chung, H., Lasserre, J., Vlahoviček, K. & Vingron, M. (2010). Histone modification levels are predictive for gene expression, *PNAS* 107(7): 2926–2931.

Kouzarides, T. (2007). Chromatin modifications and their function, *Cell* 128(4): 693–705.

Kwon, D., Vannucci, M., Song, J. J., Jeong, J. & Pfeiffer, R. M. (2008). A novel wavelet-based thresholding method for the pre-processing of mass spectrometry data that accounts for heterogeneous noise, *Proteomics* 8(15): 3019–3029.

Li., R., Zhu, H. & Ruan, J. (2010). De novo assembly of human genomes with massively parallel short read sequencing, *Nucleic Acid Research* 20(2): 265–272.

Matsumoto, M. & Nishimura, T. (1998). Mersenne twister: A 623-dimensionally equidistributed uniform pseudo-random number generator, *ACM Transactions on Modeling and Computer Simulation* 8(1): 3–30.

Meng, C. F., Zhu, X. J., Peng, G. & Dai, D. (2009). Promoter histone H3 lysine 9 di-methylation is associated with DNA methylation and aberrant expression of p16 in gastric cancer cells, *Oncology Report* 22(5): 1221–1227.

Morrison, N. (1994). *Introduction to Fourier Analysis*, Wiley-Interscience.

Murrell, A., Rakyan, V. K. & Beck, S. (2005). From genome to epigenome, *Human Molecular Genetics* 14(1): 3–10.

Raghavan, K. & Ruskin., H. J. (2011). Computational epigenetic micromodel - framework for parallel implementation and information flow., *Proceedings of the Eighth International Conference on Complex Systems*, Vol. 8, NECSI Knowledge Press, pp. 340–353.

Raghavan, K., Ruskin, H. J. & Perrin, D. (2011). Computational analysis of epigenetic information in human DNA sequences, *Proceedings of the International Conference on Bioscience, Biochemistry and Bioinformatics 2011*, Vol. 5, International Proceedings of Chemical, Biological and Environmental Engineering, pp. 383–387.

Raghavan, K., Ruskin, H. J., Perrin, D., Burns, J. & Goasmat, F. (2010). Computational micromodel for epigenetic mechanisms, *PLoS One* 5(11): e14031.

Riggs, A. D. & Xiong, Z. (2004). Methylation and epigenetic fidelity, *PNAS* 101(1): 4–5.

Salz, J. & Weinstein, S. B. (1969). Fourier transform communication system, *Proceedings of the first ACM symposium on Problems in the optimization of data communications systems*, ACM, pp. 99–128.

Schones, D. E., Cui, K., Cuddapah, S., Roh, T.-Y., Barski, A., Wang, Z., Wei, G. & Zhao, K. (2008). Dynamic regulation of nucleosome positioning in the human genome, *Cell* 132(5): 887–898.
URL: *http://linkinghub.elsevier.com/retrieve/pii/S0092867408002705*

Segal, E. & Widom, J. (2009). What controls nucleosome positions, *Trends in Genetics* 25(8): 335–343.

Silverman, B. & Linskera, R. (1986). A measure of DNA periodicity., *Journal of Theoretical Biology* 118(7): 295–300.

Strachan, T. & Read, A. P. (1999). *Human Molecular Genetics*, 2 edn, New York: Wiley-Liss.

Su, A. I., Wiltshire, T., Batalov, S., Lapp, H., Ching, K. A., Block, D., Zhang, J., Soden, R., Hayakawa, M., Kreiman, G., Cooke, M. P., Walker, J. R. & Hogenesch, J. B. (2004). A gene atlas of the mouse and human protein-encoding transcriptomes, *Proceedings of*

the National Academy of Sciences of the United States of America 101(16): 6062–6067.
  URL: *http://www.pnas.org/content/101/16/6062.abstract*

Su, J., Zhang, Y., Lv, J., Liu, H., Tang, X., Wang, F., Qi, Y., Feng, Y. & Li, X. (2010). Cpg_mi:
  a novel approach for identifying functional CpG islands in mammalian genomes,
  Nucleic Acids Research 38(1): e6.

Sun, J. M., Chen, H. Y., Espino, P. S. & Davie, J. R. (2007). Phosphorylated serine 28 of histone
  h3 is associated with destabilized nucleosomes in transcribed chromatin, *Nucleic
  Acids Research* 35(19): 6640–6647.

Sundararajan, N., Mao, D., Chan, S., Koo, T.-W., Su, X., Sun, L., Zhang, J., Sung, K.-b.,
  Yamakawa, M., Gafken, P. R., Randolph, T., McLerran, D., Feng, Z., Berlin, A. A. &
  Roth, M. B. (2006). Ultrasensitive detection and characterization of posttranslational
  modifications using surface-enhanced raman spectroscopy, *Analytical Chemistry*
  78(11): 3543–3550.

Takai, D. & Jones, P. A. (2002). Comprehensive analysis of CpG islands in human
  chromosomes 21 and 22, *PNAS* 99(6): 3740–3745.

Taplick, J. (1998). Histone H4 acetylation during interleukin-2 stimulation of mouse t cells,
  FEBS Letters 436(3): 349–352.

Tsonis, A. A., Kumar, P., Elsner, J. B. & Tsonis, P. A. (1996). Wavelet analysis of DNA sequences,
  Physical Review E 53(2): 1828–1834.

Turner, B. M. (2001). *Chromatin and Gene Regulation – Mechanisms in Epigenetics*, 2nd edition
  edn, BlackWell Science Ltd.

Ushijima, T., Watanabe, N., Okochi, E., Kaneda, A., Sugimura, T. & Miyamoto, K. (2003).
  Fidelity of the methylation pattern, its variation in the genome, *Genome Research*
  13(5): 868–874.

Waddington, C. H. (1942). The epigenotype, *Endeavour* 1: 18–20.

Whittle, P. (1955). Some distribution and moment formulae for the markov chain, *Journal of
  the Royal Statistical Society Series B (Methodological)* 17(2): 235–242.

Wyrick, J. J. & Parra, M. A. (2008). The role of histone H2A and H2B post-translational
  modifications in transcription: A genomic perspective, *Biochimica et Biophysica Acta*
  1789(1): 37–44.

Yasuhara, J. C., DeCrease, C. H. & Wakimoto, B. T. (2005). Evolution of heterochromatic genes
  of drosophila, *PNAS* 102(31): 10958–10963.

Yin, H. & Lin, H. (2007). An epigenetic activation role of piwi and a piwi associated piRNA in
  drosophila melanogaster, *Nature* 450(7167): 304–308.

Yu, H., Zhu, S., Zhou, B., Xue, H. & Han, J. (2008). Inferring causal relationships
  among different histone modifications and gene expression, *Genome Research*
  18(8): 1314–1324.

Zhang, L., Eugeni, E. E., Parthun, M. R. & Freitas, M. A. (2003). Identification of novel
  histone post-translational modifications by peptide mass fingerprinting, *Chromosoma*
  112(2): 77–86.

Zhang, L., Su, X., Liu, S., Knapp, A. R., Parthun, M. R., Marcucci, G. & Freitas, M. A.
  (2007). Histone H4 n-terminal acetylation in kasumi-1 cells treated with depsipeptide
  determined by acetic acid. urea polyacrylamide gel electrophoresis, amino acid
  coded mass tagging, and mass spectrometry, *Journal of Proteome Research* 6(q): 81–88.

Zhao, J., Yang, X. W., Li, J. P. & Tang, Y. Y. (2001). DNA sequences classification based on Wavelet package analyses, *WAA '01: Proceedings of the Second International Conference on Wavelet Analysis and Its Applications*, Springer-Verlag.

Zheng, C. & Hayes, J. J. (2003). Structure and interactions of core histone tail domains, *Bioplymers* 68(4): 539–546.

Inheritance of DNA Methylation in Plant Genome

Tomoko Takamiya[1], Saeko Hosobuchi[2],
Kaliyamoorthy Seetharam[4], Yasufumi Murakami[2]
and Hisato Okuizumi[3]
[1]School of Pharmacy, Nihon University ,
[2]Tokyo University of Sciences,
[3]Genetic Resources Center,
National Institute of Agrobiological Sciences (NIAS),
[4]Tamil Nadu Agriculture University,
[1,2,3]Japan
[4]India

1. Introduction

1.1 The role of DNA methylation in mammals and plants

Genomic DNA contains not only information of DNA sequence, but also epigenetic information that is the direct DNA modification by methylation (the addition of methyl group to the 5th carbon of pyrimidine ring of cytosine) and histone modifications (acetylation, methylation, etc). Epigenetic information is closely related to regulation of gene expression. If a methyl group is dislocated to position 5 of the pyrimidine ring of cytosine, the hydrogen bond between complementary GC bases will not be inhibited, but this methyl group is positioned so as to be exposed in the major groove of the double-helix structure of DNA, and according to the genome region/sequence undergoing modification of methylation, gene expression is inhibited by the interaction between the genome and DNA-binding molecules.

Methylated cytosine is very common in plant and mammalian genomes, and plays an important role in the regulation of many cellular processes including X inactivation, chromosome stability, chromatin structure, embryonic development and transcription. DNA methylation in most mammals occurs at cytosine on the CG sequence, a 2-base sequence lined up in the order of cytosine-guanine. In plants, on the other hand, methylation of a non-CG sequence (CNG and CHH, where N is A, G, C or T, and H is A, C or T) frequently exists in addition to methylation of the CG sequence. Moreover, there is a large difference between mammals and plants in methylation dynamics throughout the life cycle [Law & Jacobsen 2010]. In mammals, methylation patterns change dramatically during gametogenesis and early development [Monk *et al.* 1987, and Tada *et al.* 1997]. In mice, the methylation level of the genome decreases after fertilization to the lowest level at implantation, which is rapidly induced at the time of tissue differentiation after implantation. In reproductive cells, moreover, methylation of genomic DNA is eliminated once and then methylation occurs again, but the sex-specific methylation pattern is written according to the gene and is adjusted so that a specific uniparental allele may be expressed. The gene showing sex-specific expression is called the imprinted gene and plays an important role in development and differentiation [Obata *et al.* 1998, Ueda *et al.* 2000, and Davis *et al.* 2000].

In contrast, generational changes in methylation status and inheritance in plants have been unclear. The methylation statuses of some genes are stably inherited [Bender *et al.* 1995, Jacobsen *et al.* 1997, Kakutani *et al.* 1999], but, recent studies show that DNA methylation patterns are altered in F1 hybrids from interspecific or intraspecific crossing [Matzke *et al.* 1999, Wendel 2000, Shaked *et al.* 2001, Pikaard 2001, Madlung *et al.* 2002, Comai *et al.* 2003, Liu *et al.* 2004, Dong *et al.* 2006, and Akimoto *et al.* 2007]. These alterations might be caused by the interaction among alleles, and/or the change of epigenetic regulation. For example, imprinted genes that have sex-dependent methylation patterns in endosperm have been identified in plants, and they play an important role in the control of flowering or seed development [Grossniklaus *et al.* 1998, Kiyosue *et al.* 1999, Kinoshita *et al.* 1999, Choi *et al.* 2002, Xiao *et al* 2003, Kohler *et al.* 2003, Scott *et al.* 2004, Kohler *et al.* 2005, Baroux *et al.* 2006, and Gehring *et al.* 2006]. Another example, paramutation is an allele-dependent transfer of epigenetic information, which results in the heritable silencing of one allele by another [Brink 1956, and Coe 1959]. In recently, Shiba *et al.* (2006) suggested that tissue-specific monoallelic *de novo* methylation in F1 involved in determining the dominance interactions that determine the cruciferous self-incompatibility phenotype. These analyses of DNA methylation inheritance may help to identify new important genes, such as imprinted gene, and to further clarify the biological significance of DNA methylation.

1.2 Concept of RLGS as a tool for DNA methylation analysis

Restriction Landmark Genome Scanning (RLGS) [Hayashizaki *et al.* 1993, Okazaki *et al.* 1995] is a unique quantitative approach well suited for simultaneous assay of methylation status [Costello *et al.*, 2002]. The other genome wide methylation analyze methods *viz.*, Tiling microarrays [Zhang *et al.* 2006] and methylation-sensitive amplification polymorphism (MSAP) [Reyna-Lopez *et al.* 1997, Xiong *et al.* 1999] are the comprehensive or easily applied method, respectively, but RLGS excels these methods because of its reduced cost and short span of experimental time (3 days). In RLGS, intensity of the RLGS spots directly reveals the methylation level and partial methylation such as imprinting, whereas in other methods the methylation levels are not inferred directly from the results. These important advantages of RLGS rank it as an appropriate method for genome wide methylation survey. This method had been used for development of genetic linkage maps [Okuizumi *et al.* 1995A, Okuizumi *et al.* 1995B], methylation analysis in tumor tissue [Ohsumi *et al.* 1995, Miwa *et al.* 1995, Wang *et al.* 2009], and identification of imprinted genes in mammals [Hayashizaki *et al.* 1994A, Shibata *et al.* 1995, Plass *et al.* 1996] and based on which several interesting research in epigenetics and genetics such as alteration of genomic DNA methylation [Takamiya *et al.* 2008B, Takamiya *et al.* 2009] and genetic diversity study [Okuizumi *et al.* 2010] had been carried out. The "*in silico* RLGS", is a software originally developed and named by us, and now the name and the concept are spread to other researchers [Matsuyama 2008]. This software can be utilized for the RLGS analysis of the organisms for those the whole genome sequence is available. One among such organism is Rice which enables the utilization of "*in silico* RLGS" analysis

[Takamiya *et al.* 2006] because of the availability of its whole genome sequence [International Rice Genome Sequencing Project, 2005].

2. Methodology of RLGS to detect DNA methylated sites efficiently in the plant genome

2.1 Method development to use isoschizomer restriction enzyme such as *Msp*I and *Hpa*II

RLGS is a high-speed genome scanning system. It employs direct end-labeling of the genomic DNA digested with rare-cutting restriction enzyme such as *Not*I, followed by high-resolution two-dimensional (2-D) electrophoresis. Thousands of loci with high reproducibility can be detected as spots on the 2-D pattern in this method. A lot of methylated sites can be analyzed using the conventional RLGS method because the recognition site (GCGGCCGC) of the first cutter *Not*I [Hayashizaki *et al.* 1993, Watanabe *et al.* 1995], which is a methylation sensitive restriction enzyme (Fig. 1A), is often located in the CpG islands [Bird 1992]. But, this conventional method has bottle-neck in distinguishing methylation polymorphism from sequence polymorphism.

This bottle-neck of the conventional RLGS urges for the development of improved RLGS method (Figs. 1A and1B). In this improved method, isoschizomer restriction enzymes (*Msp*I and *Hpa*II) were used for the direct detection of methylated sites, that we produce 2 patterns; the "[*Msp*I] pattern" employs the restriction enzyme combination of *Not*I-*Msp*I-*Bam*HI as the 1st-2nd-3rd cutter, and another "[*Hpa*II] pattern" uses *Not*I-*Hpa*II-*Bam*HI as the 1st-2nd-3rd cutter. Rationale to utilize the isoschizomer is, they recognize the same sequence (CCGG) but with difference in methylation sensitivity. For example, the *Msp*I cleaves the CCGG at 2nd C which is methylated (C5mCGG) but *Hpa*II doesn't cleave because of its differential sensitivity for the methylation. The different methylation sensitivity between *Msp*I and *Hpa*II is reflected in difference of RLGS spot patterns (Fig 1B). Briefly, the spot which is only detected in either [*Msp*I] or [*Hpa*II] pattern show the *Msp*I/*Hpa*II site of the RLGS spot is methylated. We called this spot as "methylated spot." The additional qualities of this improved RLGS method were (1) easy identification of methylated sites and their location in genome, (2) methylation of coding regions are surveyed efficiently and (3) its ability to scan methylation sites in an individual or a tissue. Furthermore, in an un-sequenced species and even in a cloned organism, this improved method distinguishes DNA mutation from the methylation changes.

Genomic DNA was isolated from 8-week old leaf blades and leaf sheath of Nipponbare, Kasalath, and F1 plants (Crossing subsp. *japonica* cv. Nipponbare as the seed parent with subsp. *indica* cv. Kasalath as the pollen parent gave F_1 hybrids designated NKF_1) grown for 8 weeks. Isolated genomic DNA (0.2 µg) was treated with 2 U of DNA polymerase I (NIPPON GENE, Tokyo, Japan) in 10 µL of blocking buffer (10 mM Tris-HCl, pH 7.4, 10 mM $MgCl_2$, 1 mM dithiothreitol (DTT), 0.4 µM dGTP, 0.2 µM dCTP, 0.4 µM ddATP and 0.4 µM ddTTP) at 37°C for 20 min. Thereafter, the polymerase I was inactivated by incubating at 65 °C for 30 min. Then the DNA was digested with 20 U of *Not*I (NEB, Beverly, MA, USA) in volume of 20 µL. The cleavage ends were filled in and labeled with ^{32}P in the presence of 1.3 U of Sequenase ver. 2.0™ (USB, Cleveland, OH, USA), 0.33 µM [α-^{32}P] dGTP (3,000 Ci/mmol), 0.33 µM [α-^{32}P] dCTP (6,000 Ci/mmol) and 1.3 mM DTT at 37 °C for 30 min in 22.5 µL. Next, this reaction mixture was incubated at 65 °C for 30 min to inactivate the enzyme. The processed sample was divided into two tubes. One was digested with 25 U of *Msp*I (Toyobo, Tokyo, Japan) and the other was treated with 25 U of *Hpa*II (Toyobo) and incubated at 37°C for 1 h. Each sample was fractionated on an agarose disc gel (0.8% Seakem GTG™ agarose, FMC Bioproducts, Rockland, Maine, USA) in the 2.4 mm diameter × 63 cm long tube, and then electrophoresed in the 1st-dimensional (1-D) buffer (0.1 M Tris-acetate, pH 8.0, 40 mM sodium acetate, 3 mM EDTA, pH 8.0, 36 mM NaCl) at 100 V for 1 h followed by 230 V for 23 h. The size fractionated genomic DNA was carefully extruded from the tube and soaked for 30 min in the reaction buffer for *Bam*HI. Thereafter, DNA was digested in the gel with 1500 U of *Bam*HI for 2 h. The gel was fused into the top edge of a 50 cm (W) × 50 cm (H) × 0.1 cm (thickness) 5% polyacrylamide vertical gel by adding melted agarose (0.8% at 60-65°C) to connect each gel. The 2nd-dimensional (2-D) electrophoresis was carried out in Tris borate EDTA (TBE) buffer (50 mM Tris, 62 mM boric acid, 1 mM EDTA), at 100 V for 1 h followed by 150 V for 23 h. An area 35 cm × 41 cm of the original gel was excised and dried. Autoradiography was performed for 3-10 days on a film (XAR-5; Kodak, Rochester, NY, USA) at –80 °C using an intensifying screen (Quanta III; Sigma-Aldrich, St. Louis, MO USA), or for 1-3 days on an imaging plate (Fuji Photo Film, Tokyo, Japan). Finally, the imaging plate was analyzed by the BAS-2000™ (Fuji Photo Film).

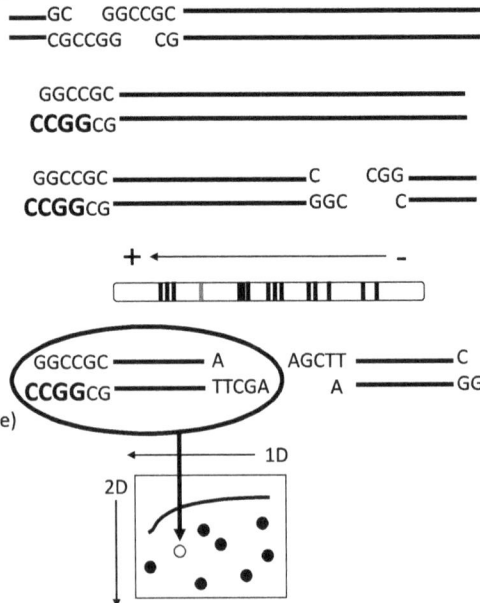

(a)

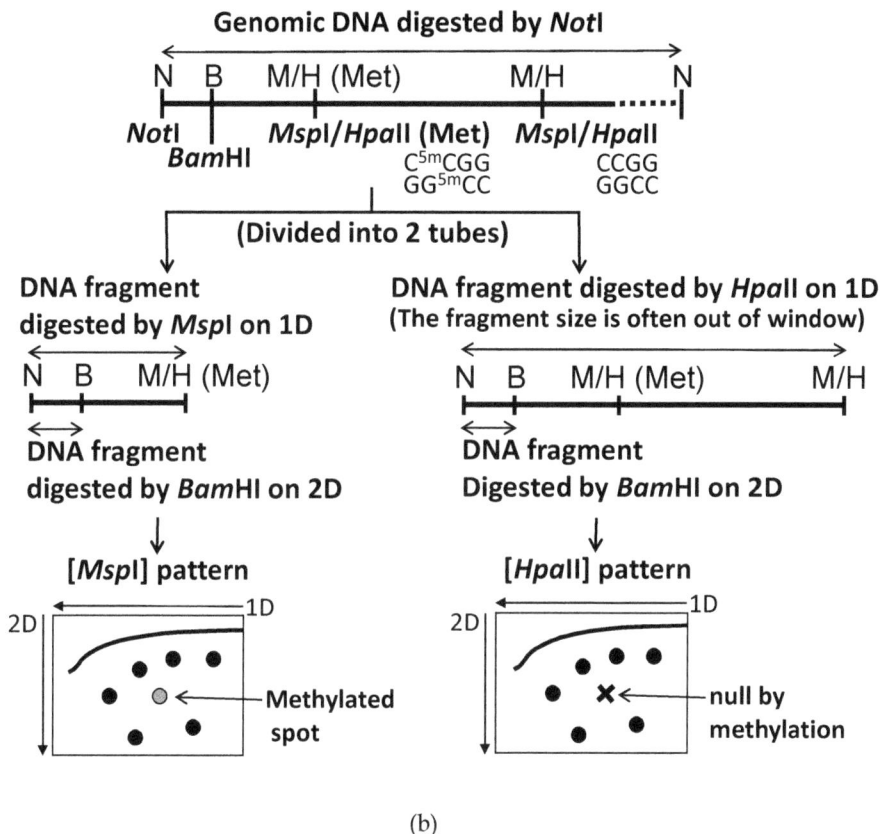

(b)

Fig. 1. Conventional (right side) and improved (left side) RLGS procedures. (a)The
conventional RLGS method is a tool that uses landmarks to directly label restriction enzyme
sites scattered on genomic DNA, expand them on a single image by high-resolution two
dimensional electrophoresis, and detect several thousands of spots at once. In the
conventional method, the differences of methylation status among samples are surveyed
using NotI methylation sensitivity. However, it has been difficult to distinguish methylation
polymorphism from sequence polymorphism, and can't adopt for methylation surveillance
of one sample. In improved method (left flow chart), isoschizomers (MspI and HpaII) are
used as second cutter. We are possible to detect methylated sites directly in even if an
individual or a tissue by comparison between [MspI] pattern and [HpaII] pattern. (b) MspI
and HpaII both recognize CCGG site, but have different methylation sensitivity. We show
one example of methylated spot: MspI can digest the methylated MspI/HpaII site (C5mCGG),
which is the nearest to NotI site, and the DNA fragment (from NotI end to MspI/HpaII end)
is electrophoresed on detectable first dimensional area. On the other hand, HpaII cannot
digest the methylated MspI/HpaII site, and digests non-methylated MspI/HpaII site at the
downstream. Therefore, longer DNA fragment (from NotI end to non-methylated
MspI/HpaII end) is electrophoresed at out of window on the first dimension. This figure was
cited from Takamiya, T. 2007.

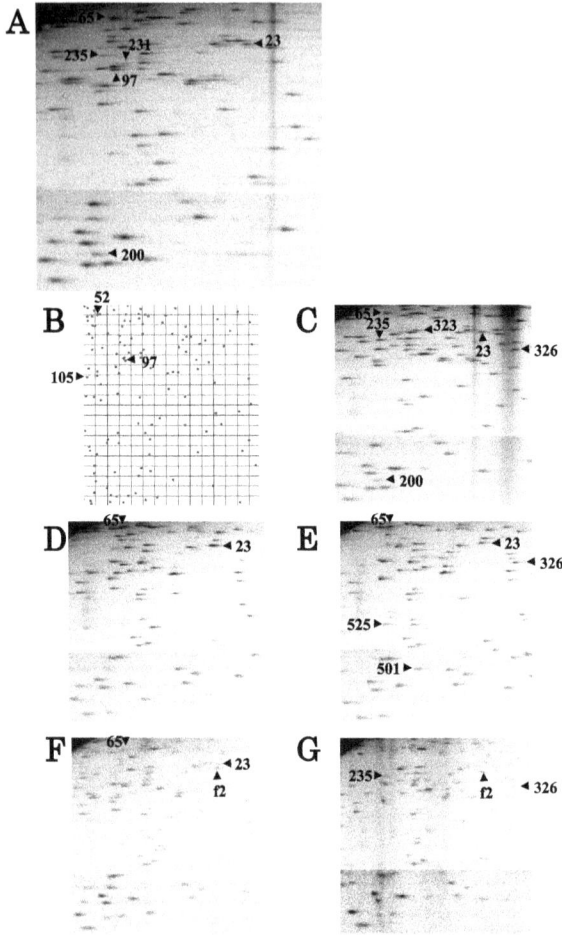

Fig. 2. RLGS patterns for detection of methylated sites. (A) RLGS pattern of Nipponbare genomic DNA with restriction enzyme combination *NotI-MspI-BamHI* ([*MspI*] pattern). (B) *in silico* RLGS pattern predicted from rice genome sequence data. (C) Nipponbare *NotI-HpaII-BamHI* ([*HpaII*]) pattern. In comparison with [*MspI*] pattern, 25 spots were specific to [*MspI*] and 18 to [*HpaII*]. (D) Kasalath [*MspI*] pattern. (E) Kasalath [*HpaII*] pattern from the genomic DNA in (D): 19 spots were specific to [*MspI*], and 13 to [*HpaII*]. (F) F1 hybrid [*MspI*] pattern. In comparison to its parents, spots 200 and 235 were absent, and spot f2 was new. (G) F1 hybrid [*HpaII*] pattern from the genomic DNA in (F): 29 spots were specific to [*MspI*], and 26 to [*HpaII*]. Compared with the parental patterns (C, E), spots 23, 65, 200, 231, 323, 501, and 525 were absent, and F1-specific spot f2 was new (G). These figures were cited from Takamiya, T., Hosobuchi, S., Noguchi, T., Asai, K., Nakamura, E., Habu, Y., Paterson, A. H., Iijima, H., Murakami, Y., Okuizumi, H. (2008). Inheritance and alteration of genome methylation in F1 hybrid rice. Electrophoresis. 29, 4088-4095. Copyright Wiley-VCH Verlag GmbH & Co. KGaA. Reproduced with permission.

2.2 *In silico* RLGS analysis oriented by whole genome sequence

A software named "*in silico* RLGS" was developed that simulates RLGS spots based on known sequence data to identify each spot in actual RLGS. The software searches for restriction enzyme sites in the entire genome sequence that were used in an actual RLGS experiment, and calculates fragment length between the restriction sites and mobility for simulating a 2-D spot pattern. In our experiment the whole rice genome sequence data was obtained from http://rgp.dna.affrc.go.jp/J/IRGSP/Build3/build3.html. *Not*I sites and *Msp*I sites that were near to the *Not*I sites were searched through the whole sequence and given ID numbers. Length of identified fragments between restriction sites (from *Not*I end to the nearest *Msp*I end or from *Not*I end to next *Not*I end) and mobility of each DNA fragment in the 1-D electrophoresis were calculated by the software according to Southern E. (1979). The exact mobility of RLGS spots were confirmed from electrophoresis of λ DNA fragments with known sequences. In addition, the *Bam*HI sites were also surveyed and DNA fragment (from *Not*I end to the nearest *Bam*HI end) length in 2-D was estimated. Based on this, a 2-D graph (*in silico* RLGS pattern) was drawn. The *in silico* RLGS pattern was compared to its corresponding autoradiographic (actual) RLGS pattern with relative spot positions. This comparison leads to the identification each RLGS spot immediatelyand precisely.

3. Results of DNA methylation analysis using improved RLGS

3.1 Detection of methylated spots in F1 hybrid and the parents

To detect the DNA methylation in plant genome, improved RLGS and *in silico* RLGS were used [Takamiya *et al.* 2008A]. Two rice individuals, *Oryza sativa* L. var Nipponbare and *O. sativa* L. var Kasalath as parent and its F1s were used as the experimental material to analyze the pattern of the DNA methylation and its inheritance in the F1 hybrid. F1 were obtained from crossing Nipponbare as the seed parent with Kasalath as the pollen parent.

The RLGS pattern with *Not*I-*Msp*I-*Bam*HI combination ([*Msp*I] pattern) and *Not*I-*Hpa*II-*Bam*HI combination ([*Hpa*II] pattern) were obtained for the parents (Nipponbare and Kasalath) and F1. The [*Msp*I] pattern showed 85 spots with Nipponbare, 77 spots with Kasalath and 111 spots with F1 hybrid. In the same way, [*Hpa*II] pattern showed 78, 71 and 108 spots with Nipponbare, Kasalath and F1 hybrid, respectively. The genome sequence of the Nipponbare was analyzed in *in silico* software and it showed 117 spots. The spot pattern obtained was compared with actual RLGS pattern and 56 spots were found to be in common. For example the spot 97 was detected in both patterns (Fig 1A and 1B),, with a locus on chromosome 9.

To detect the RLGS spots differed in methylation ("methylated spot"), the [*Msp*I] pattern and [*Hpa*II] pattern were compared. This comparison showed that 43 methylated spots in Nipponbare and 32 methylated spots in Kasalath. Next, we compared Nipponbare and Kasalath patterns to detect methylation polymorphysms between parents. Thirty five spots of [*Msp*I] pattern and 42 spots of [*Hpa*II] pattern were specific to Nipponbare and similarly the Kasalath also had 27 and 35 spots in specific with [*Msp*I] pattern and [*Hpa*II] pattern respectively. Moreover, 50 spots of [*Msp*I] pattern and 36 spots of [*Hpa*II] pattern were common between Nipponbare and Kasalath. The spots which were not identified by *in silico* RLGS were cloned from the 2D polyacrylamide gel.

In F1, though most of the spots followed the Mendelian law of inheritance, still 8 parental spots (spot numbers: 23, 65, 200, 231, 235, 323, 501 and 525) disappeared and one new spot (f2) was detected. These 9 spots (8 disappeared spots and 1 new spot) indicate the abnormal inheritance and the details were summarized in Table 1. The pattern of spot 323 in parents, F1 and selfed progenies were shown Figure 3c.

Spot No.	in silico	Nipponbare		Kasalath		F1	
		MspI	HpaII	MspI	HpaII	MspI	HpaII
23	x	o	o	o	o	o	x
65	o	o	o	o	o	o	x
200	x	o	o	x	x	x	x
231	x	o	o	x	x	x	x
235	x	o	o	x	x	x	o
323	x	x	o	x	x	x	x
501	x	x	x	x	o	x	x
525	x	x	x	x	o	x	x
f2	x	x	x	x	x	o	o
52	o	x	x	x	x	x	x
105	o	x	x	x	x	x	x

Table 1. Altered inheritance of RLGS and *in silico* RLGS spots. o, spot present; x, spot absent. One RLGS spot (f2) was newly detected in F1 hybrids, and 8 RLGS spots were absent in the NKF1. Two *in silico* RLGS spots (52 and 105) were found to be altered methylation inheritance by PCR analysis. This table was cited from Takamiya, T., Hosobuchi, S., Noguchi, T., Asai, K., Nakamura, E., Habu, Y., Paterson, A. H., Iijima, H., Murakami, Y., Okuizumi, H. (2008). Inheritance and alteration of genome methylation in F1 hybrid rice. Electrophoresis. 29, 4088-4095. Copyright Wiley-VCH Verlag GmbH & Co. KGaA. Reproduced with permission.

3.2 Mapping of methylation status of rice genome using RLGS

The mapping of methylation status of rice genome using RLGS was carried out as following steps. (1) Methylated spot was detected by comparison [*MspI*] and [*HpaII*] patterns, and its locus was identified by *in silico* RLGS. (2) The presence of restriction site of the *NotI* and the *MspI/HpaII* of methylated spot was confirmed using PCR analysis. (3)The methylation status of restriction sites of methylated spot was confirmed by another PCR analysis. (4) The parental origin of the methylated alleles in F1 was determined by using the sequence analysis. These steps are explained in detail as follows.

The presence of the restriction sites of *NotI* and *MspI/HpaII* was confirmed by PCR analysis by designing the flanking primers for the restriction sites of RLGS spots that were identified by *in silico* RLGS or spot cloning. The presence of restriction sites was confirmed in the RLGS spot 97 of Nipponbare and Kasalath (Fig. 4) as an example. The Figure 4A shows the various PCR products in lane 1 to 4. The lane 1 and 2 was the amplified genomic DNA of parents, Nipponbare and Kasalath using the flanking primers and lane 3 and 4 were the PCR products, obtained from the each parent after treatment with *MspI*. The DNA fragments of lane 3 and 4 are smaller than the fragments of lane 1 and 2, this is because the *MspI/HpaII* site were digested and divided into 456 bp (detected) and 123 bp (estimated). This difference in the fragment size confirms the presence of *MspI/HpaII* sites in both Nipponbare and Kasalath.

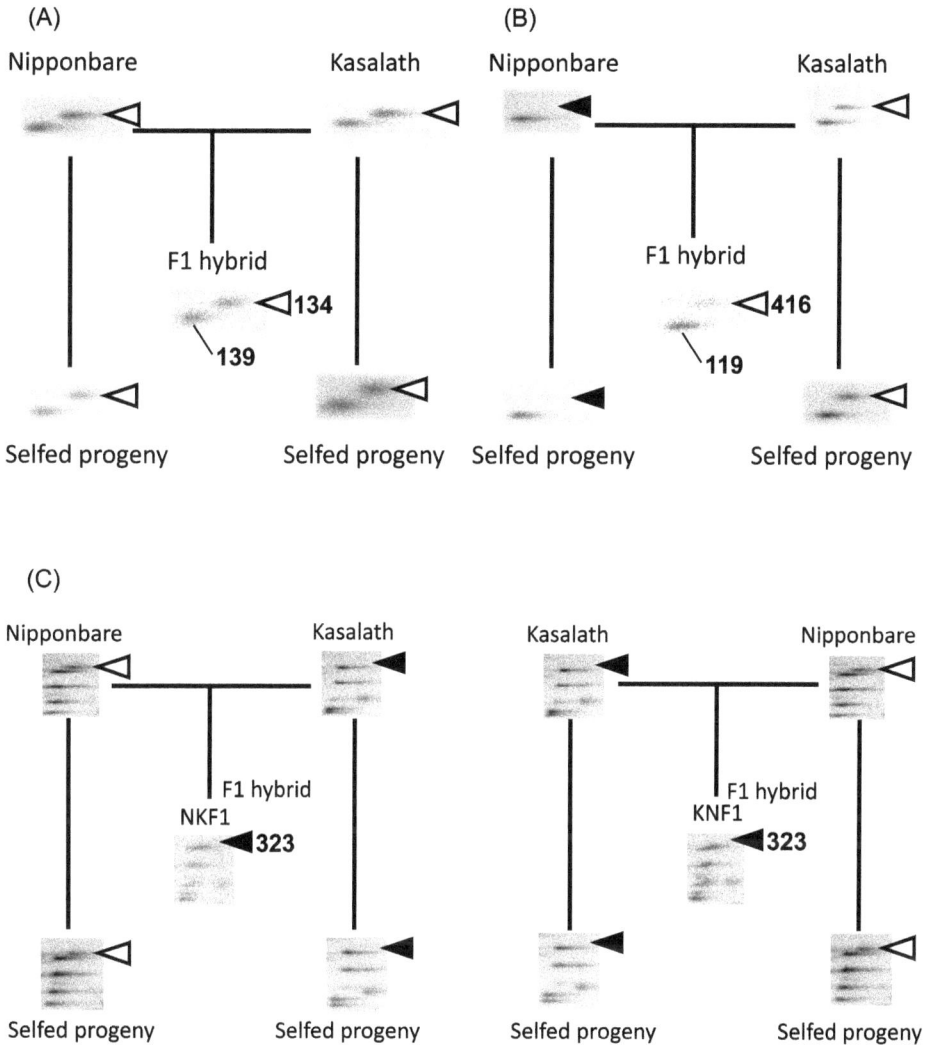

Fig. 3. Mendelian and non-Mendelian inheritance of RLGS spots pattern. (A) Spot 134 (arrowhead) was detected in both parents viz., Nipponbare and Kasalath and the same pattern of the spot was also detected in the F1. (B) Spot 416 (arrowhead) was detected in only one parent (Kasalath). This spot on the RLGS pattern of F1 was detected at diminished (half) intensity of the normal spots such as spot 119. The spots 134 and 416 showed Mendelian inheritance. (C) There was a spot 323 (arrowhead) in the pattern of parental Nipponbare, but not in parental Kasalath. It was expected to be at half intensity in F1 patterns, however, the spot 323 disappeared in all 9 patterns of NKF1 individuals and 8 patterns of KNF1s. In the patterns of all 9 selfed progenies of parental Nipponbare, the spot 323 was detected similar to parental Nipponbare. This spot 323 was indicating non-Mendelian inheritance. These figures were cited fromTakamiya, T. 2007.

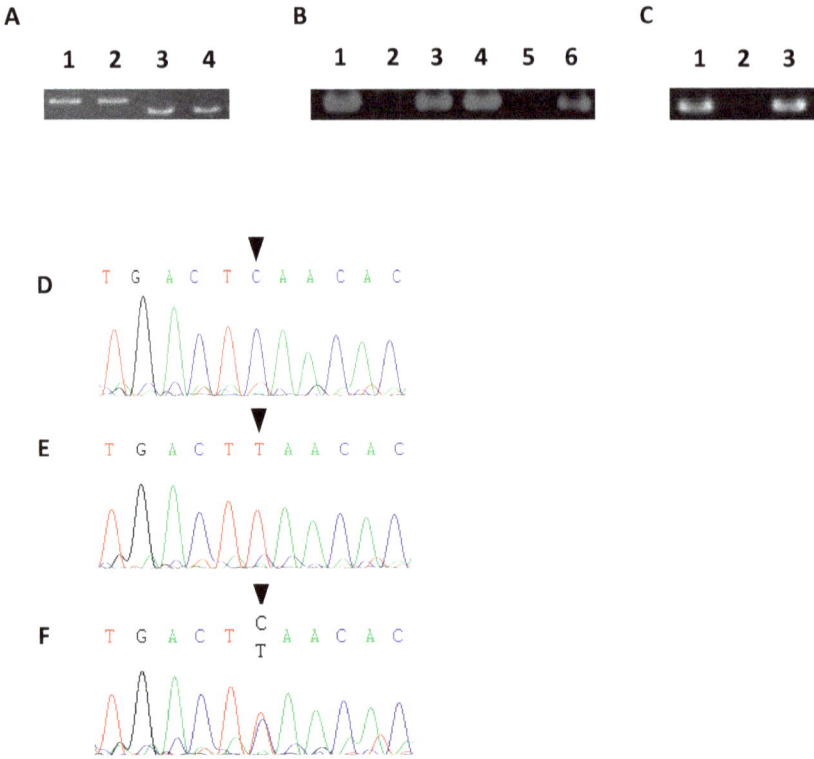

Fig. 4. Confirmation of methylation status. (A) Confirmation of restriction enzyme sites with flanking primers for the *Msp*I/*Hpa*II site of spot 97. Lanes 1 and 2 are PCR products amplified from genomic DNA of Nipponbare (N) and Kasalath (K), respectively. The PCR products were purified, and treated with *Msp*I, then, loaded into lanes 3 and 4, respectively. (B) PCR-based DNA methylation analysis of spot 97 of the parents (Nipponbare and Kasalath). Lanes 1, 2, and 3 for Nipponbare or 4, 5, and 6 for Kasalath are the PCR products for the templates (un-, *Msp*I-, and *Hpa*II-digested genomic DNA, respectively). (C) PCR-based DNA methylation analysis of spot 97 in the F1. Lane 1 is the uncut positive control. Lanes 2 and 3 are PCR products from F1 genomic DNA that was treated with *Msp*I and *Hpa*II, respectively. (D) Part of the nucleotide sequence of the DNA fragment in lane 1 shown in Fig 4B (Nipponbare). (E) Part of the nucleotide sequence of the DNA fragment in lane 4 shown in Fig 4B (Kasalath). One SNP (shown by arrow-head) was detected in this region.(F) Part of the nucleotide sequence of the DNA fragment in lane 3 shown in Fig 4C. This DNA fragment had C/T heterozygously. These figures were cited from Takamiya, T., Hosobuchi, S., Noguchi, T., Asai, K., Nakamura, E., Habu, Y., Paterson, A. H., Iijima, H., Murakami, Y., Okuizumi, H. (2008). Inheritance and alteration of genome methylation in F1 hybrid rice. Electrophoresis. 29, 4088-4095. Copyright Wiley-VCH Verlag GmbH & Co. KGaA. Reproduced with permission.

To confirm the methylation status of spots with restriction enzyme site in Nipponbare and Kasalath, different PCR products were obtained (Fig. 4B). The PCR product loaded in lane 1 genomic DNA of Nipponbare and the lane 2 and 3 were loaded with the product obtained by using *Msp*I or *Hpa*II treated genomic DNA of Nipponbare as template, respectively. The lanes 4 to 6 are loaded with the PCR products of Kasalath obtained in similar manner to that of Nipponbare. The lane 1 and 4 are used as the positive control for Nipponbare and Kasalath, respectively. The lane 2 and 5 do not have any bands where as the lane 3 and 6 had the band sizes similar to that of their positive controls. This result showed that the *Msp*I/*Hpa*II site was methylated ($C^{5m}CGG$), and it was correspondence with RLGS result. Similar to the parents, the *Msp*I/*Hpa*II site was methylated in F1 (Fig. 4C).

In Nipponbare, the methylation status of 90 *Not*I and 92 *Msp*I/*Hpa*II sites were checked. Out of 182 sites, 60 sites (33%) were methylated and 4 Nipponbare specific sites were identified. Similarly in Kasalath, 82 *Not*I and 84 *Msp*I/*Hpa*II sites were tested for methylation status resulted in identification of 59 methylation sites (36%) and 10 Kasalath specific sites.

Finally, to determine the parental origin of the methylated allele, the bands were subjected for sequence analysis. The bands of Nipponbare in lane 1 and bands of Kasalath in lane 4 (Fig 4B) were purified and sequence analysis was done. This analysis detects an SNP for spot 97 showing C in Nipponbare (Fig 4D) and T in Kasalath (Fig 4E). Similarly the band in lane 3 (Fig. 4C), which was amplified from *Hpa*II digested genomic DNA of F1, was sequenced and it was found to be heterozygous with both C and T (Fig. 4F). The sequence analysis shows that methylation prevented digestion of the *Msp*I/*Hpa*II sites of both the alleles, and the methylation status of the parents was inherited to F1 following Mendelian law. Similarly, we examined the other 25 methylated spots that showed the same appearance or absence between the parents and F1, and confirmed that all methylation status were inherited to F1.

Some of the *Not*I and *Msp*I/*Hpa*II sites that were specifically found in the *in silico* RLGS were also checked based on the PCR and sequence analysis. This analysis identified some new altered methylations in F1. The identified spots were demethylations at *Not*I of spot 52 (Chr.11) and the *Msp*I/*Hpa*II site of spot 105 (Chr.3) in F1, and their details are given in the Table 1 and Figure 5. The specific occurrence of the spots only in *in silico* RLGS is due to the methylation in *Not*I sites.

In the entire process of analysis, a total of 103 RLGS spots were identified and most of these spots were analyzed for methylation status by RLGS and PCR. The result of this analysis was summarized in methylation map (Fig 5). In the map, the numbered spot depict that those spots were methylated at one or more *Not*I or *Msp*I/*Hpa*II sites of Nipponbare, Kasalath or F1. In total, seven altered spots were mapped (shown as 'AI' in Fig 5). The other spots *viz.*, 23, 501, 525 and f2 are not yet to cloned. Out of this 103 *Not*I sites, 17 and 14 sites were located within 2.0 kb upstream and downstream of a gene respectively, and 63 sites located within the gene. The remaining 9 sites were in the intergenic regions. Similarly, 25 *Msp*I/*Hpa*II sites were in 5′ upstream region, 48 within gene and 15 sites in 3′ downstream region. Thus, 182 sites (88%) out of 206 were located between 2.0 kb upstream and downstream of genes.

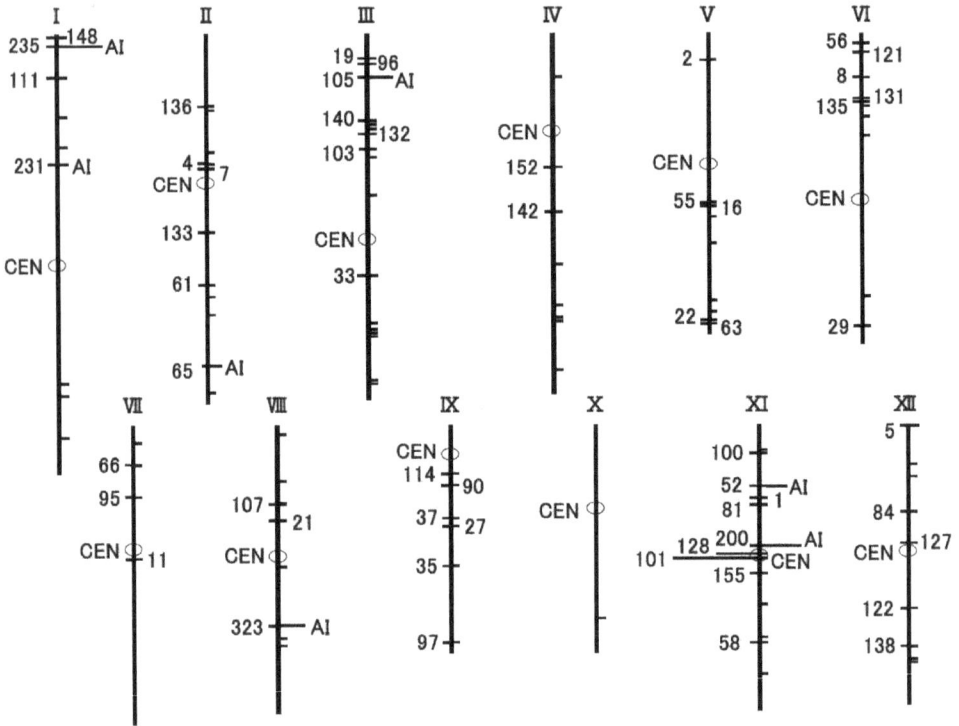

Fig. 5. Map of methylation sites. Numbered loci had at least 1 methylation in NotI or MspI/HpaII digests of Nipponbare, Kasalath, or F1s. "AI" indicates 7 loci with altered inheritance. Inheritance at the other loci appeared to be consistent. Centromeres (CEN) are indicated by ellipses. This map was cited from Takamiya, T., Hosobuchi, S., Noguchi, T., Asai, K., Nakamura, E., Habu, Y., Paterson, A. H., Iijima, H., Murakami, Y., Okuizumi, H. (2008). Inheritance and alteration of genome methylation in F1 hybrid rice. Electrophoresis. 29, 4088-4095. Copyright Wiley-VCH Verlag GmbH & Co. KGaA. Reproduced with permission.

3.3 Non-Mendelian inheritance of DNA methylation

To prove non-Mendelian inheritance of DNA methylation in altered spot loci, we analyzed reciprocal F_1 hybrids (subsp. *japonica* cv. Nipponbare × subsp. *indica* cv. Kasalath) of rice (*Oryza sativa* L.). Reciprocal hybrids were produced by crossing the same individual of each cultivar as the female parent on one culm and as the male parent on another culm. Crossing Nipponbare as the seed parent with Kasalath as the pollen parent gave F_1 hybrids designated NKF_1. The inverse cross gave KNF_1 hybrids. The seeds of Nipponbare, Kasalath, NKF_1 (nine individuals from the same parents), and KNF_1 (nine individuals from the same parents), and the selfed progeny of the parents were grown for 2 months, and the genomic DNA was isolated from the leaf blade and sheath of each individual, and the RLGS analysis was performed.

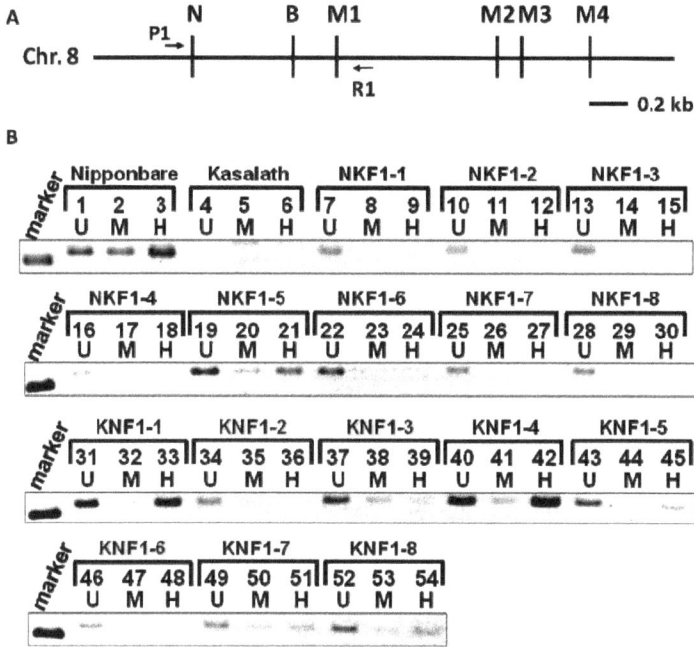

Fig. 6. Methylation analysis of an abnormally inherited RLGS spot 323. (A) DNA of spot 323 was located on Chr. 8. Arrows P1 and R1 indicate flanking primers for PCR-based methylation analysis. (B) PCR-based methylation analysis of M1 site of spot 323 in the parents and 8 NKF1 (NKF1-1~NKF1-8) and 8 KNF1 (KNF1-1~KNF1-8) hybrids. Lanes 1, 4, 7, 10, 13, 16, 19, 22, 25, 28, 31, 34, 37, 40, 43, 46, 49, and 52 are the PCR products from genomic DNA of each line as positive controls (U = uncut). M and H indicates *Msp*I and *Hpa*II digests of each line. The methylation status of M2 and M3 sites were also checked (data not shown). m: size marker, 1.0 kb band. These figures were cited from Takamiya, T., Hosobuchi, S., Noguchi, T., Asai, K., Nakamura, E., Habu, Y., Paterson, A. H., Iijima, H., Murakami, Y., Okuizumi, H. (2008). Inheritance and alteration of genome methylation in F1 hybrid rice. Electrophoresis. 29, 4088-4095. Copyright Wiley-VCH Verlag GmbH & Co. KGaA. Reproduced with permission.

Consider spot 323 of Nipponbare as an example which is a *Hpa*II specific spot but not detected in the F1. In most of the F1, the demethylation of the M1/M2 site of spot 323 was detected based on the PCR analysis (Fig 6 shown the result of M1 site). The PCR product of Kasalath genomic DNA was not amplified by the primer P1 and R1. This non-amplification of these regions by the primers may be due to some difference in base sequence between Nipponbare and Kasalath. The spots 200, 231, 235, 501 and 525 were specific to Nipponbare or Kasalath and these spots do not appeared in F1. Therefore, the regions of these spots may have some differences in DNA sequence or methylation status between Nipponbare and Kasalath. Alternatively, these polymorphic regions might have altered methylation status in the F1 hybrid [Takamiya *et al.* 2009]. The altered methylation of spot 323 suggests possible sequence-dependent demethylation, for example as a result of paramutation induced by allelic exclusion (Chandler *et al.*, 2000). For better understanding of the methylation behavior of spot 323, detailed analyses like expression analysis of the gene near to the spot may be required.

The spot 200 was detected in both the [*Msp*I] and [*Hpa*II] patterns of Nipponbare at a diminished spot intensity but disappeared in Kasalath (Fig 7). The spot is another non-Mendelian example [Takamiya *et al.*, 2009]. In RLGS analysis halved/diminished spot intensity indicates the heterozygote which was confirmed theoretically and practically in earlier studies (Hayashizaki *et al.* 1994B., Okuizumi *et al.* 1997). The DNA fragment of spot 200 was cloned and sequence analysis was carried out, which place it in the 5′ region of a non-protein coding transcript (Os11g0417300) (Fig 8A). The spot position of autoradiographic RLGS pattern of Nipponbare was compared with *in silico* RLGS pattern. Then, the DNA fragments digested at *Not*I (N) and *Msp*I (M) was separated by 1-D electrophoresis, and the N and *Bam*HI (B) sites were fractionate by 2-D electrophoresis (Fig 8A). Restriction enzyme digestion and sequencing was employed to confirm the existence of N, M and B in the parental Nipponbare, whereas in Kasalath the N and M site exist but the B site was absent. The presence or absence of spot 200 obtained based on the RLGS analysis of NKF_1 and KNF_1 shows that this spot segregated as 1:1 in both the population (Fig 7 and Table 2). The diminished spot intensity and its segregation in F1 hybrids show that the *Msp*I/*Hpa*II site of spot 200 is methylated heterozygously. Similarly, it was assumed that the detection and absence of spot 200 in F1 was due to the non-methylated and methylated M site, respectively. Moreover in all selfed progenies (9 individuals) of Nipponbare, the spot 200 was detected with halved intensity (Fig 7, Table 2). From this, it was concluded that in the selfed progeny and in parental Nipponbare, the M site was methylated heterozygous because of non-Mendelian inheritance of methylation.

Generation (Cultivar) [Sample size]	RLGS pattern of spot 200		RLGS pattern of spot 231	
	*Msp*I patterns (intensity)	*Hpa*II patterns (intensity)	*Msp*I patterns (intensity)	*Hpa*II patterns (intensity)
Parent (Nipponbare)	Present (1/2)	Present (1/2)	Present (1/2)	Present (1/2)
Parent (Kasalath)	Absent	Absent	Absent	Absent
Selfed progeny (Nipponbare) [9 individuals]	9 present (1/2)	9 present (1/2)	9 present (1/2)	9 present (1/2)
Selfed progeny (Kasalath) [4 individuals]	4 absent	4 absent	4 absent	4 absent
NKF_1 (Nipponbare × Kasalath) [9 individuals]	Segregated 5 present : 4 absent (1/2 : 0)	Segregated 4 present : 5 absent (1/2 : 0)	Segregated 2 present : 6 absent* (1/2 : 0)	Segregated 5 present : 4 absent (1/2 : 0)
KNF_1 (Kasalath × Nipponbare) [9 individuals]	Segregated 7 present : 2 absent (1/2 : 0)	Segregated 4 present : 5 absent (1/2 : 0)	Segregated 2 present : 7 absent (1/2 : 0)	Segregated 3 present : 6 absent (1/2 : 0)

This table was cited from Takamiya *et al.* 2009.

Table 2. Summary of RLGS pattern of spot 200 and 231.

♀ Parent
Nipponbare

♂ Parent
Kasalath

♀ Parent
Kasalath

♂ Parent
Nipponbare

◀ spot 200

NKF₁ hybrids

KNF₁ hybrids

Selfed progeny

Selfed progeny

Selfed progeny

Selfed progeny

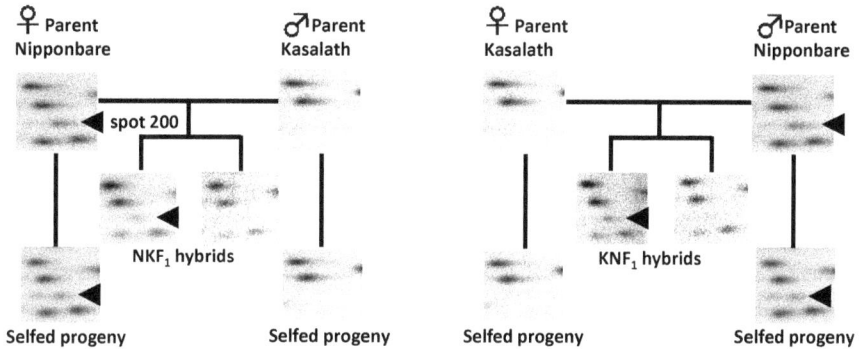

Fig. 7. Non-Mendelian pattern of spot 200. A part of RLGS [NotI-MspI-BamHI] combination patterns of the parents, their selfed progeny, and their reciprocal F₁ hybrids. Spot 200 (arrowhead) was detected in the [MspI] patterns and [HpaII] patterns (data not shown) of Nipponbare and its selfed progeny. The presence or absence of the spot segregated in both F₁ populations (NKF₁ and KNF₁). The spot intensity of this spot was half that of the others. This figure was cited from Takamiya et al. 2009.

Expression analysis of the non-protein coding transcript (Os11g0417300), which is the nearest gene to the MspI/HpaII site of spot 200 (Fig 8A) [Takamiya et al., 2009], was done to clarify whether the methylation status in correlated with expression of nearest gene. The cDNA (Genebank accessions No: AK 109537) of the non protein coding transcript expressed in flower, leaf and panicle (http://www.ncbi.nlm.nih.gov/sites/entrez?db=unigene& cmd=search&term=AK109537). Total RNA was isolated from the leaf blade and sheath of the parental Nipponbare, Kasalath, two individuals each in NKF₁ (NK5 and NK7) and KNF₁ (KN5 and KN10) hybrids. The pattern of NK5 and KN5 alone detected the spot 200 whereas in the pattern NK7 and KN10 the spot was not detected. The cDNAs' of both NK5 and KN5 were PCR amplified and separated by agarose gel electrophoresis (Fig 8B). The gene was expressed in the parents, NK5, and KN5 (Fig. 8B). The single nucleotide polymorphism (C/T) between Nipponbare and Kasalath was found and the RT-PCR products were sequence to reveal the parental origin of the expressed sequence in F₁ hybrid (Fig 8C). The RT-PCR products of NK5 and KN5 which had spot 200 in their RLGS pattern were subjected to sequence analysis. Allelic expression bias for the Nipponbare allele was found in NK5 and KN5 (Fig 8C). Similarly, the bias for Nipponbare allele was found in KN7 and NK10 which do not have spot 200 (data not shown). Strong allele bias was found in the reciprocal hybrids and this shows the monoallelic expression of the Nipponbare allele. In addition, a splicing variant (smaller transcript with lower expression than Nipponbare allele) specific to Kasalath was detected (Fig 8B), but this was absent in NKF₁ and KNF₁. This transcript was sequenced and a splicing variant that leads to a 76 bp deletion at the 3′ end of exon 2 was revealed. The reason is unknown, but, it implies that some effect of methylation is influencing the variant expression.

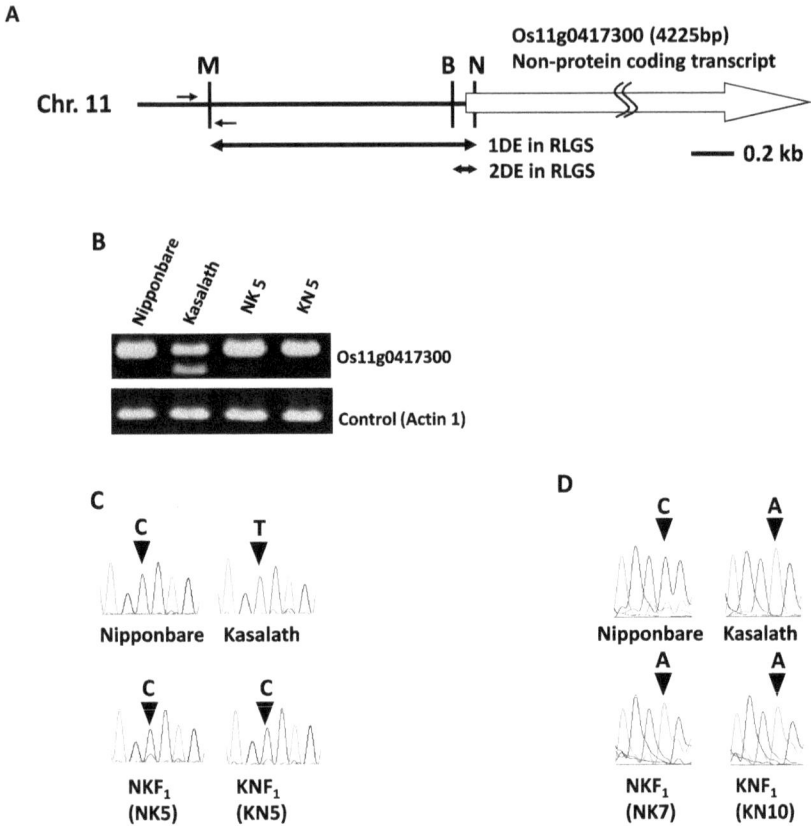

Fig. 8. Analysis of another abnormally inherited RLGS spot 200. (A) Schematic representation of the region of chromosome 11 containing the restriction enzyme sites located in the region 5′ to the transcription start site of the non-protein coding transcript (Os11g0417300). The DNA fragments were digested at the NotI (N) and MspI/HpaII (M) sites and fractionated by one dimensional electrophoresis. Next, the DNA fragments that were digested at the BamHI (B) sites were fractionated by two dimensional electrophoresis, which allowed detection of the B-N fragment as an RLGS spot. Spot 200 corresponds to the fragment between the N and B sites. (B) RT-PCR showed that a non-protein coding transcript (Os11g0417300) was expressed in leaf blade and sheath of Nipponbare, Kasalath, NKF$_1$, and KNF$_1$ plants. (C) Sequence analysis of the RT-PCR products of the expressed Os11g0417300 allele. The single nucleotide polymorphism between Nipponbare (Cytosine) and Kasalath (Tymine) is indicated in the RT-PCR products by arrowheads. Specific expression of the Nipponbare allele was confirmed by detection of base Cytosine in both NK5 and KN5 plants. (D) Sequence analysis of RT-PCR products of the expressed Os01g0327900 allele. The single nucleotide polymorphism in RT-PCR products between Nipponbare (Cytosine) and Kasalath (Adenine) is indicated by arrowheads. Specific expression of the Kasalath allele was confirmed by detection of base Adenine in both NK7 and KN10 plants. These figures were cited from Takamiya et al. 2009.

The spot 231, which behaved similar to that of non-Mendelian spot 200, had diminished spot intensity and 1:1 segregation in NKF_1 and KNF_1 (Table 2), and also it was detected in all the selfed progeny of Nipponbare. For this spot 231, the expression of the nearest gene (DUF 295 family protein Os01g0327900) was analyzed in two NKF_1 (NK5 and NK7) and two KNF_1 (KN5 and KN10) individuals. Sequence analysis of the RT-PCR products showed that only the Kasalath allele was expressed in NK5, NK7, KN5, and KN10 (Fig 8D shown the result of NK7 and KN10). In this study, two examples (spot 200 and 231) have been given for the nearest gene to a heterozygous methylated site showing allelic expression bias.

Monoallelic expression in F_1 hybrids of plants has been reported in several crops. Intraspecific maize hybrids have shown unequal expression of parental alleles [Guo et al. 2004, Springer et al. 2007A, and Springer et al. 2007B]. Besides, 17 out of 30 genes analyzed showed >1.5-fold expression bias for one of two alleles, with monoallelic expression of one gene in *Populus* interspecific hybrids [Zhuang et al. 2007]. Therefore, it is considered that allelic expression bias was caused by epigenetic status of DNA methylation and/or histone modification.

In plants, identification of more RLGS spots exhibiting non-Mendelian inheritance and simultaneously studying their methylation status of the corresponding DNA sequence and their expressed allele will explain the importance and better understanding of gene regulatory mechanisms such as monoallelic expression. Extensive and detailed expression analysis of genes in F1s with different genetic backgrounds is very much essential because the findings can be applied to other genes as well. The mechanism of allelic exclusion inducing heterosis, hybrid weakness and genome barriers might be better understood by revealing the regulation and function of the splicing variant of Kasalath (Fig. 8B) in F_1 hybrids.

3.4 Various aspects on DNA methylation roles

The RLGS method is very powerful for methylation analysis other than that for genetic analysis of DNA methylation. In the Takamiya et al. 2006, the methylated status was compared among 3 ecotypes of Arabidopsis using the RLGS method. Methylation at a total of 17 sites (*Not*I: 9 sites and *Hpa*II: 8 sites) was detected in the 3 ecotypes. Among them, there were 8 common methylation sites among the 3 ecotypes, and the 9 residual sites (53%) showed methylation polymorphism. Among all restriction enzyme sites analyzed (37 non-methylated sites and 17 methylated sites), the sites showing a different methylated status among the 3 ecotypes accounted for 17% of the total sites (9/54). In the studies so far, it has been reported that methylation is involved in inactivation and metastasis inhibition of transposon and retrotransposon [Hirochika et al. 2000]. In the pseudogene and gypsy-like retrotransposon family-like region in centromeres, both the *Not*I site and *Msp*I/*Hpa*II site were methylated in the 3 ecotypes, which may suggest the relation between the pseudogene in the centromere region and inactivation of the movable element.

Moreover, the RLGS method can be applied to the detection of genomic variation in plant tissue culture [Takamiya et al. 2008B]. The genome DNA of 2 ramets obtained from one seed of rice was extracted and analyzed for RLGS with the combination of *Not*I-*Hpa*II-*Bam*HI and with *Not*I-*Msp*I-*Bam*HI to compare their pattern. As a result, 10 different spots (6%) were detected between ramets. One spot among the 10 different spot was cloned to confirm the methylated status by PCR, it was found that the methylated status at the restriction enzyme (*Hpa*II) site was different between the 2 ramets.

In our recent RLGS analysis, we tried to detect the tissue-specific methylated status by conducting RLGS analysis of endosperm, embryo, leaf blade and leaf sheath and comparing the methylated status of each. As a result, there were 35 shared methylated spots in 3 defferent tissues among 58 methylated spots in total (Fig. 9). Fifty-six methylated spots detected in endosperm and embryo, including 19 tissue-specific spots. That is, a 34% difference was observed between endosperm and embryo. Next, 56 methylated spots were detected in endosperm and the leaf blade/sheath, including 20 tissue-spcific spots which showed a 36% difference between endosperm and leaf blade/sheath. Next, 49 methylated spots were detected in embryo and leaf blade/sheath, including 7 tissue-specific spots. That is, a 14% difference was observed between embryo and leaf blade/sheath. The result of comparison between endosperm and embryo (34%) showed a 2.4-fold larger difference with the comparison between embryo and leaf blade/sheath (14%), and the result of comparison between endosperm and blade/sheath (36%) showed a 2.6-fold larger difference than the comparison between embryo and leaf blade/sheath (14%). The leaf blade/sheath is differentiated from the embryo, but because the endosperm and embryo are independent tissues and have different functions and gene expression, the difference in methylated status is also considered large. Two recent studies show that endosperm DNA methylation is reduced genome-wide, and this reduction is likely to originate from demethylation in the central cell nucleus of the female gametophyte [Hsieh *et al.* 2009, and Gehring *et al.* 2009].

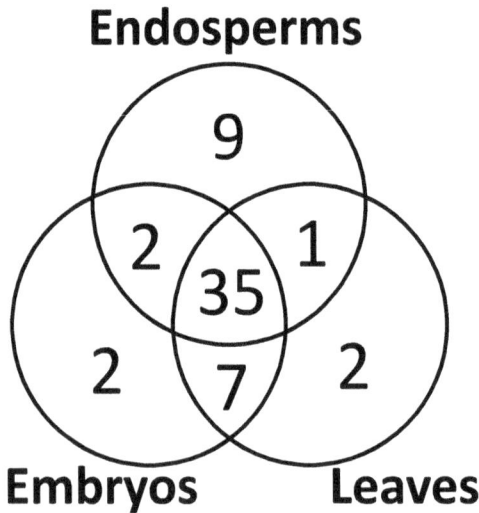

Fig. 9. Comparison of methylated spot among endosperm, embryo, and leaf blade/sheath of rice. We compared detection of methylated spots among 3 tissues. The numerals in circles indicate the number of methylated spots. This figure was cited from Hosobuchi, S. 2007.

Among the spots showing tissue-specific methylated status, expression analysis of spot 226 was conducted by RT-PCR. The *MspI/HpaII* site of spot 226 was positioned in the 5′ region of the Chr. 5 gene Zn-finger, C-x8-C-x5-C-x3-H type domain-containing protein (Os05g0497500) (Fig. 10A). When RT-PCR analysis (primer sets: rtpcr226-F, 5′-CTGGTGGAGATATGAAGAACAA-3′; rtpcr226-R: 5′-TATGTTTAACAACGGGATGTGT-3′)

A

MspI NotI

5'-527 bp | 5'-18 bp |

Zn-finger, C-x8-C-x5-C-x3-H type domain containing protein.

B

Endosperms

before after Embryos Leaves

spot 226

18S rRNA

C

Endosperms

before after Embryos Leaves

marker MboI MspI HpaII MboI MspI HpaII MboI MspI HpaII MboI MspI HpaII

300bp →

1 2 3 4 5 6 7 8 9 10 11 12 13

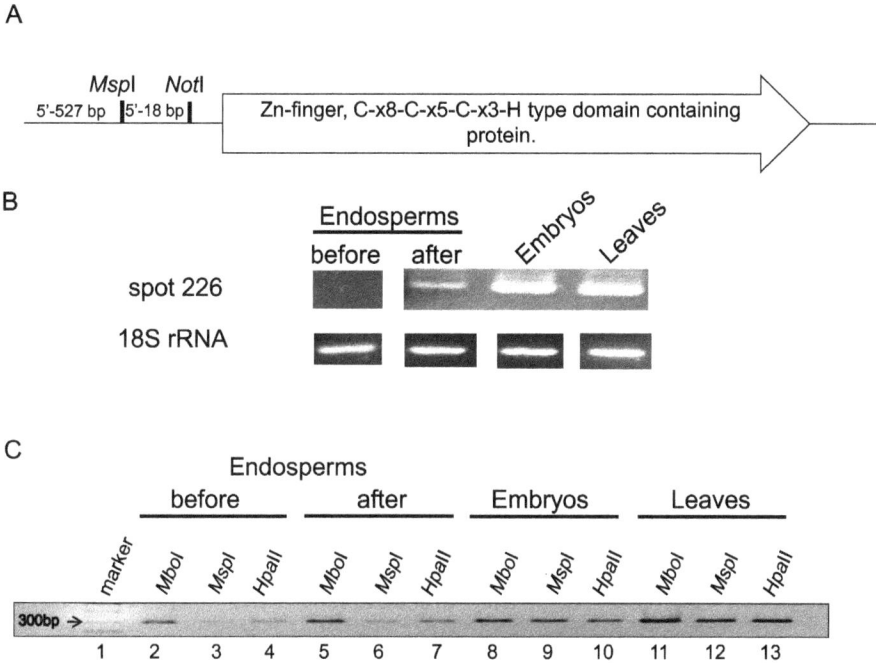

Fig. 10. Analysis of tissue-specific methylated spot 226. (A) Positions of *NotI* site and *MspI* site toward a nearby gene are shown. Both sites are located on the 5' side of the Zn-finger, C-x8-C-x5-C-x3-H type domain-containing protein (Os05g0497500). (B) The results of RT-PCR analysis of Zn-finger, C-x8-C-x5-C-x3-H type domain-containing protein (Os05g0497500). This gene was not expressed in endosperm before water absorption, but it was expressed in embryos and leaf blades/sheaths, and weakly in endosperm after water absorption. 18S rRNA is a control. (C) In regard to the *MspI/HpaII* site of spot 226, we analyzed the methylation status in the endosperm before and after water absorption and in embryos and leaf blades/sheaths by the PCR method. We used *MboI*-treated DNA as a positive control. The *MspI/HpaII* site of spot 226 was methylated in embryos and leaf blades/sheaths. These figures were cited from Hosobuchi, S. 2007.

of this gene was conducted, this gene was expressed strongly in the embryo and leaf blade /sheath in which methylation was detected at the *MspI/HpaII* site (Fig. 10B) by RLGS analysis. Moreover, this gene never expressed in the endosperm before absorption of water, but weak expression was confirmed in the endosperm after absorption of water for 10 minutes (Fig. 10B). Next, when the methylated status at the *MspI/HpaII* site was analyzed by the PCR method, methylation was confirmed in the embryo and leaf blade/sheath (Fig. 10C). When comparing the endosperm before and after absorption of water, more PCR products were amplified in the endosperm after absorption of water (Fig. 10C. Comparison between Lanes 4 and 7). In the future, quantitative analysis will be required, but from the present results, it is assumed that the partial methylation rate of the *MspI/HpaII* site is higher in the endosperm after absorption of water than in the endosperm before the absorption of water. That is, there is the possibility that induction of DNA methylation and gene expression may begin within as very a short time as 10 minutes. When the 5' region is methylated, gene expression is usually inhibited, but our results are inconsistent with the general role. However, because the methylation analysis was conducted only at the *Msp*I/*Hpa*II site of spot 226, it is necessary to analyze the DNA methylated status in the promoter region of this gene widely and quantitatively.

4. Conclusion

RLGS is useful for genome wide surveillance of epigenetic alterations which effects to gene regulation and unknown phenomena with DNA methylation, because RLGS is suitable for exploratory studies on account of its low cost, and short set-up time. It can analyze any un-sequenced living things. As mentioned above, RLGS analysis of reciprocal hybrids in rice provided new interesting observations, and this strategy will apply to study of mammalian epigenetics.

5. Acknowledgments

We thank to Paterson, A. H., Nishiguchi, M., Kakutani, T., Mizuno, Y., Yamashita, H., Nakamura, E., Asai, K., Iijima, H., Higo, K., Hirochika H., Kawase, M., Tomioka, K., Habu, Y., Ohtake, Y., Ueda, T., Takahashi, S., Nonaka, E., Nakamura, M., Noguchi, T., Saguchi, T., and Fujita, T. for their supports.

6. References

Akimoto, K., Katakami, H., Kim, H. J., Ogawa, E., *et al.* (2007). Epigenetic inheritance in rice plants. *Ann. Bot.* (Lond.), 100, 205-217.
Baroux, C., Gagliardini, V., Page, D. R., Grossniklaus, U. (2006). Dynamic regulatory interactions of Polycomb group genes: MEDEA autoregulation is required for imprinted gene expression in Arabidopsis. *Genes Dev.*, 20, 1081-1086.
Bender, J. and Fink G. R. (1995). Epigenetic control of an endogenous gene family is revealed by a novel blue fluorescent mutant of Arabidopsis. *Cell*, 83, 725-734.
Bird A. (1992). The essentials of DNA methylation. *Cell*, 70, 5-8.
Brink, A. R. (1956). A Genetic Change Associated with the R Locus in Maize Which Is Directed and Potentially Reversible. *Genetics*, 41 (6), 872–889.
Chandler, V. L., Eggleston, W. B., Dorweiler, J. E. (2000). Paramutation in maize. *Plant Mol Biol*, 43, 121-145.
Choi, Y., Gehring, M., Johnson, L., Hannon, M., *et al.* (2002). DEMETER, a DNA glycosylase domain protein, is required for endosperm gene imprinting and seed viability in arabidopsis. *Cell*, 110, 33-42.
Coe, E. H. (1959). A REGULAR AND CONTINUING CONVERSION-TYPE PHENOMENON AT THE B LOCUS IN MAIZE. *Proc. Natl. Sci. USA*, 45, 828-832.
Comai, L., Madlung, A., Josefsson, C., Tyagi, A. (2003). Do the different parental 'heteromes' cause genomic shock in newly formed allopolyploids? *Phil. Trans. R. Soc. Lond. B Biol. Sci.*, 358, 1149-1155.
Costello, J.F., D.J. Smiraglia and C. Plass. (2002). Restriction landmark genome scaning. *Methods* Vol 27, issue 2, 144-149.

Davis, T.L., Yang, G.J., McCarrey, J.R., Bartolomei, M.S. (2000). The H19 methylation imprint is erased and re-established differentially on the parental alleles during male germ cell development. *Hum. Mol. Genet.* 9, 2885-2894.

Dong, Z. Y., Wang, Y. M., Zhang, Z. J., Shen, Y., *et al.* (2006). Extent and pattern of DNA methylation alteration in rice lines derived from introgressive hybridization of rice and Zizania latifolia Griseb *Theor. Appl. Genet.*, 113, 196-205.

Gehring, M., Huh, J. H., Hsieh, T. F., Penterman, J., *et al.* (2006). DEMETER DNA glycosylase establishes MEDEA polycomb gene self-imprinting by allele-specific demethylation. *Cell*, 124, 495-506.

Gehring, M., Bubb, K. L. & Henikoff, S. (2009). Extensive demethylation of repetitive elements during seed development underlies gene imprinting. *Science,* 324, 1447–1451.

Grossniklaus, U., Vielle-Calzada, J. P., Hoeppner, M. A., Gagliano, W. B. (1998). Maternal control of embryogenesis by MEDEA, a polycomb group gene in Arabidopsis. *Science*, 280, 446-450.

Guo, M., Rupe, M.A., Zinselmeier, C., Habben, J. *et al.* (2004). Allelic variation of gene expression in maize hybrids, *Plant Cell*, 16, 1707-1716.

Hayashizaki, Y., Hirotsune, S., Okazaki, Y., Hatada, I., *et al.* (1993). Restriction landmark genomic scanning method and its various applications. *Electrophoresis*, 14, 251-258.

Hayashizaki, Y., Shibata, H., Hirotsune, S., Sugino, H., *et al.* (1994A), Identification of an imprinted U2af binding protein related sequence on mouse chromosome 11 using the RLGS method. *Nat. Genet.*, 6, 33-40.

Hayashizaki, Y., Hirotsune, S., Okazaki, Y., Shibata, H. *et al.* (1994B). A genetic linkage map of the mouse using restriction landmark genomic scanning (RLGS), *Genetics*, 138, 1207-1238.

Hirochika, H., Okamoto, H., Kakutani, T. (2000). Silencing of retrotransposons in arabidopsis and reactivation by the ddm1 mutation. *Plant Cell,* 12, 357-369.

Hosobuchi, S. (2007). Analysis of DNA methylation of rice developing stages and tissues using RLGS method. pp105. Master thesis of Tokyo University of Science. (in Japanese).

Hsieh, T. F., Ibarra, C. A., Silva, P., Zemach, A. *et al.* (2009). Genome-wide demethylation of *Arabidopsis* endosperm. *Science*, 324, 1451–1454.

International Rice Genome Sequencing Project (2005), The map-based sequence of the rice genome. *Nature*, 436, 793-800.

Jacobsen, SE. and Meyerowitz, EM. (1997). Hypermethylated SUPERMAN epigenetic alleles in arabidopsis. *Science*, 277, 1100-1103.

Kakutani, T., Munakata, K., Richards, E. J., Hirochika, H. (1999). Meiotically and mitotically stable inheritance of DNA hypomethylation induced by ddm1 mutation of Arabidopsis thaliana.*Genetics*, 151, 831-838.

Kinoshita, T., Yadegari, R., Harada, J. J., Goldberg, R. B., *et al.* (1999). Imprinting of the MEDEA polycomb gene in the Arabidopsis endosperm. *Plant Cell*, 11, 1945-1952.

Kiyosue, T., Ohad, N., Yadegari, R., Hannon, M., *et al.* (1999). Control of fertilization-independent endosperm development by the MEDEA polycomb gene in Arabidopsis. *Proc. Natl. Acad. Sci. USA*, 96, 4186-4191.

Kohler, C., Hennig, L., Spillane, C., Pien, S., *et al.* (2003). The Polycomb-group protein MEDEA regulates seed development by controlling expression of the MADS-box gene PHERES1. *Genes Dev*, 17, 1540-1553.

Kohler, C., Page, D. R., Gagliardini, V., Grossniklaus, U. (2005). The Arabidopsis thaliana MEDEA Polycomb group protein controls expression of PHERES1 by parental imprinting. *Nat. Genet.*, 37, 28-30.

Law, JA. and Jacobsen, SE. (2010). Establishing, maintaining and modifying DNA methylation patterns in plants and animals. *Nat. Rev. Genet.*, 11 (3), 204-220.

Liu, Z., Wang, Y., Shen, Y., Guo, W., *et al.* (2004). Extensive alterations in DNA methylation and transcription in rice caused by introgression from Zizania latifolia. *Plant Mol. Biol.*, 54, 571-582.

Madlung, A., Masuelli, R. W., Watson, B., Reynolds, S. H., *et al.* (2002). Remodeling of DNA methylation and phenotypic and transcriptional changes in synthetic Arabidopsis allotetraploids.*Plant Physiol.*, 129, 733-746.

Matsuyama, T. (2008) Epigenetics: use of *in silico* genome scanning by the virtual image restriction landmark method. *FEBS Journal* 275, 1607.

Matzke, M. A., Mittelsten Scheid, O., Matzke, A. J. (1999). Rapid structural and epigenetic changes in polyploid and aneuploid genomes. *Bioessays*, 21, 761-767.

Miwa, W., Yashima, K., Sekine, T., Sekiya, T., *et al.* (1995). Demethylation of a repetitive DNA sequence in human cancers. *Electrophoresis*, 16, 227-232.

Monk, M., Boubelik, M., Lehnert, S. (1987). Temporal and regional changes in DNA methylation in the embryonic, extraembryonic and germ cell lineages during mouse embryo development. *Development*, 99, 371-382.

Obata, Y., Kaneko-Ishino, T., Koide, T., Takai, Y., *et al.* (1998). Disruption of primary imprinting during oocyte growth leads to the modified expression of imprinted genes during embryogenesis. *Development*, 125, 1553-1560.

Ohsumi, T., Okazaki, Y., Okuizumi, H., Shibata, K., *et al.* (1995). Loss of heterozygosity in chromosomes 1, 5, 7 and 13 in mouse hepatoma detected by systematic genome-wide scanning using RLGS genetic map. *Biochem. Biophys. Res. Commun.*, 212, 632-639.

Okazaki, Y., Okuizumi, H., Sasaki, N., Ohsumi, T., *et al.* (1995), An expanded system of restriction landmark genomic scanning (RLGS Ver. 1.8). *Electrophoresis*, 16, 197-202.

Okuizumi, H., Okazaki, Y., Ohsumi, T., Hayashizaki, Y., *et al.* (1995A), Genetic mapping of restriction landmark genomic scanning loci in the mouse. *Electrophoresis*, 16, 233-240.

Okuizumi, H., Okazaki, Y., Ohsumi, T., Hanami, T., *et al.* (1995B), A single gel analysis of 575 dominant and codominant restriction landmark genomic scanning loci in mice interspecific backcross progeny. *Electrophoresis*, 16, 253-260.

Okuizumi, H., Okazaki, Y., Hayashizaki, Y. (1997). RLGS spot mapping method, in: Y. Hayashizaki, S. Watanabe (Eds.), Restriction Landmark Genomic Scanning (RLGS), 57-93, Springer-Verlag, Tokyo.

Okuizumi, H., Noguchi, T., Saguchi, T., Fujita, T., Nonaka, E., Yamanaka, S., Kombate, K., Sivakumar, S., Ganesamurthy, K., Murakami, Y. (2010). Application of the Restriction Landmark Genome Scanning (RLGS) method for analysis of genetic diversity between Asian and African sorghum. *Electronic Journal of Plant Breeding*, 1(4), 1144-1147.

Pikaard, C. S. (2001). Genomic change and gene silencing in polyploids. *Trends Genet.*, 17, 675-677.

Plass, C., Shibata, H., Kalcheva, I., Mullins, L., *et al.* (1996), Identification of Grf1 on mouse chromosome 9 as an imprinted gene by RLGS-M. *Nat. Genet.*, 14, 106-109.

Reyna-Lopez, GE., Simpson, J., Ruiz-Herrera, J. (1997), Differences in DNA methylation patterns are detectable during the dimorphic transition of fungi by amplification of restriction polymorphisms.*Mol. Gen. Genet.*, 253, 703-710.

Scott, R. J., & Spielman, M. (2004). Epigenetics: imprinting in plants and mammals--the same but different? *Curr. Biol.*, 14, R201-203.

Shaked, H., Kashkush, K., Ozkan, H., Feldman, M., *et al.* (2001). Sequence elimination and cytosine methylation are rapid and reproducible responses of the genome to wide hybridization and allopolyploidy in wheat. *Plant Cell*, 13, 1749-1759.

Shiba, H., Kakizaki, T., Iwano, M., Tarutani, Y., *et al.* (2006). Dominance relationships between self-incompatibility alleles controlled by DNA methylation. *Nat. Genet.*, 38 (3), 297-299.

Shibata, H., Yoshino, K., Muramatsu, M., Plass, C., *et al.* (1995). The use of restriction landmark genomic scanning to scan the mouse genome for endogenous loci with imprinted patterns of methylation. *Electrophoresis*, 16, 210-217.

Southern, E. (1979). Measurement of DNA length by gel electrophoresis. *Anal. Biochem.*, 100, 319-323.

Springer, N.M., Stupar, R.M., (2007A). Allelic variation and heterosis in maize: how do two halves make more than a whole? *Genome Res.*, 17, 264-275.

Springer, N.M., Stupar, R.M. (2007B). Allele-specific expression patterns reveal biases and embryo-specific parent-of-origin effects in hybrid maize. *Plant Cell*, 19, 2391-2402.

Tada, M., Tada, T., Lefebvre, L., Barton, S. C., *et al.* (1997). Embryonic germ cells induce epigenetic reprogramming of somatic nucleus in hybrid cells. *Embo. J.*, 16, 6510-6520.

Takamiya, T., Hosobuchi, S., Asai, K., Nakamura, E., *et al.* (2006). Restriction landmark genome scanning method using isoschizomers (*MspI/HpaII*) for DNA methylation analysis. *Electrophoresis* 2006, 27, 2846-2856.

Takamiya, T. (2007). Analysis of genome methylation using RLGS method. pp119. Doctoral thesis of Tokyo University of Science. (in Japanese).

Takamiya, T., Hosobuchi, S., Noguchi, T., Asai, K. *et al.* (2008A). Inheritance and alteration of genome methylation in F1 hybrid rice. *Electrophoresis*, 29, 4088-4095.

Takamiya, T., Ohtake, Y., Hosobuchi, S., Noguchi, T. *et al.* (2008B) Application of RLGS method for detection of alteration in tissue cultured plants. *JARQ*, 42, 151-155.

Takamiya, T., Hosobuchi, S., Noguchi, T., Paterson, A. H. *et al.* (2009). The application of restriction landmark genome scanning (RLGS) method for surveillance of non-Mendelian inheritance in F$_1$ hybrids. *Comparative and Functional Genomics*, ID 245927, 6 pages doi:10.1155/2009/245927

Ueda, T., Abe, K., Miura, A., Yuzuriha, M., et al. (2000). The paternal methylation imprint of the mouse H19 locus is acquired in the gonocyte stage during foetal testis development. *Genes Cells*, 5, 649-659.

Wang, S.S., Smiraglia, D.J., Wu, Y.Z., Ghosh, S. et al. (2009). Identification of novel methylation markers in cervical cancer using restriction landmark genomic scanning. *Cancer Res.*, 68, 2489-2497.

Watanabe, S., Kawai, J., Kawai J., Hirotsune, S., Suzuki, H., Hirose, K., *et al.* (1995). Accessibility to tissue-specific genes from methylation profiles of mouse brain genomic DNA. *Electrophoresis*, 16, 218-226.

Wendel, J. F. (2000). Genome evolution in polyploids. *Plant Mol. Biol.*, 42, 225-249.

Xiao, W., Gehring, M., Choi, Y., Margossian, L., *et al.* (2003). Imprinting of the MEA Polycomb gene is controlled by antagonism between MET1 methyltransferase and DME glycosylase. *Dev. Cell*, 5, 891-901.

Xiong, L. Z., Xu, C. G., Saghai Maroof, M. A., Zhang, Q. (1999). Patterns of cytosine methylation in an elite rice hybrid and its parental lines, detected by a methylation-sensitive amplification polymorphism technique. *Mol. Gen. Genet.*, 261, 439-446.

Zhang, X., Yazaki, J., Sundaresan, A., Cokus, S., *et al.* (2006). Genome-wide high-resolution mapping and functional analysis of DNA methylation in arabidopsis. *Cell*, 126, 1189-1201.

Zhuang, Y., Adams, K.L. (2007). Extensive allelic variation in gene expression in populus F_1 hybrids, *Genetics*, 177, 1987-1996.

4

GC$_3$ Biology in Eukaryotes and Prokaryotes

Eran Elhaik[1,2] and Tatiana Tatarinova[3]

[1]McKusick - Nathans Institute of Genetic Medicine,
Johns Hopkins University School of Medicine, Baltimore,
[2]Department of Mental Health,
Johns Hopkins University Bloomberg School of Public Health,
[3]Faculty of Advanced Technology, University of Glamorga
[1,2]USA
[3]Wales

1. Introduction

In this chapter we describe the distribution of Guanine and Cytosine (GC) content in the third codon position (GC$_3$) distributions in different species, analyze evolutionary trends and discuss differences between genes and organisms with distinct GC$_3$ levels. We scrutinize previously published theoretical frameworks and construct a unified view of GC$_3$ biology in eukaryotes and prokaryotes.

2. The wobble position

2.1 Why is GC$_3$ referred to as the wobble position

The genetic code, the set of rules by which information encoded in genetic material, is found in every cell of every living organism. This code consists of all possible combinations of tri-nucleotide sequences in coding regions, called codons. With a few exceptions, such as start- and stop-codons, a triplet codon in a DNA sequence specifies a single amino acid – protein's building block. The human genome, for example, consists of one start- and three stop-codons out of (3^4) 64 codons. With each codon promoting the binding of specific tRNA to the ribosome, the cell would theoretically need almost 64 types of tRNAs, each with different anticodons to complement the available codons. However, because only 20 amino acids are encoded, there is a significant degeneracy of the genetic code so that the third base is less discriminatory for the amino acid than the other two bases. This third position in the codon is therefore referred to as *the wobble position*. At this position U's and C's may be read by a G in the anticodon. Similarly, A's and G's may be read by a U or ψ (pseudouridine) in the anticodon.

2.2 Identification of the mRNA codons by tRNA

For most amino acids, there are specific enzymes ligating their cognate amino acid to the tRNA molecule-bearing anticodon that correspond to that amino acid. These enzymes and the unique structure of each tRNA ensure that a particular tRNA is the substrate for its cognate synthetase and not for all other syntheses present in the cell (Shaul *et al.* 2010). Like in the genetic code, there is much redundancy in the types of tRNA molecules required per cell. Because the wobble base positions are capable of binding to several codons, a minimum set of 31 tRNA are required to unambiguously translate all the codons instead of 61 types tRNA molecule required to match each codon. This redundancy in tRNA anticodons is accomplished, in part, by using inosine, which can pair with U, C, or A, at the third position of the mRNA (Ikemura 1985).

3. Biological role of GC$_3$: Codon usage, codon bias, mRNA, gene expression, gene and promoter organization, gene function, and methylation

Deviations from unimodal bell-shaped distributions of GC$_3$ appear in many species (Aota and Ikemura 1986; Belle, Smith, and Eyre-Walker 2002; Jorgensen, Schierup, and Clark 2007). This bimodality in homeotherm (termed, "warm-blooded", at that time) vertebrates was originally explained by the presence of *isochores* – long (>300,000 bp) and relatively homogeneous stretches of DNA (Mouchiroud, Fichant, and Bernardi 1987). Although there are similarities between genes in high-GC human isochores and GC$_3$-rich genes in grasses, the isochore hypothesis does not fully explain the existence of GC$_3$-rich genes in grasses: first, there is no correlation between GC contents of open reading frames (ORFs) and the flanking regions; second, most species with isochores do not have a GC$_3$-rich peak (Tatarinova *et al.* 2010). Therefore, the remaining possible causes of bimodality may be elucidated by comparing genes in different GC$_3$ classes, such as GC$_3$-rich and GC$_3$-poor classes. These classes differ in nucleotide composition and compositional gradient along coding regions (Figure 1). GC$_3$-rich class genes have a significantly higher frequency of CG dinucleotides (potential targets for methylation); therefore, there is an additional regulatory mechanism for GC$_3$-rich genes. Springer *et al.* 2005 reported that out of eight classes of methyl-CpG-binding domain proteins present in dicots, only six exist in monocots, suggesting a difference between dicots and monocots in silencing of methylated genes.

In 2010 Tatarinova, Alexandrov, Bouck and Feldmann proposed the following explanation of relationship between DNA methylation and GC$_3$ content. Two competing processes may affect the frequency of methylation targets: the GC-based mismatch repair mechanism and AT-biased mutational pressure. In recombining organisms (e.g., grasses and homeotherms vertebrates), the GC content of coding and regulatory regions is enhanced because of the action of the GC-based mismatch repair mechanism; this effect is especially pronounced for GC$_3$. Recombination has been shown to be a driving force for the increase in GC$_3$ in many organisms. Repair (recombination) happens all over the genome with a certain precision, leading to an increase in GC content. If repair did not occur in defense-related genes, the organism may fail to survive or to reproduce. However, if repair did not happen in genes that are not essential for survival, and, consequently, their GC content remained the same, it may not be detrimental to the organism. AT-biased mutational pressure, resulting from cytosine deamination or oxidative damage to C and G bases, counteracts the influence of recombination; and in most asexually-reproducing species and self-pollinating plants, AT bias is the winning process. Our analysis from aligning *Indica* and *Japonica* indicates that genomic regions under higher selective pressure are more frequently recombining and therefore increasing their GC$_3$ content (Tatarinova *et al.* 2010). This mechanism may explain the pronounced differences in GC$_3$ between *A. thaliana* and its closest relatives. Comparison of the nucleotide compositions of coding regions in *A. thaliana, R. sativus, B. rapa, and B. napus* reveals that the GC$_3$ values of *R. sativus, B. rapa,* and *B. napus* genes are on average 0.05 higher than those of the corresponding *A. thaliana* orthologs.

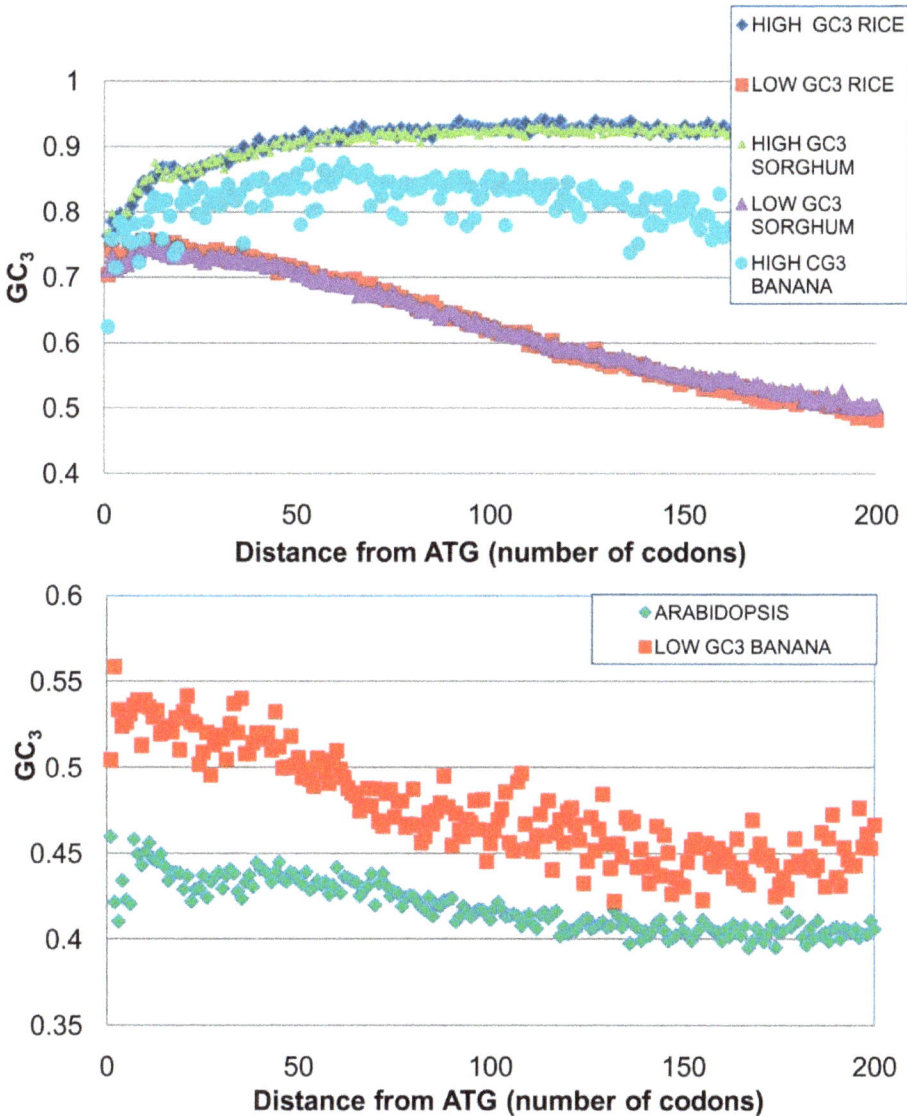

Fig. 1. GC$_3$ gradient from 5' to 3' ends of coding regions. At the 5' end of the open reading frame, high GC$_3$ genes of rice, sorghum, and banana have a slight positive gradient, whereas low GC$_3$ genes in arabidopsis, rice, sorghum, and banana become more AT$_3$-rich (From Tatarinova *et al* 2010)

An important difference between A. *thaliana* and *Brassica* and *Raphanus* is that the latter two genera are self-incompatible, whereas A. *thaliana* is self-pollinating. Self-pollination in arabidopsis keeps its recombination rates low and thus reduces the GC_3 content of its genes. Self-pollination is also reported in some grasses, such as wheat, barley and oats. Analysis of recombination in wheat showed that the genome contains areas of high and low recombination. Grasses have an efficient reproductive mechanism and high genetic variability that enables them to adapt to different climates and soil types. We hypothesize that since self-pollination generally lowers recombination rates, evolutionary pressure will selectively maintain high recombination rates for some genes. Analysis of highly recombinogenic genomic regions of wheat, barley, maize, and oat identified several genes of agronomic importance in these regions (including resistance genes against obligate biotrophs and genes encoding seed storage proteins) (Keller and Feuillet 2000). In addition to the methylation-driven growth of high-GC_3 genes, we hypothesize that the development of GC_3-richness in some genes may, if unbalanced by AT-bias, work as a feed-forward mechanism. Once GC_3-richness appears in genes under selective pressure, it provides additional transcriptional advantage. GC pairs differ from AT pairs in that guanine binds to cytosine with three hydrogen bonds, while adenine forms only two bonds with thymine. This additional hydrogen bond makes GC pairs more stable; thus GC-rich genes will have different biochemical properties from AT-rich genes. When an AT pair is replaced by a GC pair in the third position of a codon, the protein sequence remains largely unchanged, but an additional hydrogen bond is introduced. This additional bond can make transcription more efficient and reliable, change the array of RNA binding proteins or significantly alter the three-dimensional folding of the messenger RNA. In this case, those plant species that thrive and adapt successfully to harsh environments demonstrate a strong preference for GC in the third position of the codon.

High GC_3 content provides more targets for methylation. The correlation between methylation and GC_3 is supported by Stayssman *et al.* (2009), who reported a positive correlation between methylation of internal unmethylated regions and expression of the host gene. In this paper the authors have demonstrated a positive correlation between GC_3 and variability of gene expression; they also found that GC_3-rich genes are more enriched in CG than the low-GC_3-poor gene class. Therefore, GC_3-rich genes provide more targets for *de novo* methylation, which can serve as an additional mechanism of transcriptional regulation and affect the variability of gene expression. Overall, additional transcriptional regulation makes species more adaptable to external stresses.

4. Genome-wide view

4.1 GC_3 in animals

The GC_3 varies substantially within animal genomes. Animals can be divided to homeotherms-those that maintain a stable internal body temperature, like mammals and birds, and sometimes termed "warm-blooded" and poikilotherms - those whose internal temperature varies considerably and are often termed "cold-blooded." These differences were integrated into molecular evolution by Mouchiroud *et al.* (1987) who argued that in poikilotherm vertebrate, genes are mostly GC-poor and are harbored by GC-poor intergenic regions, whereas most genes of homeothermy vertebrates are GC-rich and found predominantly on the scant GC-rich intergenic regions. Because GC_3 was shown to be correlated to the GC content of the gene, it became the primary tool to study the differences between homeotherms and poikilotherms. Indeed, poikilotherms were shown to have lower GC$_3$ than homeotherms on average, although some poikilotherms exhibit higher GC$_3$ values than homeotherms. Most poikilotherms also exhibit a lower variation in GC$_3$ and a correlation to the GC content of the first two codon positions, indicating a systemic compositional variation across their genome (Belle, Smith, and Eyre-Walker 2002).

GC$_3$ exhibits a wide variation in different animals compared with the average GC content (Figure 2). In humans, GC$_3$ ranges from 22 to 97% compared with the range of GC content (32-80%), and in zebrafish the GC$_3$ range is more limited 27-92% (μ=56%,σ=8%) yet still wider than the GC content range (34-68%) (Figure 3) (Elhaik, Landan, and Graur 2009). Because GC$_3$ is mostly unconstrained by functional requirements, that is, by the need to code specific amino acids and because GC$_3$ exhibits a non-uniform distribution, the third-codon position became a natural candidate to investigate the forces that shaped the composition of the genome.

Species	No. of genes	GC$_3$			GC content		
		Mean	σ	Range	Mean	σ	Range
Homo sapiens	17,451	0.6	0.17	0.22-0.97	0.45	0.06	0.32-0.8
Bos Taurus	5,522	0.62	0.16	0.25-0.97	0.43	0.06	0.33-0.76
Mus musculus	17,009	0.59	0.11	0.21-0.96	0.43	0.05	0.27-0.76
Rattus norvegicus	8,983	0.59	0.11	0.23-0.96	0.42	0.06	0.33-0.73
Gallus gallus	3,036	0.56	0.15	0.28-0.99	0.42	0.05	0.36-0.8
Danio rerio	4,344	0.56	0.08	0.27-0.92	0.35	0.02	0.34-0.68

Fig. 2. GC$_3$ and GC content for 6 Vertebrate Taxa. From (Elhaik, Landan, and Graur 2009).

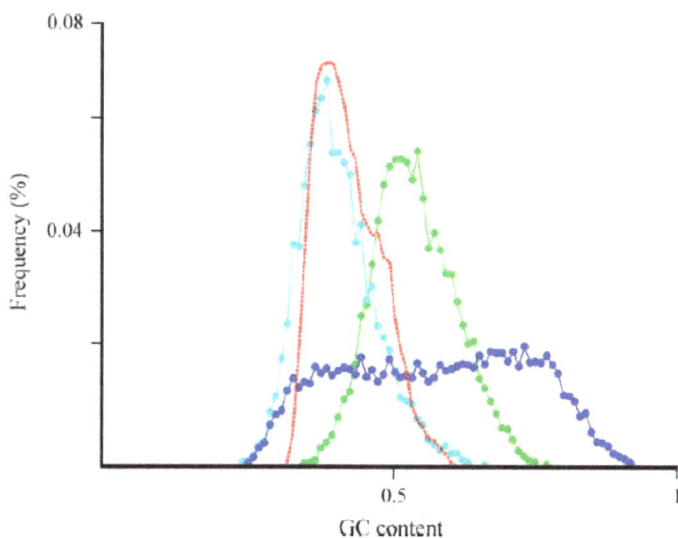

Fig. 3. GC content in codon positions: GC$_1$ (green), GC$_2$ (turquoise), GC$_3$ (blue), and 200-kb flanking regions (dashed red) in human. From (Elhaik, Landan, and Graur 2009).

4.2 GC$_3$ in yeast

The variation in GC content along yeast chromosomes was first reported in *S. cerevisiae* (Sharp and Lloyd 1993) where GC$_3$ ranged from 35% to 50% on chromosome III. Later observations confirmed similar patterns for the remaining chromosomes and linked the GC variation to recombination, that is, rich GC$_3$ regions would have a higher recombination rate and recombination hot spots would occur in high picks of GC content (Bradnam *et al.* 1999). Findings from eukaryotic genomes suggested that biased gene conversion (see below) may be the molecular mechanism that facilitated the increase of GC$_3$ in recombination hot spots.

The large variation in GC$_3$ also exists between yeast species with *C. tropicalis* having the lowest median GC$_3$ (22%) and *C. lusitaniae* with a GC$_3$ median of 49%. Overall, species that were more closely related to each other tended to have more similar GC$_3$ distributions (Figure 4). Despite of these vast differences, the locations of peaks and troughs of GC$_3$ largely coincide. In other words, the differences in base composition between the species varies systematically across all genes, with higher divergence in GC-rich genes than GC-poor genes (Lynch *et al.* 2010). The conserved nature of the GC$_3$ variation along chromosomes and the coinciding peaks and troughs led Lynch *et al.* (2010) to propose that the GC-poor troughs indicate the positions of ancient centromeres at the points where deep GC-poor regions were found.

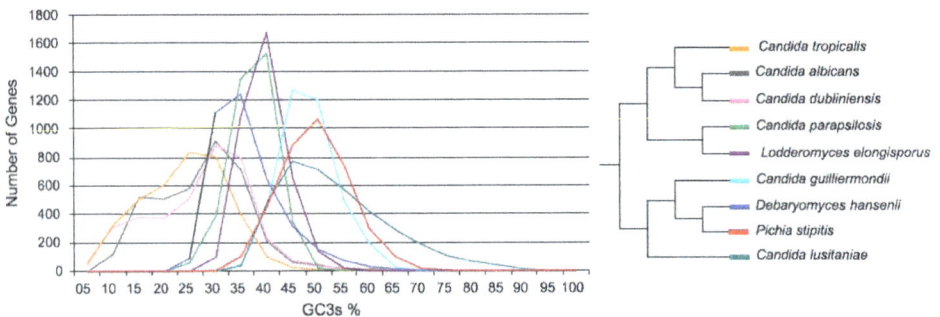

Fig. 4. Distribution of GC$_{3s}$ values among nine species in the Candida clade. A set of 3,687 orthologous genes was identified among all these species. The phylogenetic tree on the right is modified from Butler *et al.* (2009) and is derived from 644 single-gene families. From (Lynch *et al.* 2010).

4.3 GC$_3$ in plants: Adaptation to environment and codon usage (Tatarinova et al. 2010)

Pronounced differences in GC$_3$ exist both within and between plant genomes. For example, GC$_3$ in rice genes ranges from 43% to 92% (Wang, Singer, and Hickey 2004). Grasses have undergone several genome duplications. Genomic regions varied in their recombination rates and GC$_3$ contents. Since high GC$_3$ content in a gene provided an evolutionary advantage, this was frequently the sole copy retained in grasses. This may explain why GC$_3$-rich genes frequently lack paralogs. GC$_3$-rich genes provide an evolutionary advantage because of their optimized codon usage and the existence of methylation targets allowing for an additional mechanism of transcriptional regulation. Therefore, GC$_3$-rich genes were maintained in grasses for generations.

The evolutionary forces affecting development of plants are realized through introducing new mutations during meiotic recombination and fixation with the help of DNA methylation and transcriptional mechanisms. The presence of GC_3-rich genes is not likely to be a consequence of chromosomal isochores or horizontal gene transfer. Regardless of their initial origin, GC_3-rich genes in recombining species possessed a self-maintaining mechanism that over time could only increase their drift towards even higher GC_3 values. This uncompensated drift may explain the pronounced bimodality of some rapidly-evolving species. Competing forces acting in grasses make GC_3 distribution distinctly bimodal; GC_3-rich genes are more transcriptionally regulated, provide more targets for methylation and accumulate more mutations than GC_3-poor genes.

4.4 GC₃ in prokaryotes

In prokaryotic genomes the nucleotide composition varies from the extremely low GC content (15%) in obligatory intracellular bacteria to the high GC content (75%) in Proteobacteria and from 27% to 66% in Archaea. (http://www.ncbi.nlm.nih.gov/genomes/lproks.cgi). The base compositional deviations show tremendous variation even at the nucleotide level of the three codon positions. With GC_1 follows a global tendency of monotonic decrease versus the increase of the genomic GC content and the GC_2 follows a global monotonic increase, as expected. The GC_3 positions range from 10% to 90% (Muto and Osawa 1987) and exhibit a more complicated pattern that decreases first and increases last with a global minimum at about 40% of the genomic GC content (Ma and Chen 2005).

In many organisms, alternative synonymous codons are not used with equal frequency, hence the codon usage is considered biased. This bias exists not only between different organisms, but often among genes within a genome (Suzuki, Saito, and Tomita 2009). Different factors have been proposed to contribute to synonymous codon usage bias, including replication strand bias, translational selection, and GC composition (Ermolaeva 2001). Because codon third positions are largely degenerate - 70% of changes at third codon positions are synonymous and they are commonly considered correlated with synonymous codon usage bias, although in practicality, the strength of this correlation varies widely among species (Suzuki, Saito, and Tomita 2007). The large deviations in base composition of these sites were also thought to reveal the underlying mutational bias of the genome and served as the basis for the original formulation of the neutral theory (Sueoka 1962; Sueoka 1988).

4.5 The isochore theory

Prior to the publication of the draft human genome (Lander *et al*. 2001), scientists were limited to the study of genes and short (<500 bp) flanking regions. The publication of the draft human genome (Lander *et al*. 2001) in 2001 was quickly followed by the publications of fully sequenced genomes from other species (e.g., Chimpanzee, mouse, and cow), which enabled us to study the evolution of genomes whole biological entities, rather than as a collection of genes. One of the most common ways to describe a genome is by means of the nucleotide distribution, particularly the distribution of GC content. In the absence of genomic data, inferences made on short fragments were based on the assumption that these fragments represent the compositional complexity of the entire genome (e.g., Aissani *et al*. 1991; Mouchiroud and Bernardi 1993). The GC content patterns emerging from these analyses were used by Bernardi and colleagues (Macaya, Thiery, and Bernardi 1976; Thiery, Macaya, and Bernardi 1976; Bernardi *et al*. 1985) to explain the differences between the genome organization of "warm-blooded" and "cold-blooded" vertebrates (Cuny *et al*. 1981; Bernardi *et al*. 1985; Bernardi 2000) with the first described as a mosaic of GC-poor and GC-rich isochores and the later as devoid of GC-rich isochores (Bernardi *et al*. 1985).

Because GC_3 is mostly unconstrained by functional requirements, that is, by the need to code specific amino acids, the third-codon position was a natural candidate for a predictive proxy of flanking GC content. In spite of the lack of correlation between GC_3 and large flanking regions of "isochoric regions" harboring the genes (Bernardi 1993a), over time it became a common belief that such a relationship exists. Over the next two decades GC_3 was used extensively as the primary means to predict isochore structure, surprisingly enough, even after full genome sequences were made available. Many of the theories concerning the evolution of isochores are also based on studies that used GC_3 as a predictor for isochore composition or that simply assumed the existence of isochores. (Bernardi 2001; Ponger, Duret, and Mouchiroud 2001; D'Onofrio 2002; D'Onofrio, Ghosh, and Bernardi 2002; Hamada *et al*. 2003; Romero *et al*. 2003; Federico *et al*. 2004; Chojnowski *et al*. 2007; Fortes *et al*. 2007)

Two opposite explanations were proposed to explain the formation of isochores. The first view was that isochores may simply reflect variable mutation processes among genomic regions, consistent with the neutral model (Wolfe, Sharp, and Li 1989; Sueoka and Kawanishi 2000; Galtier *et al*. 2001). Alternatively, isochores were posited as results of natural selection for certain compositional environment required by certain genes (Matassi, Sharp, and Gautier 1999). It should be noted that these hypotheses are not mutually exclusive; two or more of the processes could be acting together (Eyre-Walker and Hurst 2001). For example, the most accepted hypothesis for the unequal usage of synonymous codons in bacterial genomes is that the unequal usage is the result of a very complex balance among different evolutionary forces (mutation and selection) (Suzuki, Saito, and Tomita 2007). Several hypothesis derive from the selectionist view, such as the biased gene conversion hypothesis (Galtier *et al*. 2001), the thermodynamic stability hypothesis (Bernardi and Bernardi 1986; Bernardi 1993b), the transposable elements hypothesis (Duret and Hurst 2001), the recombination hypothesis (Eyre-Walker 1993) and the cytosine deamination hypothesis (Fryxell and Zuckerkandl 2000).

The presumed relationship between GC_3 and isochores has been used numerous times in the literature to study isochore function and evolution until Elhaik *et al*. (2009) showed that no such relationship exists. By testing the relationship between GC_3 and the GC content of the flanking regions of the genes of 6 animals, the authors demonstrated that GC_3 explains a very small proportion of the variation in GC content of long genomic sequences flanking the genes. The predictive power either decreases rapidly the further one gets from the gene or does not exist at all. These findings also implied that the isochore theory cannot be discussed without further analysis of the complete genomic data. Indeed, further analyses showed that the descriptions of the human or vertebrate genomes as mosaics of isochores are erroneous (Cohen *et al*. 2005; Elsik *et al*. 2009; Elhaik, Graur, and Josić 2010; Elhaik *et al*. 2010). Due to the lack of predictive power of GC_3, new genomic studies scan the entire genomic structure using automatic algorithms rather than rely on unreliable proxies. The emerging view of the mammalian genome depicts an assortment of compositionally nonhomogeneous domains with numerous short compositionally homogeneous domains and relatively few long one (Elsik *et al.* 2009; Elhaik *et al.* 2010). Similar results were found for invertebrate genomes (Sodergren *et al.* 2006a; Sodergren *et al.* 2006b; Richards *et al.* 2008; Kirkness *et al.* 2010; Werren *et al.* 2010; Smith *et al.* 2011a; Smith *et al.* 2011b; Suen *et al.* 2011).

5. Applications of GC₃ in everyday's biology

Although the role of GC₃ as a proxy to large flanking regions was severely minimized, the question of which processes determine the GC content in 4-fold degenerate codons remains unanswered. It was therefore proposed to use GC₃ to detect sites under selection. The rationale is simple, if genomic GC-content is solely a consequence of mutation bias and the base composition is at equilibrium, then we expect equal numbers of synonymous mutations at 4-fold sites to be segregating within a species (Hildebrand, Meyer, and Eyre-Walker 2010), whereas deviation from such prediction may indicate selection. GC₃ remains a very useful tool to estimate codon usage bias and species diversity (Suzuki, Saito, and Tomita 2009).

GC₃ is useful for detection of recent horizontal gene transfer (HGT) events. Horizontal gene transfer occurs when an organism incorporates genetic material from another organism without being the offspring of that organism. Recently acquired genes retain nucleotide composition of the original genome and their identification is important for accurate reconstruction of phylogenetic trees, epidemiology, and genetic engineering.

GC₃ can also be used for gene prediction and genome annotation. In monocots, many genes demonstrate a negative GC gradient, that is, the GC content declines along the orientation of transcription. It is important to detect the presence of GC-rich sequences at the 5' end of genes because it influences the conformation of chromatin, the expression level of genes, and the recombination rate. Performance of genome annotation programs is often affected by 5'-3' gradients of nucleotide composition of coding region (Figure 1). Rare tissue-specific and stress-specific genes (that may not have mRNA support) are likely to belong to GC₃-rich class, and have a distinct pattern on the 5'-3'. If the gene-finding program is tailored to the more prevalent GC₃-poor genes, de-novo identification of these rare, but probably extremely important for stress response and adaptation, GC₃-rich genes will be hindered (Souvorov *et al* 2011).

6. References

Alvarez-Valin, F., G. Lamolle, and G. Bernardi. 2002. Isochores, GC₃ and mutation biases in the human genome. Gene. 300:161-168.

Aota, S., and T. Ikemura. 1986. Diversity in G + C content at the third position of codons in vertebrate genes and its cause. Nucleic Acids Res. 14:6345-6355.

Belle, E. M., N. Smith, and A. Eyre-Walker. 2002. Analysis of the phylogenetic distribution of isochores in vertebrates and a test of the thermal stability hypothesis. Journal of Molecular Evolution. 55:356-363.

Bernardi, G., S. Hughes, and D. Mouchiroud. 1997. The major compositional transitions in the vertebrate genome. Journal of Molecular Evolution. 44 Suppl 1:S44-51.

Carlini, D. B., Y. Chen, and W. Stephan. 2001. The Relationship Between Third-Codon Position Nucleotide Content, Codon Bias, mRNA Secondary Structure and Gene Expression in the Drosophilid Alcohol Dehydrogenase Genes Adh and Adhr. Genetics. 159:623-633.

Duret, L., D. Mouchiroud, and C. Gautier. 1995. Statistical analysis of vertebrate sequences reveals that long genes are scarce in GC-rich isochores. Journal of Molecular Evolution. 40:308-317.

Elhaik, E., G. Landan, and D. Graur. 2009. Can GC Content at Third-Codon Positions Be Used as a Proxy for Isochore Composition? Molecular Biology and Evolution. 26:1829-1833.

Graur, D., and W.-H. Li. 2000. Fundamentals of molecular evolution. Sinauer, Sunderland, Mass.

Ikemura, T. 1985. Codon usage and tRNA content in unicellular and multicellular organisms. Molecular Biology and Evolution. 2:13-34.

Lercher, M. J., N. G. C. Smith, A. Eyre-Walker, and L. D. Hurst. 2002. The Evolution of Isochores: Evidence From SNP Frequency Distributions. Genetics. 162:1805-1810.

Lynch, D. B., M. E. Logue, G. Butler, and K. H. Wolfe. 2010. Chromosomal G + C content evolution in yeasts: systematic interspecies differences, and GC-poor troughs at centromeres. Genome Biol Evol. 2:572-583.

Mouchiroud D., D'Onofrio G, Aïssani B, Macaya G, Gautier C, Bernardi G: The distribution of genes in the human genome. Gene 1991, 100:181-7.

Romiguier, J., V. Ranwez, E. J. Douzery, and N. Galtier. 2010. Contrasting GC-content dynamics across 33 mammalian genomes: Relationship with life-history traits and chromosome sizes. Genome Research.

Shaul, S., D. Berel, Y. Benjamini, and D. Graur. 2010. Revisiting the operational RNA code for amino acids: Ensemble attributes and their implications. RNA. 16:141-153.

Stayssman R, Nejman D, Roberts D, Steinfeld I, Blum B, Benvenisty N, Simon I, Yakhili Z, Cedar H: Developmental programming of CpG island methylation profiles in the human genome. Nature Structural & Molecular Biology 2009, 16(5):564-571

Tatarinova, T. V., N. N. Alexandrov, J. B. Bouck, and K. A. Feldmann. 2010. GC_3 biology in corn, rice, sorghum and other grasses. BMC Genomics. 11:308.

Aissani, B., G. D'Onofrio, D. Mouchiroud, K. Gardiner, C. Gautier, and G. Bernardi. 1991. The compositional properties of human genes. Journal of Molecular Evolution. 32:493-503.

Aota, S., and T. Ikemura. 1986. Diversity in G + C content at the third position of codons in vertebrate genes and its cause. Nucleic Acids Res. 14:6345-6355.

Belle, E. M., N. Smith, and A. Eyre-Walker. 2002. Analysis of the phylogenetic distribution of isochores in vertebrates and a test of the thermal stability hypothesis. Journal of Molecular Evolution. 55:356-363.

Bernardi, G. 1993a. The isochore organization of the human genome and its evolutionary history--a review. Gene. 135:57-66.

Bernardi, G. 2001. Misunderstandings about isochores. Part 1. Gene. 276:3-13.

Bernardi, G. 1993b. The vertebrate genome: isochores and evolution. Molecular Biology and Evolution. 10:186-204.

Bernardi, G. 2000. Isochores and the evolutionary genomics of vertebrates. Gene. 241:3-17.

Bernardi, G., and G. Bernardi. 1986. Compositional constraints and genome evolution. Journal of Molecular Evolution. 24:1-11.

Bernardi, G., B. Olofsson, J. Filipski, M. Zerial, J. Salinas, G. Cuny, M. Meunier-Rotival, and F. Rodier. 1985. The mosaic genome of warm-blooded vertebrates. Science. 228:953-958.

Bradnam, K. R., C. Seoighe, P. M. Sharp, and K. H. Wolfe. 1999. G+C content variation along and among Saccharomyces cerevisiae chromosomes. Molecular Biology and Evolution. 16:666-675.

Chojnowski, J. L., J. Franklin, Y. Katsu, T. Iguchi, L. J. Guillette, Jr., R. T. Kimball, and E. L. Braun. 2007. Patterns of vertebrate isochore evolution revealed by comparison of expressed mammalian, avian, and crocodilian genes. Journal of Molecular Evolution. 65:259-266.

Cohen, N., T. Dagan, L. Stone, and D. Graur. 2005. GC composition of the human genome: in search of isochores. Molecular Biology and Evolution. 22:1260-1272.

Cuny, G., P. Soriano, G. Macaya, and G. Bernardi. 1981. The major components of the mouse and human genomes: Preparation, basic properties and compositional heterogeneity. European Journal of Biochemistry. 115:227-233.

D'Onofrio, G. 2002. Expression patterns and gene distribution in the human genome. Gene. 300:155-160.

D'Onofrio, G., T. C. Ghosh, and G. Bernardi. 2002. The base composition of the genes is correlated with the secondary structures of the encoded proteins. Gene. 300:179-187.

Duret, L., and L. D. Hurst. 2001. The elevated GC content at exonic third sites is not evidence against neutralist models of isochore evolution. Molecular Biology and Evolution. 18:757-762.

Elhaik, E., D. Graur, and K. Josić. 2010. Comparative testing of DNA segmentation algorithms using benchmark simulations. Molecular Biology and Evolution. 27:1015-1024.

Elhaik, E., D. Graur, K. Josic, and G. Landan. 2010. Identifying compositionally homogeneous and nonhomogeneous domains within the human genome using a novel segmentation algorithm. Nucl. Acids Res.:gkq532.

Elhaik, E., G. Landan, and D. Graur. 2009. Can GC Content at Third-Codon Positions Be Used as a Proxy for Isochore Composition? Molecular Biology and Evolution. 26:1829-1833.

Elsik, C. G.R. L. TellamK. C. Worley et al. 2009. The genome sequence of taurine cattle: a window to ruminant biology and evolution. Science. 324:522-528.

Ermolaeva, M. D. 2001. Synonymous codon usage in bacteria. Curr Issues Mol Biol. 3:91-97.

Eyre-Walker, A. 1993. Recombination and mammalian genome evolution. Proc Biol Sci. 252:237-243.

Eyre-Walker, A., and L. D. Hurst. 2001. The evolution of isochores. Nat Rev Genet. 2:549-555.

Federico, C., S. Saccone, L. Andreozzi, S. Motta, V. Russo, N. Carels, and G. Bernardi. 2004. The pig genome: compositional analysis and identification of the gene-richest regions in chromosomes and nuclei. Gene. 343:245-251.

Fortes, G., C. Bouza, P. Martínez, and L. Sánchez. 2007. Diversity in isochore structure among cold-blooded vertebrates based on GC content of coding and non-coding sequences. Genetica. 129:281-289.

Fryxell, K. J., and E. Zuckerkandl. 2000. Cytosine deamination plays a primary role in the evolution of mammalian isochores. Molecular Biology and Evolution. 17:1371-1383.

Galtier, N., G. Piganeau, D. Mouchiroud, and L. Duret. 2001. GC-content evolution in mammalian genomes: the biased gene conversion hypothesis. Genetics. 159:907-911.

Hamada, K., T. Horiike, H. Ota, K. Mizuno, and T. Shinozawa. 2003. Presence of isochore structures in reptile genomes suggested by the relationship between GC contents of intron regions and those of coding regions. Genes and Genetic Systems. 78:195-198.

Hildebrand, F., A. Meyer, and A. Eyre-Walker. 2010. Evidence of Selection upon Genomic GC-Content in Bacteria. PLoS Genet. 6:e1001107.

Jorgensen, F. G., M. H. Schierup, and A. G. Clark. 2007. Heterogeneity in regional GC content and differential usage of codons and amino acids in GC-poor and GC-rich regions of the genome of Apis mellifera. Molecular Biology and Evolution. 24:611-619.

Kirkness, E. F.B. J. HaasW. Sun *et al.* 2010. Genome sequences of the human body louse and its primary endosymbiont provide insights into the permanent parasitic lifestyle. Proceedings of the National Academy of Sciences of the United States of America. 107:12168-12173.

Lander, E. S.L. M. LintonB. Birren *et al.* 2001. Initial sequencing and analysis of the human genome. Nature. 409:860-921.

Lynch, D. B., M. E. Logue, G. Butler, and K. H. Wolfe. 2010. Chromosomal G + C content evolution in yeasts: systematic interspecies differences, and GC-poor troughs at centromeres. Genome Biol Evol. 2:572-583.

Ma, B. G., and L. L. Chen. 2005. The most deviated codon position in AT-rich bacterial genomes: a function related analysis. Journal of Biomolecular Structure and Dynamics. 23:143-149.

Macaya, G., J. P. Thiery, and G. Bernardi. 1976. An approach to the organization of eukaryotic genomes at a macromolecular level. Journal of Molecular Biology. 108:237-254.

Matassi, G., P. M. Sharp, and C. Gautier. 1999. Chromosomal location effects on gene sequence evolution in mammals. Current Biology. 9:786-791.

Mouchiroud, D., and G. Bernardi. 1993. Compositional properties of coding sequences and mammalian phylogeny. Journal of Molecular Evolution. 37:109-116.

Mouchiroud, D., G. Fichant, and G. Bernardi. 1987. Compositional compartmentalization and gene composition in the genome of vertebrates. Journal of Molecular Evolution. 26:198-204.

Muto, A., and S. Osawa. 1987. The guanine and cytosine content of genomic DNA and bacterial evolution. Proceedings of the National Academy of Sciences of the United States of America. 84:166-169.

Ponger, L., L. Duret, and D. Mouchiroud. 2001. Determinants of CpG islands: expression in early embryo and isochore structure. Genome Research. 11:1854-1860.

Richards, S.R. A. GibbsG. M. Weinstock *et al*. 2008. The genome of the model beetle and pest Tribolium castaneum. Nature. 452:949-955.

Romero, H., A. Zavala, H. Musto, and G. Bernardi. 2003. The influence of translational selection on codon usage in fishes from the family Cyprinidae. Gene. 317:141-147.

Sharp, P. M., and A. T. Lloyd. 1993. Regional base composition variation along yeast chromosome III: evolution of chromosome primary structure. Nucleic Acids Res. 21:179-183.

Smith, C. D.A. ZiminC. Holt *et al*. 2011a. Draft genome of the globally widespread and invasive Argentine ant (Linepithema humile). Proceedings of the National Academy of Sciences of the United States of America.

Smith, C. R.C. D. SmithH. M. Robertson *et al*. 2011b. Draft genome of the red harvester ant Pogonomyrmex barbatus. Proceedings of the National Academy of Sciences of the United States of America.

Sodergren, E.G. M. Weinstock E. H. Davidson *et al*. 2006a. Insights into social insects from the genome of the honeybee Apis mellifera. Nature. 443:931-949.

Sodergren, E.G. M. WeinstockE. H. Davidson *et al*. 2006b. The genome of the sea urchin Strongylocentrotus purpuratus. Science. 314:941-952.

Souvorov, A; Tatusova, T; Zaslasky, L; Smith-White, B. Glycine max and Zea mays Genome Annotation with Gnomon, ISMB/ECCB 2011.

Suen, G.C. TeilingL. Li *et al*. 2011. The Genome Sequence of the Leaf-Cutter Ant *Atta ephalotes* Reveals Insights into Its Obligate Symbiotic Lifestyle. PLoS Genet. 7:e1002007.

Sueoka, N. 1988. Directional mutation pressure and neutral molecular evolution. Proceedings of the National Academy of Sciences of the United States of America. 85:2653-2657.

Sueoka, N. 1962. On the genetic basis of variation and heterogeneity of DNA base composition. Proceedings of the National Academy of Sciences of the United States of America. 48:582-592.

Sueoka, N., and Y. Kawanishi. 2000. DNA G+C content of the third codon position and codon usage biases of human genes. Gene. 261:53-62.

Suzuki, H., R. Saito, and M. Tomita. 2007. Variation in the correlation of G + C composition with synonymous codon usage bias among bacteria. EURASIP J Bioinform Syst Biol.61374.

Suzuki, H., R. Saito, and M. Tomita. 2009. Measure of synonymous codon usage diversity among genes in bacteria. BMC Bioinformatics. 10:167.

Tatarinova, T. V., N. N. Alexandrov, J. B. Bouck, and K. A. Feldmann. 2010. GC3 biology in corn, rice, sorghum and other grasses. BMC Genomics. 11:308.

Thiery, J. P., G. Macaya, and G. Bernardi. 1976. An analysis of eukaryotic genomes by density gradient centrifugation. Journal of Molecular Biology. 108:219-235.

Wang, H. C., G. A. Singer, and D. A. Hickey. 2004. Mutational bias affects protein evolution in flowering plants. Molecular Biology and Evolution. 21:90-96.

Werren, J. H.S. RichardsC. A. Desjardins *et al.* 2010. Functional and evolutionary insights from the genomes of three parasitoid Nasonia species. Science. 327:343-348.

Wolfe, K. H., P. M. Sharp, and W. H. Li. 1989. Mutation rates differ among regions of the mammalian genome. Nature. 337:283-285.

5

MethylMeter®: A Quantitative, Sensitive, and Bisulfite-Free Method for Analysis of DNA Methylation

David R. McCarthy[1], Philip D. Cotter[2],
and Michelle M. Hanna[1,3*]

[1]RiboMed Biotechnologies, Inc., Carlsbad, CA,
[2]Department of Pathology, Children's
Hospital and Research Center, Oakland, CA,
[3]Arizona Cancer Center, University of Arizona, Tucson, AZ,
USA

1. Introduction

In this chapter we present a new *bisulfite-free* method to detect and quantify DNA methylation and its application to the detection of imprinting disorders such as Prader-Willi (PWS) and Angelman (AS) syndromes. The method, called MethylMeter®, combines affinity separation of methylated and unmethylated DNA with CAP™ (Coupled Abscription®-PCR), a new quantitative and sensitive signal generation process. In order to validate MethylMeter®, we analyzed samples from 54 patients diagnosed with Prader-Willi or Angelman syndromes, as well as samples from normal patients. Results were compared to the results obtained previously on these samples using bisulfite-based TaqMan® Methylation-Specific PCR (MS-PCR). Methylation detection with CAP was as accurate as with TaqMan, but was approximately 2000 times more sensitive. Methylated DNA was separated from unmethylated DNA with the use of magnetic beads bearing a new methyl-CpG binding domain protein. The amount of the normally imprinted SNRPN promoter region present in the bound and unbound fractions was used to determine the relative amounts of methylated and unmethylated SNRPN promoter in the sample. The results were 100% concordant with previous results generated with MS-PCR, but significantly less patient DNA and time were required to obtain results, which are more quantitative than MS-PCR. CAP based detection can be accomplished without fluorescent probes and in fewer cycles than with other PCR methods. Because methylated DNA is detected based on purification of methylated DNA, rather than on chemical conversion of unmethylated DNA, the disadvantages of bisulfite treatment are avoided. DNA is not degraded allowing analysis of samples as small as 1 ng. CAP primer development is not limited by the effects of reduced sequence complexity or a requirement to overlap primers with CpG sites in the target DNA.

---

* Corresponding Author

1.1 Current methods for detecting DNA methylation in clinical samples

The most commonly used methods for analyzing the DNA methylation levels of Differentially Methylated Regions (DMRs), such as CpG islands, and individual CpG sites, utilize treatment of patient DNA with the chemical bisulfite. Bisulfite converts unmethylated cytosine (C) residues to deoxyuracil (dU), while leaving methylated cytosine (MeC) unchanged. Upon PCR amplification of bisulfite-treated DNA, the MeC is copied to C and dU is copied to thymine (T). As a result, the retention of cytosine at a specific position indicates methylation. The modified DNA can then be analyzed by methods that detect the sequence difference between the amplified DNA generated from the methylated and unmethylated, bisulfite-treated starting material.

In methylation-specific PCR (MSP), oligonucleotides that hybridize to a region on the DNA containing 1 or more potentially methylated CpG sites are added to the bisulfite-treated DNA. The oligonucleotides are designed to hybridize either to a DNA sequence corresponding to bisulfite-treated methylated DNA (still contains C), or to a sequence corresponding to bisulfite-treated unmethylated DNA (now contains dU), and PCR is performed in order to amplify the methylated and/or unmethylated target DNA (Herman, et al., 1996). Amplicons are analyzed by gel electrophoresis. Alternatively, fluorescence-based quantitative real-time PCR can be performed on bisulfite-modified DNA (Eads, et al., 2000; Zeschnigk, et al., 2004). Adaptations of quantitative real-time PCR utilize Taqman probes to generate a fluorescent signal or blocker oligonucleotides to prevent amplification of unmethylated DNA, resulting in increased assay sensitivity (Cottrell et al., 2004). Pyrosequencing is also utilized for methylation quantitation from bisulfite modified DNA (Tost et al., 2003).

An advantage of bisulfite modification is that it is a widely used and established procedure and allows analysis of methylation at single CpG sites in the DNA. Some of the disadvantages to this assay are that bisulfite treatment of DNA can destroy a large percentage of the input DNA, resulting in limited sensitivity and the requirement for large amounts of DNA. Quality control assessment of bisulfite treated DNA is necessary before performing a detection assay to avoid misleading results. Extensive degradation can introduce sampling errors when few molecules are long enough to be amplified (Ehrich et al., 2007). Bisulfite treatment also creates DNA targets that are now very A-T rich, having converted unmethylated C to T, and this complicates the design of specific probes for PCR amplification. Bisulfite treatment remains the main source of variability for methylation detection, particularly in samples containing already degraded DNA, such as Formalin Fixed Paraffin Embedded (FFPE) tumor tissues.

Other DNA methylation detection methods that do not use bisulfite rely on detection of methylated DNA by restriction enzyme analysis. The DNA is treated with either a MSRE (methylation-sensitive restriction enzyme) or a MDRE (methylation dependent restriction enzyme), amplified and then analyzed by microarray or gel analysis. MSREs are restriction enzymes, which cut at DNA only if the C in a CpG site is unmethylated. MDREs are restriction enzymes that require CpG methylation for cleavage. By treating DNA with either of these enzymes and subsequent comparison to a control sample, the methylation state of a DNA sample *at that site* can be determined. If digestion of a sample occurs after treatment with a MDRE, then the DNA is assumed to be methylated. Conversely, if the DNA is uncut when treated with a MSRE, then this sample is also assumed to be methylated. By comparing the amount of cut vs. uncut DNA, the level of methylation can be estimated. A common read out for this type of methylation analysis is the subsequent amplification and fluorescent labeling of the digested DNA. The fragments can then be hybridized to a library microarray and analyzed (Lipmann et al., 2004) or simply resolved by electrophoresis. Quantitative real-time PCR is another mode of analysis (Ordway, et al., 2006). An advantage of MSRE/MDSE digestion is that no pretreatment of the DNA is necessary, although it is often performed in conjunction with bisulfite treatment of DNA in a procedure called COBRA (Xiong & Laird, 1997). Some disadvantages with this application are that it is a rather lengthy procedure and is dependent on the presence of specific restriction enzyme recognition sequences near the region of the target DNA.

A final method that is commonly employed is chromatin immunoprecipitation (ChIP) by the use of antibodies against methyl binding proteins. Typically, cells are fixed and then methylated DNA is immunoprecipitated by the use of antibodies specific for methyl binding proteins. The resulting DNA is amplified, labeled and analyzed by hybridization in a microarray assay. The advantages of this method are that the assay can be performed from live cells with little or no DNA purification required. The assay also has increased sensitivity, as unwanted and contaminant DNA are removed prior to analysis. However, the procedure is very time-consuming, involves several steps and requires expensive reagents. Some assays may take as long as five days to complete.

1.2 Imprinting disorder diagnosis by analysis of DNA methylation

Genetic alterations in DNA methylation that affect imprinting play important roles in Prader-Willi syndrome (PWS) and Angelman syndrome (AS). The imprinted genes in chromosome 15 region 15q11.2-q13 that are associated with PWS and AS, including the SNRPN (small nuclear ribonucleoprotein peptide N) promoter region, are normally methylated and unexpressed in the maternal chromosome and unmethylated and expressed in the paternal chromosome. Loss of the unmethylated and expressed paternal copy by deletion, maternal uniparental disomy (UDP) or by imprinting errors, leaving the methylated and unexpressed maternal copy as the only version of the gene, is associated with PWS due to loss of paternal expression. Conversely, AS is associated with the loss of the maternal copy of 15q11.2-q13 which can occur through deletion, mutation in the maternally expressed gene Ubiquitin-protein ligase E3A (UBE3A) or paternal UDP. Methylation analysis of the SNRPN promoter is used to confirm diagnosis of PWS although methylation studies alone do not define the genetic basis for the diagnosis. A positive result of a methylation analysis leads to follow-up studies to define the genetic cause.

The most commonly used diagnostic methylation tests for PWS and AS are MSP and Southern blot assays using methylation sensitive restriction enzymes (Glenn et al., 1996; Kubota et al., 1997; Ramsden et al., 2010). Methylation-specific multiplex ligation dependent amplification (MS-MLA) has also been used to identify the methylation status of the PWS region and to detect copy number changes in the region (Nygren et al., 2005). The Southern blot assay and one version of MS-MLA depend on the use of methylation sensitive restriction enzymes which probe the methylation status of one or more CpG sites. Results can be affected by incomplete digestion of genomic DNA or rare SNPs affecting restriction sites (Ramsden et al., 2010). MSP and alternative versions of MS-MLA use bisulfite treated DNA.

1.3 Abscription® – Based signal amplification

Abscription, short for Abortive Transcription, is a robust and isothermal signal amplification process that utilizes an RNA polymerase, called Abscriptase®, to generate thousands of specific short RNA oligonucleotides per minute from an artificial promoter called an Abortive Promoter Cassette (APC, Figure 1). Abscription exploits the natural phenomenon of abortive transcription during the initiation of transcription where RNA polymerases synthesizes large numbers of very short RNA molecules in the range of 2-12 nucleotides (Hsu et al., 2003; Hsu et al. 2006; Vo et al., 2003a; Vo et al., 2003b). Abortive transcription occurs very rapidly because RNA polymerase does not dissociate between rounds of short RNA synthesis and, if maintained in the abortive synthesis mode, will continue making short RNAs until nucleotide substrates are consumed. Each APC is designed to produce a short oligonucleotide of a different sequence and mass at turnovers of approximately 10,000 per minute. By attaching an APC to a biomarker target and measuring the amount of abscript produced from the APC, Abscription can be use to give a quantitative measurement of the amount of the target present. The "aborted" transcripts, or abscripts, can be quantified by several methods, including mass spectrometry (MS), capillary electrophoresis (CE) and rapid Thin Layer Chromatography (rTLC).

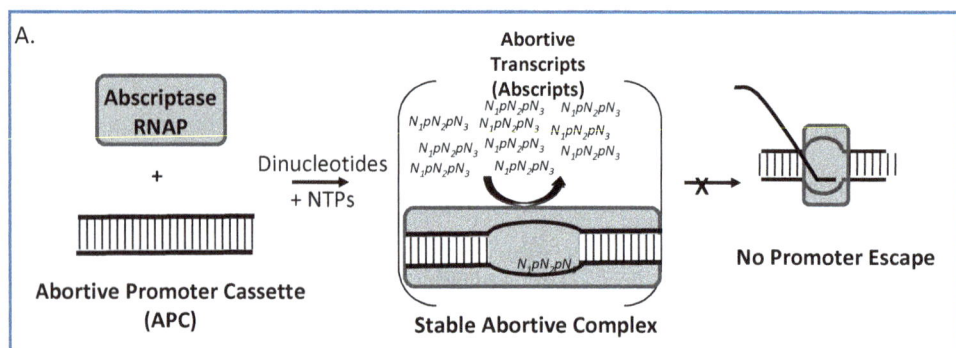

Fig. 1. Signal generation by Abscription® (Abortive transcription)

Abscription is catalyzed on a DNA template called an Abortive Promoter Cassette that encodes an RNA polymerase binding site and the start site for transcription. Abscripts are initiated with a dinucleotide complementary to the start site. Signal generation by Abscription exploits the natural phenomenon of abortive transcription which occurs during the initiation of transcription. Following promoter binding RNA polymerase temporarily synthesizes large numbers of RNA molecules in the range of 2-12 nucleotides. In normal transcription, the RNA polymerase undergoes a conformational change that allows promoter escape and entry into the processive elongation phase of the transcription cycle. Artificial promoters (APCs) and an abortive RNA polymerase, Abscriptase®, have been developed to trap the enzyme in the abortive phase and produce only abortive transcripts of specific sequence.

The highest turnovers are produced by APCs encoding trinucleotides and their synthesis from a dinucleotide initiator and a single NTP (Figure 2). For example, the trinucleotide abscript GpApG (GAG) is synthesized reiteratively by Abscriptase by an APC directed joining of the dinucleotide GpA and GTP. The sequences of the promoter and the initially transcribed segment have significant effects on the lengths and rates of synthesis of abortive transcripts (Hsu et al., 2006). Optimization of these sequences allowed the development of APCs that efficiently produce a variety of trinucleotide abortive transcripts (Hanna, 2006, 2008, 2009). There are 64 different trinucleotides that can be made from the four standard nucleotides. APCs encoding over 20 of these trinucleotides have been developed.

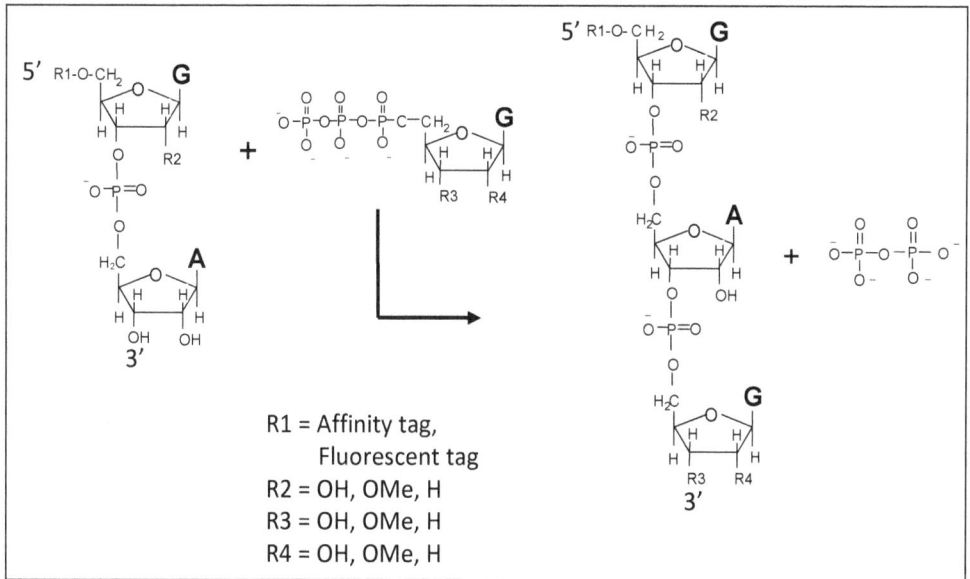

Fig. 2. Synthesis of trinucleotide abortive transcripts

Abortive transcripts can be limited to a uniform population of trinucleotides. In this example, a trinucleotide abscript is made exclusively by including a dinucleotide initiator (GpA) and a single nucleoside-triphosphate (GTP). A single polymerization step will produce a trinucleotide abortive transcript (GpApG) plus pyrophosphate (pp). The short RNA is expelled from the transcription complex and another is then made. GAG is produced at close to 10,000 copies per minute. Abscriptase is tolerant of modifications at both the 5' position of the initiator (R1) and the 3' position of the NTP (R3), allowing production of fluorescent or affinity tagged abscripts (such as biotin), if desired.

1.4 Abscription® – Based biomarker detection

Abscription is used to detect and quantify biomarkers in a sample by attaching an Abortive Promoter Cassette to that target (Figure 3). Once attached, Abscription is initiated and thousands of trinucleotides are generated per minute from the target. The amount of abscript produced can be quantified by analysis of the reaction by Liquid Chromatography – Mass Spectrometry (LC-MS). The substrates and trinucleotide product are separated by LC and then quantified by MS (Figure 4). The amount of abscript present is determined through the peak area off the MS.

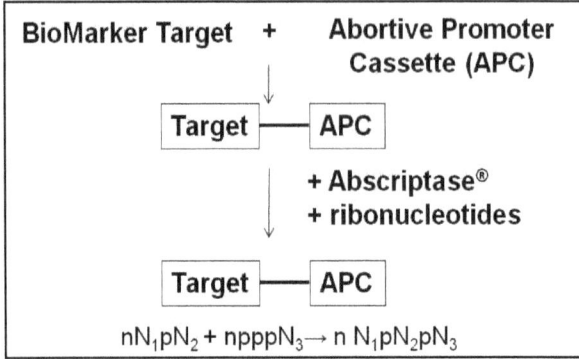

Fig. 3. Measurement of Biomarkers by Abscription®

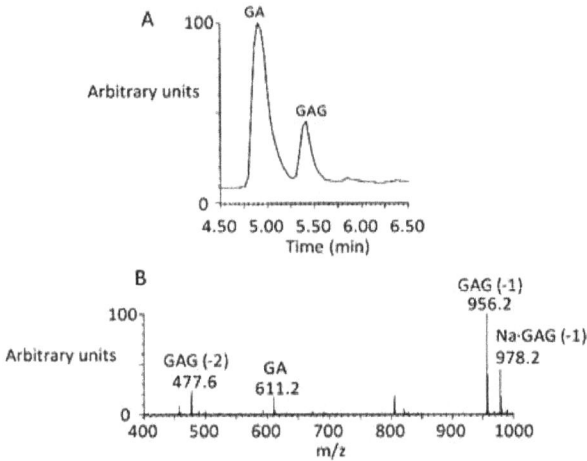

Fig. 4. Analysis of trinucleotide abscripts by LC-MS

Abortive Promoter Cassettes (APCs) are attached to biomarkers in a sample through chemical linkage to target specific antibodies (for proteins) or through single-stranded hybridization probes extending from the APCs for direct detection of nucleic acid targets. APCs also can be incorporated into nucleic acid targets during amplification procedures through the use of APC-primers as described in Fig. 7. Abscripts are then generated from the APC attached to the target by the APC-directed, enzymatic linkage of a dinucleotide initiator (N_1pN_2) to a single ribonucleoside-triphosphate ($pppN_3$) by Abscriptase to produce a uniform population of trinucleotides and pyrophosphate. The amount of abscript is determined and used as a measure of the amount of target present.

An Abscription product is quantified as the area under the trinucleotide chromatographic peak after fractionation of an Abscription reaction on a C18 column as shown in Part A for a reaction producing GpApG from GpA and GTP. GpA and GTP are separated from the trinucleotide product GpApG. The GTP passes rapidly through the column and is not shown in the elution profile in Part A. The amount of GpApG that elutes off of the column is determined very quantitatively by analyzing the chromatographic output by MS in terms of mass/charge (m/z). Part B shows the mass spectrum for a portion of the GpApG peak. The doubly charged, singly charged and the sodium adduct of GpApG are quantified and summed across the peak to give the total abscript yield.

1.5 Abscription – Based detection of DNA methylation: MethylMeter®

MethylMeter is a bisulfite-free assay for the quantitative detection of DNA methylation that combines affinity separation of methylated and unmethylated DNA with a target and signal amplification process called CAP (Coupled Abscription-PCR). The process is extremely rapid and works even on badly degraded DNA. Although the application is demonstrated here for the detection of imprinting orders from blood samples, MethylMeter has also been used to analyze small changes in the methylation levels in tumor DNA isolated from formalin-fixed paraffin embedded (FFPE) tissues.

1.5.1 MethylMagnet®: GST-MBD fusion protein with affinity for methylated CpG sites in DNA

MethylMagnet proteins are versatile tools for the study of CpG methylation in DNA. These fusion proteins contain the methyl CpG binding domain (MBD) of the mouse MBD2 protein fused to the glutathione-S-transferase protein (GST) from *S japonicum* (Figure 5A). The MBD from the MBD2b protein was chosen because MBD2b has the highest affinity among the known methyl CpG binding proteins for methylated CpG sites and the lowest cross reactivity with unmethylated CpGs (Fraga et al., 2003). Additionally, there are no sequence context effects on MBD2 CpG recognition, as there are for MeCP2, which requires a run of A-Ts near a CpG site, therefore a greater number of mCpG sites will be recognized (Klose et al., 2005). The linker between the GST and the MBD domains contains a thrombin cleavage site, so the MBD domain can be separated from the GST, if desired, (Guan & Dixon, 1001) although the fusion protein is active for binding to glutathione, GST antibodies, and specifically to methylated DNA (Figure 5B). The GST group contains surface cysteines that can be chemically modified to add reporters or affinity tags. Both the fluorescein and biotin labeled fusion proteins also retain the binding properties of the unmodified protein (data not shown).

The MethylMagnet protein is attached to glutathione modified magnetic beads and used to separate methylated DNA from unmethylated DNA in a genomic sample (Figure 6A). Intact genomic DNA should first be fragmented so that the CpG island of interest is physically separated from other regions on DNA that may be methylated, although MethylMagnet recognizes regions of CpG methylation density. Restriction with Mse I cuts DNA at the sequence TTAA and therefore is not affected by CpG methylation. DNA retrieved from formalin-fixed paraffin embedded (FFPE) tissues is already degraded and needs no restriction enzyme treatment before analysis. The fragmented DNA is incubated with the MethylMagnet protein which has been attached to glutathione magnetic beads. The DNA population will generally contain a mixture of methylated and unmethylated CpG sites. Methylated CpG islands will bind to the beads via interaction of the mCpG sites and the MBD domain. After capture, methylated DNA can be eluted from the magnetic beads several ways (Figure 6B).

A.

Thrombin Cleavage Site

GST Domain -LVPR↓GSPGISGGGGGIR- MBD

B.

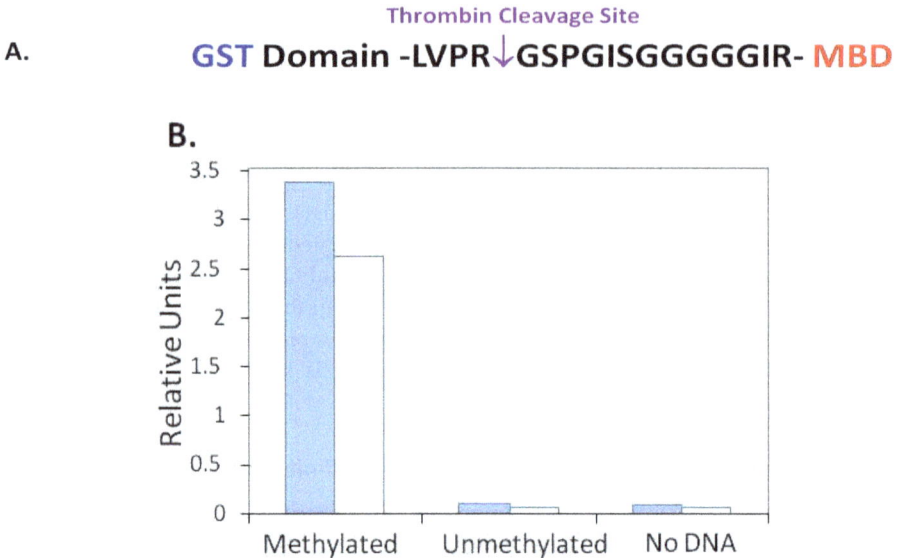

Fig. 5. Binding of the GST-MBD protein to methylated DNA A. A glutathione-S-transferase (GST) fusion protein was constructed which contains GST attached to the methyl binding domain from mouse MBD2b. The GST allows the fusion protein or its complexes with methylated DNA to be isolated on glutathione agarose or glutathione magnetic beads and eluted intact with glutathione. The GST group further allows the eluted protein-DNA complexes to be immobilized to beads or microtiter plates or visualized with GST antibodies.B. MethylMagnet proteins have much higher affinity for DNA methylated on both strands, versus hemi-methylated (data not shown) or unmethylated DNA. Purified MethylMagnet (3 nM) was incubated with 100 fmoles of a biotinylated 550 bp p16 DNA fragment that was either fully methylated or unmethylated, or with no DNA. The biotinylated DNA target was immobilized to magnetic beads. The MethylMagnet protein was incubated with the immobilized DNA. After washing, bound MethylMagnet was detected with an anti-GST-HRP conjugate. The filled and unfilled bars represent duplicate measurements. Binding was very specific for the methylated DNA fragment. The signal from unmethylated DNA was no greater than that in the minus DNA control.

1.5.2 MethylMeter™: Bisulfite-free, abscription based detection of methylated DNA

The amount of a targeted CpG island or DMR in the methylated and unmethylated DNA fractions is quantitatively detected by a new method called CAP™, for Coupled Abscription®-PCR. CAP is a combination of target amplification by PCR and linear signal generation by Abscription. Figure 7 shows the strategy for detecting DNA methylation using MethylMagnet purification followed by CAP. Methylated and unmethylated DNA molecules are separated using magnetic bead bound GST-MBD protein. Methylated DNA attaches tightly to the beads due to cooperative binding with multiple MBD domains. The unmethylated fragments remain in the supernatant fraction. The methylated DNA is eluted from the beads and then both the supernatant fraction containing unmethylated DNA, and the eluted fraction, containing methylated DNA are analyzed by CAP (Steps 2-3). CAP involves the amplification of both the methylated and unmethylated fragments (in separate tubes) using a conventional primer and a primer that contains a single-stranded abortive promoter cassette at its 5′ end. The APC is inactive for Abscription in the single-stranded form but becomes activated when it is converted into the duplex form during the amplification of the target. The PCR reaction is supplemented with Abscriptase® along with the dinucleotide and NTP substrates for abscript synthesis.

A.

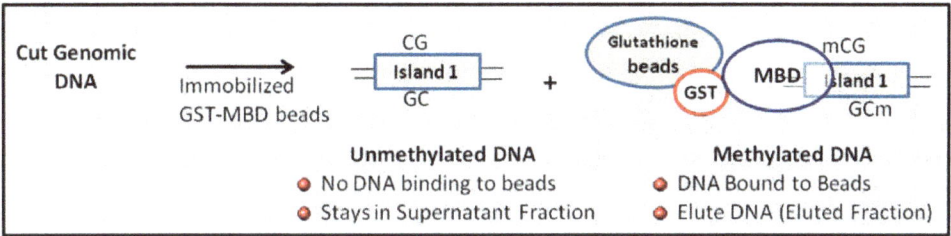

B.

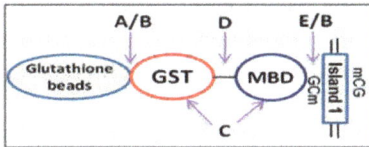

Fig. 6. Separation of Methylated and Unmethylated DNA fragments with MethylMagnet® A. Genomic DNA is incubated with GST-MBD magnetic beads, leading to the immobilization of methylated fragments. Unmethylated DNA remains in solution and is recovered in the supernatant fraction of the binding reaction. B. The methylated DNA can be eluted from the bead several ways. Treatment with glutathione releases the GST-MBD DNA complex intact. Heat releases the DNA and denatured proteins from the beads. Proteinase K digests the proteins and releases the methylated DNA and peptide fragments. Thrombin will cleave in the linker between the GST and the MBD and will release the DNA as a complex with the MBD protein. And lastly, salt can be used to fractionate the methylated DNA based on methylation density, releasing the methylated DNA from the beads and leaving the GST-MBD protein behind.

Each new amplicon now contains the APC and begins the reiterative synthesis of trinucleotide abscripts from each target. Because Abscription generates tens to hundreds of thousands of detectable trinucleotides per each amplicon, CAP is more sensitive than PCR alone, allowing detection of DNA methylation from very little DNA without fluorescent probes and in fewer cycles than with other PCR methods. Because methylated DNA is detected based on fractionation, rather than on chemical conversion, the disadvantages of bisulfite treatment are avoided. DNA is not degraded, allowing analysis of samples as small as 100 pg to 1 ng of genomic DNA. CAP primer development is not limited by the effects of reduced sequence complexity caused by bisulfite or a requirement to overlap primers with CpG sites, as in MS-PCR.

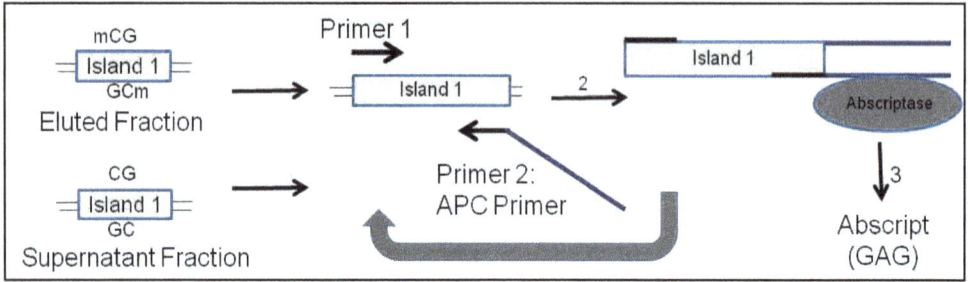

Fig. 7. Coupled Abscription®-PCR (CAP) based quantification of methylated DNA

Fragmented DNA is separated into methylated (Eluted) and unmethylated (Supernatant) fractions with the use of magnetic beads bearing the MethylMagnet® protein. The protein binds to DNA fragments with high densities of methylated CpG dinucleotides. The presence of a targeted CpG island in either the methylated DNA fraction or the unmethylated fraction is measured by coupled Abscription-PCR (CAP, steps 2-3). Targets are amplified with a conventional primer (primer 1) matched with a primer encoding an Abortive Promoter Cassette (APC) at its 5′ end (primer 2). Conversion of the APC from a single-stranded form to a double-stranded promoter activates it for Abscription. Addition of Abscriptase, a dinucleotide initiator and a single NTP allows production of the encoded trinucleotide transcript. Abortive transcripts are detected by rapid Thin Layer Chromatography (rTLC, Figure 13) or by LC-MS.

2. Results

2.1 CAP assay for detection of methylated DNA

In this study, the methylation status of a DNA segment from +234 to +539 nucleotides downstream from the start-site for SNRPN RNA variant 1 was analyzed to detect Prader-Willi and Angelman Syndrome disorders from blood. Our strategy for detecting methylated DNA involves the separation of methylated DNA fragments from the unmethylated versions through the use of the 76 amino acid mouse methyl-CpG binding domain of MBD2 fused to glutathione-S-transferase. The GST-MBD fusion protein, called MethylMagnet® was immobilized to glutathione magnetic beads to allow capture of methylated DNA fragments from a sample. After restriction, this promoter CpG island is part of a 1,237 nucleotide MseI fragment. There are 62 MseI sites separating this segment from the next upstream CpG island, two MseI sites separating it from an immediately adjacent cluster of CpG sites which probably are part of the targeted CpG island, and 179 MseI sites before the next downstream CpG island. Experiments with synthetic DNAs showed that MBD2 protein has a strong bias for densely spaced methylated sites. Synthetic DNAs with 4 methylated CpG sites spaced on average 11 nt apart were fractionated with approximately 50% efficiency. Those with six methylated CpG sites were bound quantitatively by GST-MBD2 beads while DNAs having seven to twelve methylated sites spaced between 65 to 91 nucleotides apart were bound very inefficiently (data not shown). Consequently, the linkage of widely spaced methylated CpGs that are not part of a CpG island following MseI digestion is unlikely to significantly influence the fractionation of the CpG island.

2.1.1 APC-primer development

Signal generation in the CAP assay depends on the conversion of the inactive single-stranded APC into an active duplex form. Ideally this is accomplished only when a CAP primer is copied in the course of amplifying the target. The CAP assay is potentially vulnerable to primer dimer effects that would activate the APC in the absence of target DNA. Primer-primer interactions that lead to DNA synthesis copying the APC-primer from the downstream priming segment into the double-stranded and active APC cause strong background signals independent of target DNA. The SNRPN primers were tested for this possibility by performing PCR reactions with, or without DNA over a range of annealing temperatures.

Primer	Assay	Primer Sequence (5'-3')
SNRPN F1	qPCR, CAP (APC-primer, Abscript: GAG)	TGCATAGGGATTTTAGGCGG
SNRPN R1	qPCR, CAP	CCGATCACTTCACGTACCTTC
SNRPN F2	CAP (APC-primer, Abscript: AUC)	ACCTCCGCCTAAAATCCCTATG
SNRPN R2	CAP	CTTGCTGTTGTGCCGTTCTG

Table 1. Primer sequences for CAP and TaqMan® assays

Figure 8 shows the results for one of the two well-designed SNRPN promoter primer pairs that were used to analyze patient samples (Table 1). PCR reactions were carried out for 30 cycles (2 cycles more than the standard protocol) followed by 30 min of Abscription. No background signal was seen over annealing temperatures from 62.9°C through 68.5°C (Figure 8A). Relatively strong abscript signals were generated in the presence of 1,000 copies of HeLa DNA at annealing temperatures 62.9°C and 64.9°C. Signal intensity fell at higher stringencies up to 68.5°C (Figure 8B).

Primer validation included an assessment of amplification specificity in the presence of genomic DNA. Figure 8C shows an example for the SNRPN primer pairs that were used to analyze patient samples. PCR reactions were carried out with 3,000 copies of DNA from a normal patient sample and for 32 cycles in order to allow potential non-specific amplicons to be detected by agarose gel electrophoresis. Neither primer pair produced detectable non-specific amplicons.

2.2 Analysis of the SNRPN imprinting center in patient DNAs

The MethylMeter® SNRPN assay was applied to the analysis of genomic DNA samples from 38 patients whose diagnosis of PWS and AS were confirmed based on MS-PCR of the SNRPN imprinting center (Kubota et al., 1997; Kosaki et al., 1997). A collection of 16 normal samples was also analyzed. Purified DNA samples were exhaustively cleaved with MseI in overnight digestions to unlink the targeted SNRPN promoter region from neighboring CpG islands. Treated DNAs were fractionated with GST-MBD magnetic beads without prior purification to remove the restriction enzyme. MseI was inactivated by incubating the samples at 65°C for 20 min. The supernatants of the binding reactions containing unmethylated fragments were analyzed with the paired fractions containing methylated fragments that were eluted from the beads. The eluted fractions were in the same volume and in the same binding buffer as the supernatant fractions to eliminate potential PCR

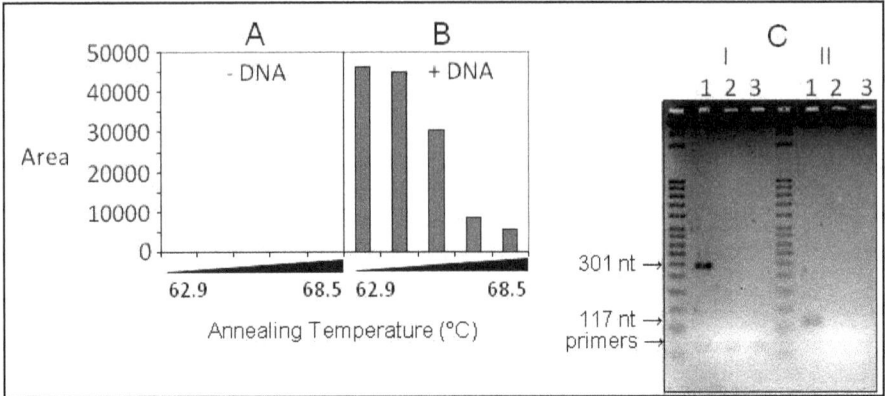

Fig. 8. APC-primer pair specificity A. APC-SNRPN F2 and SNRPN R2 were tested for specific amplification of the SNRPN and their ability to synthesize abscripts by Coupled Abscription PCR at a series of increasing stringencies in the absence and presence of HeLa DNA. Samples were analyzed by LC-MS and the total amount of trinucleotide abscript made was measured and plotted as the relative peak area. Abscription signals in the absence of DNA indicate primer dimers or self priming events that activate the APC independently of a DNA target. B. Assays in the presence of DNA indicate the optimum annealing temperature that avoids self priming without unnecessarily sacrificing abortive transcript yield in the presence of DNA. C. The SNRPN APC-primer pairs APC-SNRPN F1-SNRPN R1 (I) and APC-SNRPN F2-SNRPN-R2 (II) were used to amplify 3,000 copies of HeLa DNA for 32 cycles. Samples were fractionated in a 2% w/v agarose gel followed by staining with ethidium bromide. Lane 1 shows the amplification reaction containing DNA. Lane 2 represents the no-template control. Lane 3 shows a no-amplification control containing DNA but lacking Taq DNA polymerase. The expected amplicons contain 301 nucleotides (APC-SNRPN F1-R1) and 117 nucleotides (APC-SNRPN F2-R2)

biases due to sample buffer differences. The fractions were subjected to 28 cycles with 2 alternative sets of CAP primers with APCs that encoded either GAG or GUG. PCR reactions were followed by Abscription reactions for 15 to 30 min. The production of alternative abortive transcripts did not affect the results. The methylation status of a sample was determined by comparing abscript yields for the supernatant fraction (containing exclusively unmethylated SNRPN targets) and the eluted fraction (containing exclusively methylated SNRPN targets).

Figure 9 shows results based on abscript detection by rapid Thin Layer Chromatography (rTLC) and Liquid Chromatography- Mass Spectrometry (LC-MS). Rapid TLC had the advantage of allowing rapid processing of multiple samples in parallel. Although the results are qualitative, the imprinting disorders lend themselves to a qualitative yes-no analysis since the samples are expected to contain either all methylated, all unmethylated or a 1:1 ratio of both types of DNA. Figure 9A shows typical results for PWS (abscript GAG only in the methylated eluted fraction), AS (GAG only in the unmethylated supernatant fraction) and Normal (equal amounts of abscript GAG in both fractions). In all cases visual inspection of rTLC results for the 54 samples led to classifications of the samples in agreement with the more quantitative mass spectrometry assay (Figure 9B).

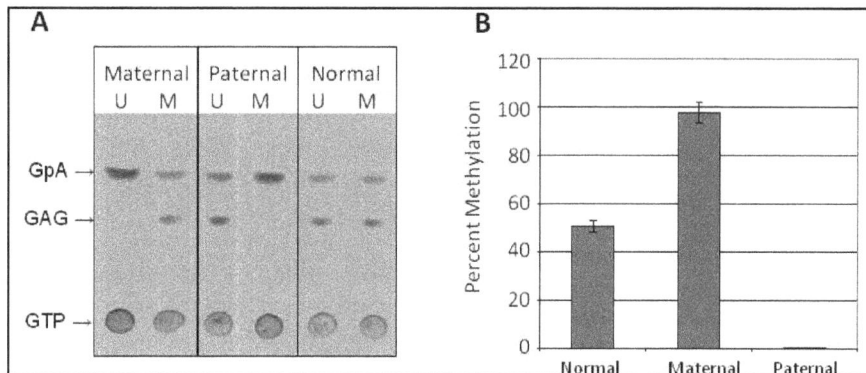

Fig. 9. CAP analysis of PWS and AS patient DNA

Patient DNAs were separated with MethylMagnet into unmethylated (U) and methylated (M) fractions. Both fractions were amplified for the SNRPN imprinting center with CAP promoter primers and then were subjected to Abscription to produce the abscript GAG. Part A shows representative rTLC results for samples showing exclusively methylated SNRPN DNA (PWS), exclusively unmethylated SNRPN (AS) or an equal representation of both methylated and unmethylated DNA (Normal). GTP remains at the origin during the development of the chromatogram while GpA and GAG are separated from each other based on their differing rates of migration in the rTLC solvent. Part B shows a quantitative summary of all the CAP assays with LC-MS detection of the abscript signals. Normal samples (n = 16) showed 50.4% ± 2.5 methylation, PWS (Maternal pattern) samples (n = 25) showed 97.5% ± 4.2 methylation and AS samples (n = 13) showed 0.23% ± 0.31 methylation. MethylMagnet inputs were between 50 ng to 150 ng of DNA.

Quantitative detection of methylated DNA was performed by LC-MS of diluted Abscription reactions. CAP reactions were fractionated by HPLC to separate the dinucleotide initiator from the trinucleotide abscript. The outflow from the HPLC column was injected into an electrospray ionization mass spectrometer (Waters LCT-Premier) to generate a chromatographic profile based on the mass/charge ratio of the abscript. The area of the abscript chromatographic peak is linearly related to the amount of abscript. Figure 9B and Table 2 show summary data for all the samples grouped into Normal, Maternal (methylated) and Paternal (unmethylated) patterns. The percent methylation was based on the abscript amount in the eluted fraction divided by the total abscript amount (the sum of the supernatant and eluted fractions). The results were highly reproducible in spite of the fact that the averages were based on multiple individuals. The large variance for the methylation level of the Paternal pattern was probably due to the statistics of sampling because the level of methylation indicates that the signals were probably generated from fewer than 5 molecules per CAP reaction based on estimates of the input amounts in calibrated CAP assays.

The analysis of patient samples also indicated that nonspecific amplification was insignificant. A non-specific amplicon that is not linked to the SNRPN target would skew the results of the normal DNA samples away from a 1:1 ratio of methylated to unmethylated target DNA. If a hypothetical secondary target was methylated the paternal methylation pattern would show a proportionately high background of methylated signal. The opposite bias would be seen in the maternal methylation pattern if the secondary target was unmethylated. As shown below, all of the patient samples fit one of the PWS, AS, or Normal methylation patterns without significant backgrounds (Table 2). The primer sequences that were the basis for the APC-primer pair SNRPN F2, R2 assay showed no apparent bias for unmethylated DNA over methylated DNA. For example the average amplification efficiencies over a 10°C annealing temperature range were 92.7% ± 7.0 (n = 11) for HeLa DNA and 92.9% ± 5.5 (n=4) for artificially methylated HeLa in TaqMan® assays.

Methylation Pattern	Percent Methylation ± SD	No. samples	95% confidence interval
Normal	50.4 ± 2.5	16	1.23
Maternal	97.5 ± 4.2	25	1.65
Paternal	0.23 ± 0.31	13	0.17

Table 2. Cumulative results for Prader-Willi/Angelman syndrome samples

The reproducibility of the MethylMagnet fractionation method was evaluated by performing between 4 to 7 independent fractionations on DNA from 4 normal samples (Table 3). All of the fractionation runs gave good reproducibility with coefficients of variation (CVs) between 1.2% and 6.3%. The use of the alternative primers did not have a significant quantitative effect on the detection results. Sample #41 was analyzed in triplicate with APC-SNRPN F2, R2 that encoded AUC instead of GAG. The average methylation level was 49.8% ± 0.69 SD in agreement with the result for sample #41 in Table 3. The single-blinded assignments of individuals to the 3 groups based on TLC and LC-MS were 100% concordant with earlier classification of the samples based on MS-PCR (data not shown).

	[1]Cumulative Normal	Individual Normal Samples			
		#36	#39	#41	#53
Percent Methylation	50.4	50.5	47.0	49.0	46.3
SD	2.5	0.58	2.6	1.02	2.9
CV	4.9%	1.2%	5.6%	2.1%	6.3%
No. Fractionations	16	4	7	7	4

Table 3. Reproducibility of MethylMagnet fractionation [1]Data from Table 2

2.3 Sensitivity of CAP verses TaqMan®

Coupled Abscription-PCR involves the addition of Abscription, a linear signal amplification step, to PCR and is therefore much more sensitive than PCR alone. The relative sensitivity of the CAP assay was compared to a TaqMan PCR assay using the same SNRPN priming sites and PCR cycling conditions (Figure 10). Figure 10A shows a plot of Ct verses DNA amount for TaqMan amplification of HeLa DNA along with analogous plots for CAP. In the case of the CAP assay, the Ct-analog for the raw data is interpreted as the minimal cycle number required for detection of a particular input DNA copy number. These DNA amounts were not equivalent to limits of detection because the signals overshot the threshold for detection (an LC-MS signal of 100) by amounts ranging up to 549. The actual limits of detection at various cycles were calculated by extrapolating the lines relating LC-MS signals and DNA copy numbers to the detection threshold of 100 for each endpoint experiment. These extrapolated DNA copy numbers were plotted with the associated end point cycles in Figure 10A (unfilled squares).

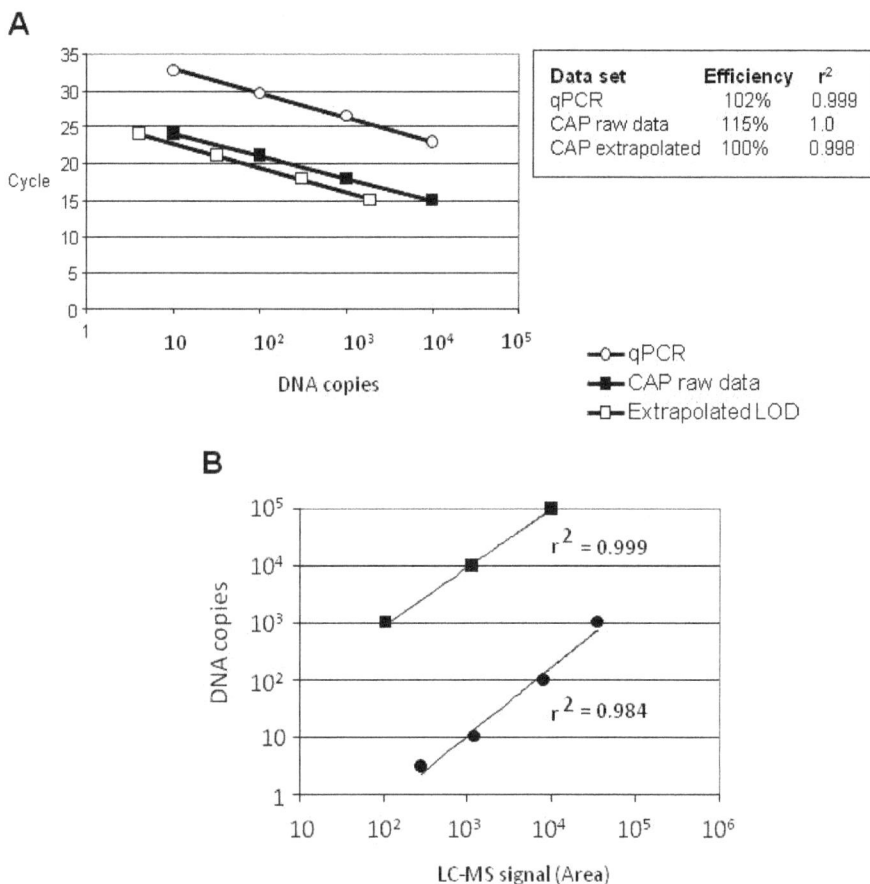

Data set	Efficiency	r^2
qPCR	102%	0.999
CAP raw data	115%	1.0
CAP extrapolated	100%	0.998

Fig. 10. Sensitivity and dynamic range of CAP. A. Calibration curves were made with dilutions of HeLa DNA. The Ct values for detection of 10, 100, 1,000 and 10,000 copies are plotted for TaqMan PCR (open circles, Efficiency: 102%; r2: 0.999). The minimal cycle number at which a plotted DNA copy number can be detected is shown for a series of end-point CAP experiments (filled rectangles, filled and open triangles). The limit of detection at a given CAP cycle number was calculated using curve fitting of the relationship between LC-MS signal and DNA copy number for each titration. The LOD was fixed to a signal intensity of 100 (unfilled squares represent the processed versions of the filled squares). Abscription reactions were for either 3 hr (squares) or for 15 min (triangles). B. HeLa DNA was titrated at the copy numbers indicated on the ordinate. Samples containing 3 to 1,000 molecules were amplified for 28 PCR cycles followed by 30 min of Abscription (circles). Samples containing 1,000 to 100,000 molecules were amplified for 20 PCR cycles followed by 15 min of Abscription (squares). Abscript signals were detected by LC-MS. Primer sequences used in the assay are listed in Table 1.

The sensitivity of the CAP assay could be adjusted with variations in the Abscription time following the PCR step. Overall a 3 hr Abscription reaction resulted in detection with a reduction of 11 cycles compared to the TaqMan assay. This translates to about a 2,000-fold improvement in sensitivity assuming 99.7% amplification efficiency. Sensitivity can also be estimated by determining the number of target copies at the detection threshold. The number of amplicons, Nt, at the Ct can be calculated by extrapolating the number of cycles to 0 in Figure 10A, where the x intercept is interpreted as the number of targets at the detection threshold (Rutledge & Cote, 2003). The qPCR assay produced an Nt of 6 $\times 10^{11}$ while the normalized CAP Nt was 4 $\times 10^7$, a 2,650-fold improvement in sensitivity with a 3 hr Abscription period. A 15 min Abscription time produced a 64-fold improvement in sensitivity over qPCR (Figure 10A).

The dynamic range of the CAP assay could be made to extend over approximately 5 orders of magnitude by performing two CAP reactions at high and low cycle numbers as shown in Figure 10B. Samples with 10,000 and 100,000 DNA copies are prone to underestimates of DNA amount at high cycle numbers due to depletion of the Abscription reagents at moderate Abscription reaction times. The high end of the dynamic range is accurately quantified by performing 20 PCR cycles coupled with 15 min of Abscription. At the low end of the range (DNA input <1,000 copies), PCR is performed for 28 cycles followed by an Abscription time between 15 to 30 min. Following this strategy we were able to consistently quantify DNA amounts between 3 molecules to 100,000 molecules.

2.4 Analysis of DNA methylation in other bodily fluids with MethylMeter

MethylMeter is an extremely robust, sensitive, rapid and quantitative method to analyze even small changes in DNA methylation levels of differentially methylated regions. It works well and reproducibly in on DNA isolated from saliva and urine (Figures A and B), in which the normal methylation pattern is 50% methylated and 50% unmethylated. In addition, the process works well and reproducibly on DNA from FFPE tissue (Figure 11C), where bisulfite based methods often fail.

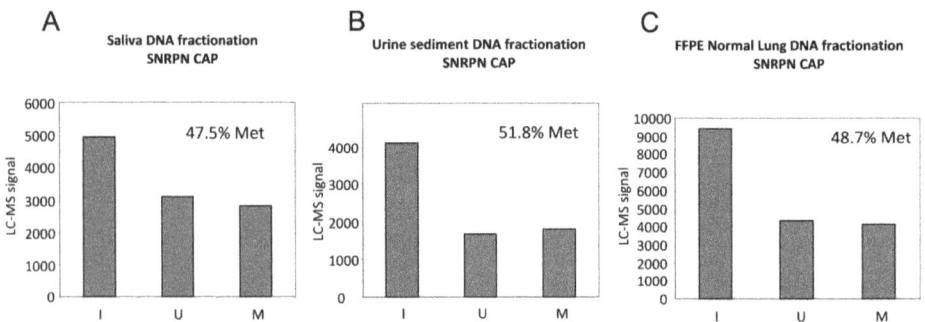

Fig. 11. Detection of methylated DNA from saliva, urine, and FFPE tumor slides. A and B. Purified normal DNAs from saliva and urine sediments (50 ng) from the same individual were separated into methylated (M) and unmethylated (U) fractions with MethylMagnet. Methylated SNRPN CpG island was detected after 28 cycles of PCR and 15 min Abscription. C. FFPE DNA from normal lung was purified from 2 glass slides (10 μM thickness). SNRPN CpG island DNA was detected after 29 cycles of amplification and 30 min of Abscription.

2.5 MethylMeter: Comparison of results to bisulfite sequencing

Although the results obtained by MethylMeter were 100% concordant with the results from MS-PCR, the method was further validated by analyzing SNRPN methylation in HeLa DNA (50%) with both MethylMeter and bisulfite sequencing. The region of SNRPN CpG island that was analyzed with bisulfite sequencing is shown in Figure 12.

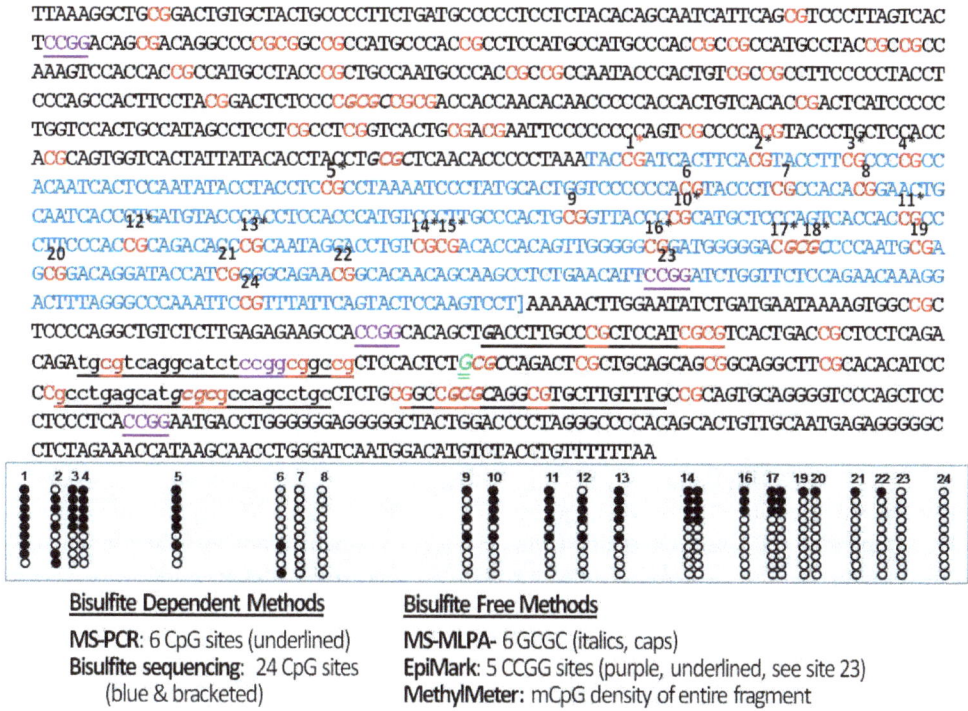

Fig. 12. Methylation of SNRPN Island: Method Comparison. The entire island probed by MethylMagnet is shown. The region that was sequenced is in the middle of the island and the CpG sites that were interrogated are numbered 1 to 24. The results of the bisulfite sequencing are shown under the sequence. A black circle indicates a methylated site. Sites that are used in MS-PCR analysis and some restriction enzyme based methods are shown. Note that site number 24, which was determined to be unmethylated by pyrosequencing, is the site used in the commercially available method EpiMark™ (NEN), which uses a restriction enzyme that cleaves differentially at this site based on CpG methylation, so this method would have incorrectly scored all of these samples as unmethylated.

The MseI generated DNA segment from +234 to +539 nucleotides downstream from the start-site for SNRPN RNA variant 1 is shown. MseI cut HeLa DNA was assayed for methylation of this SNRPN CpG island, which is imprinted and 50% methylated in normal somatic tissues. The average percent methylated DNA determined by MethylMeter was 47.7 ± 2.9%. Unfractionated HeLa DNA was sent out for bisulfite sequencing of the SNRPN segment within the MseI fragment that was probed by MethylMeter. Filled circles are methylated CpG sites (1-24). Five islands were scored as methylated and 5 unmethylated, identical to the MethylMeter results, which took 10X less DNA and far less time and money. Sites in this island that are probed by other common methods are also indicated.

3. Materials and methods

3.1 Patient samples

Deidentified, purified DNAs from peripheral blood samples were obtained from the Molecular Genetics Laboratory, Children's Hospital and Research Center at Oakland under IRB approval. The DNA samples were previously analyzed for the methylation status of the SNRPN imprinting center with MS-PCR (Kubota et al., 1997; Kosaki et al., 1997).

3.2 Separation of methylated and unmethylated DNA with a Methyl CpG binding domain protein

Patient DNAs were fragmented with restriction endonuclease MseI. MseI cuts segments between CpG islands at high frequencies but cuts CpG islands infrequently. Fragmented DNA (7.5-150 ng) samples were fractionated with the MethylMagnet® CpG DNA isolation kit following the instructions in the user manual (RiboMed, MM101K). DNA samples were diluted 5-fold into Binding Buffer and were incubated with 5 µl of GST-MBD magnetic beads for 1 hr at 22°C with shaking at 1,000 rpm in an Eppendorf Thermomixer. Supernatant fractions were recovered after collecting the beads with a magnet. The beads were washed 2 times in 400 µl of Wash Buffer 2 with 5 min incubations at 22°C and shaking at 1,000 rpm. A third wash without incubation was performed with 400 µl of 10 mM Tris·HCl pH 8, 1 mM EDTA. Methylated DNAs were eluted from GST-MBD magnetic beads by incubation in 80% (v/v) Binding Buffer:18% (v/v) ultrapure water:2%(v/v) NEB buffer #4 at 80°C for 10 min with shaking at 1,000 rpm in an Eppendorf Thermomixer. The supernatant and the eluted fractions of the binding reactions were saved for CAP analysis.

3.3 Coupled Abscription-PCR (CAP) reaction

PCR reactions were performed with Maxima Hot Start Taq DNA Polymerase (Fermentas) using the manufacturers reaction buffer [20 mM Tris·HCl pH 8.3 (25°C), 20 mM KCl, 50 mM $(NH_4)_2SO_4$] and 2 mM $MgCl_2$, and 5% (v/v) DMSO. The dNTPS were added to final concentrations of 0. 2 µM each. CAP primers for the SNRPN promoter region were present at 1 µM each. Taq polymerase was added to 2 units/20 µl reaction. Inclusion of DMSO in the PCR reaction is essential to avoid amplification bias against methylated DNA. PCR conditions involved an initial denaturation step at 95°C for 30 sec, followed by 28 cycles of 95° for 15 sec, 64°C for 15 sec and 72°C for 30 sec. A final elongation step was at 72°C for 5 min followed by indefinite incubation at 4°C. Abscription reactions were set up by supplementing 10 µl PCR reactions with 2 µl of an Abscription master mix consisting of 3.7 x Abscription buffer (1x buffer: 40 mM HEPES pH 7.5, 40 mM KCl, 10 $MgCl_2$), 6 mM dinucleotide initiator [(GpA, GpU or ApU, RiboMed I31, I34, I14], 6 mM NTP (either GTP or CTP) and 1 unit of thermostable Abscriptase (RiboMed MME-1). Abscription was performed in a thermocycler at 77.6°C for periods ranging from 15 min to 3 hours.

3.4 Primer sequences

All CAP promoter primers were based on 2 sets of primer sequences listed in Table 1. The SNRPN F1 and SNRPN F2 had 43 nucleotide single-stranded Abortive Promoter Cassettes (APCs) linked to their 5′ ends for the CAP assays. SNRPN F1 and SNRPN R1 were also used for qPCR experiments without the additional APC sequence.

3.5 qPCR

Conditions for TaqMan® qPCR reactions were essentially identical to the PCR portion of the CAP reactions except for the inclusion of ROX reference dye to 300 nM. Primers for the qPCR experiments did not include an APC extension. The primers SNRPN-F2 and SNRPN-R2 targeted the promoter region of SNRPN transcript variant 1. A set of parallel CAP primers were derived from the same primer sequences (Table 1). The probe (5′AGGTATATTGGAGTGATTGTGGCGGG3′) was labeled at the 5′ end with 6-carboxyfluorescene and at the 3′ end with Iowa Black®FQ (IDT). Titrations of HeLa DNA from 10 copies/PCR to 10,000 copies/PCR were amplified in 20 μl volumes in an ABI 7700 Sequence Analyzer.

3.6 Abscript analysis: HPLC-Mass Spectrometry (LC-MS)

For high performance liquid chromatography-mass spectrometry (LC-MS) detection of abscripts, 10 μl of the CAP reaction was diluted into 20 μl of HPLC water in a 384 well plate. Volumes of 10 μl were injected into the LC-MS.

3.7 Abscript analysis: Rapid Thin Layer Chromatography (rTLC)

Analysis of abscripts by rTLC is both fast and extremely affordable. All that is needed is a small TLC developing tank and a handheld 254 nm light to visualize the products. Dinucleotides and trinucleotides migrate differently due to their varying polarities. Products are visualized by shining a 254 nm light on the TLC plate, which contains a fluorescent indicator. Nucleotides quench this fluorescence, resulting in a dark spot (UV shadowing). The TLC is photographed with a digital or CCD camera. Alternatively, fluorescent abscripts can be synthesized with labeled dinucleotides and then separated by TLC and visualized with a fluorescent imager or a 336 nm hand-held light (for fluorescein). Analysis by TLC involves three simple steps (Figure 13).

For rTLC detection of abscripts in the patient samples, 1.5 μl of CAP reaction was spotted 1 cm from the bottom edge of a 10 cm tall silica gel plate containing a UV-excitable fluorophore (Whatman, cat# 4420 222). The sample spots were air dried before placing the TLC plate in an air-tight rapid TLC solvent chamber (RiboMed cat# TC6-01) containing 100 ml of freshly made rTLC solvent [6:3:1 (v/v) isopropanol :ammonium hydroxide: Activation buffer 1 (RiboMed, cat AB-1)]. The plate was submerged in solvent at a depth of approximately 5 mm, with the solvent just below the point at which the reaction sample was spotted on the plate. The plate was left in the tank for approximately 20 minutes to allow the solvent to flow upwards through the TLC plate by capillary action, causing separation of the components of the CAP reaction. When the solvent was approximately 1-2 cm from the top of the plate, it was removed from the tank. Developed plates were air dried and photographed with 254 nm UV-illumination.

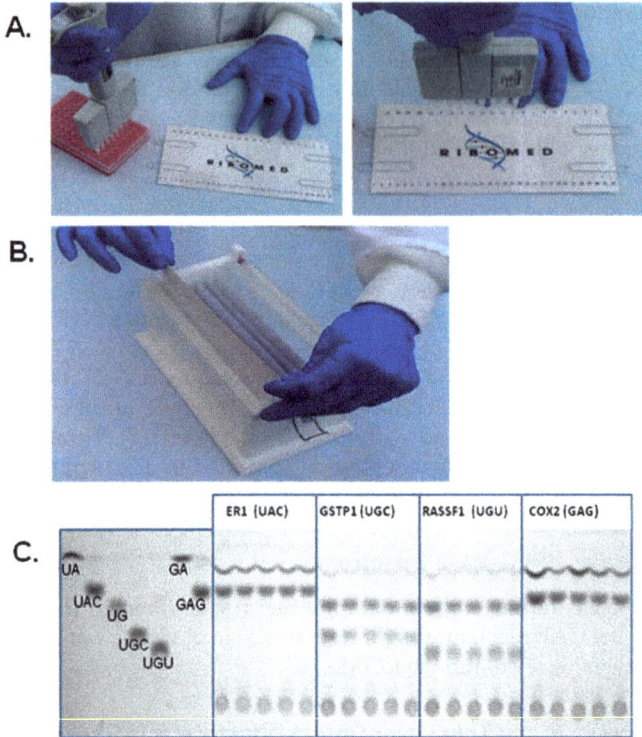

Fig. 13. Rapid Thin Layer Chromatography (rTLC) Step 1 (A). Spot samples. Samples (1 to 1.5 µl) are spotted directly onto precut TLC plates. No sample dye is required. The spotting template is designed for use with multi-channel pipettors on one side, allowing 21 samples per plate to be spotted. Even more samples can be analyzed by pipetting individually on the other side of the template, which can hold 32 samples. Step 2 (B): Put plates in the TLC developing tank. The TLC chamber holds 6 plates, so 120 samples plus 6 standards can be analyzed simultaneously using the multi-channel pipettor side. The entire process takes less than 1 hour. Up to 192 samples, or the equivalent of two 96 well plates, can be simultaneously processed when using the manual spotting template. Step 3: Irradiate the plate and to visualize the results. The TLC takes between 20 minutes to an hour to run, after which the plate is removed, air dried and visualized by UV shadowing, as shown below in (C). Spots can be quantified with the same imaging software used for gels.

4. Conclusion

Most methods for detecting methylated-CpG islands rely on chemical conversion of DNA by treatment with bisulfite. In conjunction with nucleotide sequencing this approach has the advantage that the methylation status of each CpG site in a targeted genomic region can be determined within sequence context. However, there are major disadvantages to bisulfite treatment especially when used with methods designed to determine the overall level of methylation in a genomic segment. Bisulfite conversion is carried out under relatively harsh conditions of low pH and high temperature that result in extensive degradation of the sample

DNA. This not only causes a requirement for relatively high DNA inputs, but also tends to place constraints on primer development limiting efficient amplification to small amplicons (Rutledge & Cote, 2003). Highly optimized commercial conversion kits work best with DNA samples in the range of 200 ng to 1 µg. Although samples in the range of 50 ng (15 genomic copies) can be treated, they cannot be used to generate quantitative data based on the effects of sampling statistics and losses during the post conversion clean-up (Rutledge & Cote, 2003;Ehrich el al., 2007) High cycle numbers or nested PCR are often needed to detect small amounts of bisulfite treated DNA. The reduction in sequence complexity caused by the conversion of cytosines to uracils places a further constraint on primer development by tending to homogenize the sequence compared to the untreated DNAs.

Methylation specific PCR (MSP) methods based on bisulfite treated DNA rely on sets of primers and probes that overlap CpG sites. Discrimination based on primer complementarity to CpG verses TpG is vulnerable to false positive results due to incomplete conversion (Kristensen et al., 2008). Deep sequencing of DNA prepared by a commercial kit showed over 1% unconverted methyl-CpGs which potentially could be detected as false positives give the high sensitivity of MSP based methods pushed to high cycle numbers (Taylor et al., 2007). Heterogeneous patterns of methylation are not readily detected by MSP based methods because a consensus status must exist at all of the CpG sites encompassed by the primers and the probe. One must infer the status of an entire CpG island based on a small number of sites. A single consistently unmethylated CpG site will cause a false negative for a heavily methylated CpG island (Yegnasubramanian et al., 2006)

The use of MBD proteins to fractionate methylated from unmethylated DNA fragments avoids the complications associated with bisulfite treatment. All of the CpG sites in a CpG island can contribute to the binding of a DNA by the MBD protein. Consequently it is not necessary to infer the overall methylation status based on a few CpG sites. The MBD2 binding domain shows a strong bias in favor of DNA fragments containing multiple closely spaced methylated CpGs (Yegnasubramanian et al., 2006). This specific bias in favor of methylated CpG clusters has been exploited to survey the genome for methylated CpG islands (Serre et al., 2009).

Because MBD based fractionation does not damage DNA, less sample is required than is needed for methods that rely on bisulfite treatment. We have successfully used MBD based DNA fractionation with as little as 1 ng of genomic DNA. DNA amounts in the microgram range can be processed by scaling-up the volume of MBD-magnetic beads in the binding reaction.

The linkage of a linear signal amplification to the target amplification of PCR in the CAP method greatly increases assay sensitivity without adding complexity to assay development. Modest Abscription times between 15 min to 30 min allowed a reduction of between 6-7 cycles compared to TaqMan® assays on undamaged DNA. The CAP approach is expected to show a greater relative advantage in sensitivity compared to the application of TaqMan® assays to bisulfite treated DNA. Primer development for CAP assays is free of the constraints associated with bisulfite assays. CAP primers do not have to overlap with CpG sites nor do they have any constraints on primer spacing since no intervening probe sequence is needed. Care must be taken to avoid primer-dimer effects that lead to activation of the APC in the absence of DNA. Potential problems can be identified with conventional primer development software that shows homodimer- and heterodimer interactions. The potential for priming events can be eliminated by changing nonconserved promoter sequences or by choosing a new reverse primer. Abscription provides an extremely sensitive assay to confirm the absence of primer-dimer effects and to identify optimal stringency conditions.

There are several options for abscript detection that depend on the nature of the targets. In the case of imprinting disorders where classification of samples is based on a simple yes or no result, a qualitative approach can be taken with rTLC. This method has the advantage that multiple samples can be processed rapidly in parallel allowing for greater through-put than LC-MS where samples are processed sequentially. All of the PWS AS samples gave unambiguous results that allowed them to be correctly classified based on visual inspection (Figure 9A). LC-MS is better suited for analysis of methylation in tumor cell DNA where quantitation is important or where it is important to monitor changes in methylation levels over time.

One potential complication of the DNA fractionation-CAP approach is associated with the fragmentation of the DNA sample to unlink the region of interest from neighboring dense clusters of methylated CpGs. Incomplete digestion is usually not a problem because numerous MseI sites are typically situated between neighboring CpG islands. However in the case where a CpG island is closely linked to a methylated repetitive element it might be necessary to use an alternative restriction enzyme or shear the DNA to very small sizes. Analysis of very small fragments should be feasible with the CAP assay because there are no assay specific restrictions on primer placement or minimal spacing between primers.

5. Acknowledgement

This work was supported in part by funding from the National Institutes of Health through National Cancer Institute Small Business Innovative Grant number 1R43CA132851-1 and NCI Contract number HHSN261200900047C to RiboMed.

6. References

Cottrell, S.E., Distler, J., Goodman, N.S., Mooney, S.H., Kluth, A., Olek, A., Schwope, I., Tetzner, R., Ziebarth, H. & Berlin, K. (2004) A real-time PCR assay for DNA-methylation using methylation specific blockers. Nucleic Acids Research, 32, e10. ISSN 1362-4962

Eads, C.A., Danenberg, K.D., Kawakami, K., Saltz, L.B., Blake, C., Shibata, D., Danenberg, P.V. & Laird, P.W. (2000) MethyLight: a high-throughput assay to measure DNA methylation. Nucleic Acids Research, 28, E32. ISSN 1362-4962

Ehrich, M., Zoll, S., Sur, S., & van den Boom, D. (2007). A new method for accurate assessment of DNA quality after bisulfite treatment. Nucleic Acids Research Vol.35, No.5, p. e29, ISSN 1362-4962

Fraga, M. F., Ballestar, E., Montoya, G., Taysavang, P., Wade, P. A., & Esteller, M. (2003). The affinity of different MBD proteins for a specific methylated locus depends on their intrinsic binding properties. Nucleic Acids Research Vol.31, No.6, pp. 1765-1774, ISSN 1362-4962

Free, A., Wakefield, R.I., Smith, B.O., Dryden, D.T., Barlow, P.N. and Bird, A.P. (2001) DNA recognition by the methyl-CpG binding domain of MeCP2. J Biol Chem, 276, 3353-3360.

Glenn, C. C., Saitoh, S., Jong, M. T., Filbrandt, M. M., Surti, U., Driscoll, D. J., & Nicholls, R. D. (1996). Gene structure, DNA methylation, and imprinted expression of the human SNRPN gene. American Journal of Human Genetics Vol.58, No.2, pp. 335-346, ISSN 0002-9297

Guan, K.L. and Dixon, J.E. (1991) Eukaryotic proteins expressed in Escherichia coli: an improved thrombin cleavage and purification procedure of fusion proteins with glutathione S-transferase. Analytical Biochemistry, 192, 262-267

Hanna, M. M. (2006) Molecular detection systems utilizing reiterative oligonucleotide synthesis. Patent 7,045,319

Hanna, M. M. (2008) Methods for determining nucleic acid methylation, Patent 7,470,511

Hanna, M.M. (2009) Abortive Promoter Cassettes. Patent 7,473,775

Herman, J.G., Graff, J.R., Myohanen, S., Nelkin, B.D. & Baylin, S.B. (1996) Methylation-specific PCR: a novel PCR assay for methylation status of CpG islands. Proc Natl Acad Sci U S A, 93, 9821- 9826.

Hsu, L. M., Vo, N. V., Kane, C. M., & Chamberlin, M. J. (2003). In vitro studies of transcript initiation by Escherichia coli RNA polymerase. 1. RNA chain initiation, abortive initiation, and promoter escape at three bacteriophage promoters. Biochemistry Vol.42, No.13, pp. 3777-3786, ISSN 0006-2960

Hsu, L. M., Cobb, I. M., Ozmore, J. R., Khoo, M., Nahm, G., Xia, L., Bao, Y., & Ahn, C. (2006). Initial transcribed sequence mutations specifically affect promoter escape properties. Biochemistry Vol.45, No.29, pp. 8841-8854, ISSN 0006-2960

Kane, J.F. (1995) Effects of rare codon clusters on high-level expression of heterologous proteins in Escherichia coli. Curr Opin Biotechnol, 6, 494-500.

Kosaki, K., M. McGinniss, M. J., Veraksa, A. N., McGinnis, W. J., & Jones, K. L. (1997). Prader-Willi and Angelman syndromes: diagnosis with a bisulfite-treated methylation-specific PCR method. American Journal of Human Genetics Vol.73, No.3, pp. 308-313, ISSN 0148-7299

Klose, R.J., Sarraf, S.A., Schmiedeberg, L., McDermott, S.M., Stancheva, I. and Bird, A.P. (2005) DNA binding selectivity of MeCP2 due to a requirement for A/T sequences adjacent to methyl-CpG. Mol Cell, 19, 667-678

Kristensen, L. S., Mikeska, T., Krypuy, M., & Dobrovic, A. (2008). Sensitive Melting Analysis after Real Time- Methylation Specific PCR (SMART-MSP): high-throughput and probe-free quantitative DNA methylation detection. Nucleic Acids Research Vol.36, No.7, p. e42, ISSN 1362-4962

Kubota, T., Das, S., Christian, S. L., Baylin, S. B., Herman, J. G., & Ledbetter, D. H. (1997). Methylation-specific PCR simplifies imprinting analysis. Nature Genetics Vol.16, No.1, pp. 16-17, ISSN 1061-4036

Lippman, Z., Gendrel, A.V., Black, M., Vaughn, M.W., Dedhia, N., McCombie, W.R., Lavine, K., Mittal, V., May, B., Kasschau, K.D., Carrington, J.C., Doerge, R.W., Colot, V. & Martienssen, R. (2004) Role of transposable elements in heterochromatin and epigenetic control. Nature, 430, 471-476.

Munson, K., Clark, J., Lamparska-Kupsik, K., & Smith, S. S. (2007). Recovery of bisulfite-converted genomic sequences in the methylation-sensitive QPCR. Nucleic Acids Research Vol.35, No.9, pp. 2893-2903, ISSN 1362-4962

Nan, X., Meehan, R.R. and Bird, A. (1993) Dissection of the methyl-CpG binding domain from the chromosomal protein MeCP2. Nucleic Acids Research, 21, 4886-4892. ISSN 1362-4962

Nygren, A. O., Ameziane, N., Duarte, H. M., Vijzelaar, R. N., Waisfisz, Q., Hess, C. J., Schouten, J. P., & Errami, A. (2005). Methylation-specific MLPA (MS-MLPA): simultaneous detection of CpG methylation and copy number changes of up to 40 sequences. Nucleic Acids Research Vol.33, No.14, p. e128, ISSN 1362-4962

Ohki, I., Shimotake, N., Fujita, N., Jee, J., Ikegami, T., Nakao, M. and Shirakawa, M. (2001) Solution structure of the methyl-CpG binding domain of human MBD1 in complex with methylated DNA. Cell, 105, 487-497.

Ordway, J.M., Bedell, J.A., Citek, R.W., Nunberg, A., Garrido, A., Kendall, R., Stevens, J.R., Cao, D., Doerge, R.W., Korshunova, Y., Holemon, H., McPherson, J.D., Lakey, N., Leon, J., Martienssen, R.A. & Jeddeloh, J.A. (2006) Comprehensive DNA methylation profiling in a human cancer genome identifies novel epigenetic targets. Carcinogenesis, 27, 2409-2423.

Ohki, I., Shimotake, N., Fujita, N., Nakao, M. and Shirakawa, M. (1999) Solution structure of the methyl-CpG-binding domain of the methylation-dependent transcriptional repressor MBD1. Embo J, 18, 6653-6661.

Ramsden, S. C., Clayton-Smith, J., Birch, R ., & Buiting, K. (2010). Practice guidelines for the molecular analysis of Prader-Willi and Angelman syndromes. BMC Medical Genetics Vol.11, p. 70, ISSN 1471-2350

Rutledge, R. G. & Cote, C. (2003). Mathematics of quantitative kinetic PCR and the application of standard curves. Nucleic Acids Research Vol.31, No.16, p. e93, ISSN 1362-4962

Serre, D., Lee, B. H., & Ting, A. H. (2009). MBD-isolated Genome Sequencing provides a high-throughput and comprehensive survey of DNA methylation in the human genome. Nucleic Acids Research Vol.38, No.2, pp. 391-399, ISSN 1362-4962

Taylor, K. H., Kramer, R. S., Davis, J. W., Guo, J., Duff, D. J., Xu, D., Caldwell, C. W., & Shi, H. (2007). Ultradeep bisulfite sequencing analysis of DNA methylation patterns in multiple gene promoters by 454 sequencing. Cancer Research Vol.67, No.18, pp. 8511-8518, ISSN 0008-5472

Tost, J., Kunker, J. & Gut, I.G. (2003) Analysis and quantification of multiple methylation variable positions in CpG islands by Pyrosequencing. Biotechniques, 35, 152-156.

Vo, N. V., Hsu, L. M., Kane, C. M., & Chamberlin, M. J. (2003a). In vitro studies of transcript initiation by Escherichia coli RNA polymerase. 3. Influences of individual DNA elements within the promoter recognition region on abortive initiation and promoter escape. Biochemistry Vol.42, No.13, pp. 3798-3811, ISSN 0006-2960

Vo, N. V., Hsu, L. M., Kane, C. M., & Chamberlin, M. J. (2003b). In vitro studies of transcript initiation by Escherichia coli RNA polymerase. 2. Formation and characterization of two distinct classes of initial transcribing complexes. Biochemistry Vol.42, No.13, pp. 3787-3797, ISSN 0006-2960

Wakefield, R.I., Smith, B.O., Nan, X., Free, A., Soteriou, A., Uhrin, D., Bird, A.P. and Barlow, P.N. (1999) The solution structure of the domain from MeCP2 that binds to methylated DNA. J Mol Biol, 291, 1055-1065.

Wojdacz, T.K. & Dobrovic, A. (2007) Methylation-sensitive high resolution melting (MS-HRM): a new approach for sensitive and high-throughput assessment of methylation. Nucleic Acids Research, 35, e41. ISSN 1362-4962

Xiong, Z. & Laird, P.W. (1997) COBRA: a sensitive and quantitative DNA methylation assay. Nucleic Acids Research, 25, 2532-2534. ISSN 1362-4962

Yegnasubramanian, S., Lin, X., Haffner, M. C., DeMarzo, A. M., & Nelson, W. G. (2006). Combination of methylated-DNA precipitation and methylation-sensitive restriction enzymes (COMPARE-MS) for the rapid, sensitive and quantitative detection of DNA methylation. Nucleic Acids Research Vol.34, No.3, p. e19, ISSN 1362-4962

Zeschnigk, M., Bohringer, S., Price, E.A., Onadim, Z., Masshofer, L. & Lohmann, D.R. (2004) A novel real-time PCR assay for quantitative analysis of methylated alleles (QAMA): analysis of the retinoblastoma locus. Nucleic acids research, 32, e125 ISSN 1362-4962.

Part 2

Human and Animal Health

6

DNA Methylation and Trinucleotide Repeat Expansion Diseases

Mark A. Pook
Division of Biosciences,
School of Health Sciences and Social Care,
Brunel University, Uxbridge
UK

1. Introduction

DNA methylation of CpG dinucleotides is essential for mammalian development, X inactivation, genomic imprinting, and may also be involved in immobilization of transposons and the control of tissue-specific gene expression (Bird & Wolffe, 1999). The common theme in each of these processes is gene silencing. Therefore, gene silencing is a major biological consequence of DNA methylation. As such, DNA methylation can play a very important role in human disease. For example, DNA methylation-induced silencing of tumour suppressor genes can result in cancer, while gain or loss of DNA methylation can produce loss of genomic imprinting in diseases such as Beckwith-Wiedermann syndrome (BWS), Prader-Willi syndrome (PWS) or Angelman syndrome (AS) (Robertson, 2005). Yet another group of diseases where DNA methylation has a prominent role to play in disease aetiology and pathology is that of the inherited trinucleotide repeat (TNR) expansion diseases.

TNR expansion diseases can be divided into two major subgroups: (i) those involving large non-coding repeats (typically 100-1000 repeats), and (ii) those involving short coding repeats (< 100 repeats, coding for polyglutamine or polyalanine). The majority of TNR expansion diseases that have disease-associated DNA hypermethylation are of the large non-coding repeat type. These include fragile X syndrome (FRAXA), which is caused by CGG repeat expansion in the 5'-untranslated region (UTR) of the *FMR1* gene (Verkerk et al., 1991), myotonic dystrophy type I (DM1), which is caused by CTG repeat expansion in the 3'-UTR of the *DMPK* gene (Brook et al., 1992), and Friedreich ataxia (FRDA), which is caused by GAA repeat expansion within intron 1 of the *FXN* gene (Campuzano et al., 1996). However, there is also evidence for possible involvement of DNA methylation in the short CAG repeat, polyglutamine-encoding, types of TNR expansion diseases, such as spinocerebellar ataxia type 1 (SCA1) (Dion et al., 2008) and spinocerebellar ataxia type 7 (SCA7) (Libby et al., 2008).

This review focuses on recent advances in our understanding of DNA methylation association with inherited TNR expansion diseases. It first describes the relevant TNR expansion diseases, the genes that are mutated and what is currently known about DNA methylation profiles in each case. This is followed by consideration of the potential causes of DNA methylation, the subsequent effects of DNA methylation on disease phenotype, and how understanding the mechanisms of DNA methylation may benefit efforts towards therapy for TNR expansion diseases.

2. TNR expansion diseases with associated DNA methylation

2.1 Fragile site-related mental retardation syndromes

Seven folate-sensitive fragile sites have been identified within human chromosomes: FRAXA (Verkerk et al., 1991), FRAXE (Knight et al., 1993), FRAXF (Parrish et al., 1994), FRA10A (Sarafidou et al., 2004), FRA11B (Jones et al., 1995), FRA12A (Winnepenninckx et al., 2007) and FRA16A (Nancarrow et al., 1994) (Table 1). In each case, the fragile site is associated with a large non-coding CGG repeat expansion, together with methylation of the CpG sites within the repeat expansion as well as within an adjacent upstream CpG island (Lopez Castel et al., 2010a). In the majority of cases, the CGG repeat expansion occurs within the 5'-UTR of a specific gene and the CpG island resides within the promoter region of this gene (Table 1). The effect of DNA methylation is to induce silencing of the gene, and the outcome of this, for the majority of fragile site-expressing patients, is the development of mental retardation.

Disease	Fragile site	Chromosomal position	Associated Gene	CGG repeat size	
				Normal	Disease
FRAXA Fragile X syndrome	FRAXA	Xq27.3	FMR1	6-54	>200
FRAXE Fragile X syndrome	FRAXE	Xq28	FMR2	4-39	>200
None identified at present	FRAXF	Xq28	FAM11A	7-40	>300
None identified at present	FRA10A	10q23.3	FRA10AC1	8-14	>200
Jacobsen syndrome	FRA11B	11q23.3	CBL2 (candidate)	11	>100
Mental retardation	FRA12A	12q13.1	DIP2B	6-23	>150
None identified at present	FRA16A	11q22	Unknown	16-49	>1000

Table 1. Fragile sites that are associated with aberrant methylated, expanded CGG repeats and methylated adjacent CpG sites in disease state.

2.1.1 Fragile X syndrome (FRAXA)

The most prominent of the fragile site disorders is Fragile X syndrome (FRAXA), an X linked disorder that is recognized as the most common inherited form of mental retardation (Brouwer et al., 2009). FRAXA is caused by CGG repeat expansion within the 5' UTR of the FMR1 (fragile X mental retardation 1) gene, which is located at the FRAXA fragile site on chromosome Xq27.3 (Verkerk et al., 1991) (Fig. 1). Unaffected individuals have a range of allele sizes between 6-54 CGG repeats. However, allele sizes of 55-200 CGG repeats, known as 'premutations', are unstable and can expand upon transmission to FRAXA individuals, who have alleles that exceed 200 CGG repeats, known as 'full mutations' (Fu et al., 1991). The expanded CGG repeats become methylated, as does the CpG island within the *FMR1* promoter, resulting in reduced expression of the *FMR1* gene product FMRP during development. Detailed analysis of the *FMR1* gene has revealed a distinct boundary of DNA methylation at a site between 650 and 800 nucleotides upstream of the CGG repeat in unaffected individuals that is lost in FRAXA patients (Naumann et al., 2009). This suggests that the *FMR1* promoter is normally protected from the spread of DNA methylation by a specific chromatin structure, which is somehow removed as a consequence of the expanded CGG repeat sequence.

Premutation CGG repeats ranging in size from 55-200 do not induce the typical DNA methylation and gene silencing that is seen with full mutations. Instead, unmethylated premutation CGG repeats produce overexpression of the *FMR1* gene, resulting in a toxic gain-of-function RNA that gives rise to the phenotypically distinct disorder called fragile X tremor/ataxia syndrome (FXTAS) (Jacquemont et al., 2003).

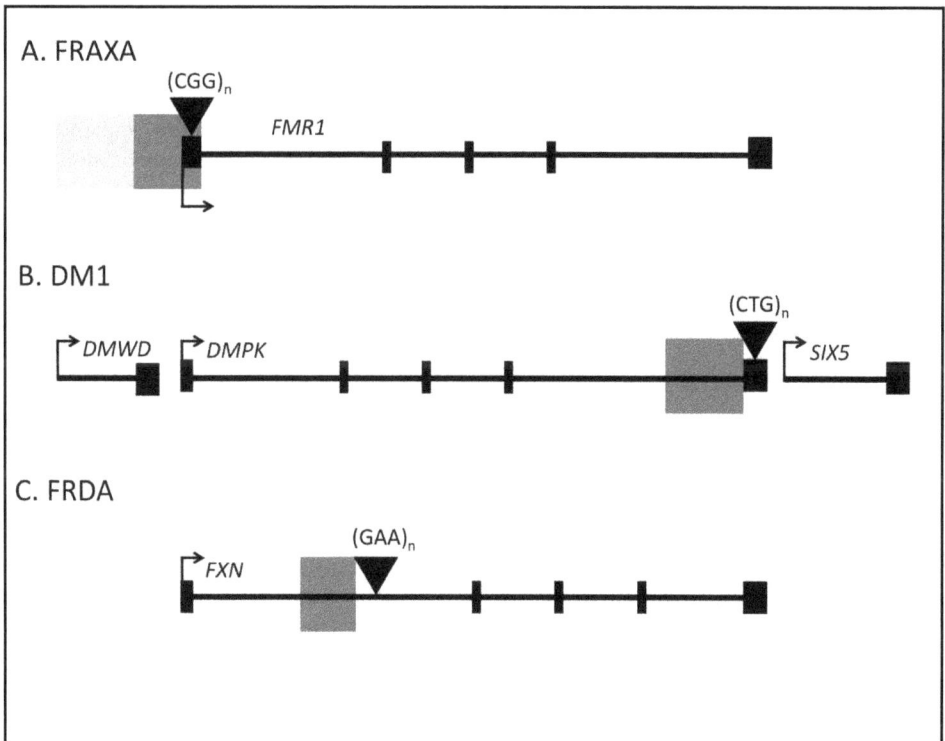

Fig. 1. Location of DNA methylation within expanded TNR loci. (A) FRAXA: The expanded CGG repeat is located in the 5'-UTR of the *FMR1* gene. A boundary of DNA methylation associated with normal CGG alleles (light grey box) shifts upon CGG expansion to enclose the CGG repeat (dark grey box). (B) DM1: The expanded CTG repeat in the 3'-UTR of the *DMPK* gene, and (C) FRDA: The expanded GAA repeat in intron 1 of the *FXN* gene are each associated with regions of DNA methylation just upstream of the expanded repeat.

2.2 Myotonic Dystrophy type 1 (DM1)

Myotonic dystrophy type 1 (DM1) is an autosomal dominant inherited multisystem disorder characterized by clinical features such as muscle weakness, myotonia and heart conduction defects (Schara & Schoser, 2006). The molecular basis for DM1 is expansion of a CTG repeat sequence within the 3'-UTR of the *DMPK* gene (Brook et al., 1992). Unaffected individuals have CTG repeat sizes of 5-37, and there is a premutation range of 34-90 CTG repeats, whereas affected individuals have expanded CTG repeat sizes that can range from 90 to thousands of units (Lopez Castel et al., 2010a). Both intergenerational and somatic instability of the CTG repeat are evident, providing a molecular basis for the anticipation phenomenon observed in DM1 families, together with tissue specific differences in disease pathology (Lavedan et al., 1993; Monckton et al., 1995). The effects of expanded CTG repeats are two-fold. Firstly, expression of an expanded CUG RNA sequence causes a toxic gain-of-function effect by altering the activity of RNA splicing factors (Ranum & Cooper, 2006). Secondly, the expanded CTG repeat induces epigenetic changes, including DNA methylation, at the DM1 locus that result in reduced expression of the *DMPK* gene and upstream and downstream genes *SIX5* and *DMWD* (Klesert et al., 1997; Alwazzan et al., 1999; Eriksson et al., 2001). Disease-associated DNA methylation was first reported to occur within a region approximately 1kb upstream of the *DMPK* gene (Steinbach et al., 1998). However, this was a rather restricted study based on the use of methylation-sensitive restriction enzymes that do not identify all CpG sites. A more recent study, which used bisulphite sequencing to characterize the DM1 locus at higher resolution, confirmed the disease-associated DNA methylation upstream of the expanded CTG repeat and further identified a distinct boundary at the expanded CTG repeat beyond which there is no DNA methylation (Lopez Castel et al., 2010b) (Fig.1).

2.3 Friedreich Ataxia (FRDA)

Friedreich ataxia (FRDA) is an autosomal recessive neurodegenerative disorder caused by homozygous GAA repeat expansion within intron 1 of the *FXN* gene (Campuzano et al., 1996). The effect of the expanded GAA repeat is to reduce expression of the essential mitochondrial protein frataxin (Campuzano et al., 1997), which results in progressive spinocerebellar neurodegeneration and cardiomyopathy (Pandolfo, 2009). Unaffected individuals have *FXN* alleles containing 5-32 GAA repeats, there is a premutation range of 33-65 GAA repeats, and affected individuals have alleles of 66-1700 GAA repeats. Both intergenerational and somatic instability of the GAA repeat are evident in FRDA, with expanded GAA repeats occurring prominently in disease-related CNS tissue (De Michele et al., 1998; De Biase et al., 2007a, 2007b). There is no disease-associated change in the DNA methylation status of the CpG island that spans the *FXN* 5'-UTR and exon 1 regions. However, disease-associated DNA hypermethylation has been identified within a region of *FXN* intron 1 immediately upstream of the expanded GAA repeat in FRDA cell culture, FRDA patient tissues and FRDA mouse models (Greene et al., 2007; Al-Mahdawi et al., 2008). Furthermore, the level of DNA methylation in this region correlates with expanded GAA repeat size and inversely correlates with age of FRDA disease onset (Castaldo et al., 2008). Interestingly, DNA hypomethylation has been identified in the *FXN* intron 1 *Alu* repeat sequence (which is normally fully methylated) immediately downstream of the expanded GAA repeat (Al-Mahdawi et al., 2008) (Fig.1). This effect of demethylation may have some, as yet unknown, relevance for GAA repeat instability and frataxin expression.

2.4 Polyglutamine-encoding TNR expansion disorders

To date, eleven inherited disorders are known to be caused by expansion of CAG repeats within the coding region of genes, resulting in the production of abnormal proteins that have long stretches of polyglutamine repeats (Lopez Castel et al., 2010a). Included within this group of polyglutamine disorders are Huntington disease (HD) and the spinocerebellar ataxias (SCAs). In each case, the severity of disease correlates with the size of the expanded CAG repeat, which is subject to both intergenerational and somatic instability (Koefoed et al., 1998; Wheeler et al., 2007). There is currently no evidence to support a disease-associated role for DNA methylation in HD (Reik et al., 1993) or SCA3 (Emmel et al., 2011). However, DNA methylation has been implicated in stability of CAG repeats in SCA1 (Dion et al., 2008) and in increased instability of CAG repeats in SCA7 (Libby et al., 2008).

3. Causes of DNA methylation

DNA methylation is involved in human diseases such as cancer, imprinting disorders and inherited TNR disorders, but at present it is not known why certain CpG sequences succumb to disease-associated aberrant DNA methylation. When considering the causes of DNA methylation in the TNR disorders, distinction must be made between DNA methylation of the expanded CGG repeat itself in the fragile site disorders, such as FRAXA and FRAXE, and DNA methylation of the flanking CpG sites in both fragile site disorders and other TNR disorders, including DM1 and FRDA. In the case of expanded CGG repeat disorders, the CGG repeat contains CpG residues that may be subject to methylation by direct effects. Thus, expanded CGG repeats have been shown to form single-stranded hairpins that lead to slippage structures during replication. Unrepaired slippage structures that contain extrahelical and mispaired cytosines may then act as substrates for direct *de novo* methylation by DNA methyltransferase enzymes (Chen et al., 1995; Laayoun & Smith, 1995; Chen et al., 1998). On the other hand, a common theme for all of the large non-coding TNR expansion diseases is DNA hypermethylation of CpG dinucleotides in the local vicinity of the TNR expansion. This suggests the action of a unified, but as yet unknown, secondary molecular mechanism. Evidence in favour of both *cis*- and *trans*-acting secondary effects has been put forward. Thus, aberrant DNA methylation of expanded CGG repeats or CpG sequences flanking expanded TNR sequences may be based upon underlying *cis*-acting DNA sequence context. For example, there is evidence to suggest that methylation can spread from core repetitive DNA sequences (Yates et al., 1999), and particular motifs have been identified as candidates for methylation-targeting DNA sequences (Feltus et al., 2006).

Another potential mechanism, which could be either *cis*- or *trans*-acting, is the induction of DNA methylation by short interfering RNAs (siRNAs). Studies of human cells have shown that long CNG repeat hairpins can be cleaved by the ribonuclease Dicer to form short double-stranded siRNAs (Krol et al., 2007), which may then induce DNA methylation as a process of transcriptional gene silencing (Kawasaki & Taira, 2004; Morris et al., 2004). Bidirectional transcription across TNRs may also produce siRNAs, which then recruit histone methyltransferases, HP1 and DNA methyltransferases to result in DNA methylation, as proposed for a general model of heterochromatin formation at repetitive elements (Grewal & Jia, 2007). Alternatively, siRNAs targeted to gene promoter CpG islands may be produced by bidirectional transcription at these regions (Morris et al., 2008), and such bidirectional transcripts have indeed been identified at several TNR loci (Cho et al.,

2005; Moseley et al., 2006; Ladd et al., 2007; De Biase et al., 2009; Chung et al., 2011). Furthermore, DNA methylation at the TNR locus may be induced by *trans*-acting siRNAs that are generated from a different locus (Watanabe et al., 2011). In each case, DNA methylation is likely to be a later long-term gene silencing effect, following on from earlier increases in histone methylation (Hawkins et al., 2009). Interestingly, siRNA targeting of the huntingtin gene promoter has failed to induce DNA methylation (Park et al., 2004), agreeing with a lack of any evidence for DNA methylation induction by CAG repeat expansion (Reik et al., 1993).

Another general mechanism that may be involved in the formation of DNA methylation at TNR loci is the loss of a methylation-sensitive chromatin insulator and subsequent spreading of DNA methylation. Of particular note is the chromatin insulator protein CTCF (CCCTC-binding protein), since CTCF binding sites have been identified in the flanking regions of *FRAXA* CGG repeats (Ladd et al., 2007), *DM1* CTG repeats (Filippova et al., 2001) and *SCA7* CAG repeats (Libby et al., 2008), and also in the upstream region of *FXN* GAA repeats (De Biase et al., 2009) (Fig.2).

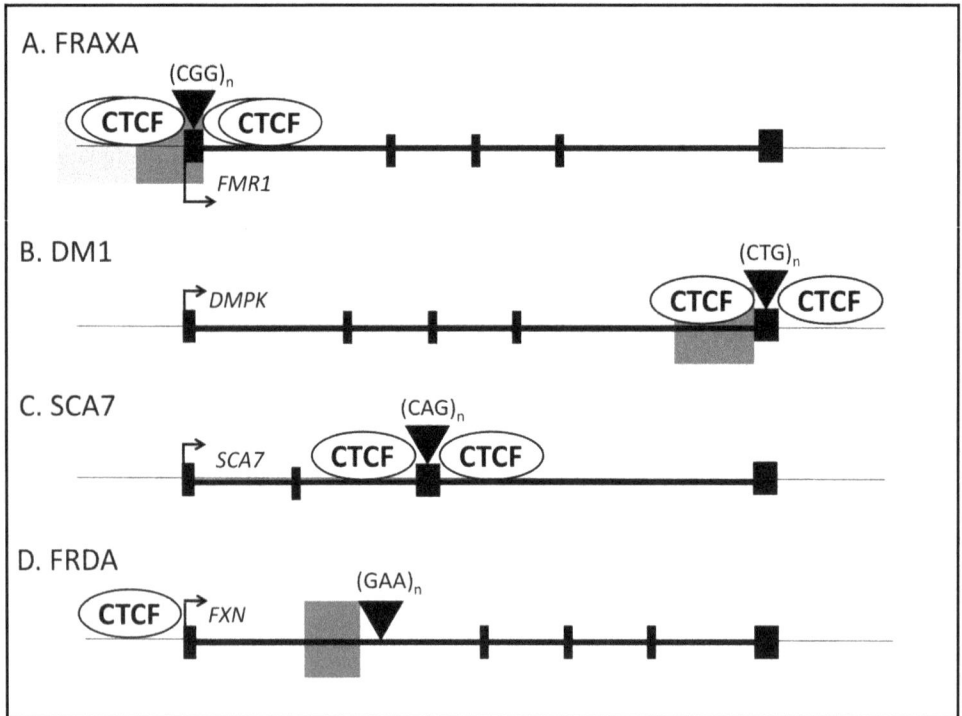

Fig. 2. The position of CTCF binding sites within TNR expansion loci (A) FRAXA, (B) DM1, (C) SCA7 and (D) FRDA. Grey boxes represent regions of DNA methylation.

A model has been proposed for the *DM1* locus whereby the normal CTG repeat allele is associated with bidirectional transcription, siRNAs, H3K9 dimethylation and HP1 recruitment in the region of the CTG repeats, but without any associated DNA methylation.

This local heterochromatin formation is limited to a small region by flanking CTCF sites and it is likely that CTCF binding protects the *DM1* CTG region from DNA methylation. However, expanded *DM1* CTG repeats are associated with loss of CTCF binding, spread of heterochromatin and regional CpG methylation (Cho et al., 2005). In FRDA, expanded GAA repeats are similarly associated with depletion of CTCF binding in the 5′-UTR of the *FXN* gene and associated hypermethylation of CpG sites just upstream of the GAA repeat. (Greene et al., 2007; Al-Mahdawi et al., 2008; De Biase et al., 2009). However, it is not currently known if depletion of CTCF actually precedes DNA methylation, or vice versa, in the context of TNR expansion diseases. For example, DNA methylation is known to inhibit CTCF binding at the *DM1* locus and other genetic loci (Filippova, 2008), whereas it does not appear to inhibit CTCF binding at the *FXN* locus (De Biase et al., 2009). Therefore, further studies will be required to determine the precise order of events that connect TNR expansions, CTCF binding and DNA methylation.

4. Effects of DNA methylation

There are two principal effects of DNA methylation on TNR expansion disorders. Firstly, silencing of gene transcription may take place, exerting profound effects on the subsequent disease phenotype. Secondly, modulation of TNR instability may occur throughout development and within different specific tissues, which will also impact upon progression of the disease phenotype.

4.1 Silencing of gene expression

The main effect of DNA methylation in TNR expansion disorders, as with other diseases such as cancer, is silencing of gene transcription. DNA methylation inhibits transcription by two general mechanisms: (i) preventing binding of basal transcription proteins or other regulatory DNA binding proteins (e.g. CTCF), and (ii) influencing nucleosome positioning or stability and reinforcing heterochromatin formation through the actions of methyl-CpG-binding proteins (MBPs), histone modifications and chromatin remodeling (Klose & Bird, 2006). There is evidence for both mechanisms at play in silencing of gene transcription in TNR expansion disorders. For FRAXA, expansion of CGG repeats in the 5′-UTR of the *FMR1* gene to greater than 200 units induces CpG methylation within the repeat tract and also the adjacent promoter region, leading to transcriptional silencing (Pieretti et al., 1991; Sutcliffe et al., 1992). Hypermethylation of the *FMR1* region is associated with histone deacetylation, H3K9 methylation and chromatin remodeling, which may impact upon *FMR1* transcription (Coffee et al., 1999; Coffee et al., 2002). However, it is suggested that DNA methylation, rather than histone modifications, is the key event for silencing of *FMR1* transcription (Pietrobono et al., 2002; Pietrobono et al., 2005). In contrast to FRAXA, the unmethylated 55-200 CGG repeats of the *FMR1* gene that characterize FXTAS produce a 2-10 fold increase of *FMR1* transcription, leading to an RNA toxic gain-of-function disease effect (Tassone et al., 2000). For DM1, the expanded CTG repeat, which is situated in the 3′-UTR of the *DMPK* gene, induces CpG methylation and H3K9 methylation flanking the repeat. This then silences transcription of *DMPK* and the neighbouring *SIX5* and *DMWD* genes (Klesert et al., 1997; Thornton et al., 1997; Alwazzan et al., 1999; Eriksson et al., 2001), likely by mechanisms that involve bidirectional transcription and siRNA formation (Cho et al., 2005). For FRDA, the expanded GAA repeat within intron 1 of the *FXN* gene induces CpG methylation, histone deacetylation and H3K9 methylation in the region immediately upstream of the GAA repeat, but the *FXN* promoter appears to be unaffected (Greene et al., 2003; Herman et al., 2006; Al-Mahdawi et al., 2008). Since DNA methylation within the body of a gene has not been linked to transcriptional silencing (Brenet et al., 2011), it is unlikely that CpG methylation within intron 1 is the primary cause of *FXN* gene silencing. The increased DNA methylation within intron 1 of the *FXN* gene is more likely to be secondary to gene silencing caused by chromatin changes (Greene et al., 2007) or changes in bidirectional transcription and CTCF binding (De Biase et al., 2009).

4.2 Modulation of TNR instability

Another major effect of DNA methylation, specifically related to TNR expansion disorders, is the ability to influence the dynamics of the trinucleotide repeat stability. Both germline and somatic instability of TNR sequences are known to play major roles in the aetiology and progression of all TNR expansion disorders (Lopez Castel et al., 2010a). For FRAXA, germline instability of the CGG repeat involves maternally derived expansions, but deletion in the gametes of full-mutation males. The CGG deletions occur during replication and are dependent on replication fork dynamics, size of repeat and CpG methylation status (Nichol Edamura et al., 2005). The period of somatic CGG instability is restricted to early stages of embryonic and foetal growth and ends when expanded CGG sequences become abnormally methylated (Devys et al., 1992; Taylor et al., 1999). Subsequent CpG methylation of FRAXA 'full mutation' expanded CGG repeats causes somatic stability (Wohrle et al., 1995). Several studies, which have examined the effect of DNA methylation on germline TNR instability using mouse models of DM1, HD and SCA1, find that instability is particularly associated with periods of demethylation in the developing germline (Kaytor et al., 1997; Kovtun & McMurray, 2001; Savouret et al., 2004). Furthermore, treatment of cultured cells from DM1 patients with DNA demethylating compounds produced destabilization of CTG repeats, with a bias towards expansion (Gorbunova et al., 2004). Thus, it appears that changes in DNA methylation patterns during germline epigenetic reprogramming may trigger the intergenerational TNR expansions that lead to disease. It is currently not known how DNA methylation stabilizes germline TNR sequences. However, several hypotheses have been proposed. For example, a recent study of heterozygous DNA methyltransferase 1 (*Dnmt1+/-*) knockout SCA1 mice revealed Dnmt1-deficient promotion of CAG intergenerational instability, together with aberrant DNA methylation and histone methylation within the CpG island adjacent to the CAG repeat, suggesting a role for local chromatin structure in germline TNR instability. However, no effect of Dnmt1 deficit was seen on somatic instability (Dion et al., 2008). Another study, which investigated somatic instability of the *DM1* CTG repeat in relation to replication and CTCF binding, has led to the suggestion that CpG methylation may regulate, in a tissue-specific manner, the role of CTCF in DNA replication and thereby CTG repeat instability (Cleary et al., 2010). Yet another connection between DNA methylation, CTCF and TNR expansion has been identified in studies of *SCA7* transgenic mice, which revealed further destabilization of unstable expanded CAG repeats by CpG methylation of CTCF binding sites (Libby et al., 2008). Furthermore, other studies suggest a potential connection between DNA methylation, transcription and TNR instability. For example, cyclical changes in promoter CpG methylation have been identified during transcription (Kangaspeska et al., 2008; Metivier et al., 2008), and at the same time, transcription through the repeat tract has been identified as a major contributor to expansion of GAA repeats (Ditch et al., 2009). Therefore, periods of active demethylation during transcription may present a window of opportunity for TNR expansion. Finally, several studies have highlighted an important role of DNA repair mechanisms, and in particular the mismatch repair (MMR) system, in TNR instability (Savouret et al., 2003; Wheeler et al., 2003; Dragileva et al., 2009), and DNMT1 deficiency has been shown to result in MMR defects that increase the rate of CAG repeat contraction (Lin & Wilson, 2009). Therefore, it may be interesting to further pursue connections between DNA methylation, DNA repair and TNR instability.

5. Demethylation therapy for TNR expansion diseases

The finding that DNA methylation of the CGG repeat and flanking CpG sequences of the *FMR1* promoter cause transcriptional silencing of the *FMR1* gene in FRAXA, has lead to consideration of DNA demethylation as a potential therapy. To date, investigations have focused on the use of the cytidine analogue DNA methyltransferase inhibitor 5-aza-2′-deoxycytidine (5-aza-CdR), which is an FDA approved drug (decitabine) used for the treatment of myelodysplastic anaemia (Oki et al., 2007). Treatment of fully methylated FRAXA patient cell lines with 5-aza-CdR leads to a decrease in promoter CpG methylation, together with increased histone acetylation, decreased H3K9 methylation and increased H3K4 methylation, that result in an increase in *FMR1* transcription (Chiurazzi et al., 1998; Pietrobono et al., 2002; Tabolacci et al., 2005). Combined treatment with 5-aza-CdR and histone deacetylase (HDAC) inhibitors have also been shown to produce a synergistic increase in *FMR1* transcription (Chiurazzi et al., 1999). However, 5-aza-CdR is a drug that induces substantial cytotoxicity, and therefore the development and testing of other less toxic DNA methylation inhibitors, such as zebularine (Cheng et al., 2003) or hydralazine (Cornacchia et al., 1988) may be necessary before treatment for FRAXA can be considered further. Another potential adverse effect of DNA methylation inhibitor treatment that will have to be considered is the finding that DNA demethylation can induce TNR instability, with a bias towards expansions (Gorbunova et al., 2004), which may then negatively impact upon gene expression.

6. Conclusions

DNA methylation is a molecular process that is clearly associated with TNR expansion disorders, particularly those of the long non-coding repeat type. Recent studies have revealed common themes for TNR gene silencing, including bidirectional transcription, siRNA formation, CTCF binding, histone modifications and chromatin remodeling. However, the exact role that DNA methylation plays within TNR expansion disease pathogenesis remains uncertain and further investigations are still needed. At the same time, DNA methylation also appears to impact upon TNR instability, which is an important part of TNR expansion disease progression. Therefore, the interplay between DNA methylation, DNA replication, DNA repair and transcription will need particular investigation if future consideration can realistically be given to DNA demethylation therapies for TNR expansion disorders.

7. References

Al-Mahdawi, S., Pinto, R. M., Ismail, O., Varshney, D., Lymperi, S., Sandi, C., Trabzuni, D. & Pook, M. (2008). The Friedreich ataxia GAA repeat expansion mutation induces comparable epigenetic changes in human and transgenic mouse brain and heart tissues. *Hum Mol Genet*, Vol.17, No.5, (Mar 1), pp. 735-46, ISSN.1460-2083

Alwazzan, M., Newman, E., Hamshere, M. G. & Brook, J. D. (1999). Myotonic dystrophy is associated with a reduced level of RNA from the DMWD allele adjacent to the expanded repeat. *Hum Mol Genet*, Vol.8, No.8, (Aug), pp. 1491-7, ISSN.0964-6906

Bird, A. P. & Wolffe, A. P. (1999). Methylation-induced repression--belts, braces, and chromatin. *Cell*, Vol.99, No.5, (Nov 24), pp. 451-4, ISSN.0092-8674

Brenet, F., Moh, M., Funk, P., Feierstein, E., Viale, A. J., Socci, N. D. & Scandura, J. M. (2011). DNA methylation of the first exon is tightly linked to transcriptional silencing. *PLoS One*, Vol.6, No.1, pp. e14524, ISSN.1932-6203

Brook, J. D., McCurrach, M. E., Harley, H. G., Buckler, A. J., Church, D., Aburatani, H., Hunter, K., Stanton, V. P., Thirion, J. P., Hudson, T. & et al. (1992). Molecular basis of myotonic dystrophy: expansion of a trinucleotide (CTG) repeat at the 3' end of a transcript encoding a protein kinase family member. *Cell*, Vol.69, No.2, (Apr 17), pp. 385, ISSN.0092-8674

Brouwer, J. R., Willemsen, R. & Oostra, B. A. (2009). The FMR1 gene and fragile X-associated tremor/ataxia syndrome. *Am J Med Genet B Neuropsychiatr Genet*, Vol.150B, No.6, (Sep 5), pp. 782-98, ISSN.1552-485X

Campuzano, V., Montermini, L., Molto, M. D., Pianese, L., Cossee, M., Cavalcanti, F., Monros, E., Rodius, F., Duclos, F., Monticelli, A., Zara, F., Canizares, J., Koutnikova, H., Bidichandani, S. I., Gellera, C., Brice, A., Trouillas, P., De Michele, G., Filla, A., De Frutos, R., Palau, F., Patel, P. I., Di Donato, S., Mandel, J. L., Cocozza, S., Koenig, M. & Pandolfo, M. (1996). Friedreich's ataxia: autosomal recessive disease caused by an intronic GAA triplet repeat expansion. *Science*, Vol.271, No.5254, (Mar 8), pp. 1423-7, ISSN.0964-6906

Campuzano, V., Montermini, L., Lutz, Y., Cova, L., Hindelang, C., Jiralerspong, S., Trottier, Y., Kish, S. J., Faucheux, B., Trouillas, P., Authier, F. J., Durr, A., Mandel, J. L., Vescovi, A., Pandolfo, M. & Koenig, M. (1997). Frataxin is reduced in Friedreich ataxia patients and is associated with mitochondrial membranes. *Hum Mol Genet*, Vol.6, No.11, (Oct), pp. 1771-80, ISSN.0036-8075

Castaldo, I., Pinelli, M., Monticelli, A., Acquaviva, F., Giacchetti, M., Filla, A., Sacchetti, S., Keller, S., Avvedimento, V. E., Chiariotti, L. & Cocozza, S. (2008). DNA methylation in intron 1 of the frataxin gene is related to GAA repeat length and age of onset in Friedreich's ataxia patients. *J Med Genet*, (Aug 12), pp. 808-812, ISSN.1468-6244

Chen, X., Mariappan, S. V., Catasti, P., Ratliff, R., Moyzis, R. K., Laayoun, A., Smith, S. S., Bradbury, E. M. & Gupta, G. (1995). Hairpins are formed by the single DNA strands of the fragile X triplet repeats: structure and biological implications. *Proc Natl Acad Sci U S A*, Vol.92, No.11, (May 23), pp. 5199-203, ISSN.0027-8424

Chen, X., Mariappan, S. V., Moyzis, R. K., Bradbury, E. M. & Gupta, G. (1998). Hairpin induced slippage and hyper-methylation of the fragile X DNA triplets. *J Biomol Struct Dyn*, Vol.15, No.4, (Feb), pp. 745-56, ISSN.0739-1102

Cheng, J. C., Matsen, C. B., Gonzales, F. A., Ye, W., Greer, S., Marquez, V. E., Jones, P. A. & Selker, E. U. (2003). Inhibition of DNA methylation and reactivation of silenced genes by zebularine. *J Natl Cancer Inst*, Vol.95, No.5, (Mar 5), pp. 399-409, ISSN.0027-8874

Chiurazzi, P., Pomponi, M. G., Willemsen, R., Oostra, B. A. & Neri, G. (1998). In vitro reactivation of the FMR1 gene involved in fragile X syndrome. *Hum Mol Genet*, Vol.7, No.1, (Jan), pp. 109-13, ISSN.0964-6906

Chiurazzi, P., Pomponi, M. G., Pietrobono, R., Bakker, C. E., Neri, G. & Oostra, B. A. (1999). Synergistic effect of histone hyperacetylation and DNA demethylation in the reactivation of the FMR1 gene. *Hum Mol Genet*, Vol.8, No.12, (Nov), pp. 2317-23, ISSN.0964-6906

Cho, D. H., Thienes, C. P., Mahoney, S. E., Analau, E., Filippova, G. N. & Tapscott, S. J. (2005). Antisense transcription and heterochromatin at the DM1 CTG repeats are constrained by CTCF. *Mol Cell*, Vol.20, No.3, (Nov 11), pp. 483-9, ISSN.1097-2765

Chung, D. W., Rudnicki, D. D., Yu, L. & Margolis, R. L. (2011). A natural antisense transcript at the Huntington's disease repeat locus regulates HTT expression. *Hum Mol Genet*, Vol.20, No.17, (Sep 1), pp. 3467-77, ISSN.1460-2083

Cleary, J. D., Tome, S., Lopez Castel, A., Panigrahi, G. B., Foiry, L., Hagerman, K. A., Sroka, H., Chitayat, D., Gourdon, G. & Pearson, C. E. (2010). Tissue- and age-specific DNA replication patterns at the CTG/CAG-expanded human myotonic dystrophy type 1 locus. *Nat Struct Mol Biol*, Vol.17, No.9, (Sep), pp. 1079-87, ISSN.1545-9985

Coffee, B., Zhang, F., Warren, S. T. & Reines, D. (1999). Acetylated histones are associated with FMR1 in normal but not fragile X-syndrome cells. *Nat Genet*, Vol.22, No.1, (May), pp. 98-101, ISSN.1061-4036

Coffee, B., Zhang, F., Ceman, S., Warren, S. T. & Reines, D. (2002). Histone modifications depict an aberrantly heterochromatinized FMR1 gene in fragile x syndrome. *Am J Hum Genet*, Vol.71, No.4, (Oct), pp. 923-32, ISSN.0002-9297

Cornacchia, E., Golbus, J., Maybaum, J., Strahler, J., Hanash, S. & Richardson, B. (1988). Hydralazine and procainamide inhibit T cell DNA methylation and induce autoreactivity. *J Immunol*, Vol.140, No.7, (Apr 1), pp. 2197-200, ISSN.0022-1767

De Biase, I., Rasmussen, A., Endres, D., Al-Mahdawi, S., Monticelli, A., Cocozza, S., Pook, M. & Bidichandani, S. I. (2007a). Progressive GAA expansions in dorsal root ganglia of Friedreich's ataxia patients. *Ann Neurol*, Vol.61, No.1, (Jan), pp. 55-60, ISSN.0364-5134

De Biase, I., Rasmussen, A., Monticelli, A., Al-Mahdawi, S., Pook, M., Cocozza, S. & Bidichandani, S. I. (2007b). Somatic instability of the expanded GAA triplet-repeat sequence in Friedreich ataxia progresses throughout life. *Genomics*, Vol.90, No.1, (Jul), pp. 1-5, ISSN.0888-7543

De Biase, I., Chutake, Y. K., Rindler, P. M. & Bidichandani, S. I. (2009). Epigenetic silencing in Friedreich ataxia is associated with depletion of CTCF (CCCTC-binding factor) and antisense transcription. *PLoS One*, Vol.4, No.11, pp. e7914, ISSN.1932-6203

De Michele, G., Cavalcanti, F., Criscuolo, C., Pianese, L., Monticelli, A., Filla, A. & Cocozza, S. (1998). Parental gender, age at birth and expansion length influence GAA repeat intergenerational instability in the X25 gene: pedigree studies and analysis of sperm from patients with Friedreich's ataxia. *Hum Mol Genet*, Vol.7, No.12, (Nov), pp. 1901-6, ISSN.0964-6906

Devys, D., Biancalana, V., Rousseau, F., Boue, J., Mandel, J. L. & Oberle, I. (1992). Analysis of full fragile X mutations in fetal tissues and monozygotic twins indicate that abnormal methylation and somatic heterogeneity are established early in development. *Am J Med Genet*, Vol.43, No.1-2, (Apr 15-May 1), pp. 208-16, ISSN.0148-7299

Dion, V., Lin, Y., Hubert, L., Jr., Waterland, R. A. & Wilson, J. H. (2008). Dnmt1 deficiency promotes CAG repeat expansion in the mouse germline. *Hum Mol Genet*, Vol.17, No.9, (May 1), pp. 1306-17, ISSN.1460-2083

Ditch, S., Sammarco, M. C., Banerjee, A. & Grabczyk, E. (2009). Progressive GAA.TTC repeat expansion in human cell lines. *PLoS Genet*, Vol.5, No.10, (Oct), pp. e1000704, ISSN.1553-7404

Dragileva, E., Hendricks, A., Teed, A., Gillis, T., Lopez, E. T., Friedberg, E. C., Kucherlapati, R., Edelmann, W., Lunetta, K. L., MacDonald, M. E. & Wheeler, V. C. (2009). Intergenerational and striatal CAG repeat instability in Huntington's disease knock-in mice involve different DNA repair genes. *Neurobiol Dis*, Vol.33, No.1, (Jan), pp. 37-47, ISSN.1095-953X

Emmel, V. E., Alonso, I., Jardim, L. B., Saraiva-Pereira, M. L. & Sequeiros, J. (2011). Does DNA methylation in the promoter region of the ATXN3 gene modify age at onset in MJD (SCA3) patients? *Clin Genet*, Vol.79, No.1, (Jan), pp. 100-2, ISSN.1399-0004

Eriksson, M., Hedberg, B., Carey, N. & Ansved, T. (2001). Decreased DMPK transcript levels in myotonic dystrophy 1 type IIA muscle fibers. *Biochem Biophys Res Commun*, Vol.286, No.5, (Sep 7), pp. 1177-82, ISSN.0006-291X

Feltus, F. A., Lee, E. K., Costello, J. F., Plass, C. & Vertino, P. M. (2006). DNA motifs associated with aberrant CpG island methylation. *Genomics*, Vol.87, No.5, (May), pp. 572-9, ISSN.0888-7543

Filippova, G. N., Thienes, C. P., Penn, B. H., Cho, D. H., Hu, Y. J., Moore, J. M., Klesert, T. R., Lobanenkov, V. V. & Tapscott, S. J. (2001). CTCF-binding sites flank CTG/CAG repeats and form a methylation-sensitive insulator at the DM1 locus. *Nat Genet*, Vol.28, No.4, (Aug), pp. 335-43, ISSN.1061-4036

Filippova, G. N. (2008). Genetics and epigenetics of the multifunctional protein CTCF. *Curr Top Dev Biol*, Vol.80, pp. 337-60, ISSN.0070-2153

Fu, Y. H., Kuhl, D. P., Pizzuti, A., Pieretti, M., Sutcliffe, J. S., Richards, S., Verkerk, A. J., Holden, J. J., Fenwick, R. G., Jr., Warren, S. T. & et al. (1991). Variation of the CGG repeat at the fragile X site results in genetic instability: resolution of the Sherman paradox. *Cell*, Vol.67, No.6, (Dec 20), pp. 1047-58, ISSN.0092-8674

Gorbunova, V., Seluanov, A., Mittelman, D. & Wilson, J. H. (2004). Genome-wide demethylation destabilizes CTG.CAG trinucleotide repeats in mammalian cells. *Hum Mol Genet*, Vol.13, No.23, (Dec 1), pp. 2979-89, ISSN.0964-6906

Greene, E., Handa, V., Kumari, D. & Usdin, K. (2003). Transcription defects induced by repeat expansion: fragile X syndrome, FRAXE mental retardation, progressive myoclonus epilepsy type 1, and Friedreich ataxia. *Cytogenet Genome Res*, Vol.100, No.1-4, pp. 65-76, ISSN.1424-859X

Greene, E., Mahishi, L., Entezam, A., Kumari, D. & Usdin, K. (2007). Repeat-induced epigenetic changes in intron 1 of the frataxin gene and its consequences in Friedreich ataxia. *Nucleic Acids Res*, Vol.35, No.10, pp. 3383-90, ISSN 1362-4962

Grewal, S. I. & Jia, S. (2007). Heterochromatin revisited. *Nat Rev Genet*, Vol.8, No.1, (Jan), pp. 35-46, ISSN.1471-0056

Hawkins, P. G., Santoso, S., Adams, C., Anest, V. & Morris, K. V. (2009). Promoter targeted small RNAs induce long-term transcriptional gene silencing in human cells. *Nucleic Acids Res*, Vol.37, No.9, (May), pp. 2984-95, ISSN.1362-4962

Herman, D., Jenssen, K., Burnett, R., Soragni, E., Perlman, S. L. & Gottesfeld, J. M. (2006). Histone deacetylase inhibitors reverse gene silencing in Friedreich's ataxia. *Nat Chem Biol*, Vol.2, No.10, (Oct), pp. 551-8, ISSN.1552-4450

Jacquemont, S., Hagerman, R. J., Leehey, M., Grigsby, J., Zhang, L., Brunberg, J. A., Greco, C., Des Portes, V., Jardini, T., Levine, R., Berry-Kravis, E., Brown, W. T., Schaeffer, S., Kissel, J., Tassone, F. & Hagerman, P. J. (2003). Fragile X premutation tremor/ataxia syndrome: molecular, clinical, and neuroimaging correlates. *Am J Hum Genet*, Vol.72, No.4, (Apr), pp. 869-78, ISSN.0002-9297

Jones, C., Penny, L., Mattina, T., Yu, S., Baker, E., Voullaire, L., Langdon, W. Y., Sutherland, G. R., Richards, R. I. & Tunnacliffe, A. (1995). Association of a chromosome deletion syndrome with a fragile site within the proto-oncogene CBL2. *Nature*, Vol.376, No.6536, (Jul 13), pp. 145-9, ISSN.0028-0836

Kangaspeska, S., Stride, B., Metivier, R., Polycarpou-Schwarz, M., Ibberson, D., Carmouche, R. P., Benes, V., Gannon, F. & Reid, G. (2008). Transient cyclical methylation of promoter DNA. *Nature*, Vol.452, No.7183, (Mar 6), pp. 112-5, ISSN.1476-4687

Kawasaki, H. & Taira, K. (2004). Induction of DNA methylation and gene silencing by short interfering RNAs in human cells. *Nature*, Vol.431, No.7005, (Sep 9), pp. 211-7, ISSN. 1476-4687

Kaytor, M. D., Burright, E. N., Duvick, L. A., Zoghbi, H. Y. & Orr, H. T. (1997). Increased trinucleotide repeat instability with advanced maternal age. *Hum Mol Genet*, Vol.6, No.12, (Nov), pp. 2135-9, ISSN.0964-6906

Klesert, T. R., Otten, A. D., Bird, T. D. & Tapscott, S. J. (1997). Trinucleotide repeat expansion at the myotonic dystrophy locus reduces expression of DMAHP. *Nat Genet*, Vol.16, No.4, (Aug), pp. 402-6, ISSN.1061-4036

Klose, R. J. & Bird, A. P. (2006). Genomic DNA methylation: the mark and its mediators. *Trends Biochem Sci*, Vol.31, No.2, (Feb), pp. 89-97, ISSN.0968-0004

Knight, S. J., Flannery, A. V., Hirst, M. C., Campbell, L., Christodoulou, Z., Phelps, S. R., Pointon, J., Middleton-Price, H. R., Barnicoat, A., Pembrey, M. E. & et al. (1993). Trinucleotide repeat amplification and hypermethylation of a CpG island in FRAXE mental retardation. *Cell*, Vol.74, No.1, (Jul 16), pp. 127-34, ISSN.0092-8674

Koefoed, P., Hasholt, L., Fenger, K., Nielsen, J. E., Eiberg, H., Buschard, K. & Sorensen, S. A. (1998). Mitotic and meiotic instability of the CAG trinucleotide repeat in spinocerebellar ataxia type 1. *Hum Genet*, Vol.103, No.5, (Nov), pp. 564-9, ISSN.0340-6717

Kovtun, I. V. & McMurray, C. T. (2001). Trinucleotide expansion in haploid germ cells by gap repair. *Nat Genet*, Vol.27, No.4, (Apr), pp. 407-11, ISSN.1061-4036

Krol, J., Fiszer, A., Mykowska, A., Sobczak, K., de Mezer, M. & Krzyosiak, W. J. (2007). Ribonuclease dicer cleaves triplet repeat hairpins into shorter repeats that silence specific targets. *Mol Cell*, Vol.25, No.4, (Feb 23), pp. 575-86, ISSN.1097-2765

Laayoun, A. & Smith, S. S. (1995). Methylation of slipped duplexes, snapbacks and cruciforms by human DNA(cytosine-5)methyltransferase. *Nucleic Acids Res*, Vol.23, No.9, (May 11), pp. 1584-9, ISSN.0305-1048

Ladd, P. D., Smith, L. E., Rabaia, N. A., Moore, J. M., Georges, S. A., Hansen, R. S., Hagerman, R. J., Tassone, F., Tapscott, S. J. & Filippova, G. N. (2007). An antisense transcript spanning the CGG repeat region of FMR1 is upregulated in premutation carriers but silenced in full mutation individuals. *Hum Mol Genet*, Vol.16, No.24, (Dec 15), pp. 3174-87, ISSN.0964-6906

Lavedan, C., Hofmann-Radvanyi, H., Shelbourne, P., Rabes, J. P., Duros, C., Savoy, D., Dehaupas, I., Luce, S., Johnson, K. & Junien, C. (1993). Myotonic dystrophy: size- and sex-dependent dynamics of CTG meiotic instability, and somatic mosaicism. *Am J Hum Genet*, Vol.52, No.5, (May), pp. 875-83, ISSN.0002-9297

Libby, R. T., Hagerman, K. A., Pineda, V. V., Lau, R., Cho, D. H., Baccam, S. L., Axford, M. M., Cleary, J. D., Moore, J. M., Sopher, B. L., Tapscott, S. J., Filippova, G. N., Pearson, C. E. & La Spada, A. R. (2008). CTCF cis-regulates trinucleotide repeat instability in an epigenetic manner: a novel basis for mutational hot spot determination. *PLoS Genet*, Vol.4, No.11, (Nov), pp. e1000257, ISSN.1553-7404

Lin, Y. & Wilson, J. H. (2009). Diverse effects of individual mismatch repair components on transcription-induced CAG repeat instability in human cells. *DNA Repair (Amst)*, Vol.8, No.8, (Aug 6), pp. 878-85, ISSN.1568-7856

Lopez Castel, A., Cleary, J. D. & Pearson, C. E. (2010a). Repeat instability as the basis for human diseases and as a potential target for therapy. *Nat Rev Mol Cell Biol*, Vol.11, No.3, (Mar), pp. 165-70, ISSN.1471-0080

Lopez Castel, A., Nakamori, M., Tome, S., Chitayat, D., Gourdon, G., Thornton, C. A. & Pearson, C. E. (2010b). Expanded CTG repeat demarcates a boundary for abnormal CpG methylation in myotonic dystrophy patient tissues. *Hum Mol Genet*, Vol.20, No.1, (Jan 1), pp. 1-15, ISSN.1460-2083

Metivier, R., Gallais, R., Tiffoche, C., Le Peron, C., Jurkowska, R. Z., Carmouche, R. P., Ibberson, D., Barath, P., Demay, F., Reid, G., Benes, V., Jeltsch, A., Gannon, F. & Salbert, G. (2008). Cyclical DNA methylation of a transcriptionally active promoter. *Nature*, Vol.452, No.7183, (Mar 6), pp. 45-50, ISSN.1476-4687

Monckton, D. G., Wong, L. J., Ashizawa, T. & Caskey, C. T. (1995). Somatic mosaicism, germline expansions, germline reversions and intergenerational reductions in myotonic dystrophy males: small pool PCR analyses. *Hum Mol Genet*, Vol.4, No.1, (Jan), pp. 1-8, ISSN.0964-6906

Morris, K. V., Chan, S. W., Jacobsen, S. E. & Looney, D. J. (2004). Small interfering RNA-induced transcriptional gene silencing in human cells. *Science*, Vol.305, No.5688, (Aug 27), pp. 1289-92, ISSN.1095-9203

Morris, K. V., Santoso, S., Turner, A. M., Pastori, C. & Hawkins, P. G. (2008). Bidirectional transcription directs both transcriptional gene activation and suppression in human cells. *PLoS Genet*, Vol.4, No.11, (Nov), pp. e1000258, ISSN.1553-7404

Moseley, M. L., Zu, T., Ikeda, Y., Gao, W., Mosemiller, A. K., Daughters, R. S., Chen, G., Weatherspoon, M. R., Clark, H. B., Ebner, T. J., Day, J. W. & Ranum, L. P. (2006). Bidirectional expression of CUG and CAG expansion transcripts and intranuclear polyglutamine inclusions in spinocerebellar ataxia type 8. *Nat Genet*, Vol.38, No.7, (Jul), pp. 758-69, ISSN.1061-4036

Nancarrow, J. K., Kremer, E., Holman, K., Eyre, H., Doggett, N. A., Le Paslier, D., Callen, D. F., Sutherland, G. R. & Richards, R. I. (1994). Implications of FRA16A structure for the mechanism of chromosomal fragile site genesis. *Science*, Vol.264, No.5167, (Jun 24), pp. 1938-41, ISSN.0036-8075

Naumann, A., Hochstein, N., Weber, S., Fanning, E. & Doerfler, W. (2009). A distinct DNA-methylation boundary in the 5'- upstream sequence of the FMR1 promoter binds nuclear proteins and is lost in fragile X syndrome. *Am J Hum Genet*, Vol.85, No.5, (Nov), pp. 606-16, ISSN.1537-6605

Nichol Edamura, K., Leonard, M. R. & Pearson, C. E. (2005). Role of replication and CpG methylation in fragile X syndrome CGG deletions in primate cells. *Am J Hum Genet*, Vol.76, No.2, (Feb), pp. 302-11, ISSN.0002-9297

Oki, Y., Aoki, E. & Issa, J. P. (2007). Decitabine--bedside to bench. *Crit Rev Oncol Hematol*, Vol.61, No.2, (Feb), pp. 140-52, ISSN.1040-8428

Pandolfo, M. (2009). Friedreich ataxia: the clinical picture. *J Neurol*, Vol.256 Suppl 1, (Mar), pp. 3-8, ISSN.0340-5354

Park, C. W., Chen, Z., Kren, B. T. & Steer, C. J. (2004). Double-stranded siRNA targeted to the huntingtin gene does not induce DNA methylation. *Biochem Biophys Res Commun*, Vol.323, No.1, (Oct 8), pp. 275-80, ISSN.0006-291X

Parrish, J. E., Oostra, B. A., Verkerk, A. J., Richards, C. S., Reynolds, J., Spikes, A. S., Shaffer, L. G. & Nelson, D. L. (1994). Isolation of a GCC repeat showing expansion in FRAXF, a fragile site distal to FRAXA and FRAXE. *Nat Genet*, Vol.8, No.3, (Nov), pp. 229-35, ISSN.1061-4036

Pieretti, M., Zhang, F. P., Fu, Y. H., Warren, S. T., Oostra, B. A., Caskey, C. T. & Nelson, D. L. (1991). Absence of expression of the FMR-1 gene in fragile X syndrome. *Cell*, Vol.66, No.4, (Aug 23), pp. 817-22, ISSN.0092-8674

Pietrobono, R., Pomponi, M. G., Tabolacci, E., Oostra, B., Chiurazzi, P. & Neri, G. (2002). Quantitative analysis of DNA demethylation and transcriptional reactivation of the FMR1 gene in fragile X cells treated with 5-azadeoxycytidine. *Nucleic Acids Res*, Vol.30, No.14, (Jul 15), pp. 3278-85, ISSN.1362-4962

Pietrobono, R., Tabolacci, E., Zalfa, F., Zito, I., Terracciano, A., Moscato, U., Bagni, C., Oostra, B., Chiurazzi, P. & Neri, G. (2005). Molecular dissection of the events leading to inactivation of the FMR1 gene. *Hum Mol Genet*, Vol.14, No.2, (Jan 15), pp. 267-77, ISSN.0964-6906

Ranum, L. P. & Cooper, T. A. (2006). RNA-mediated neuromuscular disorders. *Annu Rev Neurosci*, Vol.29, pp. 259-77, ISSN.0147-006X

Reik, W., Maher, E. R., Morrison, P. J., Harding, A. E. & Simpson, S. A. (1993). Age at onset in Huntington's disease and methylation at D4S95. *J Med Genet*, Vol.30, No.3, (Mar), pp. 185-8, ISSN.0022-2593

Robertson, K. D. (2005). DNA methylation and human disease. *Nat Rev Genet*, Vol.6, No.8, (Aug), pp. 597-610, ISSN.1471-0056

Sarafidou, T., Kahl, C., Martinez-Garay, I., Mangelsdorf, M., Gesk, S., Baker, E., Kokkinaki, M., Talley, P., Maltby, E. L., French, L., Harder, L., Hinzmann, B., Nobile, C., Richkind, K., Finnis, M., Deloukas, P., Sutherland, G. R., Kutsche, K., Moschonas, N. K., Siebert, R. & Gecz, J. (2004). Folate-sensitive fragile site FRA10A is due to an expansion of a CGG repeat in a novel gene, FRA10AC1, encoding a nuclear protein. *Genomics*, Vol.84, No.1, (Jul), pp. 69-81, ISSN.0888-7543

Savouret, C., Brisson, E., Essers, J., Kanaar, R., Pastink, A., te Riele, H., Junien, C. & Gourdon, G. (2003). CTG repeat instability and size variation timing in DNA repair-deficient mice. *EMBO J*, Vol.22, No.9, (May 1), pp. 2264-73, ISSN.0261-4189

Savouret, C., Garcia-Cordier, C., Megret, J., te Riele, H., Junien, C. & Gourdon, G. (2004). MSH2-dependent germinal CTG repeat expansions are produced continuously in spermatogonia from DM1 transgenic mice. *Mol Cell Biol*, Vol.24, No.2, (Jan), pp. 629-37, ISSN.0270-7306

Schara, U. & Schoser, B. G. (2006). Myotonic dystrophies type 1 and 2: a summary on current aspects. *Semin Pediatr Neurol*, Vol.13, No.2, (Jun), pp. 71-9, ISSN.1071-9091

Steinbach, P., Glaser, D., Vogel, W., Wolf, M. & Schwemmle, S. (1998). The DMPK gene of severely affected myotonic dystrophy patients is hypermethylated proximal to the largely expanded CTG repeat. *Am J Hum Genet*, Vol.62, No.2, (Feb), pp. 278-85, ISSN.0002-9297

Sutcliffe, J. S., Nelson, D. L., Zhang, F., Pieretti, M., Caskey, C. T., Saxe, D. & Warren, S. T. (1992). DNA methylation represses FMR-1 transcription in fragile X syndrome. *Hum Mol Genet*, Vol.1, No.6, (Sep), pp. 397-400, ISSN.0964-6906

Tabolacci, E., Pietrobono, R., Moscato, U., Oostra, B. A., Chiurazzi, P. & Neri, G. (2005). Differential epigenetic modifications in the FMR1 gene of the fragile X syndrome after reactivating pharmacological treatments. *Eur J Hum Genet*, Vol.13, No.5, (May), pp. 641-8, ISSN.1018-4813

Tassone, F., Hagerman, R. J., Taylor, A. K., Gane, L. W., Godfrey, T. E. & Hagerman, P. J. (2000). Elevated levels of FMR1 mRNA in carrier males: a new mechanism of involvement in the fragile-X syndrome. *Am J Hum Genet*, Vol.66, No.1, (Jan), pp. 6-15, ISSN.0002-9297

Taylor, A. K., Tassone, F., Dyer, P. N., Hersch, S. M., Harris, J. B., Greenough, W. T. & Hagerman, R. J. (1999). Tissue heterogeneity of the FMR1 mutation in a high-functioning male with fragile X syndrome. *Am J Med Genet*, Vol.84, No.3, (May 28), pp. 233-9, ISSN.0148-7299

Thornton, C. A., Wymer, J. P., Simmons, Z., McClain, C. & Moxley, R. T., 3rd (1997). Expansion of the myotonic dystrophy CTG repeat reduces expression of the flanking DMAHP gene. *Nat Genet*, Vol.16, No.4, (Aug), pp. 407-9, ISSN.1061-4036

Verkerk, A. J., Pieretti, M., Sutcliffe, J. S., Fu, Y. H., Kuhl, D. P., Pizzuti, A., Reiner, O., Richards, S., Victoria, M. F., Zhang, F. P. & et al. (1991). Identification of a gene (FMR-1) containing a CGG repeat coincident with a breakpoint cluster region exhibiting length variation in fragile X syndrome. *Cell*, Vol.65, No.5, (May 31), pp. 905-14, ISSN.0092-8674

Watanabe, T., Tomizawa, S., Mitsuya, K., Totoki, Y., Yamamoto, Y., Kuramochi-Miyagawa, S., Iida, N., Hoki, Y., Murphy, P. J., Toyoda, A., Gotoh, K., Hiura, H., Arima, T., Fujiyama, A., Sado, T., Shibata, T., Nakano, T., Lin, H., Ichiyanagi, K., Soloway, P. D. & Sasaki, H. (2011). Role for piRNAs and noncoding RNA in de novo DNA methylation of the imprinted mouse Rasgrf1 locus. *Science*, Vol.332, No.6031, (May 13), pp. 848-52, ISSN.1095-9203

Wheeler, V. C., Lebel, L. A., Vrbanac, V., Teed, A., te Riele, H. & MacDonald, M. E. (2003). Mismatch repair gene Msh2 modifies the timing of early disease in Hdh(Q111) striatum. *Hum Mol Genet*, Vol.12, No.3, (Feb 1), pp. 273-81, ISSN.0964-6906

Wheeler, V. C., Persichetti, F., McNeil, S. M., Mysore, J. S., Mysore, S. S., MacDonald, M. E., Myers, R. H., Gusella, J. F. & Wexler, N. S. (2007). Factors associated with HD CAG repeat instability in Huntington disease. *J Med Genet*, Vol.44, No.11, (Nov), pp. 695-701, ISSN.1468-6244

Winnepenninckx, B., Debacker, K., Ramsay, J., Smeets, D., Smits, A., FitzPatrick, D. R. & Kooy, R. F. (2007). CGG-repeat expansion in the DIP2B gene is associated with the fragile site FRA12A on chromosome 12q13.1. *Am J Hum Genet*, Vol.80, No.2, (Feb), pp. 221-31, ISSN.0002-9297

Wohrle, D., Kennerknecht, I., Wolf, M., Enders, H., Schwemmle, S. & Steinbach, P. (1995). Heterogeneity of DM kinase repeat expansion in different fetal tissues and further expansion during cell proliferation in vitro: evidence for a casual involvement of methyl-directed DNA mismatch repair in triplet repeat stability. *Hum Mol Genet*, Vol.4, No.7, (Jul), pp. 1147-53, ISSN.0964-6906

Yates, P. A., Burman, R. W., Mummaneni, P., Krussel, S. & Turker, M. S. (1999). Tandem B1 elements located in a mouse methylation center provide a target for de novo DNA methylation. *J Biol Chem*, Vol.274, No.51, (Dec 17), pp. 36357-61, ISSN.0021-9258

7

DNA Methylation in Mammalian and Non-Mammalian Organisms

Michael Moffat[1], James P. Reddington[1],
Sari Pennings[2] and Richard R. Meehan[1*]
*[1]MRC Human Genetics Unit, IGMM,
Western General Hospital, Edinburgh,
[2]Queen's Medical Research Institute,
University of Edinburgh, Edinburgh,
UK*

1. Introduction

Of particular interest in biology is how different chromatin states contribute to the complex regulation of gene transcription that is necessary to establish and maintain multi-cellular organisms. This is because (with few exceptions) each cell within an organism contains the same genomic sequence, meaning that the diversity of expression states is not due to variability in the underlying genetic sequence but could result from differences in chromatin landscape. This area of research comes under the umbrella of 'epigenetics', which is concerned with molecular processes involved in regulating gene expression that are transmittable and independent of changes in DNA sequence. However, perhaps owing to the ambiguity of this definition, the term often means different things to different people. A more precise definition of 'epigenetic' mechanisms is 'the structural adaptation of chromosomal regions so as to register, signal or perpetuate altered activity states' (Bird, 2007). Differentiated cells are said to have an 'epigenetic memory' imparted by epigenetic processes that can maintain a pattern of gene transcription through time and cell divisions. Understanding the mechanisms involved in setting up and maintaining these processes is an exciting area of research investigation.

The nucleosome is the basic repeating unit of chromatin and consists of 146bp of DNA wrapped 1.7 times around the histone octamer, which is composed of two molecules of each of four types of histone protein: H2A, H2B, H3 and H4 (An et al., 1998; Zlatanova et al., 2009). The core histones each have a distinct C-terminal, a structured globular domain and a flexible, unstructured N-terminal tail, which protrudes from the nucleosome (Luger et al., 1997; Schroth et al., 1990). The repressive effect of nucleosomes on transcription can be enhanced or reduced by combinations of histone post-translational modifications (PTM). Histone acetylation, methylation, ubiquitylation and other modifications play crucial roles in diverse biological processes, such as embryogenesis, development and maintenance of genome integrity. Recently, an integrated, mass spectrometry-based proteomics approach

* Corresponding Author

resulted in the identification of over 67 new PTM sites in histones including lysine crotonylation, expanding the total number of known histone PTMs by about 70% (Tan et al., 2011). Histone PTMs create docking sites for non-histone effector proteins that that can subsequently modify chromatin structure (Andrews & Luger, 2011). However in some cases, charge changes resulting from modifications can alter chromatin structure directly by disrupting the DNA-histone interaction. Linker histone modifications can also influence higher order chromatin structure and thus alter gene expression states.

Modification of DNA can alter its biological properties and involves enzymatic mechanisms that are sustained through cellular replication. Particular molecular signatures of DNA together with histone modifications are associated with active and repressed chromatin states (Barski et al., 2007). DNA methylation patterns are developmentally regulated and are thought to define tissue states in plants and animals (Feng et al., 2010b). In cancer as well as in embryos generated through somatic cell nuclear transfer, normal patterns of DNA methylation are altered implying that precise molecular pathways are involved in setting up and maintaining diverse patterns of modification (Hochedlinger & Jaenisch, 2006).

1.2 New modifications in the genome

The major form of epigenetic information in mammalian genomes is centred on DNA methylation, which now comes in the forms of 5-methylcytosine (5mc), 5-hydroxymethylcytosine (5hmC) and the more recently discovered 5-carboxylcytosine (5caC) and 5-formylcytosine (5fC) (He et al., 2011; Ito et al., 2011). The observed levels of both 5caC and 5fC are extremely low in ES cells (~3 5caC and ~18 5fC for every 10^6 C) (Ito et al., 2011) and these may represent transient intermediates in a demethylation pathway. In contrast to these low levels, far greater levels of the 5hmC modification were observed in ES cells - indicating that this mark may have additional functional roles. The presence of 5hmC, 5caC and 5fC in genomes is dependent on 5mC which is the substrate for conversion by the TET family (1-3) of Fe(II) and α-KG-dependent dioxygenases, which utilise molecular oxygen to convert 5mC to 5hmC, 5fC and 5caC (He et al., 2011; Ito et al., 2010; Ito et al., 2011; Ko et al., 2010; Koh et al., 2011; Tahiliani et al., 2009; Wossidlo et al., 2011).

1.3 The DNA methylation machinery

To generate 5mC, a methyl group is added covalently to the 5 position of cytosine by DNA cytosine methyltransferases (DNMTs), mostly within the context of CpG dinucleotides in somatic cells; however, non-CpG methylation also occurs at a high frequency in mouse and human embryonic stem (ES) cells (Lister et al., 2009;Ramsahoye et al., 2000). Non-CpG methylation may be a feature of the pluripotent state, as it is present in induced Pluripotent Stem (iPS) cells generated by transduction of a non-pluripotent somatic cell with stem cell-associated genes, which results in reprogramming of the recipient cell's epigenetic profile (Takahashi et al., 2007). The importance of 5hmC and its cousins in epigenetics is that the hydroxymethyl group is suggested to alter the biological properties of methylated DNA (Ndlovu et al., 2011). The rediscovery of 5hmC also presents an unanticipated experimental problem, as conventional techniques were originally unable to distinguish between 5mC and 5hmC in DNA (Nestor et al., 2010). Recent technical developments can now distinguish prominent 5hmC sites in the genome (C.X. Song et al., 2011). However, it is clear that 5hmC is less abundant than 5mC, and the latter is still the most prominent modification in vertebrate DNA in many tissues. 5mC values are stable at a typical value of around 4.5% of all cytosine in tissues, whereas 5hmC values vary significantly (Munzel et al., 2010). This suggests that 5hmC has a specific function that is not absolutely correlated with 5mC levels. Initial analysis suggests that 5hmC is predominantly associated with the gene bodies of highly expressed genes (C.X. Song et al., 2011).

The presence of DNA methylation at regulatory sequences in somatic cells is generally associated with transcriptional repression, and potentially has a long term impact on the stability of gene expression states and on genome integrity (Sharma et al., 2010). Alterations in genomic methylation patterns underpin imprinting syndromes such as Beckwith-Wiedemann, Prader-Willi and Angelman, and have been implicated in a number of other disease conditions including cancer (Goll & Bestor, 2005). The enzymes responsible for targeting and maintaining global DNA methylation in mammals are constructed from a complex set of functional modules, broadly divided into the N-terminal 'regulatory' domain and the C-terminal 'catalytic' domain. The regulatory domain acts as an interaction platform for protein interactions, DNA binding and mediates its differential nuclear targeting during the cell cycle (Goll & Bestor, 2005). Not surprisingly, the localisation of the maintenance methyltransferase, Dnmt1, in mammals is co-ordinated with DNA replication so that newly synthesised hemi-methylated DNA is rapidly and fully methylated. Three methyltransferase enzymes, Dnmt1 along with the *de novo* methyltransferases Dnmt3a and Dnmt3b, coordinate the establishment and maintenance of DNA methylation patterns in mammals. The C-terminal domain of each enzyme comprises ten motifs responsible for the enzyme's catalytic activity; six of these motifs are conserved in nearly all cytosine methyltransferases from bacteria to mammals. Dnmt3a and Dnmt3b target cytosine methylation to previously unmethylated CpG dinucleotides, which the Dnmt1 preserves during cell division. Dnmt3a and 3b are thought to act with an equal preference for hemimethylated and unmethylated DNA *in vitro*, but *in vivo* they have differential targets which may be mediated by partner proteins such as transcription factors (Hervouet et al., 2009). They are necessary for *de novo* methylation of the genome during development and potentially newly integrated retroviral sequences (Okano et al., 1998b, 1999).

The N-terminal domain of Dnmt1 interacts with many chromatin-associated proteins including the *de novo* methyltransferases, methyl-CpG binding proteins (MeCPs) and histone modifying enzymes (Qin et al., 2011). It also contains a replication targeting region and a cysteine-rich Zn2+-binding domain that can potentially bind non-methylated CG rich DNA. Binding of the CXXC domain to unmethylated CpG DNA is thought to result in a repositioning of the CXXC-BAH1 linker between the DNA and the active site of DNMT1, thereby preventing *de novo* methylation (J. Song et al., 2011). In addition, a loop projecting from the BAH2 domain interacts with the target recognition domain (TRD), stabilising it in a retracted position so it cannot access the DNA major groove. Hemimethylated CpG dinucleotides that do not bind the CXXC domain can gain access to the active site of Dnmt1 by bypassing this molecular mechanism. Biochemical and molecular analyses of Dnmt1 suggest that it participates in multiple complex networks involved in gene regulation, epigenetic signalling and genome stability via the mismatch repair pathway. Dnmt1 is also post-translationally modified by the protein lysine methyltransferase SET7 which regulates its stability (Esteve et al., 2009). This modification on Lysine 142 is mutually exclusive with phosphorylation on Ser143; phosphorylated Dnmt1 is more stable than its methylated version (Esteve et al., 2011).

Dnmt3b is specialised in methylation of specific regions of the genome, such as pericentromeric repeats and CpG islands on the inactive X-chromosome, whereas Dnmt3a is required for maternal imprints of differentially methylated regions (DMRs), in addition to their general *de novo* roles (Kim et al., 2009). The PWWP domain of Dnmt3a specifically recognises the histone 3 lysine 36 trimethylation mark and this may be important for its subnuclear localisation (Dhayalan et al., 2010). Deletion of Dnmt3a in primordial germ cells disrupts paternal and maternal imprinting, whereas Dnmt3b is dispensable for mouse gametogenesis and imprinting (Kaneda et al., 2010;Kato et al., 2007). Protein interaction domains in the regulatory N-termini of Dnmt3a and Dnmt3b also mediate binding to transcriptional co-repressors (Qiu et al., 2002). Unlike Dnmt1 and Dnmt3a/b, the DNA methyltransferase Dnmt2 has only weak activity *in vitro* towards DNA, and its inactivation does not result in alterations to global CpG methylation levels (Okano et al., 1998b). A cofactor, Dnmt3L (DNMT3-Like), is expressed only in germ and ES cells. It is not a methyltransferase but enhances the *de novo* methyl transferase activity of Dnmt3a and 3b in mouse ES cells (Ooi et al., 2010).

1.4 Role of DNA methylation in mammals

In general, repression by DNA methylation is considered to occur downstream of other epigenetic or trans-acting factors that signal the initial inactivation event. For example, initial repression of Pou5f1 during differentiation of mouse ES cells is mediated by sequence-specific transcription repressors such as GCNF leading to conversion of the 'active' histone modification state to an inactive one that is subsequently followed by *de novo* DNA methylation at its promoter (Cedar & Bergman, 2009). The number of potential genes that can be directly regulated by DNA methylation in a tissue and developmental specific manner may be quite small corresponding to 100–200 of annotated CpG island (CGI) genes in somatic cells (Meissner et al., 2008). However, new data suggests there are approximately 23,000 and 25,500 CGIs in the mouse and human genomes respectively, about half of which are associated with annotated transcription start sites for mainly constitutively expressed genes (Illingworth et al., 2010). The non-annotated or 'orphan' CGI's show higher levels of tissue specific methylation (14-20%) and may be directly regulated by DNA methylation in different tissues and developmental stages. The importance of the preservation of these patterns is highlighted by the observation that *de novo* methylation of promoter CGIs associated with tumour suppressor genes occurs in many neoplastic cells (Sharma et al., 2010). At the same time as *de novo* methylation of CGIs, global methylation levels associated with satellite repeats and retroposons are often reduced in cancers (Sharma et al., 2010). Recent results suggest that retrotransposons mobilise to protein-coding genes that are differentially expressed and active in the brain; suggesting that retrotransposition may result in somatic genome mosaicism and alteration in the genetic circuitry that underpins normal and abnormal neurobiological processes (Baillie et al., 2011).

1.5 Histone modifications and gene regulation

The combination of histone PTMs and their resulting effects on gene expression is often referred to as the "Histone Code" (Turner, 2007). Trimethylation of lysine 4 on histone H3 (H3K4me3) is enriched at transcriptionally active gene promoters, whereas trimethylation of H3K9 (H3K9me3) and H3K27 (H3K27me3) is present at inactive gene promoters. H3K9me3 can function in concert with DNA methylation whereas H3K27me3 may be exclusive of DNA methylation. Genome-wide studies of these histone marks in the genome have increased our understanding of how these diverse modifications act in a cooperative manner to regulate gene expression (Sharma et al., 2010). The polycomb complex (PRC), which mediates trimethylation of lysine 27 on histone H3 (H3K27me3) appears to be targeted specifically to genes involved in development and differentiation (Mikkelsen et al., 2007). Heterochromatin protein (HP1) binds H3K9me2/3 containing chromatin through its chromodomain (Dialynas et al., 2008). Interaction partners for HP1 include DNMT1, the histone H3K9 methyltransferases Suvar39H1 and G9a (Esteve et al., 2006; Smallwood et al., 2007), which may coordinate DNA and H3K9 methylation at genomic loci.

Binding of the PRC2 complex to specific genes, such as the Hox cluster, results in trimethylation of histone H3K27 by the histone methyltransferase EZH2 (Morey & Helin, 2010). However, polycomb target genes in ES cells can have a bivalent chromatin signature, being also marked by the activating modification H3K4me3 (Barski et al., 2007; Mikkelsen et al., 2007). These marks may be resolved as development proceeds leading to developmental and tissue specific patterns of gene expression (Barski et al., 2007; Boyer et al., 2006). Mature heterochromatin HP1alpha and H4K20me3 signatures do not arise until late in development (Wongtawan et al., 2011). Like DNA methylation, gene silencing via histone modification can be maintained *in vivo* through multiple cell divisions. It has been reported that CGIs that are aberrantly methylated in cancer cells coincide with sites targeted by polycomb in human ES cells (Schlesinger et al., 2007). Approximately 50% of tumor-specific methylated CGIs are H3K27 trimethylated in ES cells (Illingworth et al., 2010). These findings suggest that the mechanisms governing tumour-specific and normal directed CGI methylation are distinct.

It becomes clear that as our knowledge of the DNA methylation regulatory system in mammals and their component parts deepens, that this system functions out with the enzymatic modification of cytosine to its modified forms. In other words many of the enzymes involved have non-catalytic functions. Potentially this occurs in ways that we cannot predict, which is why study of DNA modification systems in other animal model systems may add mechanistic and biological insight into the role of DNA modification pathways in development and disease. In the following we undertake a short review of the role of DNA modification (primarily 5mC) in 4 organisms: frog, zebrafish, chicken, and honeybee. Now is a particularly relevant time to review 5mC in these organisms, with each of them benefitting from at least one recently completed genome-wide methylome.

2. Zebrafish – *Danio rerio*

Initial evidence for the existence of 5mC in zebrafish came from transgenesis studies using zebrafish. The expression of a chloramphenicol acetyltransferase (CAT) transgene was found to be variegating; an expression pattern that would be consistent with transgene methylation (Stuart et al., 1990). It was subsequently shown that treatment with 5-azacytidine (an anologue of cytidine, with methyltransferase inhibitory action (Friedman, 1979)) significantly increased the expression of transgenes, strongly implying that transgene repression could be mediated by DNA methylation. The above experiment suggested that a working *de novo* methylation process was present in zebrafish, and this was confirmed when the CpG island of the *ntl* (notail) gene of zebrafish larvae was found to undergo *de novo* methylation (Yamakoshi & Shimoda, 2003). As discussed earlier, the DNMT enzymes can be categorised as either *de novo* or maintenance in function. Surprisingly, the zebrafish genome contains at least 8 potential DNMTs (see Table 1 and Figure 1), including the tRNA methylase Dnmt2, but does not have an obvious DNMT3L homologue.

DNMT proteins

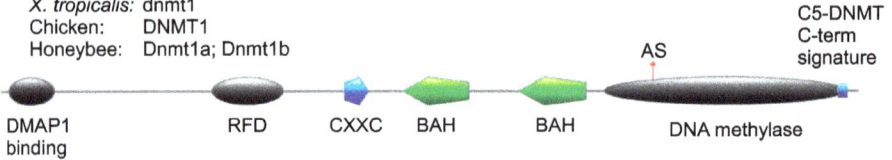

DNMT1
Zebrafish: Dnmt1
X. tropicalis: dnmt1
Chicken: DNMT1
Honeybee: Dnmt1a; Dnmt1b

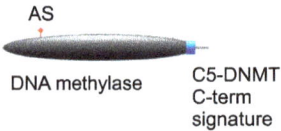

TRDMT1 (DNMT2)
Zebrafish: Trdmt1
X. tropicalis: trdmt1
Chicken: TRDMT1
Honeybee: Dnmt2

DNMT3L
Zebrafish: not present
X. tropicalis: not present
Chicken: not present
Honeybee: not present

DNMT3A
Zebrafish: Dnmt6; Dnmt8
X. tropicalis: dnmt3a
Chicken: DNMT3A
Honeybee: not present

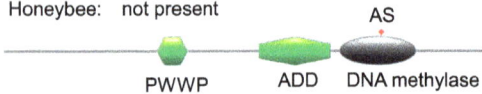

DNMT3B
Zebrafish: Dnmt3; Dnmt4; Dnmt5; Dnmt7
X. tropicalis: not present
Chicken: DNMT3B
Honeybee: Dnmt3

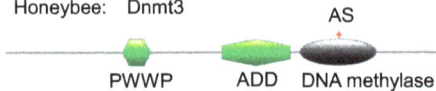

Fig. 1. **DNMT proteins.** Domains were identified by searching PROSITE (http://prosite.expasy.org/) and the Sanger pfam database (http://pfam.sanger.ac.uk/) using the sequences belonging to the accession numbers shown. Diagrams were constructed using the output from PROSITE searches, and pfam output added to the diagrams using the MyDomains tool in PROSITE. ADD: ATRX-DNMT3-DNMT3L domain; AS: active site; BAH: Bromo-adjacent homology; CXXC: CXXC zinc finger domain; PWWP: Pro-Trp-Trp-Pro domain; RFD: Cytosine specific DNA methyltransferase replication foci domain

MBD proteins

MBD1
Zebrafish: Mbd1
X. tropicalis: mbd1
Chicken: not present
Honeybee: not present

MBD   CXXC CXXC

MBD2
Zebrafish: Mbd2
X. tropicalis: mbd2
Chicken: MBD2
Honeybee: not present

MBD

MBD3
Zebrafish: Mbd3a; Mbd3b
X. tropicalis: mbd3
Chicken: MBD3
Honeybee: Mbd3

MBD

MBD4
Zebrafish: Mbd4
X. tropicalis: mbd4
Chicken: MBD4
Honeybee: not present

MBD   HhH-GPD

MBD5
Zebrafish: Mbd5
X. tropicalis: mbd5
Chicken: MBD5
Honeybee: not present

MBD

PWWP

MBD6
Zebrafish: Mbd6
X. tropicalis: mbd6
Chicken: not present
Honeybee: not present

MBD

MeCP2
Zebrafish: MeCP2
X. tropicalis: mecp2
Chicken: MeCP2
Honeybee: not present

MBD

Fig. 2. **MBD proteins in mouse.** Domains were identified by searching PROSITE (http://prosite.expasy.org/) and the Sanger pfam database (http://pfam.sanger.ac.uk/) using the sequences belonging to the accession numbers shown. Diagrams were constructed using the output from PROSITE searches, and pfam output added to the diagrams using the MyDomains tool in PROSITE. CXXC: CXXC zinc finger domain; HhH-GPD: Helix-hairpin-helix-gly-pro-asp superfamily base excision repair domain. MBD: methyl binding domain; PWWP: Pro-Trp-Trp-Pro domain

TET proteins

TET1
Zebrafish: Tet1
X. tropicalis: not present
Chicken: TET1
Honeybee: Tet1

Oxygenase

CXXC   Cys   DSBH   DSBH
                        rich
                        domain

TET2
Zebrafish: Tet2
X. tropicalis: tet2
Chicken: TET2
Honeybee: not present

Oxygenase

Cys   DSBH   DSBH
rich
domain

TET3
Zebrafish: Tet3
X. tropicalis: tet3
Chicken: TET3
Honeybee: not present

Oxygenase

Potential   Cys
CXXC   rich   DSBH
                        domain

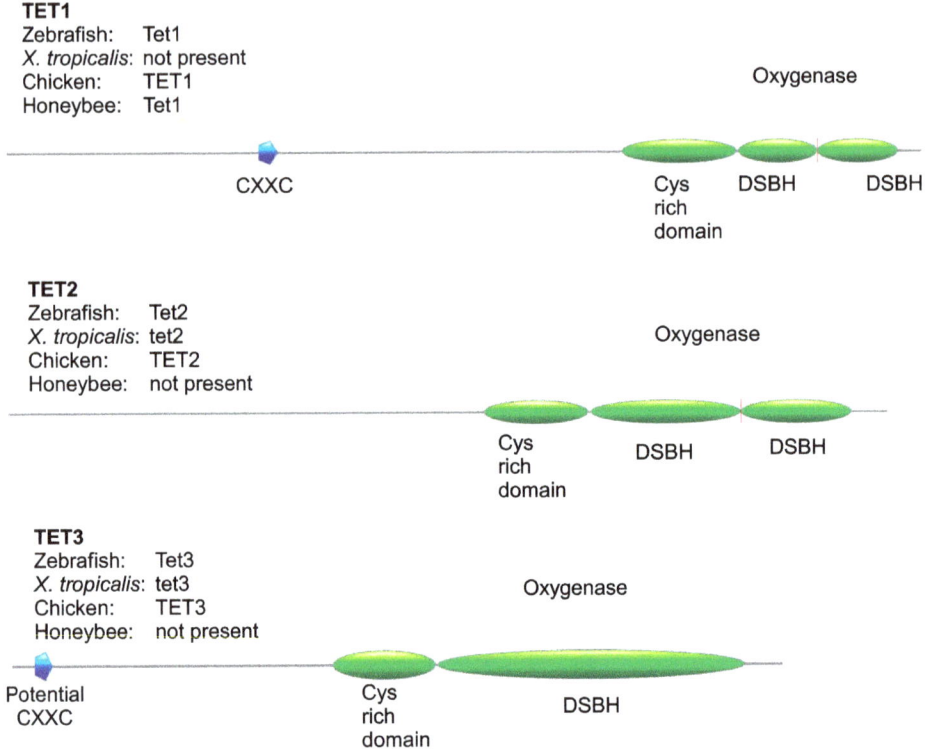

Fig. 3. **TET proteins in mouse.** Domains were identified by searching PROSITE
(http://prosite.expasy.org/) and the Sanger pfam database (http://pfam.sanger.ac.uk/)
using the sequences belonging to the accession numbers shown. Diagrams were constructed
using the output from PROSITE searches, and pfam output added to the diagrams using the
MyDomains tool in PROSITE. CXXC: CXXC zinc finger domain; DSBH: double stranded
beta helix fold of oxygenase domain of TET proteins.

A homology search revealed that 6 of the zebrafish DNMTs exhibit a high degree of
similarity to DNMT3A/B, termed Dnmt3-8 (Shimoda et al., 2005). Interestingly, knockdown
experiments suggested that Dnmt7 was responsible for *de novo* methylation of *ntl* but not
other forms of *de novo* methylation, such as at transgenes (Shimoda et al., 2005). The
homologues most similar to Dnmt3A are Dnmt6 and Dnmt8, and it is of note that Dnmt3A
is expressed as 2 major isoforms in mammals. The expression profile of zebrafish Dnmt6
and Dnmt8 has been shown to be similar to DNMT3A; while Dnmt3-4,7 is similar to
DNMT3B (Smith et al., 2011). This suggests that the 6 potential *de novo* methyltransferases
are functionally equivalent to DNMT3A/B in mammals.

In mammals, *de novo* methylation at imprinted regions is dependent upon DNMT3L
(Bourc'his et al., 2001). Zebrafish do not possess any DNMT3L homologue, but are also not

known to use imprinting extensively. Interestingly, it has been reported that a transgene in the zebrafish shows parent-of-origin dependent DNA methylation (Martin & McGowan, 1995). In line with this, parent-of-origin effects have been reported in interspecific crosses of

ZBTB proteins

ZBTB33 (Kaiso)
Zebrafish: Zbtb33
X. tropicalis: zbtb33
Chicken: ZBTB33
Honeybee: not present

BTB   3x C2H2 ZF

ZBTB4
Zebrafish: Zbtb4
X. tropicalis: not present
Chicken: not present
Honeybee: not present

BTB   C2H2 ZF 3x C2H2 ZF   2x C2H2 ZF

ZBTB38
Zebrafish: Zbtb38
X. tropicalis: zbtb38
Chicken: ZBTB38
Honeybee: not present

3x C2H2 ZF

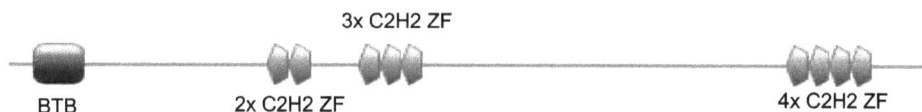

BTB   2x C2H2 ZF   4x C2H2 ZF

Fig. 4. **methyl binding ZBTB proteins in mouse.** Domains were identified by searching PROSITE (http://prosite.expasy.org/) and the Sanger pfam database (http://pfam.sanger.ac.uk/) using the sequences belonging to the accession numbers shown. Diagrams were constructed using the output from PROSITE searches.

birds, frogs, and fishes. However, in zebrafish and frogs it is possible to create viable uniparental diploids (Cheng & Moore, 1997). Recent work (Gertz et al., 2011) suggests that in humans 8% of heterozygous SNPs are associated with differential methylation in *cis*. In these cases, the vast majority of differential methylation between homologous chromosomes (>92%) occurs on a particular haplotype, as opposed to being associated with the gender of the parent of origin. This indicates that genotype affects DNA methylation far more than gametic imprinting does. Overall, this suggests that the influence of genotype on patterns of DNA methylation is widespread in the genome, and greatly exceeds the influence of imprinting on genome-wide methylation patterns. DNMT3L has also been shown to be important for the methylation of transposable elements (Bourc'his & Bestor, 2004), which are reported to be methylated in zebrafish (Feng et al., 2010a). It would be of interest to find whether any of the previously mentioned zebrafish Dnmts can methylate transposable elements, and whether their perturbation results in lack of methylation and reactivation of such elements. The existence of alternative methylation mechanisms in zebrafish is made more appealing by the existence of apparent calponin homology domains in Dnmt3 and Dnmt7 (Figure 5), although what these mechanisms could be is currently a mystery and the role (if any) of these domains in Dnmt function is currently unknown.

Organism	DNMTs	MBDs	TETs	ZBTBs
Mouse	DNMT1 [ENSMUSP00000004202] TRDMT1 (DNMT2) [ENSMUSP00000114572] DNMT3A [ENSMUSP00000020991] DNMT3B [ENSMUSP00000051830] DNMT3L [ENSMUSP00000121562]	MBD1 [ENSMUSP00000025446] MBD2 [ENSMUSP00000073701] MBD3 [ENSMUSP00000089948] MBD4 [ENSMUSP00000032469] MBD5 [ENSMUSP00000036847] MBD6 [ENSMUSP00000026476] MeCP2 [ENSMUSP00000033770]	TET1 [ENSMUSP00000133279] TET2 [ENSMUSP00000043977] TET3 [ENSMUSP00000087049]	ZBTB4 [ENSMUSP00000104279] ZBTB33 [ENSMUSP00000110795] ZBTB38 [ENSMUSP00000121753]
Zebrafish	Dnmt1 [ENSDARP00000013243] Trdmt1 (Dnmt2) [ENSDARP00000048632] Dnmt3 [ENSDARP00000110904] Dnmt4 [ENSDARP00000053417] Dnmt5 [ENSDARP00000118597] Dnmt6 [ENSDARP00000104553] Dnmt7 [ENSDARP00000108732] Dnmt8 [ENSDARP00000029456]	Mbd1 [ENSDARP00000031774] Mbd2 [ENSDARP00000104037] Mbd3a [ENSDARP00000082941] Mbd3b [ENSDARP00000048774] Mbd4 Mbd5 [ENSDARP00000077389] Mbd6 [ENSDARP00000105745] MeCP2 [ENSDARP00000109399]	Tet1 [ENSDARP00000104252] Tet2 [ENSDARP00000101295] Tet3 [ENSDARP00000115297]	Zbtb4 [ENSDARP00000068451] Zbtb33 [ENSDARP00000096464] Zbtb38 [ENSDARP00000102342]
Chicken	DNMT1 [Q92072] TRDMT1 (DNMT2) [ENSGALP00000014121] DNMT3A [ENSGALP00000006352] DNMT3B [Q4W5Z3]	MBD2 [NP_001012403.1] MBD3 [ENSGALP00000001652] MBD4 [F1P366] MBD5 [F1NE18] MeCP2 [O42403]	TET1 [XP_421571.2] TET2 [ENSGALP00000017216] TET3 [ENSGALP00000021873]	Zbtb33 [ENSGALP00000038290] Zbtb38 [ENSGALP00000004524]
Xenopus tropicalis	dnmt1 [ENSXETP00000047703] trdmt1 (dnmt2) [ENSXETP00000057609] dnmt3a [ENSXETP00000002659]	mbd1 [Q66JI6] mbd2 [ENSXETP00000029954] mbd3 [ENSXETP00000029788] mbd4 [Q28HB5] mbd5 [ENSXETP00000062167] mbd6 [ENSXETP00000053901] mecp2 [ENSXETP00000003337]	tet2 [ENSXETP00000030770] tet3 [ENSXETP00000054061]	Zbtb33 [ENSXETP00000009735] Zbtb38 [ENSXETP00000056311]
Honeybee	Dnmt1a [GB19865-PA] Dnmt1b [GB15130-PA] Dnmt2 [GB10767-PB] Dnmt3 [D7RIF7]	Mbd3 [XP_392422.2]	Tet1 [GB13880-PA]	(none)

Table 1. DNMTs, MBDs, TETs and ZBTBs found in mouse, zebrafish, chicken, *X. tropicalis* and honeybee.

Maintenance of DNA methylation is generally more widely studied than *de novo* methylation. In mice, hemimethylated DNA is recognised and bound by UHRF1, which is required for the subsequent recruitment of DNMT1 (Sharif et al., 2007). Both of these proteins are conserved in zebrafish, and mutation of either results in a global reduction of 5mC levels (Goll et al., 2009;Tittle et al., 2011). Furthermore, Uhrf1 mutation phenocopies certain aspects of Dnmt1 mutation. However, it is not clear if there are similar methylation dependant patterns of misexpression in these mutants.

Sequencing of the zebrafish genome allowed a genome wide profile of 5mC to be obtained. Bisulfite sequencing (BS-seq) was used to profile the distribution and abundance of 5mC within 5dpf zebrafish embryos (Feng et al., 2010a). Methylation was found in all 3 sequence contexts – CpG (80.3%), CHG (1.22%), CHH (0.91%); where H stands for C, T or A. These levels were higher than is seen in mouse embryos for each category (CpG, 74.2%; CHG, 0.30%; and CHH, 0.29%); but roughly equivalent to E14 mouse ES cells (Feng et al., 2010a).

Non-Mammalian Methylation Protein Anomolies

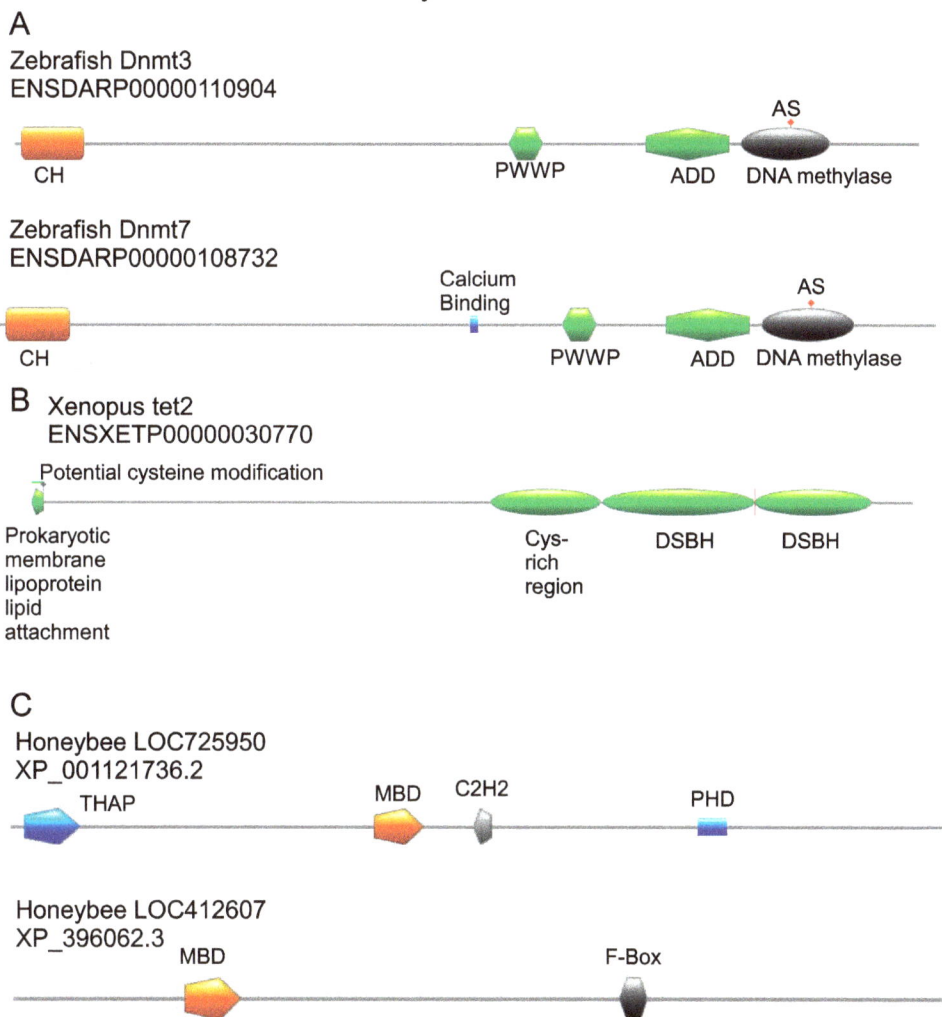

A

Zebrafish Dnmt3
ENSDARP00000110904

CH   PWWP   ADD   DNA methylase   AS

Zebrafish Dnmt7
ENSDARP00000108732

CH   Calcium Binding   PWWP   ADD   DNA methylase   AS

B Xenopus tet2
ENSXETP00000030770

Potential cysteine modification

Prokaryotic membrane lipoprotein lipid attachment   Cys-rich region   DSBH   DSBH

C

Honeybee LOC725950
XP_001121736.2

THAP   MBD   C2H2   PHD

Honeybee LOC412607
XP_396062.3

MBD   F-Box

Fig. 5. **Non-mammalian methylation protein variants.** Domains were identified by searching PROSITE (http://prosite.expasy.org/) and the Sanger pfam database (http://pfam.sanger.ac.uk/) using the sequences belonging to the accession numbers shown. Diagrams were constructed using the output from PROSITE searches, and pfam output added to the diagrams using the MyDomains tool in PROSITE. A: Zebrafish Dnmt3 and Dnmt7 each contain a calponin homology (CH) domain of unknown function, not seen in other species. Dnmt7 also contains a calcium binding domain. B: *Xenopus tropicalis* tet2 contains a potential lipid attachment site normally seen in prokaryotes. C: In *Apis mellifera*, MBDs are present in architectures found in other flying insects, but not other animals (MBD-FBOX and THAP-MBD / THAP-MBD-PHD)

Especially interesting is the increase in non-CpG methylation seen in zebrafish embryos compared to mouse embryos, equivalent to a 3-4 fold increase; but similar to mouse ES cells. BS-seq showed that 5mC is modestly enriched at repetitive elements and within gene bodies, but depleted at CpG islands covering TSSs. Overall, the organisation of the 5mC methylome is similar in all vertebrates, which have methylation throughout the genome except at CpG islands (Feng et al., 2010a). Gene body methylation is conserved with clear preference for exons in most organisms.

Mammalian development is known to be highly dependent upon methylation and there are many studies which also highlight the importance of methylation in zebrafish. First, the apparent presence of maternal Dnmt1 in zebrafish oocytes points towards an essential role of 5mC in zebrafish development (Goll et al., 2009). Treatment of embryos with 5-azacytidine resulted in abnormal tail development, and incorrect patterning of somites (Martin et al., 1999). Morpholino knockout (zdnmt1MO) of Dnmt1 in zebrafish embryos has been shown to interfere with terminal differentiation of the intestine, exocrine pancreas and retina; but not the liver or endocrine pancreas (Rai et al., 2006). This is interesting as many embryos survive past gastrulation, unlike in mice. Surviving zdnmt1MO embryos exhibit developmental defects including curled tails, pericardial oedema, and jaw defects; with eyes appearing normal. This may suggest altered dependence on DNA methylation or Dnmts in zebrafish, or alternative mechanisms of DNA methylation maintenance in non-mammalian animals. It chimes with the situation in *Xenopus*, where certain key roles of DNMT1 do not depend on it methylating DNA (Dunican et al., 2008; Stancheva et al., 2001). Genetic analysis reinforces this difference, as zebrafish mutants in uhrf1 and dnmt1 also have defects in lens development and maintenance and they are not embryonic lethal, despite being hypomethylated (Anderson et al., 2009; Tittle et al., 2011). This suggests that attributing developmental defects in these mutants to hypomethylation may be premature, as the biology may be more complex. Here it is worth noting that Lsh mutants in mice are as globally hypomethylated as Dnmt1 mutants, yet certain Lsh -/- mutant mice exhibit growth retardation and a premature aging phenotype (Sun et al., 2004). This may be partly due to the different classes of gene misexpression compared to Dnmt1 mutants in mice (Myant et al., 2011). This implies that the role of DNA methylation and its associated components is still unclear, although DNA methylation patterns at promoters are highly correlated with transcription state.

Demethylation of DNA may occur either actively or passively, with passive demethylation occurring via the absence of methylation maintenance upon cell division and active methylation via direct enzymatic action resulting in the reversion of 5mC to C (Wu & Zhang, 2010). Active demethylation in the mouse is most obvious during early development, where detection of 5mC is rapidly lost in the male pronucleus (Santos et al., 2002), and also appears to occur during PGC development (Monk et al., 1987). Whether an equivalent process occurs within zebrafish is still unclear. Early experiments looking for 5mC changes during early zebrafish development did not detect any (Macleod et al., 1999), though this investigation was not comprehensive; while a more recent report did (Mhanni & McGowan, 2004). Promising results were obtained using immunological methods in zebrafish, and it was seen that methylation is lost around 1.5-2 hpf (cleavage and early blastula stages), and increases by 4 hpf (MacKay et al., 2007). This result is particularly encouraging, as the study made use of a similar technique to that used to obtain positive results in mouse (Santos et al., 2002). The authors suggest possible explanations for negative results reported by other groups; including the relevant time point being missed, and potential contamination of samples with hypermethylated mitochondrial DNA (MacKay et al., 2007). In any case, the lack of large scale demethylation events would not prevent the existence of active demethylation, and there is evidence that such non-global processes occur in zebrafish. It has been shown that a methylated plasmid is demethylated when inserted into a one-cell zebrafish embryo, independently of DNA replication (Collas, 1998). It was also reported that the demethylase activity was dependent upon RNA, due to its inhibition by RNaseH. Upregulation of both MBD4 and AID results in global DNA demethylation, and both are found to be recruited to methylated DNA by Gadd45a (Rai et al., 2008). AID is a 5mC-deaminase which converts 5mC to T; and MBD4 has G:T mismatch thymine glycosylase. These enzymes are thus expected to promote demethylation via the conversion of 5mC to T, and the subsequent substitution of T to C. Interestingly, Gadd45a has been shown to bind neither single or double stranded DNA whether methylated or unmethylated (Sytnikova et al., 2011). However, RNA binding activity was observed, indicating that recruitment of the demethylation couplet may proceed via an RNA intermediate, explaining the RNA dependency reported.

Mammalian demethylation studies have recently benefited from a flurry of papers detailing 5hmC and its potential role in demethylation. This base is produced from the further modification of 5mC by TET enzymes (TET1-3), and appears to allow demethylation to occur through a series of intermediate conversions (He et al., 2011; Ito et al., 2011; Kriaucionis & Heintz, 2009; Tahiliani et al., 2009). It has been suggested that 5hmC is converted to 5-formylcytosine (5fC), then 5-carboxylcytosine (5caC); which is subsequently substituted for cytosine by the action of TDG (Ito et al., 2011; He et al., 2011). This pathway is yet to be studied in zebrafish, though at least one unpublished report exists claiming the presence of 5hmC in zebrafish (Yen & Jia, 2010). All 3 mammalian TET proteins have homologous sequences in zebrafish (see Table 1). Mammalian 5hmC could be suggested to have a dual functional role, as it does not appear to be immediately converted, and is present at considerably greater levels than 5fC and 5caC. Such roles could include repression of transcription SIN3A recruitment (Williams et al., 2011) and hypomethylation of target regions allowing PRC2 recruitment (Wu et al., 2011). It will be interesting to explore the distribution of 5hmC and TET binding/expression profiles in zebrafish. Immunostaining of mitotic chromosome spreads of mouse pre-implantation embryos demonstrated that paternal 5hmC is gradually lost during pre-implantation development (Inoue & Zhang, 2011). Here it is suggested that although the conversion of 5mC to 5hmC in zygotes is an enzyme-catalysed process, loss of 5hmC during pre-implantation may be a DNA replication dependent but passive process.

3. Chicken – *Gallus gallus*

DNA methylation has long been studied in the chicken, with the globin genes being the primary model of study; and is complimented by structural studies (Heitmann et al., 2003; Scarsdale et al., 2011).

The first evidence for functionally relevant DNA methylation in chicken came when methylation sensitive restriction enzymes were used to study the methylation profile around the chicken beta globin gene (McGhee & Ginder, 1979). Lack of methylation was observed in tissues expressing the gene; while tissues not expressing it showed high methylation in the region. Soon after this initial study, it was found that around the genes encoding ovalbumin, conalbumin, and ovomucoid; DNA methylation was reduced in the oviduct when they are expressed (Mandel & Chambon, 1979). It was also noted that the highest levels of methylation were found in sperm. The observation that methylation is particularly high in sperm was replicated in a study reporting a negative correlation between methylation and expression of the adult and embryonic alpha globin genes during chicken development (Haigh et al., 1982). Despite these results linking methylation to expression for globin genes, *in vivo* demethylation by 5-azacytidine treatment in adult chickens did not result in reactivation of embryonic alpha globin (Ginder et al., 1983); it was also subsequently shown that a difference in methylation profile does not necessarily correspond to a difference in expression level (Cooper et al., 1983). The link between methylation and expression was nevertheless enhanced by experiments showing by HPLC and restriction analysis that chicken methylation profiles appear to change in an age- and tissue-dependent manner (Harasawa & Mitsuoka, 1984). This idea is strengthened by a more recent study showing significant differences in global methylation levels between different chicken tissues (Xu et al., 2007), but this may only be true for a subset of genes, as no correlation was observed between methylation of the lysozyme promoter and its transcriptional activity (Wolfl et al., 1991).

Structural evidence for a link between methylation and expression of target genes in chicken came when it was shown that the presence of methylation at just 3 CpGs in the beta globin promoter was sufficient to exclude binding of histone proteins (Davey et al., 1997). The finding that a small number of CpGs could be the major determinant in gene expression changes was also seen when chickens were transfected with GFP under the control of the RSV promoter (Park et al., 2010). It was found that GFP expression varied between tissues, but didn't appear to have inserted into a tissue specific gene cluster. To test if methylation was involved in this, the methylation status of the promoter was tested, and was seen to be slightly lower in tissues with higher GFP expression, but mainly at the set of CpGs at the very start of the promoter, where the majority of CpGs were unmethylated.

The methylation of the GFP reporter construct demonstrates *de novo* methylation in chicken. The chicken genome contains homologues of each mammalian DNMT (Table 1) with the exception of DNMT3L (Yokomine et al., 2006) each having high conservation with mammalian proteins; with DNMT1 having 94% identical amino acid sequence (Tajima et al., 1995).

The chicken genome consists of 39 chromosomes, 33 of which are classed as microchromosomes (McQueen et al., 1998). These microchromosomes are gene rich, early replicating and enriched for CpG islands (McQueen et al., 1996, 1998); but remain incompletely sequenced due technical difficulties (Dodgson et al., 2011). The sequencing of the chicken genome (International Chicken Genome Sequencing Consortium, 2004) allowed a genome-wide methylation profile to be obtained. Liver and muscle tissue was analysed from two breeds of chicken, and found to be enriched within gene bodies and at repetitive sequences but depleted at TSS and TTSs. The majority of CpG islands were in an unmethylated state, and promoter methylation correlated with gene expression. No differentially methylated regions were found, consistent with the previous lack of evidence for imprinting in birds and the lack of a DNMT3L homologue in the chicken genome sequence. UHRF1 appears in the chicken genome, as do several MBDs (Table 1 and Figure 2).

A potential active demethylation system in chicken was first seen when chicken embryonic nuclear extracts were shown to be capable of demethylation, and that the extent of demethylation was different across developmental time points (Jost, 1993). This action was subsequently suggested to involve the combined action of glycosylases, (Jost et al., 1995) and repair pathways (Jost et al., 1995; Zhu et al., 2000) in an RNA dependant manner (Fremont et al., 1997; Jost et al., 1997). There is potential for 5hmC to be involved in this pathway, as the chicken genome contains potential homologues for all three TET proteins, which share all domains with their mouse equivalents (Table 1 and Figure 3).

A particularly interesting region found in the chicken is the chicken male hypermethylation region (cMHM). This region is hypermethylated in males but hypomethylated in females where the region is transcribed into ncRNAs which accumulate at DMRT1 – a gene required for chicken testis differentiation; and in a region enriched for dosage compensated genes. A recent study investigated this region in chickens subject to sex-reversal (Yang et al., 2011). Female chickens were sex reversed by the injection of fadrozole into eggs, and grouped into different states of sex reversal: slightly sex reversed, or highly sex reversed and compared to standard males and females. The cMHM in gonad cells was seen to be highly methylated in males and hypomethylated in standard females, as expected. Methylation was slightly increased in the slightly sex-reversed group of females, and there was no significant difference in cMHM methylation between highly sex-reversed chickens and males. However, in liver the cMHM was hypermethylated in males, and low in females but remained low in each of the sex-reversed groups. DMRT1 expression was seen to increase towards male levels upon sex-reversal, but the expression of other sex specific genes changed less dramatically. A link between chicken DNA methylation and development was further seen by the interesting regulation of DNMTs by miRNAs during PGC development (Rengaraj et al., 2011), a finding similar to that seen in mouse PGCs (Takada et al., 2009).

Due to the extensive use of chick in agriculture, study of disease in the organism is relevant not only as models for human diseases; but also due to potential economic benefits. A recent study sought to find links between DNA methylation and neoplastic diseases in chickens using two lines, one susceptible to tumours and one resistant (Yu et al, 2008a, 2008b). Possible links between methylation of viral DNA (Yu et al., 2008a) and DNMT genes (Yu et al., 2008b) were reported. The general importance of methylation to chicken health was seen when *in vivo* treatment with 5-azacytidine negatively affected lymphatic organs, and increased the prevalence of autoimmune diseases in chicks (Schauenstein et al., 1991). Findings particularly relevant to humans came from experiments which used betaine as a dietary supplement (Xing et al., 2011) - a methyl donor known to reduce fat deposition. A modest reduction and change in pattern of promoter methylation was noted for the LPL gene, involved in lipoprotein catabolism; which coincided with a change in LPL levels.

4. Frog – *Xenopus tropicalis / Xenopus laevis*

Xenopus laevis has a solid history in methylation research, and was used in the experiments which pioneered the use of methylation-sensitive restriction enzymes to analyse gene expression (Bird & Southern, 1978). This method allowed extensive discoveries to be made and permitted functional DNA methylation studies to begin.

Although both X. *laevis* and X. *tropicalis* have a DNMT3 homologue, *de novo* methylation in *Xenopus* is not well studied, and it is notable that they do not appear to have a DNMT3B homologue (Table 1). When *Xenopus* eggs were injected with *Xenopus* globin transgenes, no *de novo* methylation was observed (Bendig & Williams, 1983; Harland, 1982). *In vitro* methylation of the transgenes did, however, result in the maintenance of transgene methylation, and importantly, low transgene expression was observed regardless of methylation status (Bendig & Williams, 1983).

It has been shown that when *in vitro* methylated plasmids are injected into *Xenopus* embryos, the methylation is maintained (Harland, 1982) and that adenoviral genes injected into *Xenopus* were repressed by methylation (Vardimon et al., 1982b). Further studies went on to show that methylation is only repressive at certain CpG sites (Langner et al., 1984; Vardimon et al., 1982a, 1983). An attempt to link gene induction by estrogen to methylation was unsuccessful, as no methylation change was detected (Folger et al., 1983).

Dnmt1 was confirmed to be expressed in *Xenopus* (Kimura et al., 1996) and is accumulated in oocytes (Kimura et al., 1999), suggesting an *in vivo* importance during development. Functional evidence for such a role came when dnmt1 depletion in embryos resulted in temporal misexpression of genes, with various developmental markers being expressed early (Stancheva & Meehan, 2000); and misexpression of dnmts was linked to activation of apoptosis by various studies (Kaito et al., 2001;Kimura et al., 2002; Stancheva et al., 2001). However, it has recently been shown that xDnmt1 can alter gene expression in a methylation independent manner (Dunican et al., 2008), making it possible that the apoptotic phenotypes were also methylation independent. Further, when the function of dnmt1 oocyte accumulation was probed using monoclonal antibodies for xDnmt1, the resulting inhibition of cell division was also found to be unrelated to methylation (Hashimoto et al., 2003). Nevertheless, several studies link methylation to expression in *Xenopus*, such as the finding that methylation directly inhibits transcription (Harvey & Newport, 2003; Lopes et al., 2008); 5mC promotes HDAC mediated transcriptional repression (Jones et al., 1998; Wade et al., 1999); and the finding that the methyl CpG binding activity of the transcription factor Kaiso is specifically required during development; although some of the Kaiso targets do not appear to be directly regulated by promoter methylation (Ruzov et al., 2009a, 2009b). Kaiso can also interact directly with components of the Wnt signalling pathways in development and cancer, which may expand its functional repertoire (Ruzov et al., 2009a). It was demonstrated that the zinc finger regions (ZF1-3) of xKaiso (*Xenopus*), dKaiso (*Drosophila*) and gKaiso (chicken) are sufficient for direct interaction with a terminal component of the canonical Wnt pathway, xTcf3, via its HMG domain *in vitro* (Ruzov et al., 2009a); suggesting that the interaction between Kaiso and Tcf3/4 is mutually exclusive of their DNA binding. Over-expression of xKaiso in developing *Xenopus* embryos actually mimics certain aspects of xTcf3 depletion, such as ectopic Siamois expression. A potential intersection of Kaiso with Wnt signalling pathways may occur in cancer, where over-expression of Kaiso could attenuate constitutive Wnt signalling, while at the same time promoting cancer progression through silencing of *de novo* methylated tumour suppressor genes (Lopes et al., 2008). Recently published work shows that disruption of the Tcf4:Kaiso interaction in human colon cancer cell lines releases Tcf4, enabling its mutual association with β-catenin and the formation of a transcriptional complex (Del Valle-Perez et al., 2011). This also permits Kaiso binding to the methylated CDKN2A promoter in cells, leading to its decreased expression.

xMBD3 is highly expressed in *X. laevis* embryos, in a spatially distinctive pattern which overlaps significantly with DNMT1 expression (Iwano et al., 2004). Knockdown of xMBD3 resulted in the eye developmental transcription factor Pax6, consistent with the eye being one of the regions with particularly high xMBD3 expression. This is in contrast with Mbd3 being indispensible for mouse development (Hendrich et al., 2001) The *X. tropicalis* genome appears to contain multiple MBD proteins (Table 1).

The xDnmt1 sequence has all the hallmarks of a maintenance methyltransferase which can propagate pre-existing patterns of methylation that are present in the early *Xenopus* embryos (Stancheva & Meehan, 2000). Transient anti-sense RNA depletion of xDnmt1 levels by 90% results in DNA hypomethylation and premature activation of gene expression before the mid-blastula transition (MBT), a developmental landmark that coincides with general zygotic gene activation (Dunican et al., 2008; Newport & Kirschner, 1982; Stancheva & Meehan, 2000). Loss of xDnmt1 results in embryonic lethality due to activation of a programmed cell death pathway (Jackson-Grusby et al., 2001; Stancheva & Meehan, 2000). DNA methylation levels recover after anti-sense RNA depletion of xDnmt1 suggesting that a functional de novo methylation pathway is present ion *Xenopus laevis* (Stancheva & Meehan, 2000). Morphant knockdown of xDNMT1 (xDMO) matched the anti-sense RNA phenotype by exhibiting p53 dependant apoptotic embryo lethality and premature activation of zygotic transcription before the MBT. Very surprisingly, this occurred without global changes in DNA methylation levels. The underlying explanation was that a moderate reduction in xDnmt1p levels was sufficient to prematurely activate gene expression in *X. laevis* embryos independently of changes in DNA methylation levels or histone modifications (Dunican et al., 2008). Crucially, repression of target genes was re-imposed in xDMO morphants by co-injection of mRNA encoding a catalytically inactive form of DNMT1. In addition, it was observed that histone modifications (H3K4me3, H3K9me2 and H4K20me3) accumulate after the MBT and are not prematurely accrued when xDnmt1 levels were reduced and transcription occurs *before* the MBT (Akkers et al., 2009; Dunican et al., 2008). Here there is separation between the transcriptional outcome and specification by the histone code. A model was proposed in which xDnmt1 has a major silencer role in early *Xenopus* development as a chromatin bound non-enzymatic protein that regulated the timing of zygotic gene activation (Dunican et al., 2008). xDnmt1p may serve as a titratable repressor component that has been previously invoked for *Xenopus* embryos (Almouzni & Wolffe, 1995; Newport & Kirschner, 1982; Prioleau et al., 1994)

The study of DNA methylation in amphibians is particularly interesting due to the possibility of regenerative mechanisms being elucidated. Methylation of the enhancer of *Shh* - the ZRS (Lettice et al., 2003), was found to be low in tadpoles, which have full limb regeneration ability, but high in adult frogs capable of only limited regeneration (Yakushiji et al., 2007). Re-expression of *Shh* has been shown to occur during limb regeneration in other amphibians (Imokawa & Yoshizato, 1997; Torok et al., 1999).

Although the majority of frog DNA methylation studies have been performed in *Xenopus laevis*, the recent sequencing of the *Xenopus tropicalis* genome (Hellsten et al., 2010) has allowed the 5mC profile of this organism to be investigated. Genome wide 5mC profiling in embryos revealed that 5mC is mainly present in the CpG context, 45% of which are methylated, but also as CpA methylation. It appeared particularly enriched within repetitive regions (except for CpG depleted microsatellites), gene bodies and promoters; with TSSs being hypomethylated (Bogdanovic et al., 2011).

H3K4/27 trimethylation and RNA polymerase II (RNAPII) maps identify promoters and transcribed regions in *Xenopus tropicalis* (Akkers et al., 2009). Spatial differences in H3K27me3 deposition are predictive of localised gene expression. In agreement with a previous study, the appearance of K3K4me3 coincides with zygotic gene activation, whereas H3K27me3 is predominantly deposited upon subsequent spatial restriction or repression of transcriptional regulators (Akkers et al., 2009; Dunican et al., 2008).

Deep sequencing of purified methylated DNA obtained from early *X. tropicalis* embryos demonstrates that its genome is heavily methylated during blastula and gastrula stages (Bogdanovic et al., 2011). DNA methylation is absent in large H3K27me3 domains, indicating that these two repression pathways may be mutually antagonistic. Strikingly, genes that are highly expressed in *X. tropicalis* embryos but not in differentiated cells exhibit relatively high DNA methylation. Direct testing with reporter template demonstrates that methylated promoters are robustly transcribed in blastula- and gastrula-stage embryos, but not in oocytes or late embryos. This complements the situation in *X. laevis*, in which depletion of xDnmt1 leads to premature zygotic activation on a background of global methylation and a *tabula rasa* of histone modifications (Dunican et al., 2008). These findings have implications for epigenetic regulation of gene expression in early embryos and subsequent differentiation. It is noteworthy that mouse ES cells that lack *Dnmt1*, *Dnmt3a* and *Dnmt3b* survive in culture very well, but cannot differentiate properly (Tsumura et al., 2006).

It has been shown that global loss of 5mC does not occur during early *Xenopus* development (Stancheva et al., 2002; Veenstra & Wolffe, 2001). However, evidence exists for active demethylation; for example, demethylation of exogenous mouse *Oct4* is required for its transcription when micro injected into *X. laevis* oocytes (Simonsson & Gurdon, 2004), and gadd45a recruitment occurs at the demethylated region, with demethylation proceeding via action of the repair enzyme XPG (Barreto et al., 2007). It has subsequently been shown that *Xenopus* demethylation makes use of the NER system rather than BER (Schafer et al., 2010). Interestingly, gadd45a has been implicated in the switch from pluripotency to differentiation in the early *Xenopus* embryo (Kaufmann & Niehrs, 2011). *X. tropicalis* appears to possess homologues for at least TET2 and TET3 (Table 1 and Fig 3), but again the 5hmC system has not yet been investigated in *Xenopus*; though the results of such investigation would certainly be of interest. The apparent absence of a TET1 homologue (the only TET with a definite CXXC domain), is particularly interesting and could represent a demethylation mechanism absent in this species; although the function of the TET1 CXXC domain is still unclear (Frauer et al., 2011). Also notable is the presence of a potential lipid attachment site in Tet2, which is normally found in prokaryotes (Figure 1B).

5. Honeybee – *Apis mellifera*

The sequencing of the genome of the honeybee *Apis mellifera* (Honeybee Genome Sequencing Consortium, 2006) allowed potential DNA methylation genes to be identified. This was a significant turning point in the study of insect DNA methylation, with honeybee being the first insect shown to contain at least one homologue of DNMT1, 2 and 3 (Wang et al., 2006). In contrast with mammals, the honeybee genome contains two homologues of the DNMT1 maintenance methylase, and one for the DNMT3 *de novo* methylases (Table 1); with each having high sequence conservation with the mammalian genes (Wang et al., 2006). The *Apis mellifera* genome also contains UHRF1, possibly indicating a conserved DNA maintenance methylation mechanism. Inevitably, once the genome sequence was available, a methylome was obtained (Zemach et al., 2010). This initial analysis was performed on DNA obtained from whole bodies of worker bees, and revealed that overall methylation levels in honeybee are low, with CpG methylation being 0.51%; CHG methylation 0.11%; and CHH methylation 0.16%. Hypomethylation of transposable elements was observed, with methylation mainly being observed within gene bodies, and no enrichment seen at promoters. Interestingly, when transcriptional activity was compared to CpG methylation it was seen that the methylation profile peaked at around the 50th percentile of transcriptional activity, with lowly expressed genes and highly expressed genes sharing a relatively low amount of CpG methylation. Due to this study being from whole bodies of honeybees, it is possible that the results are not biologically relevant, and are an artefact of the amalgamation of all cell types and tissues being analysed together. More informative honeybee methylomes were obtained when BS-seq was used to map the 5mC profile of *Apis mellifera* in brains from both worker and queen bees (Lyko et al., 2010). This allowed both the study of 5mC within a distinct tissue; and the comparison of 5mC profiles between two genomically identical castes with massive behavioural and physiological differences. This study replicated the finding that global levels of DNA methylation are low in honeybees, with approximately 70,000 cytosines methylated from a total of 60,000,000 in the genome (although it remains possible that methylation levels are high in other tissues), and showed that honeybee cytosine methylation is almost exclusive to CpG dinucleotides. Further, the vast majority of 5mC (over 85% in both workers and queens), was within gene bodies; most abundant in exons (over 75% in both castes) and found significantly at splice sites. The lack of methylation of repeats and transposons was also confirmed by this study, with the authors suggesting that it may represent that these elements do not confer genome instability in honeybees. Interestingly, the genes found to be methylated showed higher sequence conservation across a wide range of species than non-methylated genes. Finally, the authors found a significant group of genes to be differentially methylated between worker and queen bees, pointing towards DNA methylation being one mechanism of genetic control contributing to the differences between these castes. Interestingly, when Dnmt3 is knocked down in honeybees using siRNA, they are born with queen-like characteristics such as developed ovaries, mimicking the effects of a royal jelly diet (Kucharski et al., 2008). This study was complimented by the finding that DNMT3 activity reduced with each day that larvae were fed royal jelly (tested at days 3,4 and 5) (Shi et al., 2011). The existence of MBDs in the honeybee is unclear, with Mbd3 being the only classical-type MBD apparent in the genome (Table 1). However, MBD domains are seen in other contexts (Figure 1C). The hypothetical honeybee protein LOC725950 contains an MBD domain and shares most amino acid similarity in mouse with the PHF20L1 protein, which lacks any MBD. Another hypothetical protein in Apis, LOC412607, shares most amino acid similarity in mouse with the FBXL7 protein, with which it shares an F-BOX domain but no MBD. In fact, the MBD:F-BOX architecture appears to be unique to insects, based on the Sanger pfam database. In both of these examples, the compared mouse proteins are probably not homologues, with similarity only being due to the shared domain. It is therefore likely that DNA methylation in honeybees is acted on via mechanisms unseen in non-insects. Further evidence for novel epigenetic systems in honeybee include the finding that hypermethylated genes tend to be significantly longer than hypomethylated genes (Zeng & Yi, 2010). Other potential MBD-containing proteins in honeybee include BAZ2B-like, and SETDB1, both of which appear to have legitimate homologues in mouse, which they share MBD and other domains with. Interestingly, honeybee appears to have a TET1 homologue complete with CXXC domain, and may therefore may have an active demethylation system making use of this protein, but not TET2/3 homologues (Table 1). Alternatively, TET1 may function non-catalytically to recruit/block other proteins.

6. Conclusions

Although interspecific 5mC distributions can vary, methylation of cytosine offers a selective potential as a mechanism for regulating gene expression across a wide range of species. The fundamental mechanism of DNA methylation mediated repression seems to be conserved, with active DNMT enzymes appearing in each of the species discussed, and each having homologues of UHRF1 and MBDs. Additionally, there is genomic evidence for active demethylation mechanisms via 5hmC in each of the species discussed, with each having at least one TET homologue present; although it is unclear whether large-scale active demethylation events such as those seen in mammalian development also occur in non-mammalian animals. It will be particularly interesting to find roles of active demethylation in non-mammalian animals; and to see whether TETs/5hmC are involved. Additionally, it will be interesting to investigate potential roles for domains apparently unique to non-mammalian methylation enzymes; such as the CH domain of zebrafish dnmt3/7; and the potential lipid attachment of *Xenopus* Tet2. However, attribution of the molecular pathology of mutants in components of DNA methylation machinery is not always so clear, especially as non-catalytic versions (e.g. in xDnmt1) can rescue phenotypes. For example, a model for the activation of xDnmt1-mediated apoptosis in *X. laevis* embryos has been proposed in which a chromatin associated complex of xDnmt11, xMbd4 and xMlh1 in embryos responds to DNA damage or replication stress by either repairing the lesion or to activate an apoptotic response through the activation/release of Mbd4/Mlh1 from chromatin-bound xDnmt1 (Ruzov et al., 2009c). The Mbd4/Mlh1 complex signals, perhaps via the DNA-damage kinases ATM and ATR, to activate the p53-dependent programmed cell death pathway.

A recent report has validated this model in human cells by demonstrating that DNMT1 is rapidly but transiently recruited to double stranded breaks (DSBs) (Ha et al., 2011), dependent on its ability to interact with both PCNA and CHK1, but independent of its catalytic activity. What is the potential importance of the Dnmt1 signalling mechanism? Cancers arise from the sequential acquisition of genetic alterations in specific genes leading to cellular transformation in step with epigenetic alterations (Fang et al., 2011; Figueroa et al., 2010). One possibility is that Dnmt1 is part of a signalling cascade that activates a barrier against tumour progression. Cellular sensitivity to changes in DNMT1 levels are lost when components of the signalling cascade are either absent or mutated, contributing to the generation of an altered cancer epigenome in tumours and cell lines. In a recent paper it was shown that depleting DNMT1 in proliferating human fibroblasts is sufficient to cause mismatch repair defects and increased mutation rates at a CA17 microsatellite (Loughery et al., 2011). This is associated with decreases in mismatch repair protein levels, including MBD4 following activation of the DNA damage response (DDR). Blocking the DDR, and in particular PARP over-activation, also increases survival of the DNMT1 knockdowns.

Thanks to the study of DNA modification patterns, their generation and the consequences of inactivating components of the epigenetic pathways in different model organisms we are gaining a fuller understanding of their function in mammalian and especially human disease models.

7. Acknowlegements

Work in the SP and RM labs are supported by the MRC, BBSRC and Breakthrough Breast Cancer.

8. References

Akkers, R.C., S.J.van Heeringen, U.G.Jacobi, E.M.Janssen-Megens, K.J.Francoijs, H.G.Stunnenberg, & G.J.Veenstra. 2009. A hierarchy of H3K4me3 and H3K27me3 acquisition in spatial gene regulation in Xenopus embryos. *Dev. Cell* 17: 425-434.

Almouzni, G. & A.P.Wolffe. 1995. Constraints on transcriptional activator function contribute to transcriptional quiescence during early Xenopus embryogenesis. *EMBO J.* 14: 1752-1765.

An, W., H.K.van, & J.Zlatanova. 1998. The non-histone chromatin protein HMG1 protects linker DNA on the side opposite to that protected by linker histones. *J. Biol. Chem.* 273: 26289-26291.

Anderson, R.M., J.A.Bosch, M.G.Goll, D.Hesselson, P.D.Dong, D.Shin, N.C.Chi, C.H.Shin, A.Schlegel, M.Halpern, & D.Y.Stainier. 2009. Loss of Dnmt1 catalytic activity reveals multiple roles for DNA methylation during pancreas development and regeneration. *Dev. Biol.* 334: 213-223.

Andrews, A.J. & K.Luger. 2011. Nucleosome structure(s) and stability: variations on a theme. *Annu. Rev. Biophys.* 40: 99-117.

Baillie, J.K., M.W.Barnett, K.R.Upton, D.J.Gerhardt, T.A.Richmond, S.F.De, P.Brennan, P.Rizzu, S.Smith, M.Fell, R.T.Talbot, S.Gustincich, T.C.Freeman, J.S.Mattick, D.A.Hume, P.Heutink, P.Carninci, J.A.Jeddeloh, & G.J.Faulkner. 2011. Somatic retrotransposition alters the genetic landscape of the human brain. *Nature*.

Barreto, G., A.Schafer, J.Marhold, D.Stach, S.K.Swaminathan, V.Handa, G.Doderlein, N.Maltry, W.Wu, F.Lyko, & C.Niehrs. 2007. Gadd45a promotes epigenetic gene activation by repair-mediated DNA demethylation. *Nature* 445: 671-675.

Barski, A., S.Cuddapah, K.Cui, T.Y.Roh, D.E.Schones, Z.Wang, G.Wei, I.Chepelev, & K.Zhao. 2007. High-resolution profiling of histone methylations in the human genome. *Cell* 129: 823-837.

Bendig, M.M. & J.G.Williams. 1983. Replication and expression of Xenopus laevis globin genes injected into fertilized Xenopus eggs. *Proc. Natl. Acad. Sci. U. S. A* 80: 6197-6201.

Bird, A. 2007. Perceptions of epigenetics. *Nature* 447: 396-398.

Bird, A.P. & E.M.Southern. 1978. Use of restriction enzymes to study eukaryotic DNA methylation: I. The methylation pattern in ribosomal DNA from Xenopus laevis. *J. Mol. Biol.* 118: 27-47.

Bogdanovic, O., S.W.Long, S.J.van Heeringen, A.B.Brinkman, J.L.Gomez-Skarmeta, H.G.Stunnenberg, P.L.Jones, & G.J.Veenstra. 2011. Temporal uncoupling of the DNA methylome and transcriptional repression during embryogenesis. *Genome Res.* 21: 1313-1327.

Bourc'his, D. & T.H.Bestor. 2004. Meiotic catastrophe and retrotransposon reactivation in male germ cells lacking Dnmt3L. *Nature* 431: 96-99.

Bourc'his, D., G.L.Xu, C.S.Lin, B.Bollman, & T.H.Bestor. 2001. Dnmt3L and the establishment of maternal genomic imprints. *Science* 294: 2536-2539.

Boyer, L.A., K.Plath, J.Zeitlinger, T.Brambrink, L.A.Medeiros, T.I.Lee, S.S.Levine, M.Wernig, A.Tajonar, M.K.Ray, G.W.Bell, A.P.Otte, M.Vidal, D.K.Gifford, R.A.Young, and R.Jaenisch. 2006. Polycomb complexes repress developmental regulators in murine embryonic stem cells. *Nature* 441: 349-353.

Cedar, H. & Y.Bergman. 2009. Linking DNA methylation and histone modification: patterns and paradigms. *Nat. Rev. Genet.* 10: 295-304.

Cheng, K.C. & J.L.Moore. 1997. Genetic dissection of vertebrate processes in the zebrafish: a comparison of uniparental and two-generation screens. *Biochem. Cell Biol.* 75: 525-533.

Collas, P. 1998. Modulation of plasmid DNA methylation and expression in zebrafish embryos. *Nucleic Acids Res.* 26: 4454-4461.

Cooper, D.N., L.H.Errington, & R.M.Clayton. 1983. Variation in the DNA methylation pattern of expressed and nonexpressed genes in chicken. *DNA* 2: 131-140.

Davey, C., S.Pennings, & J.Allan. 1997. CpG methylation remodels chromatin structure in vitro. *J. Mol. Biol.* 267: 276-288.

Del Valle-Perez, B., D.Casagolda, E.Lugilde, G.Valls, M.Codina, N.Dave, A.G.de Herreros, & M.Dunach. 2011. Wnt controls the transcriptional activity of Kaiso through CK1epsilon-dependent phosphorylation of p120-catenin. *J. Cell Sci.* 124: 2298-2309.

Dhayalan, A., A.Rajavelu, P.Rathert, R.Tamas, R.Z.Jurkowska, S.Ragozin, & A.Jeltsch. 2010. The Dnmt3a PWWP domain reads histone 3 lysine 36 trimethylation and guides DNA methylation. *J. Biol. Chem.* 285: 26114-26120.

Dialynas, G.K., M.W.Vitalini, & L.L.Wallrath. 2008. Linking Heterochromatin Protein 1 (HP1) to cancer progression. *Mutat. Res.* 647: 13-20.

Dodgson, J.B., M.E.Delany, & H.H.Cheng. 2011. Poultry genome sequences: progress and outstanding challenges. *Cytogenet. Genome Res.* 134: 19-26.

Dunican, D.S., A.Ruzov, J.A.Hackett, & R.R.Meehan. 2008. xDnmt1 regulates transcriptional silencing in pre-MBT Xenopus embryos independently of its catalytic function. *Development* 135: 1295-1302.

Esteve, P.O., Y.Chang, M.Samaranayake, A.K.Upadhyay, J.R.Horton, G.R.Feehery, X.Cheng, & S.Pradhan. 2011. A methylation and phosphorylation switch between an adjacent lysine and serine determines human DNMT1 stability. *Nat. Struct. Mol. Biol.* 18: 42-48.

Esteve, P.O., H.G.Chin, J.Benner, G.R.Feehery, M.Samaranayake, G.A.Horwitz, S.E.Jacobsen, & S.Pradhan. 2009. Regulation of DNMT1 stability through SET7-mediated lysine methylation in mammalian cells. *Proc. Natl. Acad. Sci. U. S. A* 106: 5076-5081.

Esteve, P.O., H.G.Chin, A.Smallwood, G.R.Feehery, O.Gangisetty, A.R.Karpf, M.F.Carey, & S.Pradhan. 2006. Direct interaction between DNMT1 and G9a coordinates DNA and histone methylation during replication. *Genes Dev.* 20: 3089-3103.

Fang, F., S.Turcan, A.Rimner, A.Kaufman, D.Giri, L.G.Morris, R.Shen, V.Seshan, Q.Mo, A.Heguy, S.B.Baylin, N.Ahuja, A.Viale, J.Massague, L.Norton, L.T.Vahdat, M.E.Moynahan, & T.A.Chan. 2011. Breast cancer methylomes establish an epigenomic foundation for metastasis. *Sci. Transl. Med.* 3: 75ra25.

Feng, S., S.J.Cokus, X.Zhang, P.Y.Chen, M.Bostick, M.G.Goll, J.Hetzel, J.Jain, S.H.Strauss, M.E.Halpern, C.Ukomadu, K.C.Sadler, S.Pradhan, M.Pellegrini, & S.E.Jacobsen. 2010a. Conservation and divergence of methylation patterning in plants and animals. *Proc. Natl. Acad. Sci. U. S. A* 107: 8689-8694.

Feng, S., S.E.Jacobsen, & W.Reik. 2010b. Epigenetic reprogramming in plant and animal development. *Science* 330: 622-627.

Figueroa, M.E., O.Abdel-Wahab, C.Lu, P.S.Ward, J.Patel, A.Shih, Y.Li, N.Bhagwat, A.Vasanthakumar, H.F.Fernandez, M.S.Tallman, Z.Sun, K.Wolniak, J.K.Peeters, W.Liu, S.E.Choe, V.R.Fantin, E.Paietta, B.Lowenberg, J.D.Licht, L.A.Godley, R.Delwel, P.J.Valk, C.B.Thompson, R.L.Levine, & A.Melnick. 2010. Leukemic IDH1 and IDH2 mutations result in a hypermethylation phenotype, disrupt TET2 function, and impair hematopoietic differentiation. *Cancer Cell* 18: 553-567.

Folger, K., J.N.Anderson, M.A.Hayward, & D.J.Shapiro. 1983. Nuclease sensitivity and DNA methylation in estrogen regulation of Xenopus laevis vitellogenin gene expression. *J. Biol. Chem.* 258: 8908-8914.

Frauer, C., A.Rottach, D.Meilinger, S.Bultmann, K.Fellinger, S.Hasenoder, M.Wang, W.Qin, J.Soding, F.Spada, & H.Leonhardt. 2011. Different binding properties and function of CXXC zinc finger domains in Dnmt1 and Tet1. *PLoS. One.* 6: e16627.

Fremont, M., M.Siegmann, S.Gaulis, R.Matthies, D.Hess, & J.P.Jost. 1997. Demethylation of DNA by purified chick embryo 5-methylcytosine-DNA glycosylase requires both protein & RNA. *Nucleic Acids Res.* 25: 2375-2380.

Friedman, S. 1979. The effect of 5-azacytidine on E. coli DNA methylase. *Biochem. Biophys. Res. Commun.* 89: 1328-1333.

Gertz, J., K.E.Varley, T.E.Reddy, K.M.Bowling, F.Pauli, S.L.Parker, K.S.Kucera, H.F.Willard, & R.M.Myers. 2011. Analysis of DNA methylation in a three-generation family reveals widespread genetic influence on epigenetic regulation. *PLoS. Genet.* 7: e1002228.

Ginder, G.D., M.Whitters, K.Kelley, & R.A.Chase. 1983. In vivo demethylation of chicken embryonic beta-type globin genes with 5-azacytidine. *Prog. Clin. Biol. Res.* 134: 501-510.

Goll, M.G., R.Anderson, D.Y.Stainier, A.C.Spradling, & M.E.Halpern. 2009. Transcriptional silencing and reactivation in transgenic zebrafish. *Genetics* 182: 747-755.

Goll, M.G. & T.H.Bestor. 2005. Eukaryotic cytosine methyltransferases. *Annu. Rev. Biochem.* 74: 481-514.

Ha, K., G.E.Lee, S.S.Palii, K.D.Brown, Y.Takeda, K.Liu, K.N.Bhalla, & K.D.Robertson. 2011. Rapid and transient recruitment of DNMT1 to DNA double-strand breaks is mediated by its interaction with multiple components of the DNA damage response machinery. *Hum. Mol. Genet.* 20: 126-140.

Haigh, L.S., B.B.Owens, O.S.Hellewell, & V.M.Ingram. 1982. DNA methylation in chicken alpha-globin gene expression. *Proc. Natl. Acad. Sci. U. S. A* 79: 5332-5336.

Harasawa, R. & T.Mitsuoka. 1984. DNA methylation in chicken brain and liver. *Experientia* 40: 390-392.

Harland, R.M. 1982. Inheritance of DNA methylation in microinjected eggs of Xenopus laevis. *Proc. Natl. Acad. Sci. U. S. A* 79: 2323-2327.

Harvey, K.J. & J.Newport. 2003. CpG methylation of DNA restricts prereplication complex assembly in Xenopus egg extracts. *Mol. Cell Biol.* 23: 6769-6779.

Hashimoto, H., I.Suetake, & S.Tajima. 2003. Monoclonal antibody against dnmt1 arrests the cell division of xenopus early-stage embryos. *Exp. Cell Res.* 286: 252-262.

He, Y.F., B.Z.Li, Z.Li, P.Liu, Y.Wang, Q.Tang, J.Ding, Y.Jia, Z.Chen, L.Li, Y.Sun, X.Li, Q.Dai, C.X.Song, K.Zhang, C.He, & G.L.Xu. 2011. Tet-mediated formation of 5-carboxylcytosine and its excision by TDG in mammalian DNA. *Science* 333: 1303-1307.

Heitmann, B., T.Maurer, J.M.Weitzel, W.H.Stratling, H.R.Kalbitzer, & E.Brunner. 2003. Solution structure of the matrix attachment region-binding domain of chicken MeCP2. *Eur. J. Biochem.* 270: 3263-3270.

Hellsten, U., R.M.Harland, M.J.Gilchrist, D.Hendrix, J.Jurka, V.Kapitonov, I.Ovcharenko, N.H.Putnam, S.Shu, L.Taher, I.L.Blitz, B.Blumberg, D.S.Dichmann, I.Dubchak, E.Amaya, J.C.Detter, R.Fletcher, D.S.Gerhard, D.Goodstein, T.Graves, I.V.Grigoriev, J.Grimwood, T.Kawashima, E.Lindquist, S.M.Lucas, P.E.Mead, T.Mitros, H.Ogino, Y.Ohta, A.V.Poliakov, N.Pollet, J.Robert, A.Salamov, A.K.Sater, J.Schmutz, A.Terry, P.D.Vize, W.C.Warren, D.Wells, A.Wills, R.K.Wilson, L.B.Zimmerman, A.M.Zorn, R.Grainger, T.Grammer, M.K.Khokha, P. M. Richardson, & D. S. Rokhsar. 2010. The genome of the Western clawed frog Xenopus tropicalis. *Science* 328: 633-636.

Hendrich, B., J.Guy, B.Ramsahoye, V.A.Wilson, & A.Bird. 2001. Closely related proteins MBD2 and MBD3 play distinctive but interacting roles in mouse development. *Genes Dev.* 15: 710-723.

Hervouet, E., F.M.Vallette, & P.F.Cartron. 2009. Dnmt3/transcription factor interactions as crucial players in targeted DNA methylation. *Epigenetics.* 4: 487-499.

Hochedlinger, K. & R.Jaenisch. 2006. Nuclear reprogramming and pluripotency. *Nature* 441: 1061-1067.

Honeybee Genome Sequencing Consortium. 2006. Insights into social insects from the genome of the honeybee Apis mellifera. *Nature* 443: 931-949.

Illingworth, R.S., U.Gruenewald-Schneider, S.Webb, A.R.Kerr, K.D.James, D.J.Turner, C.Smith, D.J.Harrison, R.Andrews, & A.P.Bird. 2010. Orphan CpG islands identify numerous conserved promoters in the mammalian genome. *PLoS. Genet.* 6.

Imokawa, Y. & K.Yoshizato. 1997. Expression of Sonic hedgehog gene in regenerating newt limb blastemas recapitulates that in developing limb buds. *Proc. Natl. Acad. Sci. U. S. A* 94: 9159-9164.

Inoue, A. & Y.Zhang. 2011. Replication-dependent loss of 5-hydroxymethylcytosine in mouse preimplantation embryos. *Science* 334: 194.

International Chicken Genome Sequencing Consortium. 2004. Sequence and comparative analysis of the chicken genome provide unique perspectives on vertebrate evolution. *Nature* 432: 695-716.

Ito, S., A.C.D'Alessio, O.V.Taranova, K.Hong, L.C.Sowers, & Y.Zhang. 2010. Role of Tet proteins in 5mC to 5hmC conversion, ES-cell self-renewal and inner cell mass specification. *Nature* 466: 1129-1133.

Ito, S., L.Shen, Q.Dai, S.C.Wu, L.B.Collins, J.A.Swenberg, C.He, & Y.Zhang. 2011. Tet proteins can convert 5-methylcytosine to 5-formylcytosine and 5-carboxylcytosine. *Science* 333: 1300-1303.

Iwano, H., M.Nakamura, & S.Tajima. 2004. Xenopus MBD3 plays a crucial role in an early stage of development. *Dev. Biol.* 268: 416-428.

Jackson-Grusby, L., C.Beard, R.Possemato, M.Tudor, D.Fambrough, G.Csankovszki, J.Dausman, P.Lee, C.Wilson, E.Lander, & R.Jaenisch. 2001. Loss of genomic methylation causes p53-dependent apoptosis and epigenetic deregulation. *Nat. Genet.* 27: 31-39.

Jones, P.L., G.J.Veenstra, P.A.Wade, D.Vermaak, S.U.Kass, N.Landsberger, J.Strouboulis, & A.P.Wolffe. 1998. Methylated DNA and MeCP2 recruit histone deacetylase to repress transcription. *Nat. Genet.* 19: 187-191.

Jost, J.P. 1993. Nuclear extracts of chicken embryos promote an active demethylation of DNA by excision repair of 5-methyldeoxycytidine. *Proc. Natl. Acad. Sci. U. S. A* 90: 4684-4688.

Jost, J.P., M.Fremont, M.Siegmann, & J.Hofsteenge. 1997. The RNA moiety of chick embryo 5-methylcytosine- DNA glycosylase targets DNA demethylation. *Nucleic Acids Res.* 25: 4545-4550.

Jost, J.P., M.Siegmann, L.Sun, & R.Leung. 1995. Mechanisms of DNA demethylation in chicken embryos. Purification and properties of a 5-methylcytosine-DNA glycosylase. *J. Biol. Chem.* 270: 9734-9739.

Kaito, C., M.Kai, T.Higo, E.Takayama, H.Fukamachi, K.Sekimizu, & K.Shiokawa. 2001. Activation of the maternally preset program of apoptosis by microinjection of 5-aza-2'-deoxycytidine and 5-methyl-2'-deoxycytidine-5'-triphosphate in Xenopus laevis embryos. *Dev. Growth Differ.* 43: 383-390.

Kaneda, M., R.Hirasawa, H.Chiba, M.Okano, E.Li, & H.Sasaki. 2010. Genetic evidence for Dnmt3a-dependent imprinting during oocyte growth obtained by conditional knockout with Zp3-Cre and complete exclusion of Dnmt3b by chimera formation. *Genes Cells.*

Kato, Y., M.Kaneda, K.Hata, K.Kumaki, M.Hisano, Y.Kohara, M.Okano, E.Li, M.Nozaki, & H.Sasaki. 2007. Role of the Dnmt3 family in de novo methylation of imprinted and repetitive sequences during male germ cell development in the mouse. *Hum. Mol. Genet.* 16: 2272-2280.

Kaufmann, L.T. & C.Niehrs. 2011. Gadd45a and Gadd45g regulate neural development and exit from pluripotency in Xenopus. *Mech. Dev.*

Kim, J.K., M.Samaranayake, & S.Pradhan. 2009. Epigenetic mechanisms in mammals. *Cell Mol. Life Sci.* 66: 596-612.

Kimura, H., G.Ishihara, & S.Tajima. 1996. Isolation and expression of a Xenopus laevis DNA methyltransferase cDNA. *J. Biochem.* 120: 1182-1189.

Kimura, H., I.Suetake, & S.Tajima. 1999. Xenopus maintenance-type DNA methyltransferase is accumulated and translocated into germinal vesicles of oocytes. *J. Biochem.* 125: 1175-1182.

Kimura, H., I.Suetake, & S.Tajima. 2002. Exogenous expression of mouse Dnmt3 induces apoptosis in Xenopus early embryos. *J. Biochem.* 131: 933-941.

Ko, M., Y.Huang, A.M.Jankowska, U.J.Pape, M.Tahiliani, H.S.Bandukwala, J.An, E.D.Lamperti, K.P.Koh, R.Ganetzky, X.S.Liu, L.Aravind, S.Agarwal, J.P.Maciejewski, & A.Rao. 2010. Impaired hydroxylation of 5-methylcytosine in myeloid cancers with mutant TET2. *Nature* 468: 839-843.

Koh, K.P., A.Yabuuchi, S.Rao, Y.Huang, K.Cunniff, J.Nardone, A.Laiho, M.Tahiliani, C.A.Sommer, G.Mostoslavsky, R.Lahesmaa, S.H.Orkin, S.J.Rodig, G.Q.Daley, & A.Rao. 2011. Tet1 and Tet2 regulate 5-hydroxymethylcytosine production and cell lineage specification in mouse embryonic stem cells. *Cell Stem Cell* 8: 200-213.

Kriaucionis, S. & N.Heintz. 2009. The nuclear DNA base 5-hydroxymethylcytosine is present in Purkinje neurons and the brain. *Science* 324: 929-930.

Kucharski, R., J.Maleszka, S.Foret, & R.Maleszka. 2008. Nutritional control of reproductive status in honeybees via DNA methylation. *Science* 319: 1827-1830.

Langner, K.D., L.Vardimon, D.Renz, & W.Doerfler. 1984. DNA methylation of three 5' C-C-G-G 3' sites in the promoter and 5' region inactivate the E2a gene of adenovirus type 2. *Proc. Natl. Acad. Sci. U. S. A* 81: 2950-2954.

Lettice, L.A., S.J.Heaney, L.A.Purdie, L.Li, B.P.de, B.A.Oostra, D.Goode, G.Elgar, R.E.Hill, & G.E.de. 2003. A long-range Shh enhancer regulates expression in the developing limb and fin and is associated with preaxial polydactyly. *Hum. Mol. Genet.* 12: 1725-1735.

Lister, R., M.Pelizzola, R.H.Dowen, R.D.Hawkins, G.Hon, J.Tonti-Filippini, J.R.Nery, L.Lee, Z.Ye, Q.M.Ngo, L.Edsall, J.ntosiewicz-Bourget, R.Stewart, V.Ruotti, A.H.Millar, J.A.Thomson, B.Ren, & J.R.Ecker. 2009. Human DNA methylomes at base resolution show widespread epigenomic differences. *Nature* 462: 315-322.

Lopes, E.C., E.Valls, M.E.Figueroa, A.Mazur, F.G.Meng, G.Chiosis, P.W.Laird, N.Schreiber-Agus, J.M.Greally, E.Prokhortchouk, & A.Melnick. 2008. Kaiso contributes to DNA methylation-dependent silencing of tumor suppressor genes in colon cancer cell lines. *Cancer Res.* 68: 7258-7263.

Loughery, J.E., P.D.Dunne, K.M.O'Neill, R.R.Meehan, J.R.McDaid, & C.P.Walsh. 2011. DNMT1 deficiency triggers mismatch repair defects in human cells through depletion of repair protein levels in a process involving the DNA damage response. *Hum. Mol. Genet.* 20: 3241-3255.

Luger, K., A.W.Mader, R.K.Richmond, D.F.Sargent, & T.J.Richmond. 1997. Crystal structure of the nucleosome core particle at 2.8 A resolution. *Nature* 389: 251-260.

Lyko, F., S.Foret, R.Kucharski, S.Wolf, C.Falckenhayn, & R.Maleszka. 2010. The honey bee epigenomes: differential methylation of brain DNA in queens and workers. *PLoS. Biol.* 8: e1000506.

MacKay, A.B., A.A.Mhanni, R.A.McGowan, & P.H.Krone. 2007. Immunological detection of changes in genomic DNA methylation during early zebrafish development. *Genome* 50: 778-785.

Macleod, D., V.H.Clark, & A.Bird. 1999. Absence of genome-wide changes in DNA methylation during development of the zebrafish. *Nat. Genet.* 23: 139-140.

Mandel, J.L. & P.Chambon. 1979. DNA methylation: organ specific variations in the methylation pattern within and around ovalbumin and other chicken genes. *Nucleic Acids Res.* 7: 2081-2103.

Martin, C.C., L.Laforest, M.A.Akimenko, & M.Ekker. 1999. A role for DNA methylation in gastrulation and somite patterning. *Dev. Biol.* 206: 189-205.

Martin, C.C. & R.McGowan. 1995. Genotype-specific modifiers of transgene methylation and expression in the zebrafish, Danio rerio. *Genet. Res.* 65: 21-28.

McGhee, J.D. & G.D.Ginder. 1979. Specific DNA methylation sites in the vicinity of the chicken beta-globin genes. *Nature* 280: 419-420.

McQueen, H.A., J.Fantes, S.H.Cross, V.H.Clark, A.L.Archibald, & A.P.Bird. 1996. CpG islands of chicken are concentrated on microchromosomes. *Nat. Genet.* 12: 321-324.

McQueen, H.A., G.Siriaco, & A.P.Bird. 1998. Chicken microchromosomes are hyperacetylated, early replicating, and gene rich. *Genome Res.* 8: 621-630.

Meissner, A., T.S.Mikkelsen, H.Gu, M.Wernig, J.Hanna, A.Sivachenko, X.Zhang, B.E.Bernstein, C.Nusbaum, D.B.Jaffe, A.Gnirke, R.Jaenisch, & E.S.Lander. 2008. Genome-scale DNA methylation maps of pluripotent and differentiated cells. *Nature* 454: 766-770.

Mhanni, A.A. & R.A.McGowan. 2004. Global changes in genomic methylation levels during early development of the zebrafish embryo. *Dev. Genes Evol.* 214: 412-417.

Mikkelsen, T.S., M.Ku, D.B.Jaffe, B.Issac, E.Lieberman, G.Giannoukos, P.Alvarez, W.Brockman, T.K.Kim, R.P.Koche, W.Lee, E.Mendenhall, A.O'Donovan, A.Presser, C.Russ, X.Xie, A.Meissner, M.Wernig, R.Jaenisch, C.Nusbaum, E.S.Lander, & B.E.Bernstein. 2007. Genome-wide maps of chromatin state in pluripotent and lineage-committed cells. *Nature* 448: 553-560.

Monk, M., M.Boubelik, & S.Lehnert. 1987. Temporal and regional changes in DNA methylation in the embryonic, extraembryonic and germ cell lineages during mouse embryo development. *Development* 99: 371-382.

Morey, L. & K.Helin. 2010. Polycomb group protein-mediated repression of transcription. *Trends Biochem. Sci.* 35: 323-332.

Munzel, M., L.Lercher, M.Muller, & T.Carell. 2010. Chemical discrimination between dC and 5MedC via their hydroxylamine adducts. *Nucleic Acids Res.* 38: e192.

Myant, K., A.Termanis, A.Y.Sundaram, T.Boe, C.Li, C.Merusi, J.Burrage, J.I.de Las Heras, & I.Stancheva. 2011. LSH and G9a/GLP complex are required for developmentally programmed DNA methylation. *Genome Res.* 21: 83-94.

Ndlovu, M.N., H.Denis, & F.Fuks. 2011. Exposing the DNA methylome iceberg. *Trends Biochem. Sci.* 36: 381-387.

Nestor, C., A.Ruzov, R.Meehan, & D.Dunican. 2010. Enzymatic approaches and bisulfite sequencing cannot distinguish between 5-methylcytosine and 5-hydroxymethylcytosine in DNA. *Biotechniques* 48: 317-319.

Newport, J. & M.Kirschner. 1982. A major developmental transition in early Xenopus embryos: II. Control of the onset of transcription. *Cell* 30: 687-696.

Okano, M., D.W.Bell, D.A.Haber, & E.Li. 1999. DNA methyltransferases Dnmt3a and Dnmt3b are essential for de novo methylation and mammalian development. *Cell* 99: 247-257.

Okano, M., S.Xie, & E.Li. 1998a. Cloning and characterization of a family of novel mammalian DNA (cytosine-5) methyltransferases. *Nat. Genet.* 19: 219-220.

Okano, M., S.Xie, & E.Li. 1998b. Dnmt2 is not required for de novo and maintenance methylation of viral DNA in embryonic stem cells. *Nucleic Acids Res.* 26: 2536-2540.

Ooi, S.K., D.Wolf, O.Hartung, S.Agarwal, G.Q.Daley, S.P.Goff, & T.H.Bestor. 2010. Dynamic instability of genomic methylation patterns in pluripotent stem cells. *Epigenetics. Chromatin.* 3: 17.

Park, S.H., J.N.Kim, T.S.Park, S.D.Lee, T.H.Kim, B.K.Han, & J.Y.Han. 2010. CpG methylation modulates tissue-specific expression of a transgene in chickens. *Theriogenology* 74: 805-816.

Prioleau, M.N., J.Huet, A.Sentenac, & M.Mechali. 1994. Competition between chromatin and transcription complex assembly regulates gene expression during early development. *Cell* 77: 439-449.

Qin, W., H.Leonhardt, & G.Pichler. 2011. Regulation of DNA methyltransferase 1 by interactions and modifications. *Nucleus.* 2: 392-402.

Qiu, C., K.Sawada, X.Zhang, & X.Cheng. 2002. The PWWP domain of mammalian DNA methyltransferase Dnmt3b defines a new family of DNA-binding folds. *Nat. Struct. Biol.* 9: 217-224.

Rai, K., I.J.Huggins, S.R.James, A.R.Karpf, D.A.Jones, & B.R.Cairns. 2008. DNA demethylation in zebrafish involves the coupling of a deaminase, a glycosylase, and gadd45. *Cell* 135: 1201-1212.

Rai, K., L.D.Nadauld, S.Chidester, E.J.Manos, S.R.James, A.R.Karpf, B.R.Cairns, & D.A.Jones. 2006. Zebra fish Dnmt1 and Suv39h1 regulate organ-specific terminal differentiation during development. *Mol. Cell Biol.* 26: 7077-7085.

Ramsahoye, B.H., D.Biniszkiewicz, F.Lyko, V.Clark, A.P.Bird, & R.Jaenisch. 2000. Non-CpG methylation is prevalent in embryonic stem cells and may be mediated by DNA methyltransferase 3a. *Proc. Natl. Acad. Sci. U. S. A* 97: 5237-5242.

Rengaraj, D., B.R.Lee, S.I.Lee, H.W.Seo, & J.Y.Han. 2011. Expression patterns and miRNA regulation of DNA methyltransferases in chicken primordial germ cells. *PLoS. One.* 6: e19524.

Ruzov, A., J.A.Hackett, A.Prokhortchouk, J.P.Reddington, M.J.Madej, D.S.Dunican, E.Prokhortchouk, S.Pennings, & R.R.Meehan. 2009a. The interaction of xKaiso with xTcf3: a revised model for integration of epigenetic and Wnt signalling pathways. *Development* 136: 723-727.

Ruzov, A., E.Savitskaya, J.A.Hackett, J.P.Reddington, A.Prokhortchouk, M.J.Madej, N.Chekanov, M.Li, D.S.Dunican, E.Prokhortchouk, S.Pennings, & R.R.Meehan. 2009b. The non-methylated DNA-binding function of Kaiso is not required in early Xenopus laevis development. *Development* 136: 729-738.

Ruzov, A., B.Shorning, O.Mortusewicz, D.S.Dunican, H.Leonhardt, & R.R.Meehan. 2009c. MBD4 and MLH1 are required for apoptotic induction in xDNMT1-depleted embryos. *Development* 136: 2277-2286.

Santos, F., B.Hendrich, W.Reik, & W.Dean. 2002. Dynamic reprogramming of DNA methylation in the early mouse embryo. *Dev. Biol.* 241: 172-182.

Scarsdale, J.N., H.D.Webb, G.D.Ginder, & D.C.Williams, Jr. 2011. Solution structure and dynamic analysis of chicken MBD2 methyl binding domain bound to a target-methylated DNA sequence. *Nucleic Acids Res.* 39: 6741-6752.

Schafer, A., L.Schomacher, G.Barreto, G.Doderlein, & C.Niehrs. 2010. Gemcitabine functions epigenetically by inhibiting repair mediated DNA demethylation. *PLoS. One.* 5: e14060.

Schauenstein, K., A.Csordas, G.Kromer, H.Dietrich, & G.Wick. 1991. In-vivo treatment with 5-azacytidine causes degeneration of central lymphatic organs and induces autoimmune disease in the chicken. *Int. J. Exp. Pathol.* 72: 311-318.

Schlesinger, Y., R.Straussman, I.Keshet, S.Farkash, M.Hecht, J.Zimmerman, E.Eden, Z.Yakhini, E.Ben-Shushan, B.E.Reubinoff, Y.Bergman, I.Simon, & H.Cedar. 2007. Polycomb-mediated methylation on Lys27 of histone H3 pre-marks genes for de novo methylation in cancer. *Nat. Genet.* 39: 232-236.

Schroth, G.P., P.Yau, B.S.Imai, J.M.Gatewood, & E.M.Bradbury. 1990. A NMR study of mobility in the histone octamer. *FEBS Lett.* 268: 117-120.

Sharif, J., M.Muto, S.Takebayashi, I.Suetake, A.Iwamatsu, T.A.Endo, J.Shinga, Y.Mizutani-Koseki, T.Toyoda, K.Okamura, S.Tajima, K.Mitsuya, M.Okano, & H.Koseki. 2007. The SRA protein Np95 mediates epigenetic inheritance by recruiting Dnmt1 to methylated DNA. *Nature* 450: 908-912.

Sharma, S., T.K.Kelly, & P.A.Jones. 2010. Epigenetics in cancer. *Carcinogenesis* 31: 27-36.

Shi, Y.Y., Z.Y.Huang, Z.J.Zeng, Z.L.Wang, X.B.Wu, & W.Y.Yan. 2011. Diet and cell size both affect queen-worker differentiation through DNA methylation in honey bees (Apis mellifera, Apidae). *PLoS. One.* 6: e18808.

Shimoda, N., K.Yamakoshi, A.Miyake, & H.Takeda. 2005. Identification of a gene required for de novo DNA methylation of the zebrafish no tail gene. *Dev. Dyn.* 233: 1509-1516.

Simonsson, S. & J.Gurdon. 2004. DNA demethylation is necessary for the epigenetic reprogramming of somatic cell nuclei. *Nat. Cell Biol.* 6: 984-990.

Smallwood, A., P.O.Esteve, S.Pradhan, & M.Carey. 2007. Functional cooperation between HP1 and DNMT1 mediates gene silencing. *Genes Dev.* 21: 1169-1178.

Smith, T.H., T.M.Collins, & R.A.McGowan. 2011. Expression of the dnmt3 genes in zebrafish development: similarity to Dnmt3a and Dnmt3b. *Dev. Genes Evol.* 220: 347-353.

Song, C.X., K.E.Szulwach, Y.Fu, Q.Dai, C.Yi, X.Li, Y.Li, C.H.Chen, W.Zhang, X.Jian, J.Wang, L.Zhang, T.J.Looney, B.Zhang, L.A.Godley, L.M.Hicks, B.T.Lahn, P.Jin, & C.He. 2011. Selective chemical labeling reveals the genome-wide distribution of 5-hydroxymethylcytosine. *Nat. Biotechnol.* 29: 68-72.

Song, J., O.Rechkoblit, T.H.Bestor, & D.J.Patel. 2011b. Structure of DNMT1-DNA complex reveals a role for autoinhibition in maintenance DNA methylation. *Science* 331: 1036-1040.

Stancheva, I., O.El-Maarri, J.Walter, A.Niveleau, & R.R.Meehan. 2002. DNA methylation at promoter regions regulates the timing of gene activation in Xenopus laevis embryos. *Dev. Biol.* 243: 155-165.

Stancheva, I., C.Hensey, & R.R.Meehan. 2001. Loss of the maintenance methyltransferase, xDnmt1, induces apoptosis in Xenopus embryos. *EMBO J.* 20: 1963-1973.

Stancheva, I. & R.R.Meehan. 2000. Transient depletion of xDnmt1 leads to premature gene activation in Xenopus embryos. *Genes Dev.* 14: 313-327.

Stuart, G.W., J.R.Vielkind, J.V.McMurray, & M.Westerfield. 1990. Stable lines of transgenic zebrafish exhibit reproducible patterns of transgene expression. *Development* 109: 577-584.

Sun, L.Q., D.W.Lee, Q.Zhang, W.Xiao, E.H.Raabe, A.Meeker, D.Miao, D.L.Huso, & R.J.Arceci. 2004. Growth retardation and premature aging phenotypes in mice with disruption of the SNF2-like gene, PASG. *Genes Dev.* 18: 1035-1046.

Sytnikova, Y.A., A.V.Kubarenko, A.Schafer, A.N.Weber, & C.Niehrs. 2011. Gadd45a is an RNA binding protein and is localized in nuclear speckles. *PLoS. One.* 6: e14500.

Tahiliani, M., K.P.Koh, Y.Shen, W.A.Pastor, H.Bandukwala, Y.Brudno, S.Agarwal, L.M.Iyer, D.R.Liu, L.Aravind, & A.Rao. 2009. Conversion of 5-methylcytosine to 5-hydroxymethylcytosine in mammalian DNA by MLL partner TET1. *Science* 324: 930-935.

Tajima, S., H.Tsuda, N.Wakabayashi, A.Asano, S.Mizuno, & K.Nishimori. 1995. Isolation and expression of a chicken DNA methyltransferase cDNA. *J. Biochem.* 117: 1050-1057.

Takada, S., E.Berezikov, Y.L.Choi, Y.Yamashita, & H.Mano. 2009. Potential role of miR-29b in modulation of Dnmt3a and Dnmt3b expression in primordial germ cells of female mouse embryos. *RNA.* 15: 1507-1514.

Takahashi, K., K.Tanabe, M.Ohnuki, M.Narita, T.Ichisaka, K.Tomoda, & S.Yamanaka. 2007. Induction of pluripotent stem cells from adult human fibroblasts by defined factors. *Cell* 131: 861-872.

Tan, M., H.Luo, S.Lee, F.Jin, J.S.Yang, E.Montellier, T.Buchou, Z.Cheng, S.Rousseaux, N.Rajagopal, Z.Lu, Z.Ye, Q.Zhu, J.Wysocka, Y.Ye, S.Khochbin, B.Ren, & Y.Zhao. 2011. Identification of 67 histone marks and histone lysine crotonylation as a new type of histone modification. *Cell* 146: 1016-1028.

Tittle, R.K., R.Sze, A.Ng, R.J.Nuckels, M.E.Swartz, R.M.Anderson, J.Bosch, D.Y.Stainier, J.K.Eberhart, & J.M.Gross. 2011. Uhrf1 and Dnmt1 are required for development and maintenance of the zebrafish lens. *Dev. Biol.* 350: 50-63.

Torok, M.A., D.M.Gardiner, J.C.Izpisua-Belmonte, & S.V.Bryant. 1999. Sonic hedgehog (shh) expression in developing and regenerating axolotl limbs. *J. Exp. Zool.* 284: 197-206.

Tsumura, A., T.Hayakawa, Y.Kumaki, S.Takebayashi, M.Sakaue, C.Matsuoka, K.Shimotohno, F.Ishikawa, E.Li, H.R.Ueda, J.Nakayama, & M.Okano. 2006. Maintenance of self-renewal ability of mouse embryonic stem cells in the absence of DNA methyltransferases Dnmt1, Dnmt3a and Dnmt3b. *Genes Cells* 11: 805-814.

Turner, B.M. 2007. Defining an epigenetic code. *Nat. Cell Biol.* 9: 2-6.

Vardimon, L., U.Gunthert, & W.Doerfler. 1982a. In vitro methylation of the BsuRI (5'-GGCC-3') sites in the E2a region of adenovirus type 2 DNA does not affect expression in Xenopus laevis oocytes. *Mol. Cell Biol.* 2: 1574-1580.

Vardimon, L., A.Kressmann, H.Cedar, M.Maechler, & W.Doerfler. 1982b. Expression of a cloned adenovirus gene is inhibited by in vitro methylation. *Proc. Natl. Acad. Sci. U. S. A* 79: 1073-1077.

Vardimon, L., D.Renz, & W.Doerfler. 1983. Can DNA methylation regulate gene expression? *Recent Results Cancer Res.* 84: 90-102.

Veenstra, G.J. & A.P.Wolffe. 2001. Constitutive genomic methylation during embryonic development of Xenopus. *Biochim. Biophys. Acta* 1521: 39-44.

Wade, P.A., A.Gegonne, P.L.Jones, E.Ballestar, F.Aubry, & A.P.Wolffe. 1999. Mi-2 complex couples DNA methylation to chromatin remodelling and histone deacetylation. *Nat. Genet.* 23: 62-66.

Wang, Y., M.Jorda, P.L.Jones, R.Maleszka, X.Ling, H.M.Robertson, C.A.Mizzen, M.A.Peinado, & G.E.Robinson. 2006. Functional CpG methylation system in a social insect. *Science* 314: 645-647.

Williams, K., J.Christensen, M.T.Pedersen, J.V.Johansen, P.A.Cloos, J.Rappsilber, & K.Helin. 2011. TET1 and hydroxymethylcytosine in transcription and DNA methylation fidelity. *Nature* 473: 343-348.

Wongtawan, T., J.E. Taylor, K.A. Lawson, I. Wilmut & S. Pennings. 2011. Histone H4K20me3 and HP1α are late heterochromatin markers in development but present in undifferentiated ES cells. *J. Cell Science,* 124, 1878-1890.

Wolfl, S., M.Schrader, & B.Wittig. 1991. Lack of correlation between DNA methylation and transcriptional inactivation: the chicken lysozyme gene. *Proc. Natl. Acad. Sci. U. S. A* 88: 271-275.

Wossidlo, M., T.Nakamura, K.Lepikhov, C.J.Marques, V.Zakhartchenko, M.Boiani, J.Arand, T.Nakano, W.Reik, & J.Walter. 2011. 5-Hydroxymethylcytosine in the mammalian zygote is linked with epigenetic reprogramming. *Nat. Commun.* 2: 241.

Wu, H., A.C.D'Alessio, S.Ito, K.Xia, Z.Wang, K.Cui, K.Zhao, Y.E.Sun, & Y.Zhang. 2011. Dual functions of Tet1 in transcriptional regulation in mouse embryonic stem cells. *Nature* 473: 389-393.

Wu, S.C. & Y.Zhang. 2010. Active DNA demethylation: many roads lead to Rome. *Nat. Rev. Mol. Cell Biol.* 11: 607-620.

Xing, J., L.Kang, & Y.Jiang. 2011. Effect of dietary betaine supplementation on lipogenesis gene expression and CpG methylation of lipoprotein lipase gene in broilers. *Mol. Biol. Rep.* 38: 1975-1981.

Xu, Q., Y.Zhang, D.Sun, Y.Wang, & Y.Yu. 2007. Analysis on DNA methylation of various tissues in chicken. *Anim Biotechnol.* 18: 231-241.

Yakushiji, N., M.Suzuki, A.Satoh, T.Sagai, T.Shiroishi, H.Kobayashi, H.Sasaki, H.Ide, & K.Tamura. 2007. Correlation between Shh expression and DNA methylation status of the limb-specific Shh enhancer region during limb regeneration in amphibians. *Dev. Biol.* 312: 171-182.

Yamakoshi, K. & N.Shimoda. 2003. De novo DNA methylation at the CpG island of the zebrafish no tail gene. *Genesis.* 37: 195-202.

Yang, X., J.Zheng, L.Qu, S.Chen, J.Li, G.Xu, & N.Yang. 2011. Methylation status of cMHM and expression of sex-specific genes in adult sex-reversed female chickens. *Sex Dev.* 5: 147-154.

Yen, J.L & X.Y Jia. 2010 Rapid Enzymatic DNA degradation for Quantitation of 5-methylcytosine and 5-hydroxymethylcytosine. Zymo Research Group. url: http://www.zymoresearch.com/zrc/pdf/10xgen-poster1.pdf

Yokomine, T., K.Hata, M.Tsudzuki, & H.Sasaki. 2006. Evolution of the vertebrate DNMT3 gene family: a possible link between existence of DNMT3L and genomic imprinting. *Cytogenet. Genome Res.* 113: 75-80.

Yu, Y., H.Zhang, F.Tian, L.Bacon, Y.Zhang, W.Zhang, & J.Song. 2008a. Quantitative evaluation of DNA methylation patterns for ALVE and TVB genes in a neoplastic disease susceptible and resistant chicken model. *PLoS. One.* 3: e1731.

Yu, Y., H.Zhang, F.Tian, W.Zhang, H.Fang, & J.Song. 2008b. An integrated epigenetic and genetic analysis of DNA methyltransferase genes (DNMTs) in tumor resistant and susceptible chicken lines. *PLoS. One.* 3: e2672.

Zemach, A., I.E.McDaniel, P.Silva, & D.Zilberman. 2010. Genome-wide evolutionary analysis of eukaryotic DNA methylation. *Science* 328: 916-919.

Zeng, J. & S.V.Yi. 2010. DNA methylation and genome evolution in honeybee: gene length, expression, functional enrichment covary with the evolutionary signature of DNA methylation. *Genome Biol. Evol.* 2: 770-780.

Zhu, B., Y.Zheng, D.Hess, H.Angliker, S.Schwarz, M.Siegmann, S.Thiry, & J.P.Jost. 2000. 5-methylcytosine-DNA glycosylase activity is present in a cloned G/T mismatch DNA glycosylase associated with the chicken embryo DNA demethylation complex. *Proc. Natl. Acad. Sci. U. S. A* 97: 5135-5139.

Zlatanova, J., T.C.Bishop, J.M.Victor, V.Jackson, & H.K.van. 2009. The nucleosome family: dynamic and growing. *Structure.* 17: 160-171.

8

Aberrant DNA Methylation of Imprinted Loci in Male and Female Germ Cells of Infertile Couples

Takahiro Arima[1], Hiroaki Okae[1],
Hitoshi Hiura[1], Naoko Miyauchi[1],
Fumi Sato[1], Akiko Sato[2] and Chika Hayashi[2]
*[1]Department of Informative Genetics,
Environment and Genome Research Center,
Tohoku University Graduate School of Medicine,
2-1 Seiryo-cho, Aoba-ku, Sendai,
[2]Departments of Obstetrics and Gynecology,
Tohoku University Graduate School of Medicine, Sendai,
Japan*

1. Introduction

Recent studies have identified an increased incidence of the normally rare imprinting disorders, Beckwith-Wiedemann syndrome (BWS; NIM130650) and Angelman syndrome (AS; NIM105830), in ART babyes (DeBaun et al., 2003; Gosden et al., 2003; Maher, 2005). The identification of epigenetic changes at imprinted loci in ART babyes has led to the suggestion that the technique itself may predispose embryos to acquire imprinting errors. Both *in vitro* fertilization (IVF) and intracytoplasmic sperm injection (ICSI) are associated with the increased risk of imprinting disorders and it is not clear at what point these imprinting errors arise (Bowdin et al., 2007; Doornbos et al., 2007).

Genomic imprinting confers different functions on the two parental genomes during development by silencing one allele of each imprinted gene in a parent-of-origin dependent manner (Ohlsson et al., 1998; Reik and Walter, 1998; Surani, 1998; Tilghman, 1999). Imprinting accounts for the requirement of both maternal and paternal genomes in normal development and plays significant roles in regulating embryonic growth, placental function and neurobehavioral processes (McGrath and Solter, 1984; Surani et al., 1984). Aberrant expression of some imprinted genes has been linked to a number of human diseases, developmental abnormalities and malignant tumors (Paulsen and Ferguson-Smith, 2001). The epigenetic modifications that are imposed during gametogenesis act as primary imprint markers to distinguish the maternal and paternal alleles (Surani, 1998). The most likely candidate for the gametic mark is DNA methylation. Allele-specific DNA methylation has been observed in the vicinity of most imprinted genes. In some instances, the methylation is present on the inactive gene, suggesting a role for DNA methylation in silencing of the gene

(Figure 1). DNA methylation is both a heritable and reversible epigenetic modification that is stably propagated after DNA replication. In order to transmit this epigenetic mark from one generation to the next, the imprints have to be erased in the primordial germ cells (PGCs) (Hajkova et al., 2002; Lee et al., 2002) and re-established during gametogenesis in a sex-specific manner (Figure 2).

Major epigenetic events take place during female and male germ cell development and the preimplantation stages of embryonic development (Lucifero et al., 2004a). *In vitro* culture may expose the genome to environmental factors that prevent the proper establishment of the DNA methylation . However, the risks cannot easily be evaluated for ART because patients who receive ART may differ both demographically and genetically from the general people. Usually, patients requesting ART have a low fertility rate, an increased reproductive loss rate and are of advanced age, all of which are associated with various fetal and neonatal abnormalities. These confounding factors make it difficult to evaluate the risk and safety of ART procedures.

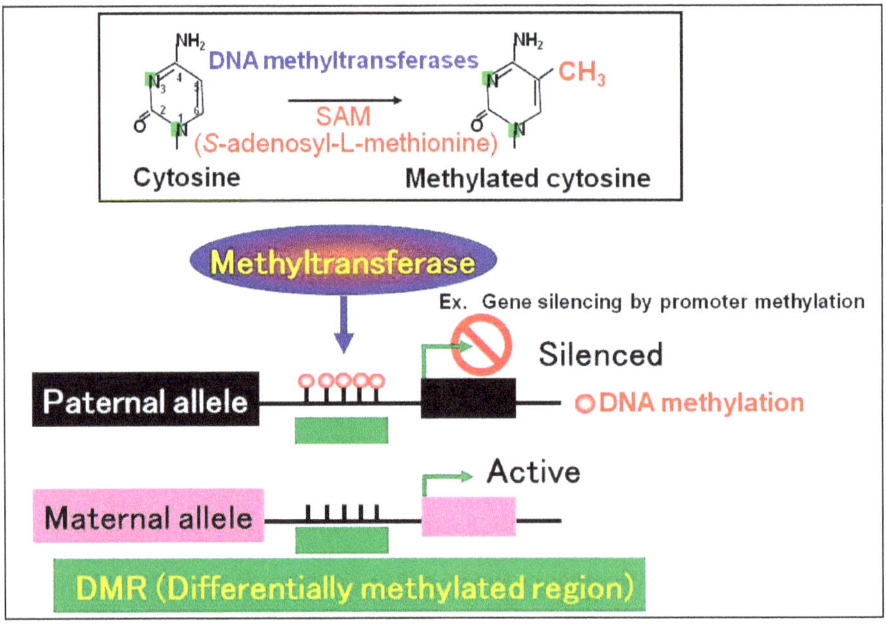

Fig. 1. The regulation of imprinted gene by DNA methylation.

DNA cytosine methylation (methyltransferase) regulates imprinted gene expression. Differentially methylated regions (DMRs) are commonly associated with imprinted genes.

Genomic imprinting is a gamete-specific modification (DNA methylation) that causes differential expression of the two parental alleles. An imprint starts by a gametogenesis process, and it is maintained for stability up to a somatic cell .In addition, it is erased in a primordial germ cell.

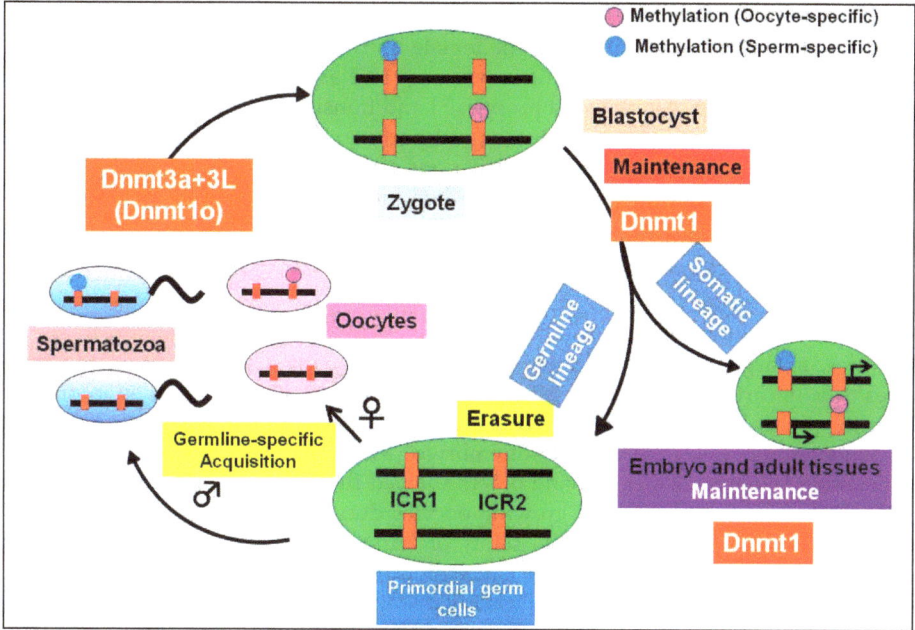

Fig. 2. Imprinted process through the life cycle.

ART involve the isolation, handling and culture of gametes and early embryos, generally after hormone stimulation protocols (superovulation), at a time when the epigenetic marks at imprinted loci are relatively malleable and therefore potentially vulnerable to external influences. Epigenetic marks appear to be at risk during several stages of the ART procedures including the superovulation, various culture mediums, cryopreservation and embryo transfer. Our recent work, and that of others, suggests the possibility that infertile men, particularly those with oligozoospermia, carry imprinting errors in their sperm (Kobayashi et al., 2007; Marques et al., 2004; Marques et al., 2008). Therefore the increase in the incidence of imprinting disorders in individuals born by ART may be due, in some cases, to the use of this sub-optimal sperm.

2.1 Aberrant of DNA methylation imprint in superovulation oocytes

Imprinted genes are particularly vulnerable targets for numerous human pathologies since single genetic or epigenetic changes can deregulate their function. Epigenetic mark, DNA methylation is found associated with only one parental allele within discrete locations known as differentially methylated regions (DMRs) (Figure 1)(Surani, 1998). The acquisition of DNA methylation at these key regions occurs primarily in the parental germ line during male and female gametogenesis and is thought to direct the imprinting process (Lucifero et al., 2002; Obata and Kono, 2002).

The acquisition of the imprint methylation marks is significantly different between the two germ lines. In the male germ line, *H19*, *Rasgrf1* and *Gtl2* methylation imprints are initiated prenatally during embryonic germ cell development and are complete by the pachytene phase of postnatal spermatogenesis in mice (Davis et al., 1999; Davis et al., 2000; Li et al., 2004; Ueda et al., 2000). In contrast, in the female line, the maternal methylations, such as *Igf2r*, *Snrpn*, *Peg1*, *Peg3* etc. methylations, are acquired asynchronously in a gene-specific manner, while oocytes are arrested at prophase I and transitioned from primordial to antral follicles during the postnatal growth phase (post-pachytene) (Lucifero et al., 2004b). Nuclear transplantation using postnatal oocytes at various stages of maturation point to this same window of oocyte development as the time when functional imprints are acquired (Bao et al., 2000; Obata and Kono, 2002).

Much debate has recently surrounded the issues of possible epigenetic alterations brought about by human ART (Lucifero et al., 2004a). One of the important issues is artificial induction of ovulation with high doses of gonadotrophins (superovulation). In the ART procedures, a large amount of gonadotrophins are usually used to obtain the mature oocytes. It is uncertain whether exogenous gonadotrophins alter the maturation process of eggs or the physiological environment of the uterus. Studies in animals have suggested that superovulation decreases the viability of embryos (McKiernan and Bavister, 1998; Van der Auwera and D'Hooghe, 2001). We and others showed the occurrence of methylation errors on several imprinted genes in full growing oocytes due to superovulation in humans and mice, which will help to estimate the safety of artificial induction of ovulation (Market-Velker et al., 2010; Sato et al., 2007). Under controlled ovarian stimulations, immature oocytes are collected. These oocytes are usually discarded due to the possibility of abnormal embryonic development or an increased rate of abortion (Smith et al., 2000). However in cases of poor responders and in patients with an unsynchronized cohort of follicles, where the presence of immature oocytes is frequent after stimulation the use of immature oocytes for IVF is important in order to incease the number of embryos obtained in each cycle. In addition, *in vitro* matured (IVM) oocytes and some devices of maturation might be a significant risk of the imprinting diseases.

2.2 Aberrant of DNA methylation imprint in oligospermic patients

In mice, paternally methylation imprints are initiated prenatally during embryonic germ cell development. In humans, limited information is available on the methylation status of imprinted genes during gametogenesis and embryogenesis. During normal spermatogenesis, the erasure of methylation marks of the maternally imprinted gene *SNRPN* (Manning et al., 2001) and the resetting of the paternally imprinted gene *H19* (Kerjean et al., 2000; Marques et al., 2004) have been reported to be completed before germ cells enter meiosis. We and other reported that there was abnormal imprinting in oligospermic patients and in a small number of the normospermic patients (Marques et al., 2004). We examined the DNA methylation status of seven imprinted genes using a combined bisulphite polymerase chain reaction (PCR) restriction analysis and sequencing technique on

We examined the DNA methylation status of several imprinted genes using a combined bisulphite (PCR) restriction analysis and sequencing technique on ejaculated sperm DNA obtained from infertile men. We found abnormal methylation of the paternal imprint in 14.4% and an abnormal maternal imprint in 20.6%. The majority of these doubly defective samples were in men with moderate or severe oligospermia. These abnormalities were specific to imprinted loci since we found that global DNA methylation was normal in these samples (Kobayashi et al., 2007). In our sperm analysis, we found incomplete methylation; normospermia in 15.2%, moderate oligospermia in 37.5% and severe oligospermia in 80% (Figure 3). These results reveal that abnormal spermatogenesis (leading to low sperm counts) is associated with a defective imprint methylation.14.4% and an abnormal maternal imprint in 20.6%. The majority of these doubly defective samples were in men with moderate or severe oligospermia. These abnormalities were specific to imprinted loci since we found that global DNA methylation was normal in these samples (Kobayashi et al., 2007). In our sperm analysis, we found incomplete methylation; normospermia in 15.2%, moderate oligospermia in 37.5% and severe oligospermia in 80% (Figure 3). These results reveal that abnormal spermatogenes(leading to low sperm counts) is associated with a defective imprint methylation.

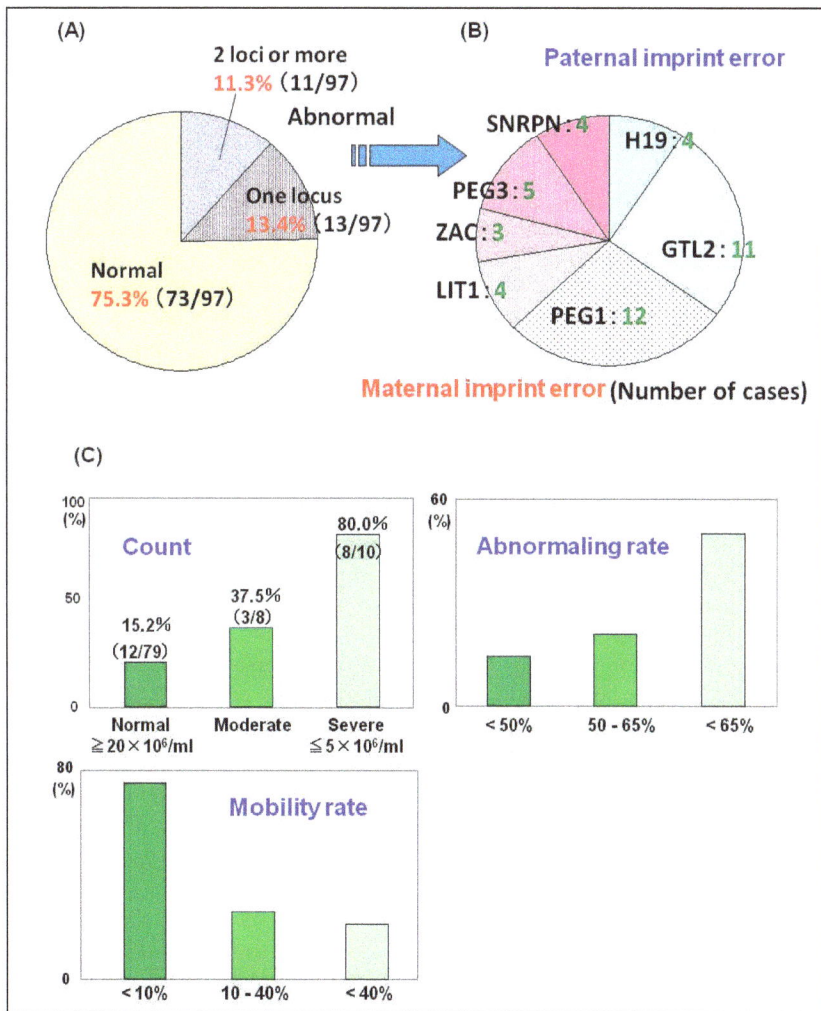

Fig. 3. Aberrant DNA methylation of imprinted loci in sperm from infertile male.

(A) Frequency of the imprint methylation error (B) Abnormal imprinted loci (C) Abnormal methylation imprint and sperm concentrations. Methylation errors at maternal and paternal imprinted loci that were specific to oligospermic men.

The most frequent methylation error was seen in the *PEG1* DMR. In our previous report, we showed that demethylation of *PEG1* was present in the growing oocytes from superovulated infertile women (Sato et al., 2007). This *PEG1* DMR may be especially vulnerable to errors. In both humans and mice, the *PEG1* DMR located in the promoter, the first exon and is unmethylated on the active paternal allele (Kobayashi et al., 1997; Lefebvre et al., 1997). Paternal transmission of a methylated *Peg1* gene results in growth-retarded embryos. Abnormal behavior has also been noted in *Peg1*-deficient females (Lefebvre et al., 1998). Generally, ART babies are characterized by low-weight birth.

Recently, we demonstrated that in a few cases, the methylation errors in the sperm were present in the ART aborted conceptus (Kobayashi et al., 2009).

2.3 Evidence of imprint defects associated with ART

Major epigenetic events take place during this time and the process of ART may expose the epigenome to external influences preventing the proper establishment and maintenance of genomic imprints (DeBaun et al., 2003; Maher et al., 2003). Except for superovulation, another issue is related to the culture conditions. Some studies have shown that exposure of mouse embryos to different culture conditions can alter the expression and imprinting of various genes which could result in abnormal development (DeBaun et al., 2003; Gicquel et al., 2003; Maher et al., 2003). The third issue is the potential effect of embryo cryopreservation (Emiliani et al., 2000; Honda et al., 2001). The timing of embryo transfer may also present issues. Some studies on monochorionic dizygotic twins and conjoined twins with BWS resulting from transfer of embryos at the blastocyst stage revealed demethylation of *LIT1* (*KCNQ1OT1*) (Miura and Niikawa, 2005; Shimizu et al., 2004), suggesting that this demethylation occurs at a critical stage of pre-implantation development. Furthermore, there may be other serious issues causing yet unknown risks of ART.

We previously reported cases of a mosaic methylation pattern in ART-babies (Kobayashi et al., 2007). Nutrients, including methyl-substrates such as vitamins B6 and folic acid, influence DNA methylation and histone modification at gene promoters. Imprinting errors in sperm at the paternally-methylated DMRs may be reversed using a similar approach. This approach has already been applied to treat autism (Chen et al., 2003).

2.4 Automated high-throughput procedure for the detection of alterations in DNA methylation

Southern blotting was the original technique used routinely to analyse DNA methylation (J.Sambrook and D.W.Russell, 2001). This technique requires a relatively large quantity of DNA (5-10 µg) that is usually digested with two restriction enzymes, one of which is methylation sensitive. The DNA is fractionated on an agarose gel, transferred a positively charged nylon membrane, hybridized with a radioisotope labeled probe, washed and exposed to autoradiography. This method has largely been superceded by methods involving The methylation status of a specific sequence can then be measured by combined bisulphite PCR restriction analysis (COBRA) or by the sequencing of the PCR product. The sodium bisulphite treatment of genomic DNA converts unmethylated cytosine to uracil leaving the methylated cytosines unconverted. The combination of COBRA and the sequencing method provides accuracy and sensitivity but nonetheless, there are still limitations with this method particularly the expertise required to obtain accurate results, the time taken to achieve a result and the relative cost, rendering it unsuitable for clinical diagnosis. Recently, we developed a new method of DNA methylation analysis , PCR-Luminex (Sato et al., 2011). This method combines PCR and sequence-specific oligonucleotide probe (SSOP) protocols with the Luminex 100 xMAP flow cytometry dual-laser system to quantitate fluorescently labeled oligonucleotides attached to color-coded microbeads (Figure 4).

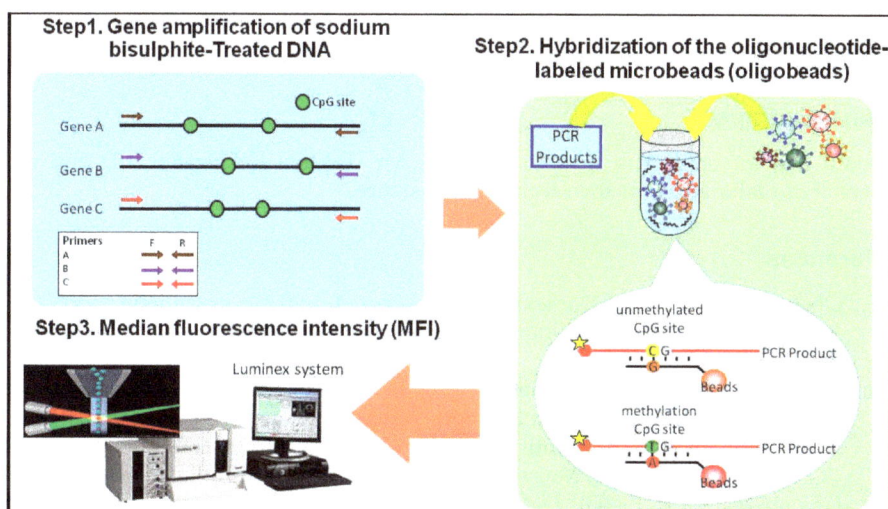

Fig. 4. BPL methylation assay.

Bisulphite PCR-Luminex (BPL) method involved PCR amplification of bisulphite-DNA, hybridization, a streptavidin-phycoerythrin (SA-PE) reaction and identification of the fluorescent microbeads by determining the preset ratio of internalized dyes to distinguish between cytosine and uracil (methylation and non-methylation). This is one of new high-throughput, high-resolution DNA methylation analysis methods.

The PCR-Luminex method can identify one base substitution by specific hybridization and therefore could potentially be applied to the identification of DNA methylation using the bisulphite conversion technique to essentially generate the two different bases, cytosine and uracil (methylation and non-methylation), called BPL. We applied these techniques to examine the methylation imprints of 8 DMRs (paternally methylated DMRs: *ZDBF2, H19* and *GTL2,* maternally methylated DMRs: *PEG1, ZAC, SNRPN, PEG3* and *LIT1*) in the sperm DNA to assess the quality of these samples with respect to imprint status (Sato et al., 2011). BPL In the mean future, new methylation analyses will prove to be a simple, accurate and rapid approach and therefore suitable for clinical application.

3. Conclusion

As a result of our studies and the work of others in this area, we recommend that imprint methylation analyses be added to the routine sperm examination (counting, mobility, abnormality analyses) to identify pre-existing imprint mutations. It may be possible, in the future, to reverse aberrant DNA methylation and the present analysis will provide useful information in that regard. Altered expression and methylation of imprinted genes is a frequent event in adult cancers (Feinberg et al., 2006). In addition to determining the frequency of classic imprinting disorders, it will be important to determine cancer occurrence in ART offspring. A retrospective examination of imprinted loci and the constitution of children born after each ART method will reveal the safest and most ethical approach to use. We believe these studies will be valuable for the development of standard ART. In our opinion, before translating new techniques into practice, more research, particularly in animals, is desirable. In addition, better ART child follow-up and a fresh approach to regulation are also needed.

4. Acknowledgment

We would like to thank Mis. Yukiko Abe RA, Mr. Takuya Koshizaka PhD and all the members of our laboratory for their technical assistance.

5. References

Bao, S., Obata, Y., Carroll, J., Domeki, I. and Kono, T. (2000). Epigenetic modifications necessary for normal development are established during oocyte growth in mice. *Biol Reprod* 62, 616-621.

Bowdin, S., Allen, C., Kirby, G., Brueton, L., Afnan, M., Barratt, C., Kirkman-Brown, J., Harrison, R., Maher, E. R. and Reardon, W. (2007). A survey of assisted reproductive technology births and imprinting disorders. *Hum Reprod* 22, 3237-3240.

Chen, W. G., Chang, Q., Lin, Y., Meissner, A., West, A. E., Griffith, E. C., Jaenisch, R. and Greenberg, M. E. (2003). Derepression of BDNF transcription involves calcium-dependent phosphorylation of MeCP2. *Science* 302, 885-889.

Davis, T. L., Trasler, J. M., Moss, S. B., Yang, G. J. and Bartolomei, M. S. (1999). Acquisition of the H19 methylation imprint occurs differentially on the parental alleles during spermatogenesis. *Genomics* 58, 18-28.

Davis, T. L., Yang, G. J., McCarrey, J. R. and Bartolomei, M. S. (2000). The H19 methylation imprint is erased and re-established differentially on the parental alleles during male germ cell development. *Hum Mol Genet* 9, 2885-2894.

DeBaun, M. R., Niemitz, E. L. and Feinberg, A. P. (2003). Association of in vitro fertilization with Beckwith-Wiedemann syndrome and epigenetic alterations of LIT1 and H19. *Am J Hum Genet* 72, 156-160.

Doornbos, M. E., Maas, S. M., McDonnell, J., Vermeiden, J. P. and Hennekam, R. C. (2007). Infertility, assisted reproduction technologies and imprinting disturbances: a Dutch study. *Hum Reprod* 22, 2476-2480.

Emiliani, S., Van den Bergh, M., Vannin, A. S., Biramane, J. and Englert, Y. (2000). Comparison of ethylene glycol, 1,2-propanediol and glycerol for cryopreservation of slow-cooled mouse zygotes, 4-cell embryos and blastocysts. *Hum Reprod* 15, 905-910.

Feinberg, A. P., Ohlsson, R. and Henikoff, S. (2006). The epigenetic progenitor origin of human cancer. *Nat Rev Genet* 7, 21-33.

Gicquel, C., Gaston, V., Mandelbaum, J., Siffroi, J. P., Flahault, A. and Le Bouc, Y. (2003). In vitro fertilization may increase the risk of Beckwith-Wiedemann syndrome related to the abnormal imprinting of the KCN1OT gene. *Am J Hum Genet* 72, 1338-1341.

Gosden, R., Trasler, J., Lucifero, D. and Faddy, M. (2003). Rare congenital disorders, imprinted genes, and assisted reproductive technology. *Lancet* 361, 1975-1977.

Hajkova, P., Erhardt, S., Lane, N., Haaf, T., El-Maarri, O., Reik, W., Walter, J. and Surani, M. A. (2002). Epigenetic reprogramming in mouse primordial germ cells. *Mech Dev* 117, 15-23.

Honda, S., Weigel, A., Hjelmeland, L. M. and Handa, J. T. (2001). Induction of telomere shortening and replicative senescence by cryopreservation. *Biochem Biophys Res Commun* 282, 493-498.

J.Sambrook and D.W.Russell. (2001). Molecular Cloning,third edition: Cold Spring Harbor Laboratory Press.

Kerjean, A., Dupont, J. M., Vasseur, C., Le Tessier, D., Cuisset, L., Paldi, A., Jouannet, P. and Jeanpierre, M. (2000). Establishment of the paternal methylation imprint of the human H19 and MEST/PEG1 genes during spermatogenesis. *Hum Mol Genet* 9, 2183-2187.

Kobayashi, H., Hiura, H., John, R. M., Sato, A., Otsu, E., Kobayashi, N., Suzuki, R., Suzuki, F., Hayashi, C., Utsunomiya, T. et al. (2009). DNA methylation errors at imprinted loci after assisted conception originate in the parental sperm. *Eur J Hum Genet* 17, 1582-1591.

Kobayashi, H., Sato, A., Otsu, E., Hiura, H., Tomatsu, C., Utsunomiya, T., Sasaki, H., Yaegashi, N. and Arima, T. (2007). Aberrant DNA methylation of imprinted loci in sperm from oligospermic patients. *Hum Mol Genet* 16, 2542-2551.

Kobayashi, S., Kohda, T., Miyoshi, N., Kuroiwa, Y., Aisaka, K., Tsutsumi, O., Kaneko-Ishino, T. and Ishino, F. (1997). Human PEG1/MEST, an imprinted gene on chromosome 7. *Hum Mol Genet* 6, 781-786.

Lee, J., Inoue, K., Ono, R., Ogonuki, N., Kohda, T., Kaneko-Ishino, T., Ogura, A. and Ishino, F. (2002). Erasing genomic imprinting memory in mouse clone embryos produced from day 11.5 primordial germ cells. *Development* 129, 1807-1817.

Lefebvre, L., Viville, S., Barton, S. C., Ishino, F., Keverne, E. B. and Surani, M. A. (1998). Abnormal maternal behaviour and growth retardation associated with loss of the imprinted gene Mest. *Nat Genet* 20, 163-169.

Lefebvre, L., Viville, S., Barton, S. C., Ishino, F. and Surani, M. A. (1997). Genomic structure and parent-of-origin-specific methylation of Peg1. *Hum Mol Genet* 6, 1907-1915.

Li, J. Y., Lees-Murdock, D. J., Xu, G. L. and Walsh, C. P. (2004). Timing of establishment of paternal methylation imprints in the mouse. *Genomics* 84, 952-960.

Lucifero, D., Chaillet, J. R. and Trasler, J. M. (2004a). Potential significance of genomic imprinting defects for reproduction and assisted reproductive technology. *Hum Reprod Update* 10, 3-18.

Lucifero, D., Mann, M. R., Bartolomei, M. S. and Trasler, J. M. (2004b). Gene-specific timing and epigenetic memory in oocyte imprinting. *Hum Mol Genet* 13, 839-849.

Lucifero, D., Mertineit, C., Clarke, H. J., Bestor, T. H. and Trasler, J. M. (2002). Methylation dynamics of imprinted genes in mouse germ cells. *Genomics* 79, 530-538.

Maher, E. R. (2005). Imprinting and assisted reproductive technology. *Hum Mol Genet* 14 Spec No 1, R133-138.

Maher, E. R., Brueton, L. A., Bowdin, S. C., Luharia, A., Cooper, W., Cole, T. R., Macdonald, F., Sampson, J. R., Barratt, C. L., Reik, W. et al. (2003). Beckwith-Wiedemann syndrome and assisted reproduction technology (ART). *J Med Genet* 40, 62-64.

Manning, M., Lissens, W., Weidner, W. and Liebaers, I. (2001). DNA methylation analysis in immature testicular sperm cells at different developmental stages. *Urol Int* 67, 151-155.

Market-Velker, B. A., Zhang, L., Magri, L. S., Bonvissuto, A. C. and Mann, M. R. (2010). Dual effects of superovulation: loss of maternal and paternal imprinted methylation in a dose-dependent manner. *Hum Mol Genet* 19, 36-51.

Marques, C. J., Carvalho, F., Sousa, M. and Barros, A. (2004). Genomic imprinting in disruptive spermatogenesis. *Lancet* 363, 1700-1702.

Marques, C. J., Costa, P., Vaz, B., Carvalho, F., Fernandes, S., Barros, A. and Sousa, M. (2008). Abnormal methylation of imprinted genes in human sperm is associated with oligozoospermia. *Mol Hum Reprod* 14, 67-74.

McGrath, J. and Solter, D. (1984). Completion of mouse embryogenesis requires both the maternal and paternal genomes. *Cell* 37, 179-183.

McKiernan, S. H. and Bavister, B. D. (1998). Gonadotrophin stimulation of donor females decreases post-implantation viability of cultured one-cell hamster embryos. *Hum Reprod* 13, 724-729.

Miura, K. and Niikawa, N. (2005). Do monochorionic dizygotic twins increase after pregnancy by assisted reproductive technology? *J Hum Genet* 50, 1-6.

Obata, Y. and Kono, T. (2002). Maternal primary imprinting is established at a specific time for each gene throughout oocyte growth. *J Biol Chem* 277, 5285-5289.

Ohlsson, R., Tycko, B. and Sapienza, C. (1998). Monoallelic expression: 'there can only be one'. *Trends Genet* 14, 435-438.

Paulsen, M. and Ferguson-Smith, A. C. (2001). DNA methylation in genomic imprinting, development, and disease. *J Pathol* 195, 97-110.

Reik, W. and Walter, J. (1998). Imprinting mechanisms in mammals. *Curr Opin Genet Dev* 8, 154-164.

Sato, A., Hiura, H., Okae, H., Miyauchi, N., Abe, Y., Utsunomiya, T., Yaegashi, N. and Arima, T. (2011). Assessing loss of imprint methylation in sperm from subfertile men using novel methylation polymerase chain reaction Luminex analysis. *Fertil Steril* 95, 129-134, 134 e1-4.

Sato, A., Otsu, E., Negishi, H., Utsunomiya, T. and Arima, T. (2007). Aberrant DNA methylation of imprinted loci in superovulated oocytes. *Hum Reprod* 22, 26-35.

Shimizu, Y., Fukuda, J., Sato, W., Kumagai, J., Hirano, H. and Tanaka, T. (2004). First-trimester diagnosis of conjoined twins after in-vitro fertilization-embryo transfer (IVF-ET) at blastocyst stage. *Ultrasound Obstet Gynecol* 24, 208-209.

Smith, S. D., Mikkelsen, A. and Lindenberg, S. (2000). Development of human oocytes matured in vitro for 28 or 36 hours. *Fertil Steril* 73, 541-544.

Surani, M. A. (1998). Imprinting and the initiation of gene silencing in the germ line. *Cell* 93, 309-312.

Surani, M. A., Barton, S. C. and Norris, M. L. (1984). Development of reconstituted mouse eggs suggests imprinting of the genome during gametogenesis. *Nature* 308, 548-550.

Tilghman, S. M. (1999). The sins of the fathers and mothers: genomic imprinting in mammalian development. *Cell* 96, 185-193.

Ueda, T., Abe, K., Miura, A., Yuzuriha, M., Zubair, M., Noguchi, M., Niwa, K., Kawase, Y., Kono, T., Matsuda, Y. et al. (2000). The paternal methylation imprint of the mouse H19 locus is acquired in the gonocyte stage during foetal testis development. *Genes Cells* 5, 649-659.

Van der Auwera, I. and D'Hooghe, T. (2001). Superovulation of female mice delays embryonic and fetal development. *Hum Reprod* 16, 1237-1243.

Could Tissue-Specific Genes Be Silenced in Cattle Carrying the Rob(1;29) Robertsonian Translocation?

Alicia Postiglioni[1], Rody Artigas[1], Andrés Iriarte[1],
Wanda Iriarte[1], Nicolás Grasso[1] and Gonzalo Rincón[2]
[1]Depto. de Genética y Mejora Animal, Facultad de Veterinaria,
Universidad de la República, Montevideo,
[2]Department of Animal Science,
University of California, Davis,
[1]Uruguay
[2]USA

1. Introduction

Robertsonian translocations (rob) are frequent chromosome rearrangements observed in human (Nielsen & Wohlert, 1991), plants (Friebe et al., 2005), mice (Saferali et al., 2010) and domestic animals (Fries & Popescu, 1999), that are considered in arthropods classical mechanisms of chromosome evolution (Robertson, 1916). In cattle, this type of translocation is the most widespread chromosomal abnormality in European and American Creole cattle breeds. In Uruguayan Creole cattle the monocentric Robertsonian translocation rob(1;29) presents a frequency of 4%, whereas a frequency between 4 to 10% is described in beef and dairy cattle breeds (King, 1991; Postiglioni et al., 1996). In these populations there is a high incidence of heterozygous carriers and a very small number or even absence of homozygous individuals (Kastelic & Mapletoft, 2003). This situation has also been observed in the majority of cattle populations in which rob(1;29) has been detected (Ducos et al., 2008).

1.1 The finding of rob(1;29) when trying to solve a sub-fertility problem in an Artificial Insemination Center

In 1969, I. Gustavsson described for the first time the Robertsonian translocation rob(1;29) in cattle. This finding, together with the discoveries of pioneers in the field such as Ohno et al., (1962) and Herzog & Hohn (1968), have been important milestones in the development of domestic animal cytogenetics. Particularly, rob(1;29) was found in the Swedish Artificial Insemination Center of Black and White cattle, associated with decreased fertility. This aneuploid alteration causes a reduction in reproductive efficiency, increasing calving intervals, non-return to service and culling rates; thus leading to important economic losses in commercial cattle herds (Bonnet-Garnier et al., 2006, 2008).

1.2 Chromosome rearrangements in the bovine genome – The origin of rob(1;29)

Bovine presents a complex genome due to the presence of high number of repetitive sequences involved in macro and micro-rearrangements that are essential in the evolution of bovids. Schibler et al., (1998) presented a high resolution integrated bovine comparative map where the analysis of break point regions revealed specific repeated density patterns, suggesting that transposons (TEs) may have played a significant role in chromosome evolution and genome plasticity. However, bovines have a morphologically simple karyotype, corresponding to 2n=60, XX; XY. All the autosomes pairs are acrocentrics and the sex chromosomes are the large submetacentric X and the tiny metacentric Y. In females the sex chromosomes correspond two large submetacentric X´s.

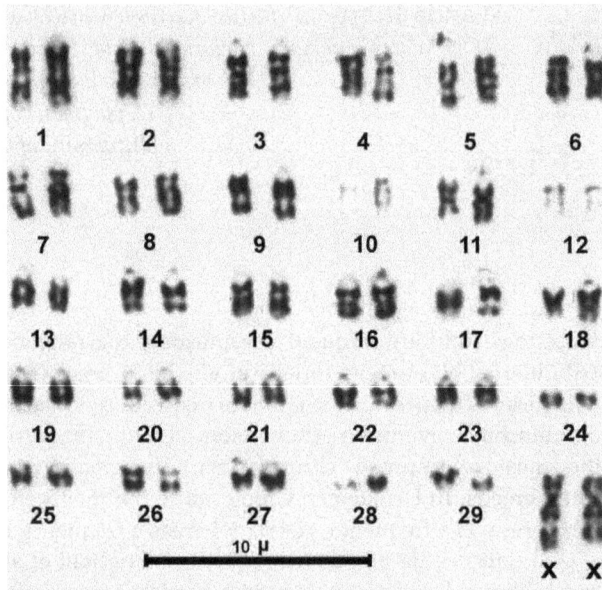

Fig. 1. Karyotype of cattle female, treated to G-banding. Observe all the acrocentric autosome pairs and the large submetacentric X ´s. Observe light G-banding in BTA1 and X`s chromosomes (courtesy: Postiglioni,A., 1987).

Theorically, all the acrocentric autosomes have the same probability to be involved in centric fusions, but the only known monocentric Robertsonian translocation corresponds to rob(1;29), being considered as a chromosome polymorphism (Di Meo et al., 2006). Lymphocyte culture in cattle carriers of rob(1;29) clearly present all the metaphases with this chromosome rearrangement.

Today, it is known that this chromosome translocation may have originated both from a centric fusion of chromosomes 1 and 29 and from chromosome rearrangements like pericentric inversions, in which a DNA probe (INRA143) normally mapping to BTA29, appeared proximally to rob(1;29)q-arm. Recently, Di Meo et al.,(2006) suggested a transposition as a complex rearrangement that could also be involved in the origin of this chromosome alteration.

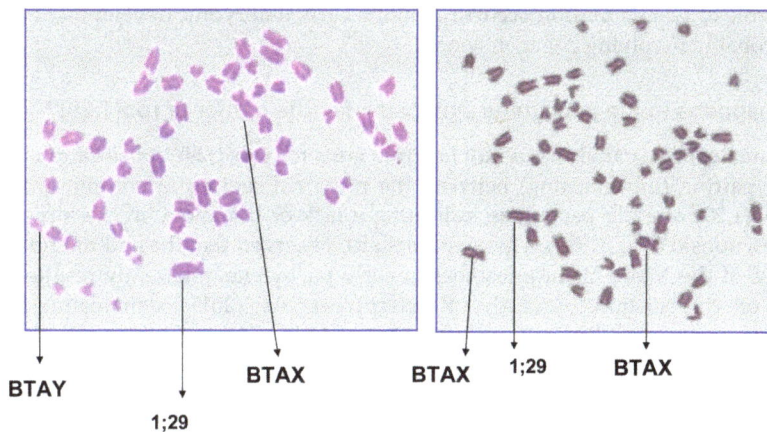

Fig. 2. Lymphocyte mephases of male and females cattle carriers of rob(1;29) (courtesy: Iriarte,W., 2008).

Treatments on metaphase chromosomes obtained from lymphocyte culture have been done (C-banding, G banding, Restriction-banding, R-banding, FISH), where the chromatin showed differential expression in the trivalent: chromosome 1, chromosome 29 and the submetacentric rob(1;29). G-banding is obtained with Giemsa stain following digestion of chromosomes with trypsin. It shows clearly dark banding along the rob(1;29) (Fig. 3A).

When metaphases of cattle carriers of rob(1;29) were treated with restriction enzymes (MspI) and C-banding, and counter-staining with propidium yodide, all the autosome centromeres were brilliant in contrast to pale regions in rob(1;29) and X chromosomes (Fig. 3B) (Postiglioni et al., 2002).

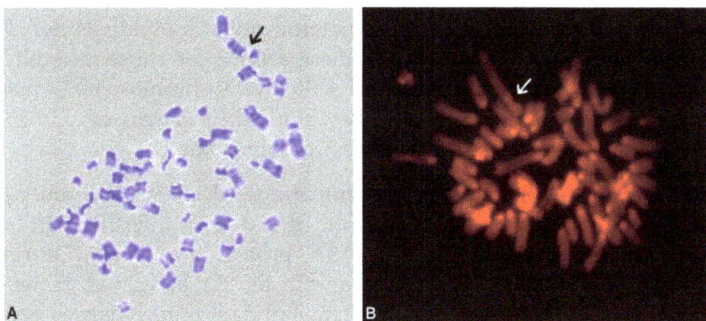

Fig. 3. Lymphocyte metaphases of cattle carriers of rob(1;29). A) treatment to G-banding. B) treatment to restriction-enzyme (MspI) and C-banding, counter staining with propidium yodide. Arrow signed the pale pericentromeric region in rob(1;29) and X chromosomes, in contrast all the autosome centromeres (courtesy: Postiglioni, A., 2002).

Changes in chromatin structure may occur during the establishment of rob(1;29), altering nucleotide sequences that could be involved in embryo development and enhancing epigenetic changes (Postiglioni et al., 2011). Furthermore, King (1991) suggested an important role of gene mutation occurring on the early embryonic development of rob(1;29) carriers, probably involving collagen genes.

1.3 What happens in the pachytene synapsis of cattle carrier of rob(1;29)?

Synaptonemal complex analysis in bull heterozygous for rob(1;29) translocation revealed an absence of pairing (unsynapsing) between the proximal regions of chromosome 1 and 29 and the rob(1;29) at early pachytene indicating a lack of homology in these regions of the trivalent (Switonski et al., 1987). These authors also observed that the end-to-end association of segments of the X and Y chromosomes at early pachytene apparently resulted in a high incidence of dissociation. Recently, Raudsepp et al., (2011) demonstrated that the pseudoautosomic region (PAR) in cattle has moved to the end of the long arm (Xq) due to X chromosome rearrangements. The location of the PAR on the long or short arm of the sex chromosomes does not affect X-Y pairing.

However, these kind of transpositions might have genetic implications in the case that these structural chromosomal rearrangements were associated with the gene content of the involved regions, that may critically implicate embryonic survival. A similar situation could be involved in the chromosome rearrangements that generated the rob(1;29), where a gene of the collagen multifamily (COL8A1) is proximal to this affected region (Postiglioni et al., 2011).

Genetic causes of these effects are not yet well understood, but there are indications that the PAR and the proximal region of rob(1;29) have tissue specific genes that might be critically involved in placental and trophoblastic embryo membranes formation, and that could have a genomic imprinting effect during early embryo development (Postiglioni et al., 2009, 2011; Raudsepp et al., 2011). This statement implies that these functional regions are not only limited to morphological segregation in male meiosis, but encourared silenced of expression of tissue specific genes involved in different stages of development in carriers of rob(1;29).

Similar events have also been shown in heterozygous individuals for Robertsonian translocations in human and mice. Noreover, these actions can cause infertility if unpaired autosomes or autosomal segments are paired with the X chromosome. A clear case was shown in a cross-breed, female Limousin-Jersey studied by Basrur et al., (2001) in which chromosome 1 was associated with the inactive X chromosome.

Recently a meiotic silencing of unsynapsed chromatin (MSUC) occurs in mice germ cells of rob(8;12) translocation carriers causing meiotic arrest and infertility. In normal mice spermatogenesis, an accumulation of DNMT3A protein was shown during the mid-pachytene stage and also a distinct association with the XY body. But in the carriers of a translocation, this protein was proportionally less abundant in unsynapsed pericentromeric regions of chromosomes 8 and 12. This event was associated to silencing of Dnmt3a gene, with the consequence of incomplete methylation of impronted genes (Saferali et al., 2010). If *Dnmt3a* is silenced, it will lead to abnormal epigenetic marking of the spermatocyte genome, including those imprinted regions that gain methylation during spermatogenesis. In mice, heterozygous carriers of the rob(8;12) translocation are fertile but fail to completely establish methylation imprints in a proportion of their sperm. Saferali et al. (2010), proposed that this imprinting defect is due to meiotic silencing of a gene or genes located in the pericentric regions of chromosome 8 and 12 during pachytene.

So, the hypothesis tested in mice was that the carriers of the translocation had lost expression of tissue specific genes from the pericentromeric region of chromosome 12 in a proportion of pachytene spermatocytes due to lack of synapsis.

Considering that DNA methylation is a potent transcriptional repressor, these authors demonstrated that lack of synapsis of chromosomes 8 and 12 during the early pachytene stage of meiosis interfere with the proper establishment of gene methylation imprinting and suggests transient silencing of genes in the unsynapsed regions. These findings support the notion that imprinting establishment extends into the prophase I of meiosis and that a similar mechanism could take place in carriers of rob(1;29).

2. An approach to understand the causes of genetic sub-fertility in rob(1;29) carriers

Cattle chromosomes BTA1 and BTA29 have genes that could be involved in early embryonic mortality, the main reproductive cause of sub-fertility of rob(1;29) carriers. The tissue-specific gene referred to as collagen type VIII alpha 1 (Col8A1) is related to extracellular matrix proteins that could be associated with placental mammal-specific gene groups. This gen is located in rob($1q_{13/21}$;29), next to the microsatellite BMS4015. Chromosome 1 has also another member of the collagen family, Col6A1 gene located in rob($1q_{4.3}$;29). Besides, the well-known and imprinted gene insulin -like growth factor 2 (IGF2) is located in the short arm of rob(1;29).

As this is a gene that has an important role in embryonic and fetal gestation, as it promotes growth of early embryos, it could also be involved in early embryonic mortality (Postiglioni et al., 2009)

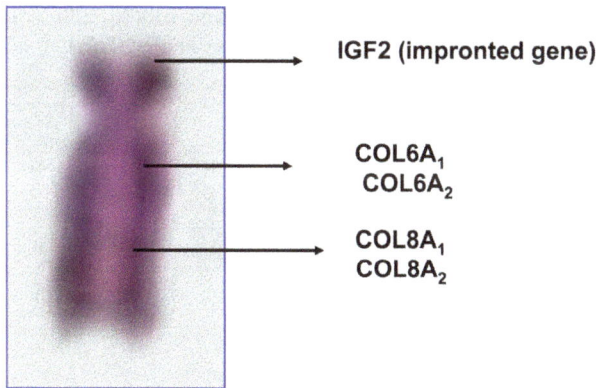

Fig. 4. The sub-metacentric chromosome rob(1;29), with the location of possible gene involved in genetic sub-fertility (courtesy: Postiglioni,A., 2009).

2.1 Chromosomal regions and clastogenic agents

In a recent study we analysed the pericentromeric region where chromosomes 1 and 29 might have suffered chromatin rearrangements due to the Robertsonian translocation polymorphism constitution. After applying different banding techniques, such as C, G and

RE (Fig. 3), clastogenic agents, like aphidicolin (APC), 5-azacytidine-C(5-AZA) and 5-bromodeoxyuridine (5-BrdU) were selected as tools to study chromatin structure alterations (Di Berardino *et al*, 1983; Sutherland & Hecht, 1985; Verma & Babu, 1995). Particularly, aphidicolin (APC) inhibits DNA polimerase α during replication. This fact allows the identification of regions rich in dCTP due to competition and spreading of the enzyme. The methodology used was the incorporation of APC (0,3μM) in one cell cycle (24 hrs.) of lymphocyte cultures, as was previously done in humans (Glover et al., 1984); cattle (Postiglioni et al., 2001; Rodriguez et al., 2002) and other domestic animals. The purpose was to find regions that are rich in cytocines and sensible to APC, and to control the response to its action in both normal and rearranged chromosomes. Other clastogenic agent used was the demetilant agent 5-azacytidine, which at high concentrations produces an undercondensation of late replicating X chromosome in G_2, due to DNA demethylation (Haaf & Schmid, 2000). A similar effect of decondensation in hypomethylated DNA was found in this important heterochromatin region of rob(1;29), after 2 hrs. of lymphocyte induction with high concentration of 5-aza-C (10mM). These findings contribute to support our hypothesis that proximal to rob(1;29) centromere exist dynamic heterochromatic regions where multiple microrearrangements could have occurred during chromosome evolution and that are possibly affecting gene expression. This hypothesis is also supported if we consider this chromosome rearrangements as possible defense mechanisms against transposons, which perpetuate in animal population as well as in microorganisms (Doolittle & Sapienza, 1980; Yoder et al., 1997).

2.2 Aphidicolin action – The response of chromatin rearrangement rob(1;29)

Inducing one cell cycle of lymphocyte cultures with APC (0,3μM), we found a new break point: (relative distance: p2c/p1=0,45), proximal to the location of IGF2 gene, and two fragile sites (c-fra) in the long arm of rob($1q_{13/21}$;29) and rob($1q_{43}$;29), corresponding to the location of the collagen genes mentioned above (Artigas et al., 2008a).

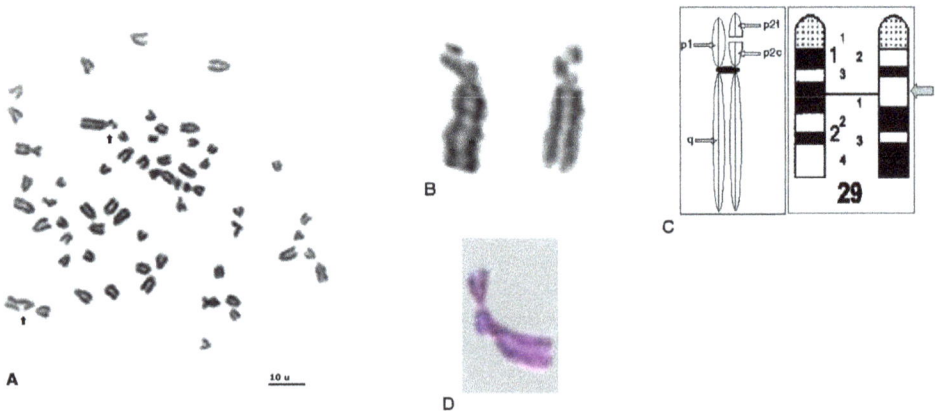

Fig. 5. Lymphocyte metaphases treated with aphidicolin (0,3μM). Observed the break point in rob($1;29p_{13/21}$) and rob($1q_{13/21}$;29). The scheme illustrates the regions meassured (courtesy: Artigas, R., 2008).

Chromosomes of Creole cows carrying rob(1;29) revealed heterozygosity in the expression of the fragile site rob($1q_{13/21}$;29), when processed with the clastogenic agent aphidicolin.

Comparing APC effects on BTA1$q_{13/21}$ and rob($1q_{13/21}$;29), the high damage found in this specific region of the chromosome rearrangements, revealed an heterozygous expression behavior in the sensitive chromatin region rob($1q_{13/21}$;29) (Table 1).

Creole cattle	(a) BTA 1;29	(b) BTA 1	(axb)	Total	Authors
Cow 1	55	13	0,16	68	Tellechea
	80%	20%	16%	100%	et al., 2004
Cow 2	44	17	0,20	61	Tellechea
	72%	28%	20%	100%	et al., 2004
Cow 3	28	21	0,25	49	Artigas
	57%	43%	25%	100%	et al., 2008

Homozygosity for fragility expression ⇐

Table 1. Comparing APC effects on BTA1$q_{13/21}$ and rob($1q_{13/21}$;29). Simultaneous expresion (a x b) corresponds to homozygosity for fragility expression.

It is important to connect this results to those that were discussed in: *"What happens in the pachytene synapsis of cattle carrier of rob(1;29)?"* (1.1.3). Two important facts, an unsynapsed region of the trivalent and the heterozygous expression of APC, indirectly suggest the incorporation of a nucleotide sequence rich in CTP in this region of rob(1;29), as the action of APC was higher than its "homologous" BTA1 chromosome. This particular APC effect could be related to epigenetic mechanisms (Wang, 2006). So, this particular region of late replication has been transformed to a heterozygous segregation region (Sutherland & Hecht, 1985).

In this particular region, we identified the microsatellite BMS4015 and the tissue-specific gene collagen type VIII alpha 1 (Col8A1) related to extracellular matrix proteins that could be associated with placental mammal-specific gene groups (Artigas et al., 2008a; Di Meo et al., 2006; Postiglioni et al., 2011).

To prove the existence of a foreign nucleotide sequence in this particular chromatin region we could incorporate the research of Joerg et al., (2001). The analysis of the microsatellite BMS4015 allowed them to find a specific allele in all rob(1;29) carriers, that clearly differenciated normal cattle that acted as a control of the experience.

2.3 5-azacytidine-C action – The response of chromatin rearrangement rob(1;29)

A pronounced chromatin despiralization of rob(1;29) similar to the inactive X chromosome of female mammals, was demonstrated when lymphocyte cultures were exposed to a DNA demethylating agent (5-azacytidine-C) (Artigas et al., 2008b; Artigas et al., 2010). This demethylating agent is considered a useful tool to study chromatin decondensation (Verma & Babu, 1995; Haaf, 1995; Haaf & Schmid, 2000).

This agent is a cytidine analog with a nitrogen atom replacing the carbon at the 5th position of the pyrimidine ring. When incorporated into DNA, the cytocine cannot be methylated and causes almost complete demethylation of genomic DNA, expressing a despiralization of the condensed chromatin of metaphase chromosomes. 5-aza-C has the ability to induce inhibition of chromatin condensation in autosomic heterochromatin, constitutive and facultative, and in the genetically inactive late replicating X chromosome (Haaf, 1995; Haaf & Schmid, 2000). The effects on chromatin structure are highly dependent on 5-aza-C concentration and treatment time. In females, a concentration of 5-aza-C of 1×10^{-3}M for 2 h inhibits condensation of the inactive X chromosome, without affecting the active X chromosome and autosomes.

Our first experiment consisted on using this demetilant agent in lymphocyte cell cultures of normal and carriers of rob(1;29) female Creole. This inductor revealed decondensation of the region proximal to the centromere of rob(1;29) and complete despiralization of the inactive X chromosome (Artigas et al., 2008b). It has been demonstrated that DNA methylation is essential for the control of gene activity in a large number of normal and pathological cellular processes, including differentiation, genomic imprinting, X inactivation and silencing of intragenomic parasitic sequence elements (Jaenisch, 1997; Yoder et al., 1997; Ng & Bird, 1999). Besides, it is used as a drug against cancer that has already been approved by the US Food and Drug Administration (FDA) for the treatment of myelodysplastic syndromes (MDS) and chronic myelomonocytic leukemia (CML) (Kulis & Esteller, 2010).

An important despiralization of the inactive X chromosome resulted similar to the rob(1;29), while the other X chromosome and the autosome chromosomes 1 and 29, maintained their condensation (Fig. 6).

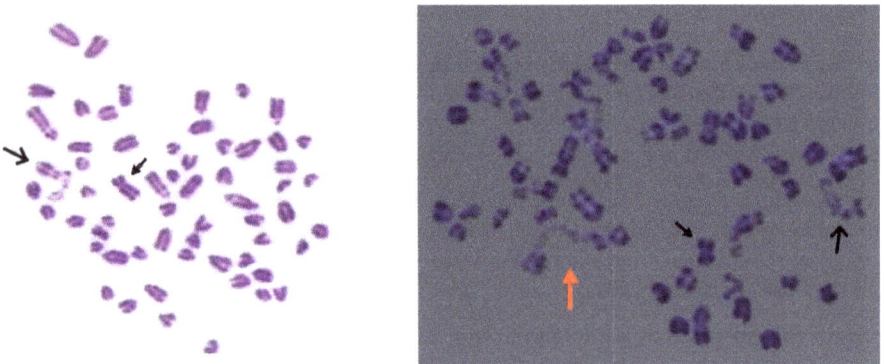

Fig. 6. Lymphocyte metaphases of cattle female normal and carriers of rob(1;29), treated to 5-aza-C (1×10^{-3}M, 2 h). Red arrow shows the demetilate rob(1;29); black arrows signed the inactive and active X´s chromosomes (cortesy: Artigas, R & Iriarte, W., 2009).

To demostrate this observation, we measured 28 metaphases using the program UTHSCSA ImageTool Version 3.0 to account for the decondensing effects of 5-aza-C on chromatin structure. Both despiralized chromosomes presented an average value and standard error of 0.75 ± 0.11 and 0.75 ± 0.083, respectively. Data from the active X chromosome clearly differed, with an average of 0.54 ± 0.09 and 1.07 ± 0.2, respectively for centromeric and biarmed despiralization (Fig. 7).

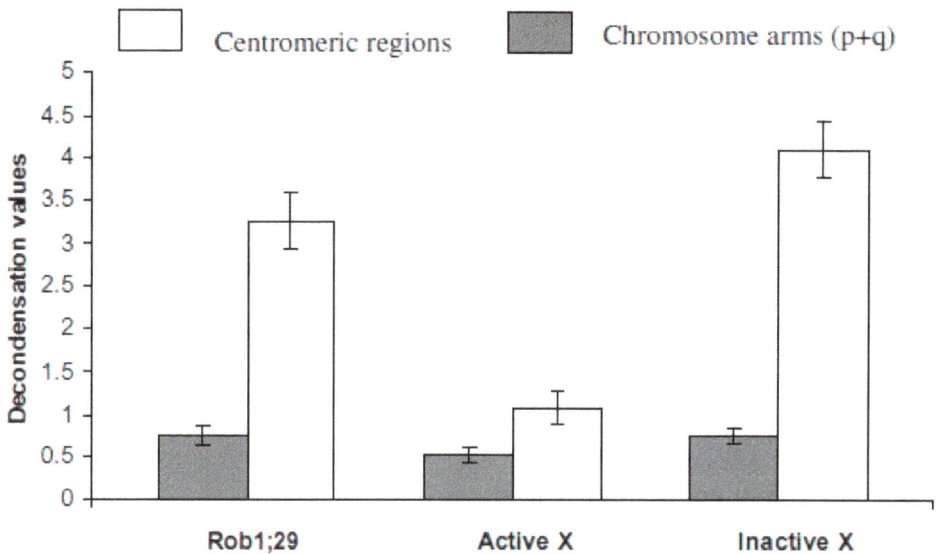

Fig. 7. Effects of 5-Aza-C on biarmed chromosome of rob(1;29) carrier cow. Histograms depict average decondensation values and standard errors in centromeric regions and chromosome arms (p + q) for rob(1;29); active and inactive X chromosomes.

Using no-parametric tests (Kruskal–Wallis global test) we demostrated our observations with level of significance (H = 32.11; p 1.07 x 10-7) and the Mann–Whitney U-test permitted to demostrate differences taking in pairs, among the three samples (Artigas et al., 2010).

We also observed a single block of condensed chromatin in the inactive X chromosome which contrasted with the high level of condensation inhibition in the rest of the chromosome. This region (Xq_{13}) corresponds to the early-replication segment associated with the center of inactivation referred to as Xist (Haaf, 1995). A similar region of condensation was observed in the rob($1q_{43}$;29). According to the International System fro Cytogenetic Nomenclature of Domestic Animals (ISCNDB, 2001), both regions correspond to R+ bands. Recently, an important gene related to placenta structure, named TRO, has been studied at the molecular level (Llambi et al., 2010). This gene, that encodes the protein trophinine, has been mapped in region $BTAXq_{25-33}$, where a fragile site of APC was found and probably proximal to the PAR region (Rodriguez et al., 2002; Raudsepp et al., 2011).

The nature and pattern of expression of genes located in condensed regions that despiralize with the action of demetilating agents, are considered facultative heterochromatin and correspond to G-bands, which may involve tissue-specific genes (Holmquist & Ashley, 2006). The rob($1q_{13/21}$;29) region was hypomethylated after treatment with 5-aza-C (1 x 10-^{3}M) for 2 h. This effect could reflect a dynamic process of the heterochromatin where multiple micro-rearrangements have occurred affecting the expression of genes co-habiting the same chromatin domain or "genomic neighborhood". These regions provide a new model for studying epigenetic changes in the bovine genome (Artigas et al., 2010).

3. An approach to silencing tissue-specific genes proximal to the rearrangement region of rob(1;29)

Based on our previous results and keeping in mind that imprinted genes are essential in embryonic development, we decided to query genes located on BTA1 and BTA29 that could undergo genome imprinting.

We have just hypothesized that the expression of tissue-specific genes co-habiting the same chromatin domain or "genomic neighborhood" could be affected by multiple micro-rearrangements that have occurred in this regions. Sequences of the region proximal to BTA1 centromere are known to have abundant CpG islands that could suffer methylations. The heritable modification of cytosine residues within CpG dinucleotides represents an important epigenetic mark that affects gene expression in diverse species (Vrana, 2007; Biliya & Bulla, 2010).

Particularly on BTA1 we begun to study genes related to structural proteins of the extracellular matrix that could be associated to a placental mammal-specific orthologous group including genes collagen typeVI-α_1 (Col6A$_1$), collagen typeVI-α_2 (Col6A$_2$), collagen typeVIII-α_1 (Col8A$_1$) and collagen typeVIII-α_2 (Col8A$_2$) (Elsik et al., 2009). As it was mentioned above, on BTA29 there is the imprinted gene insulin like growth factor II (IGF$_2$), which plays a key role in mammalian growth and is located proximal to the sensible chromatin region to dCTP, expressed as a break point to APC-induction (Schmutz et al. 1996; Vrana, 2007; Artigas et al., 2008a).

So, our next experience was to perform an *in silico* screening for CpG islands in the collagen typeVIII-α_1 (Col8A$_1$) promoter located on BTA1q$_{13/21}$, to uncover putative targets of methylation. DNA bisulfite conversion and sequencing methods were used to compare differential methylation patterns in the bovine Col8A$_1$ promoter. For the bisulfite conversion, we used the protocol referred to as "Sodium Bisulfite Conversion of Unmethylated Cytosines in DNA from solutions with low concentration of DNA" (Qiagen, Alameda, CA, USA). Besides, we used the "MethPrimer" program to design the primers to amplify the rich CpG promoter region for methylated (M) and unmethylated (U) cytosines (Li & Dahiya, 2002) (Fig. 8).

The target sequence was amplified from genomic DNA of lymphocyte cells. The eight sequence traces were aligned with sequence NM_001101176, *Bos taurus* collagen, typeVIII-α_1 (Col8A$_1$), position from 2068 to 2255 (Table 2). The alignment of the Col8α_1 sequence revealed 37 converted cytosines (75,51%), 10 methylated cytosines (20,41%) that correspond to CpG islands, and 2 inespecific alignment (4,08%). Two positions (2165 and 2186 from NM_001101176) were inespecific since some individuals showed unconverted cytosines, but others showed two peaks corresponding to a converted cytosine and an unconverted cytosine.

BiQ-Analizer software was used to show the selective conversion of unmethylated cytosines to uracils obtaining the following results: unmethylated CpGs: 0.000 (0 cases), methylated CpGs: 0.802 (77 cases) and CpGs not present: 0.198 (19 cases). All CpG islands in the promoter region of the gen Col8α_1 were methylated in the eight bovine samples (Bock et al., 2005, Luedi et al., 2007; Postiglioni et al., 2011) (Fig. 10).

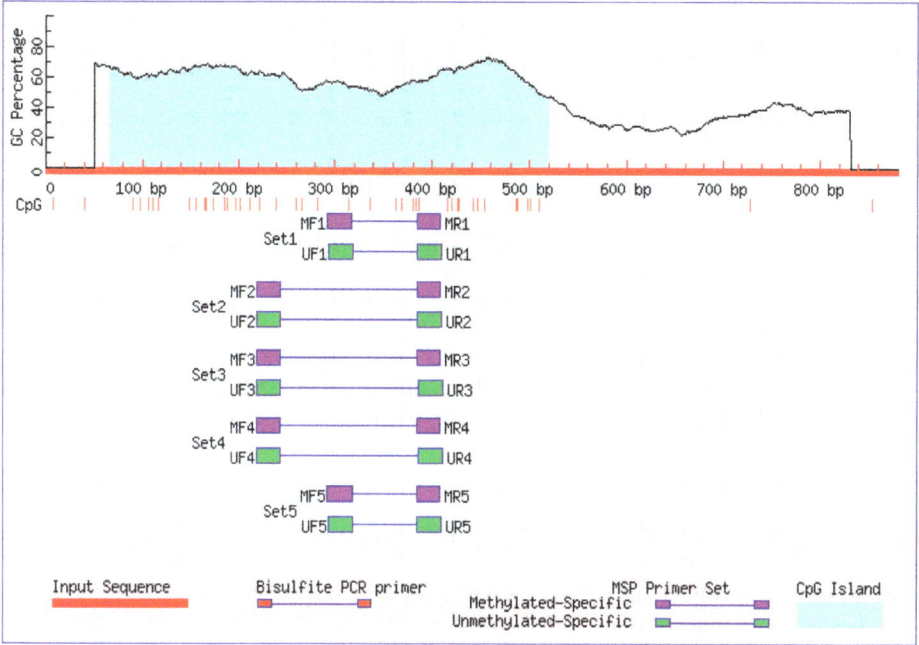

Fig. 8. Promotor region of Col8A1 gene reach in CpG island. The Methyl Primer Program permitted to select the pair primers for PCR.

Types of cytosines	NM_001101176 positions
Unmethylated cytosines (converted by MSP into uracils)	2075, 2079, 2082, 2084, 2093, 2097, 2102, 2105, 2108, 2109, 2112, 2120, 2123, 2126, 2128, 2129, 2138, 2139, 2140, 2147, 2150, 2153, 2162, 2163, 2169, 2171, 2174, 2183, 2197, 2199, 2210, 2217, 2218, 2228, 2243, 2252, 2255
Methylated cytosines	2072, 2090, 2110, 2116, 2132, 2213, 2219, 2230, 2234, 2237

Table 2. Position of methylated and unmethylated cytosines in a sample of Creole cattle traces with respect to sequence NM_001101176.

```
                    10          20          30          40          50          60
           ....|....|  ....|....|  ....|....|  ....|....|  ....|....|  ....|....|
Reference  AGTTCGACAA  ACTGCTCTAT  AACGGCAGAC  AGAACTACAA  CCCGCAGACG  GGCATCTTCA
40         .......T..  .T..T.T...  .....T...T  ....T..T..  TT..T.....  ..T..T..T.
50         .......T..  .T..T.T...  .....T...T  ....T..T..  TT..T.....  ..T..T..T.
60         .......T..  .T..T.T...  .....T...T  ....T..T..  TT..T.....  ..T..T..T.
734        .......T..  .T..T.T...  .....T...T  ....T..T..  TT..T.....  ..T..T..T.
767        .......T..  .T..T.T...  .....T...T  ....T..T..  TT..T.....  ..T..T..T.
808        .......T..  .T..T.T...  .....T...T  ....T..T..  TT..T.....  ..T..T..T.
4530       .......T..  .T..T.T...  .....T...T  ....T..T..  TT..T.....  ..T..T..T.
8692       .......T..  .T..T.T...  .....T...T  ....T..T..  TT..T.....  ..T..T..T.
```

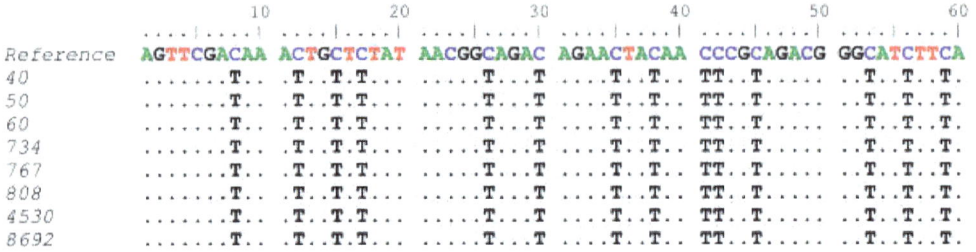

Fig. 10. Parcial consense alignment of Col8A1 gene, after treatment with Sodium Bisulfite Conversion of Unmethylated Cytosines. Observe a total of eighteen *blue* cytosines; four of them (joined to guanine) are remained as 5-methylcytosine (methylated). The resting fourteen are recognized as thymine (unmethylated).

This technique was successfully applied in this study, showing that this is a straightforward methodology that can be used to evaluate gene expression in different tissues.

4. Conclusion

We presented a research approach to the problem of changes in tissue-specific gene expression related to genetic sub-fertility problems in cattle carriers of Robertsonian translocations (early embryo mortality, slow embryonic development). The most relevant results suggest that this chromosome rearrangement rob(1;29), in heterozygous conditions, changes the nucleotide sequences around the pericentromeric region, compared with its homologous chromosomes. As this is a region rich in CpG islands, probably foreign sequences like transposons could be involved in this particular region, which also showed a despiralization of its chromatin when exposed to a demetilating agent. The bisulfite methodology applied to CpG islands of collagen promoters should be done on tissue-specific genes, like fibroblasts, to have a consistent answer to our question. Taking into account that Robertsonian translocations are the most common chromosomal rearrangements in humans, specially the rob(13q14q) and rob(14q21q), these experiences could contribute to the advance in this area. Besides, our results could be taken in consideration as a model in human cases.

Future research will have to be done to demonstrate that methylation of tissue-specific genes CpG islands occur in animals carrying the rob(1;29) Robertsonian translocation. Cattle fibroblasts, semen, early embryos and trophoblastic membranes will have to be analyzed to determine the genetic causes of embryo mortality. Our cyto-molecular approach will allow the use of this knowledge as an animal biotechnology tool in reproduction programs. New methodologies, such as high throughput sequencing and transcriptome analysis, could be incorporated to the present work to improve the understanding of questions such as: which are the precise functions of collagen genes and other genes involved in embryonic implantation and development? Is it really the rob(1;29) acting as a defensive mechanism to regulate gametic selection? As said by many other researchers: "These are just some of the questions that arise from the above summarized observations. Yet, there are very few answers".

5. Acknowledgment

We thank Lic. Ms. Eileen Armstrong for helping to improve our writing in English language. We also thank Br. Pablo Costa for improving the quality of all the figures. This work was supported by grants from PEDECIBA (Programa de Desarrollo de las Ciencias Básicas), CSIC (I + D Proyect, Universidad de la República) and CIDEC (Facultad de Veterinaria)

6. References

Artigas, R., Iriarte, A., Tellechea, B., Llambí, S., De Bethencourt, M. & Postiglioni, A. (2008a.) Aphidicolin induced break points in heterozygous Robertsonian translocation (rob1;29) in Creole cattle. *BAG. J. Basic Appl. Genet.* Vol. 19, N°. 1,pp. 1–10, ISSN 1852-6233

Artigas, R., Iriarte, W., de Soto, L., Iriarte, A., Llambí, S., De Bethencourt, M. & Postiglioni, A. (2008b). Heterochromatin decondensation in Creole cattle carrier of Robertsonian translocation (rob1;29). Action of 5-azacytidine-C. *Acta Agron Palmira* Vol. 57 N°. 1, pp 65-69, ISSN 0120-2812

Artigas, R., Iriarte, W., Iriarte, A., De Bethencourt, M., Llambí, S. & Postiglioni A. (2010). Effects of 5-azacytidine in lymphocyte-metaphases of Creole cows carrying the rob(1;29). *Research in Veterinary Science*, Vol. 88, no. 2, pp.263–266, ISSN 0034-5288

Basrur, PK., Reyes, ER., Farazmand, A., King, WA. & Popescu, PC. (2001). X-autosome ranslocation and low fertility in a family of crossbred cattle. *Anim Reprod Sci* Vol. 67, pp, 1-16, ISSN 0378-432

Biliya, S. & Bulla, LA. (2010). Genomic imprinting: the influence of differential methylation in the two sexes. *Experimental Biology and Medicine* Vol. 235, pp.139–147, ISSN 0037- 9727

Bock, CS., Reither, T., Mikeska, M., Paulsen, J. & Walter, TL. (2005). BiQ Analyzer: visualization and quality control for DNA methylation data from bisulfite sequencing. *Bioinformatics*, Vol. 21, N° 21, pp.4067-8. ISSN 1460-2059

Bonnet-Garnier, A., Pinton, A., Berland, HM., Khireddline, B., Eggen, A., Yerle, M., Darré, R. & Ducos,A. (2006). Sperm nuclei analysis of 1/29 Robertsonian translocation carrier bulls using fluorescence in situ hybridization. *Cytogenetic and Genome Research*, Vol. 112, pp. 241-247, ISSN 1424-859X

Bonnet-Garnier, A., Lacaze Beckers, JF., Berland, H., Pinton, A., Yerle, M. & Ducos, A. (2008). Meiotic segregation analysis in cows carrying the t(1;29) Robertsonian translocation. *Cytogenetic and Genome Research* Vol. 120 pp. 91-96, ISSN 1424-859X

Di Berardino, D., Iannuzzi, L. & Di Meo, G. (1983). Localization of BrdU-induced break sites in bovine chromosomes. *Caryologia* Vol 36 N° 3, pp. 285-292, ISSN 0008-7114

Di Berardino, D., Di Meo, GP., Gallagher, DS., Hayes, H. & Iannuzzi, L. (2001). ISCNDB 2000: International System fro Cytogenetic Nomenclature of Domestic Animals. *Cytogenet. Cell Genet.*Vol. 92, pp. 283–299. ISSN 0011-4537

Di Meo, GP., Perucatti, A., Chaves, R., Adega, F., De Lorenzi, L., Monteni, L., De Giovanni, A., Incamato, D., Guedes-Pinto, H., Eggen, A. & Iannuzzi, L. (2006). Cattle rob(1;29) originating from complex chromosome rearrangements as revealed by both banding and FISH-mapping techniques. *Chromosome Res.* Vol 14, pp. 649–655, ISSN 1573-6849

Doolittle, WF., & Sapienza, C. (1980). Selfish genes, the phenotpe paradigm and genome evolution. *Nature* Vol. 284, pp. 601-607, ISSN 1476-4687

Ducos, A., Revay, T., Kovacs, A., Hidas, A., Pinton, A., Bonnet-Garnier, A., Molteni, L., Slota, E., Switonski, M., Arruga, MV., Van Haeringe, WA., Nicolae, I., Chaves, A., Guedes-Pinto, H., Andersson, M. & Iannuzzi, L. (2008). Cytogenetic screening of livestock populations in Europe: an overview. *Cytogenetic and Genome Research*,Vol.120, pp. 26-41, ISSN 1424-859X

Elsik, ChG., Tellam, RL., Worley, KC. & the Bovine Genome Sequencing and Analysis Consortium. (2009). The genome sequence of taurine cattle: A window to Ruminant biology and evolution. *Science* Vol. 324, pp. 522-528, ISSN 1095-9203

Friebe, P., Zhang, B., Lincb, G., & Gill, BS.(2005). Robertsonian translocations in wheat arise by centric misdivision of univalents at anaphase I and rejoining of broken centromeres during interkinesis of meiosis II. *Cytogenet Genome Res* Vol.,109, pp.293–297.

Fries, R. & Popescu, P. (1999). Cytogenetics and physical chromosome maps, In: *The Genetics of Cattle*, Fries R & Ruvinsky A. (Eds.), Vol. VIII. ISSN 0103-9016

Glover, TW., Berger, C., Coyle, J. & Echo, B. (1984). DNA polymerase a inhibition by aphidicolin induces gaps and breaks at common fragile sites in human chromosomes. *Hum Genet* Vol. 67, 136–142 ISSN 1018-4813

Gustavsson, I. (1969). Cytogenetics, distribution and phenotypic effects of a translocation in Swedish cattle. *Hereditas* Vol. 63, pp.68–169, ISSN 1601-5223

Haaf, T. (1995). The effects of 5-azacytidine and 5-azadeoxicytidine on chromosome structure and function: implications for methylation-associated cellular process. *Pharmc. Thr.* Vol. 65, pp. 19–46, ISSN: 2042-7158

Haaf, T. & Schmid, M. (2000). Experimental condensation inhibition in constitutive and facultative heterochromatin of mammalian chromosomes. *Cytogenet. Cell Genet.* Vol 91, pp. 113–123, ISSN 0011-4537

Herzog, A. & H6hn, H. (1968). Autosomale Trisomie bei einem Kalb mit Brachygnathia Inferior und Ascites congenitus. *Dtsch Tieraerztl Wocheuschr* 75, 604-606, ISSN 0341-6593

Holmquist, G. & Ashley, T. (2006). Chromosome organization and chromatin modification: influence on genome function and evolution. *Cytogenet. Genome. Res.* Vol. 114, pp.96–125, ISSN 1424-859X

Jaenisch, R. (1997). DNA methylation and imprinting. Why bother? *Trends Genet.* Vol. 13, 323, ISSN 0168-9525.

Joerg H, Garner D, Rieder S, Suwattana D & Stranzinger G. (2001). Molecular genetic characterization of Robertsonian translocations in cattle. *J Anim Breed Genet* 118:371-377, ISSN 0931-2668

Kastelic, JK. & Mapletoft, RJ. (2003). Causas no infecciosas de muerte embrionaria en Ganado bovino. Quinto Simposio Internacional de Reproducción Animal. *IPAC* 149-159, ISSN 1031-3613

King, WA. (1991). Embryo-mediated pregnancy failure in cattle. *Can Vet J* Vol 32, pp.99-103, ISSN 0008-5286

Kulis, M. & Esteller, M.(2010). DNA methylation and Cancer. *Advances in Genetics*. 70: 27-56, ISSN, 10: 0123748968

Li, LC. & Dahiya, R. (2002). MethPrimer: designing primers for methylation PCRs. *Bioinformatics*, Vol. 18, N° 11, pp. 1427-31. ISSN 1471-2105

Luedi, PP., Dietrich, FS., Weidman, JR., Bosko, JM., Jirtle, RL. & Hartemink, AJ. (2007). Computational and experimental identification of novel human imprinted genes. *Genome Research*, Vol.17, pp.1723-1730, ISSN: 1088-9051

Llambí, S., Iriarte, A., Gagliardi, R. & Silveira, M. (2010). Análisis de secuencias del gen trofinina en bovinos Holando uruguayo. *Agrociencia* Vol XIV, N° 3, pp. 184, ISSN 1510-0839

Nielsen, J, & Wohlert, M. (1991). Chromosome abnormalities found among 34,910 newborn children: results from a 13-year incidence study in Arhus, Denmark. *Hum Genet* Vol. 87: pp. 81–83, ISSN: 0340 -6717

Ng, HH. & Bird, A. (1999). DNA methylation and chromatin modification. *Curr. Opin. Genet. Dev.* Vol. 9, pp. 158–163, ISSN 0959-437X

Ohno, S., Trujillo, J., Stenius, C., Christian, LC. & Teplitz, R. (1962). Possible germ cell chimeras among newborn dizygotic twin calves (Bos taurus). *Cytogercetics* Vol 1, pp. 258-265, ISSN 1755-8166

Postiglioni, A., Llambí, S., Gagliardi, R. & De Bethencourt, M. (1996). Genetic characterization of Uruguayan Creole cattle. I. Cytogenetic Characterization of a sample of Uruguayan Creole cattle. Archivos de Zootecnia Vol 45, N° 170, pp. 209-213, ISSN 1885-4494.

Postiglioni, A., Llambí, S. & Arruga, MV. (2001). Fragilidad cromosómica inducida por afidicolina en reordenamientos cromosómicos (1/29) bovinos. *BAG. J. Basic Appl. Genet.* N° 2, pp. 91. ISSN 1852-6233

Postiglioni, A., Llambí, S., Guevara, K., Rincón, G. & Arruga, MV. (2002). Common fragile sites expression in rob(1;29) and their relation with chromosome rearrangements. 15th European Colloquium on Cytogenetics of Domestic Animals and gene Mapping. Sorrento, Italia. *Chromosome Research* Vol 10 N° 1, pp. 5, ISSN 1573-6849.

Postiglioni, A., Arruga, MV., García, CB. & Rincón, G. (2009) ¿Qué es la impronta genética? Albéitar N° 130 noviembre,pp. 42 – 43 ISSN 1699-7883

Postiglioni, A., García, CB., Rincón, G. & Arruga, MV. (2011). Methylation specific PCR analysis in COL8A1 promoter in Creole cattle carrier of rob(1;29) *Electronic Journal of Biotechnology.* DOI: 10.2225/Vol. 14, N°3 12 ISSN 0717-3458

Raudsepp, PJ., Das, F., Avila, BP. & Chowdhary, B. (2011). The Pseudoautosomal Region and Sex Chromosome Aneuploidies in Domestic Species. *Sex Dev* DOI: 10.1159/000330627, ISSN 1661-5433

Robertson, WRB. (1916). Chromosome studies.I. Taxonomic relationships shown in the chromosomes of Tettigidae and Acrididae, V-chaped chromosomes and their significance in Acrididae, Locustidae and Gryllidae: chromosomes and variation. *J. Morphol.* Vol. 27: pp. 179-331, ISSN 1097-4687

Rodríguez, V., Llambí, S., Postiglioni, A., Guevara, K., Rincón, G., Fernández, G., Mernies, B. & Arruga. MV. (2002). Localization of aphidicolin induced break points in Holstein-Friesian cattle (Bos taurus) using RBG-banding. *Genet Sel. Evol* Vol. 34, pp.649-656, ISSN 0999-193X

Saferali, A., Berlivet, S., Schimenti, J., Bartolomei, MS., Taketo, T. & Naumova, AK. (2010) Defective imprint resetting in carriers of Robertsonian translocation Rb 8.12. Mamm Genome Vol. 21, pp.377-387, ISSN 1432-1777

Schibler, L., Vaiman, D., Oustry, A., Giraud-Delville, C. & Cribiu, EP. (1998) Comparative gene mapping: A fine-scale survey of chromosome rearrangements between ruminants and humans. *Genome Research* pp. 901-915, ISSN1549-5469

Schmutz, S., Moker, JS., Gallagher, DS., Kappes, SM. & Womack, JE. (1996). In situ hybridization mapping of LDHA and IGF2 to cattle chromosome 29. *Mammalian Genome* Vol. 7, N°6, pp. 473, ISSN: 0938-8990

Sutherland, G. & Hecht, F. (1985). *Fragile sites on Human chromosomes.* Oxford University Press, ISSN 0022-3530

Switonski, M., Gustavsson, I. & Ploen, L. (1987). The nature of the 1;29 translocation in cattle as revealed by synaptonemal complex analysis using electron microscopy. *Cytogenet Cell Genet* Vol.44, pp. 103-111, ISSN 0301-0171.

Tellechea, B., Llambí, S., De Bethencourt, M., Rincón, G. & Postiglioni, A. (2004). Avances en el estudio de la heterocromatina en rob(1;29). Un reordenamiento que produce mortalidad embrionaria temprana. XXXII Jornadas Uruguayas de Buiatría pp. 229–231.

Verma, R. & Babu, A. (1995). *Human chromosomes. Principles and techniques.* Mc Graw-Hill. ISBN 007 105432-4

Vrana, PB. (2007). Genomic imprinting as a mechanism of reproductive isolation in mammals. *Journal of Mammalogy*, Vol. 88 N° 1, pp. 5-23, ISSN 1545-1542

Wang, YH. (2006). Chromatin structure of human chromosomal fragile sites. *Cancer Letters* Vol. 232, pp. 70-78, ISSN 0304-3835

Yoder, J, Walsh, C. & Bestor, T. (1997). Cytosine methylation and the ecology of intragenomic parasites. *Trends Genet.* Vol.13, N° 8, pp. 335–340, ISSN 0168-9525

Epigenetic Defects Related Reproductive Technologies: Large Offspring Syndrome (LOS)

Makoto Nagai[1], Makiko Meguro-Horike[2,3] and Shin-ichi Horike[2,*]
[1]Ishikawa Prefectural Livestock Research Center,
[2]Frontier Science Organization, Kanazawa University,
[3]JSPS Research Fellow
Japan

1. Introduction

Assisted reproductive technologies (ART), such as somatic cell nuclear transfer (SCNT) and *in vitro* fertilization (IVF), have been used to produce genetically superior livestock. Currently, embryos from IVF are commercially available from public or private corporations. However, calves derived by ART techniques frequently suffer with pathological changes in the fetal and placental phenotype, the so-called large offspring syndrome (LOS), and this has significant consequences for development both before and after birth (Behboodi et al., 1995; Constant et al., 2006; Wilmut et al., 2002; Young et al., 1998).

Although the etiology of LOS is not fully understood, these abnormalities may arise from disruptions in expression of developmentally important genes, in particular imprinted genes (Abu-Amero et al., 2006; Amor & Halliday, 2008; Angiolini et al., 2006; Coan et al., 2005; Fowden et al., 2006; Hitchins and Moore, 2002). Genomic imprinting is an important epigenetic mechanism in mammalian development, and is thought to influence the transfer of nutrients to the fetus and the newborn from the mother (Reik & Walter, 2001). Indeed, many imprinted genes are involved in fetal and placental development. Moreover, these imprinting defects cause various developmental disorders in humans, such as Beckwith–Wiedemann syndrome (BWS) (OMIM:130650), Russell–Silver syndrome (OMIM 180860), and Prader–Willi/Angelman syndrome (OMIM 105830) (Enklaar et al., 2006; Horike et al., 2009; Horsthemke & Wagstaff, 2008; Weksberg et al., 2003, 2005). In ruminants, the current study suggests that ART techniques, particularly *in vitro* culture of preimplantation embryos, have been associated with aberrant imprinted gene expression (Tveden-Nyborg et al., 2008). However, the exact mechanisms that lead to aberrant genomic imprinting after ART remain unknown.

The most likely explanation for the aberrant genomic imprinting in SCNT and IVF cattle may be failures in epigenetic reprogramming and/or maintenance (Bertolini et al., 2002; Beyhan et al., 2007; Blelloch et al., 2006; Everts et al., 2008; Hashizume et al., 2002; Herath et al. 2006; Hochedlinger et al. 2006; Oishi et al. 2006; Pfister-Genskow et al., 2005; Smith et al.,

---

* Corresponding Author

2005; Somers et al., 2006). Genome-wide epigenetic reprogramming in germ cells is essential in order to reset the parental-origin specific marking of imprinted genes. DNA methylation is one of the most important epigenetic marks for the allele-specific silencing of imprinted genes, and its genome-wide profiles undergo drastic changes during gametogenesis (Dupont et al., 2009; Arnaud & Feil, 2005; Bao et al., 2000). Indeed, the genome-wide DNA methylation patterns of the parental genomes are erased and a new methylation pattern is established by *de novo* methylation during gametogenesis (Arnaud & Feil, 2005; Bao et al., 2000). Therefore, the failures of epigenetic reprogramming could lead to loss of imprinting for many but not all imprinted genes (Reik & Walter, 2001).

A few reports to date have described the aberrant expression of imprinted genes in LOS animals produced by ART techniques. Interestingly, LOS phenotypes are reminiscent of BWS in humans, a loss-of-imprinting pediatric overgrowth syndrome associated with congenital malformations and tumor predisposition (Amor & Halliday, 2008; DeBaun et al., 2003; Maher et al., 2003; Maher, 2005; Manipalviratn et al., 2009; Shiota & Yamada, 2005, 2009). Because the majority of sporadic BWS patients show loss of DNA methylation at KvDMR1, which may function as an imprinting control region (ICR) on the *KCNQ1OT1/CDKN1C* domain (Mitsuya et al., 1999; Weksberg et al., 2003, 2005), it is possible that LOS is related to the loss of DNA methylation at KvDMR1, leading to diminished expression of *Cdkn1c*.

In this chapter we highlight some of the epigenetic defects identified in SCNT and IVF cattle and discuss the potential role that imprinted genes may play.

2. Assisted Reproductive Technologies (ART) and Large Offspring Syndrome (LOS)

LOS calves were first described by Willadsen et al. (1991) following ART technique; the fusion of blastomeres from embryos and enucleated eggs. Since then, oversized neonates and fetuses born after various manipulations of the embryo have been reported not only in calves, but also in sheep (Wilmut et al., 1997, 2002) and mouse (Eggan et al., 2001; Fernández-Gonzalez et al., 2004; Wakayama et al., 1998). Up to 40% of SCNT-derived full-term calves and lambs have LOS, which is characterized by large size at birth, enlarged umbilical cord, enlarged organs, hydrops of the fetus, lethargy, respiratory distress, muscle fiber composition, cerebellar dysplasia and skeletal and facial malformations (Chavatte-Palmer et al., 2002; Constant et al., 2006; Fletcher et al., 2007; Loi P et al., 2006; Maxfield et al., 1997; Schmidt et al., 1996; Walker et al., 1996; Young et al., 1998). Also, it is well known that in high frequency of LOS is also frequently observed in calves that developed from *in vitro* maturation (IVM) and IVF-derived embryos (Behboodi et al., 1995; Reichenbach et al., 1992; Bertolini et al., 2004).

The most remarkable feature of LOS is large size at birth. Increases in birth weight vary widely; twice the normal birth weight is not uncommon (Young et al., 1998). In our experiments, all calves derived by SCNT (n=7) and IVF (n=2) were shown to be a large size at birth, 1.3 to 2.3 times the normal birth weight. Enlarged umbilical cord was found in almost all of the calves (five of SCNT-derived and two of IVF-derived) (Fig.1), though abnormality of organs was found only in one SCNT-derived calf in our cases.

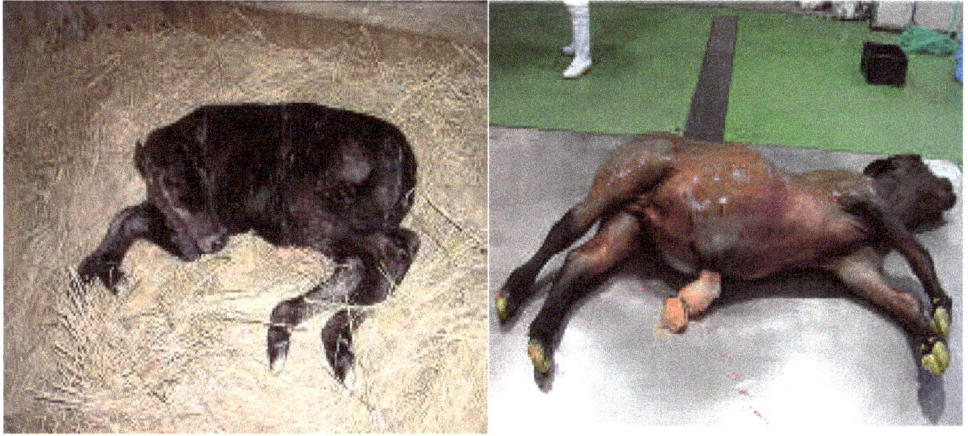

Fig. 1. Phenotype of LOS calf. Left, normal Japanese black calf produced by artificial insemination (body weight at birth: 27kg). Right, LOS Japanese black calf with enlarged umbilical cord produced by SCNT (body weight at birth: 51kg).

Placental anomalies, such as a reduced number of placentomes and increased weight of placentomes, lack of placental vascular development, reduced vascularization and poorly developed caruncles were also observed in all LOS cases in SCNT and IVF animals, and are thought to be associated with a high mortality rate and some fetal abnormalities (Bertolini & Anderson, 2002; Chavatte-Palmer et al., 2002; Constant at al., 2006; De Sousa et al., 2001; Hashizume et al., 2002; Hill et al., 2000, 2001).

While some investigations have previously suggested that reprogramming errors of the donor nucleus following SCNT could affect the fetal and placental development, the etiology of LOS remains unknown (Bertolini et al., 2002; Beyhan et al., 2007; Blelloch et al., 2006; Everts et al., 2008; Hashizume et al., 2002; Herath et al., 2006; Hochedlinger et al., 2006; Oishi et al., 2006; Pfister-Genskow et al., 2005; Smith et al., 2005; Somers et al., 2006).

Marques et al. (2004) have previously reported that paternal-allele-specific DNA methylation of the *H19* gene was significantly disrupted in spermatozoa from oligozoospermic patients. Although this result strongly suggests that transmission of paternal imprinting errors could affect embryo development, it is not likely that imprinting defects are associated with abnormal spermatogenesis in cattle, since commercially available sperms from healthy bulls are used for IVF.

3. ART culture may cause epigenetic changes

ART-derived animals can severely influence fetal growth, resulting in LOS, and any disturbance during germ cell development or early embryogenesis has the potential to alter epigenetic reprogramming and/or maintenance (Dupont et al., 2009). The birth of LOS was initially thought to associate with the procedure of ART but it is now recognized that enhanced fetal growth can also result from *in vitro* culture of oocytes or embryos (Behboodi et al., 1995; Farin et al., 2004; Farin & Farin, 1995; Maxfield et al., 1997; Smith et al., 2009; Walker et al., 1996).

Very limited information is currently available on the effects of *in vitro* culture; IVM, IVF or SCNT and *in vitro* development (IVD) on the establishment of imprinting in oocytes or embryos. The influences of *in vitro* culture on the epigenetic changes are investigated mainly in mouse. The culture medium influences the kinetics of embryo cleavage and embryo morphology up to the blastocyst stage, and can affect the imprinted expression of the *H19* gene as well as the DNA methylation status of ICR1, controlling its imprinted manner (Fauque et al., 2007). The presence of serum in culture medium for preimplantation embryos can influence the regulation of multiple growth-related imprinted genes and lead to aberrant fetal growth and development (Khosla et al., 2001). Some researchers reported that ammonium accumulates in culture medium have been linked to aberrant imprinting of *H19* and *Igf2r* (Gardner et al., 2005; Kerjean et al., 2003), however, other researchers have refuted these suggestion that follicle culture system under high ammonia levels showed normal DNA methylation patterns at regulatory sequences of *Snprn*, *Igf2r* and *H19* (Anckaert et al., 2009a, 2009b). Mineral oil, which is widely used in *in vitro* culture, has also been associated with delayed nuclear maturation and reduced development capacity in pig IVM (Shimada et al., 2002). Oil overly extracts steroid hormones in culture medium and reduces steroid hormone level by 55-70% (Anckaert et al., 2009b). Reduced steroid hormones, estrogens or xenobiotic substances with estrogenic effects in culture medium may interfere with normal imprinting establishment (Ho et al., 2006).

4. LOS in animals is reminiscent of BWS in human

The phenotypes of LOS in animals, such as large size at birth, enlarged umbilical cord and enlarged organs, are reminiscent of BWS in human. Therefore, LOS is speculated to occur primarily as the result of the misregulation of BWS-associated imprinted genes (Fig.2), while the genomic regions associated with LOS have not yet been determined. BWS is associated with epigenetic alterations at either one of two imprinting control regions on human chromosome 11p15.5, ICR1 and KvDMR1 (Enklaar et al., 2006; Ideraabdullah et al., 2008; Delaval et al., 2006; Mitsuya et al., 1999; Smith et al., 2007; Weksberg et al., 2003, 2005; Owen & Segars, 2009). The domain controlled by ICR1 includes the paternally expressed insulin-like growth factor 2 (*IGF2*) and the maternally expressed *H19* genes (Thorvaldsen et al., 1998; Owen & Segars, 2009). *IGF2* is known to be involved in regulation of fetal growth and development (Guo.et al., 2008). *H19* is also associated with embryogenesis and fetal growth in mouse (Pachnis et al., 1984), human (Goshen et al., 1993), and sheep (Lee et al., 2002). Several studies have shown that epigenetic alterations in the *Igf2/H19* domain are associated with LOS in cattle, sheep, and mice produced by ART techniques (Curchoe et al., 2009; DeChiara et al., 1991; Doherty et al., 2000; Khosla et al., 2001; Li et al., 2005; Moore et al., 2007; Yang et al., 2005; Young et al., 2000, 2003; Zhang et al., 2004). On the other hand, the domain controlled by KvDMR1 contains several maternally expressed genes including *CDKN1C*, that encodes a cyclin-dependent kinase inhibitor belong to the CIP/KIP family (Yan et al., 1997; Fitzpatrick et al., 2002; Horike et al., 2000). KvDMR1 is a maternally methylated CpG island and includes the promoter of a paternally expressed non-coding RNA (*KCNQ1OT1*) (Beatty et al., 2006; Mitsuya et al., 1999;). Interestingly, previous studies revealed that KvDMR1 is demethylated in about half of the individuals affected by BWS, and this is associated with the biallelic expression of *KCNQ1OT1* and subsequent repression of *CDKN1C* (Higashimoto et al., 2006; Lee et al., 1999; Mitsuya et al., 1999; Owen & Segars, 2009). Thus, while the *Igf2-H19* and *Cdkn1c-Kcnq1ot1* gene pairs are good LOS candidates, the phenotypic similarities between LOS and human BWS remain suggestive and deregulation of imprinting remains a plausible candidate mechanism for LOS.

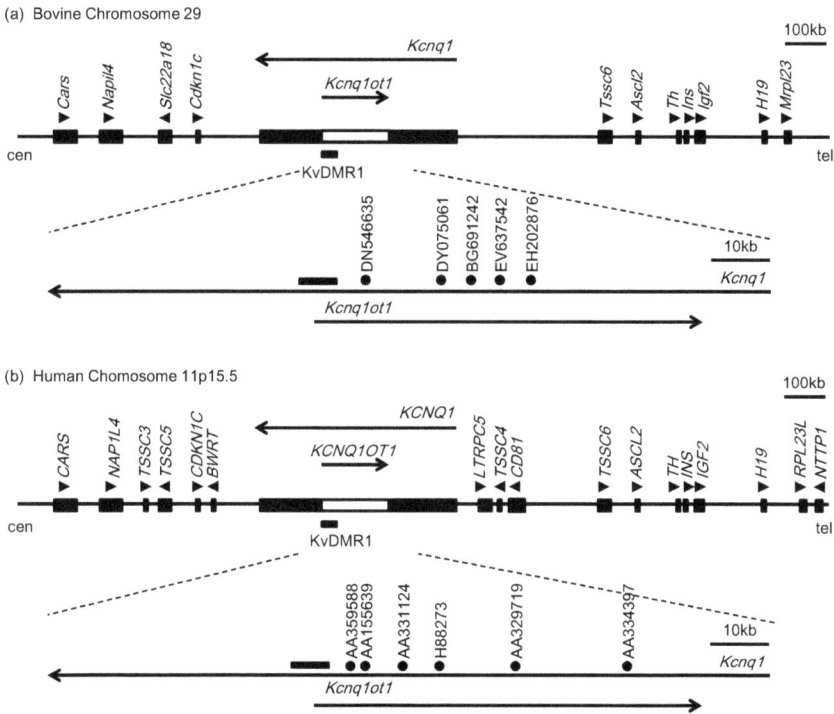

Fig. 2. Physical map of imprinting clusters in (a) bovine chromosome 29 and bovine KvDMR1, and (b) human chromosome 11p15.5 and human KvDMR1. Previously identified genes or transcripts (boxes) are drawn approximately to scale. Transcriptional orientation is indicated by arrows and arrowheads. Five and six expressed sequence tags of bovine and human are indicated by filled circle.

5. Assessment of the risk of imprinting defects in cattle born following ART

To assess of the risk of imprinting defects in cattle produced by SCNT and IVF, we analyzed DNA methylation status of the *Cdkn1c* promoter region, KvDMR1 and ICR1, and three promoter regions of other imprinted genes; *Peg1/Mest*, *Klf14* and *Gtl2* using CpG methylation sensitive restriction enzymes and bisulfite sequencing (Hori et al., 2010).

Since the use of two restriction enzymes with complementary methylation sensitivities, HpaII and McrBC, is unsurpassed as a simple, rapid method for the analysis of methylation status (Yamada et al., 2004), the HpaII-MspI-McrBC PCR assay is used for screening. HpaII and MspI recognize the CCGG sequence, but HpaII digestion is inhibited by CpG methylation at the internal cytosine while MspI is not. McrBC cleaves DNA containing a methylated cytosine and does not act upon unmethylated DNA (Fiona et al., 2000; Panne et al., 1999). In the case of a fully methylated sequence, amplification would be obtained only from the HpaII-digested template. In contrast, an unmethylated sequence is digested only with HpaII but not with McrBC, and hence amplification would be obtained only from the McrBC-digested DNA. If the target sequence is differentially methylated, such as the imprinting control region, amplification will be obtained from both HpaII- and McrBC-digested DNA. Digestion profiles visualized by PCR amplification from the main organs of seven SCNT-derived and two IVF-derived calves were compared with those of three artificial insemination-derived calves. Lastly, the HpaII–MspI–McrBC PCR assays revealed aberrant KvDMR1 hypomethylation in two of seven SCNT-derived and one of two IVF-derived calves. For other imprinting control regions such as ICR1, *Peg1/Mest* and *Gtl2* promoter, PCR amplification was obtained from both HpaII- and McrBC-digested DNA from all samples, indicating that this region is differentially methylated in both normal and SCNT- and IVF-derived calves (Fig. 3). For the *Cdkn1c* and *Klf14* promoter, PCR amplification was obtained only from the McrBC-digested DNA, as indicating that both maternal and paternal alleles are unmethylated in all samples. In addition, bisulfite sequencing analyses were demonstrated to confirm the results obtained by HpaII–MspI–McrBC PCR analyses. Bisulfite sequencing is widely recognized to be the gold standard technique to analyze CpG methylation. Finally, these bisulfate sequencing analyses showed strong concordance with the HpaII–MspI–McrBC PCR results.

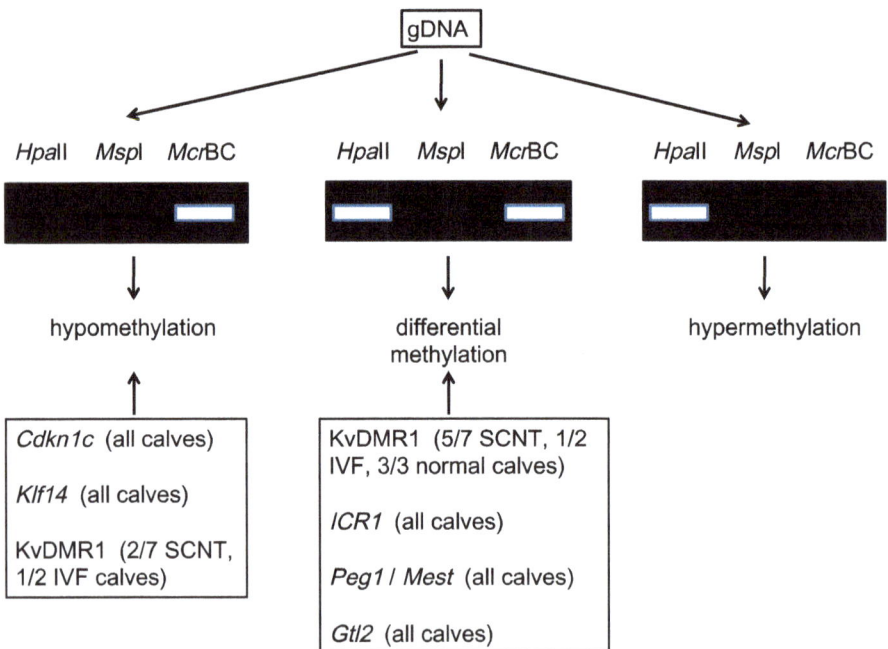

Fig. 3. A schematic gel pattern of HpaII-MspI-McrBC PCR products in hypomethylation, differentially methylation and hypermethylation cases and HpaII-MspI-McrBC PCR results of the selected six genes; Cdkn1c, Klf14, Peg1/Mest, KvDMR1, ICR1 and Gtl2 from seven SCNT-derived, Two IVF-derived and three normal calves.

To determine whether hypomethylation at KvDMR1 was linked to the aberrant expression of *Kcnq1ot1*, *Cdkn1c*, *Igf2*, or *H19*, we performed RT-PCR analysis on samples from two SCNT- and one IVF-derived calves, which showed hypomethylation status at KvDMR1, and compared gene expression patterns with those of a normal calf. In comparison to the normal calf, *Kcnq1ot1* transcript levels were increased in three ART-derived calves (two SCNT and one IVF derived calves), whereas the *Cdkn1c* transcript levels were reduced. No significant differences between three ART-derived calves and the normal calf were detected in *H19* or *Igf2* expression (Fig. 4(a)). The putative epigenetic regulation at *Kcnq1ot1/Cdkn1c* and *Igf2/H19* domains of normal and LOS cattle is shown in Fig.4 (b). These findings are consistent with the epigenetic alteration in the *Kcnq1ot1/Cdkn1c* domain of human chromosome 11p15.5 that has been observed in 50-60% of BWS patients. The biallelic expression of *Kcnq1ot1* and diminished expression of *Cdkn1c* observed in NT- and IVF-derived calves suffering with LOS in this study suggest that aberrant imprinting of the bovine *Kcnq1ot1/Cdkn1c* domain may contribute to LOS calves derived from ART techniques.

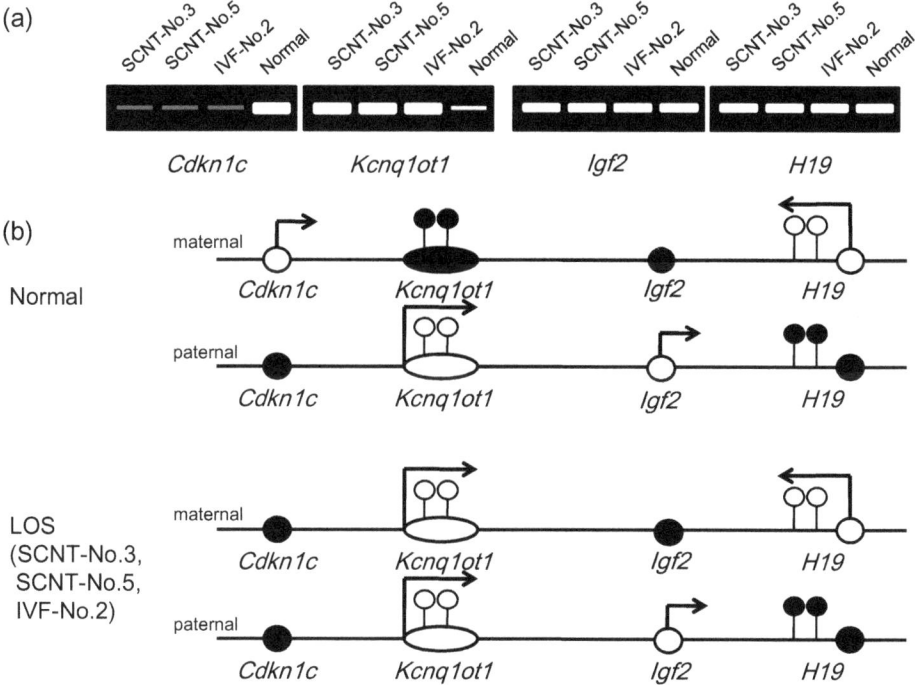

Fig. 4. (a) Scheme of RT-PCR amplification of *Cdkn1c*, *Kncq1ot1* and *H19* from SCNT-No.3 and 5, IVF-No.2 and normal cattle. (b) Putative epigenetic regulation at *Kcnq1ot1/Cdkn1c* and *Igf2/H19* domains of normal and LOS cattle. Transcription is indicated by arrows. Open and filled lollipop indicate unmethylated and methylated CpG site of KvDMR1 and ICR1.

6. Consideration and prospects

ART-derived embryos, particularly in the cow and sheep, can severely influence fetal growth, resulting in LOS. Disruptions in expression of developmentally important genes, in particular imprinted genes, were found in ART animals, suggesting that any disturbance during germ cell development or early embryogenesis may lead to altering of epigenetic changes. Aberrant gene expression is thought to associate with not only the procedure of ART, asynchronous embryo transfer or progesterone treatment but also *in vitro* culture of embryos.

The phenotypes of LOS are reminiscent of BWS in humans, an overgrowth syndrome associated with congenital malformations and tumor predisposition. Half of sporadic BWS cases show loss of DNA methylation at KvDMR1, which may function as an ICR on the *Kcnq1ot1/Cdkn1c* domain. Therefore we examined DNA methylation status of the bovine KvDMR1 in ART cattle. Abnormal hypomethylation status at an imprinting control region of *Kcnq1ot1/Cdkn1c* domain was observed in two of seven SCNT-derived calves and one of two IVF-derived calves. Moreover, abnormal expression of *Kcnq1ot1* and *Cdkn1c* were observed by RT-PCR analysis. There are very few papers which report KvDRM1 in ART-derived cattle. Coulrey and Lee (2010) reported hypomethylation of KvDMR1 in mid-gestation bovine fetuses produced by SCNT. Imprinting disruption of KvDMR1 and aberrant expression of *Kcnq1ot1* and *Cdkn1c* identified in SCNT and IVF calves may contribute to LOS in animals conceived using ART techniques. Our findings and those of Couldrey and Lee (2010) suggest that ART techniques might induce an increased risk of epigenetic defects, such as hypomethylation of KvDMR1, because epigenetic changes can be caused by embryo culture itself or the constituents of the culture medium. In humans, a significant deficit in DNA methylation at *Kcnq1ot1* in matured oocytes from stimulated cycles matured in *in vitro* culture (Khoueiry et al., 2008). This paper suggested that hyperstimulation likely recruits young follicles that are unable to acquire imprinting at KvDMR1 during the short *in vitro* maturation process. In cattle, it is unknown whether hyperstimulation is associated with acquiring imprinting at KvDMR1 of oocytes. A more thorough understanding of the stability of DNA methylation will be important for the continued safeguarding of ART techniques.

7. Acknowledgements

We thank the members of the laboratory for valuable suggestions. This work was supported by the Program for Improvement of Research Environment for Young Researchers from the Special Coordination Funds for Promoting Science and Technology (SCF) (to SH).

8. References

Abu-Amero, S.; Monk, D.; Apostolidou, S.; Stanier, P. & Moore, G. (2006). Imprinted genes and their role in human fetal growth. *Cytogenetic and genome research* 113, 262–270.

Amor, DJ. & Halliday, J. (2008). A review of known imprinting syndromes and their association with assisted reproduction technologies. *Human Reproduction* Vol.23 (No.12): 2826–2834.

Anckaert, E.; Adriaenssens, T.; Romero, S.; Dremier, S. & Smitz, J. (2009). Unaltered imprinting establishment of key imprinted genes in mouse oocytes after in vitro follicle culture under variable follicle-stimulating hormone exposure. *The International journal of developmental biology* Vol.53 (No.4): 541-548.

Anckaert, E.; Adriaenssens, T.; Romero, S. & Smitz, J. (2009). Ammonium accumulation and use of mineral oil overlay do not alter imprinting establishment at three key imprinted genes in mouse oocytes grown and matured in a long-term follicle culture. *Biology of reproduction* Vol.81 (No.4): 666-673.

Angiolini, E.; Fowden, A.; Coan, P.; Sandovici, I.; Smith, P.; Dean, W.; Burton, G.; Tycko, B.; Reik, W.; Sibley, C. & Constância, M. (2006). Regulation of placental efficiency for nutrient transport by imprinted genes. *Placenta* Vol.27 (Suppl. A): S98–S102.

Arnaud, P. & Feil, R. (2005). Epigenetic deregulation of genomic imprinting in human disorders and following assisted reproduction. *Birth defects research. Part C, Embryo today : reviews* Vol.75 (No.2): 81-97.

Bao, S.; Obata, Y.; Carroll, J.; Domeki, I. & Kono, T. (2000). Epigenetic modifications necessary for normal development are established during oocyte growth in mice. *Biology of reproduction* Vol.62 (No.3): 616-21.

Beatty, L.; Weksberg, R. & Sadowski, PD. (2006). Detailed analysis of the methylation patterns of the KvDMR1 imprinting control region of human chromosome 11. *Genomics* Vol.87 (No.1): 46–56.

Behboodi, E.; Anderson, G.B.; BonDurant, RH.; Cargill, SL.; Kreuscher, BR.; Medrano, JF. & Murray, JD. (1995). Birth of large calves that developed from in vitro-derived bovine embryos. *Theriogenology* Vol.44 (No.2): 227–232.

Bertolini, M. & Anderson, GB. (2002). The placenta as a contributor to production of large calves. *Theriogenology* Vol.57 (No.1): 181-187.

Bertolini, M.; Beam, SW.; Shim, H.; Bertolini, LR.; Moyer, AL.; Famula, TR. & Anderson, GB. (2002) Growth, development, and gene expression by in vivo- and in vitro-produced day 7 and 16 bovine embryos. *Molecular reproduction and development* Vol. 63 (No.3): 318–328.

Bertolini M, Moyer AL, Mason JB, Batchelder CA, Hoffert KA, Bertolini LR, Carneiro GF, Cargill SL, Famula TR, Calvert CC, Sainz RD, Anderson GB. (2004). Evidence of increased substrate availability to in vitro-derived bovine foetuses and association with accelerated conceptus growth. *Reproduction* Vol.128 (No.3): 341-354.

Beyhan, Z.; Forsberg, EJ.; Eilertsen, KJ.; Kent-First, M. & First, NL. (2007). Gene expression in bovine nuclear transfer embryos in relation to donor cell efficiency in producing live offspring. *Molecular reproduction and development* Vol. 74 (No.1): 18–27.

Blelloch, R, Wang Z.; Meissner, A.; Pollard, S.; Smith, A. & Jaenisch, R. (2006). Reprogramming efficiency following somatic cell nuclear transfer is influenced by the differentiation and methylation state of the dono nucleus. *Stem Cells* Vol.24 (No.6): 2007–2013.

Chavatte-Palmer, P.; Heyman, Y.; Richard, C.; Monget, P.; Le Bourhis, D.; Kann, G.; Chilliard, Y.; Vignon, X. & Renard, JP. (2002). Clinical, hormonal, and hematologic characteristics of bovine calves derived from nuclei from somatic cells. *Biology of reproduction* Vol.66 (No.6): 1596–1603.

Coan, PM.; Burton, GJ. & Ferguson-Smith, AC. (2005). Imprinted genes in the placenta – a review. *Placenta* Vol.26 (Suppl. A): S10–S20.

Constant, F.; Guillomot, M.; Heyman, Y.; Vignon, X.; Laigre, P.; Servely, JL.; Renard, JP. & Chavatte-Palmer P. (2006). Large offspring or large placenta syndrome? Morphometric analysis of late gestation bovine placentomes from somatic nuclear transfer pregnancies complicated by hydrallantois. *Biology of reproduction* Vol.75 (No.1): 122–130.

Couldrey, C. & Lee, RS. (2010). DNAmethylation patterns in tissues from midgestation bovine fetuses produced by somatic cell nuclear transfer show subtle abnormalities in nuclear reprogramming. *BMC developmental biology* Vol.10: 27.

Curchoe, CL.; Zhang, S.; Yang, L.; Page, R. & Tian, XC. (2009). Hypomethylation trends in the intergenic region of the imprinted IGF2 and H19 genes in cloned cattle. *Animal reproduction science* Vol.116 (No.3-4): 213–225.

DeBaun, MR.; Niemitz, EL. & Feinberg, AP. (2003). Association of in vitro fertilization with Beckwith-Wiedemann syndrome and epigenetic alterations of LIT1 and H19. *American journal of human genetics* Vol.72 (No.1): 156–160.

DeChiara, TM.; Robertson, EJ.; Efstratiadis, A.; (1991). Parental imprinting of the mouse insulin-like growth factor II gene. *Cell* Vol.64 (No.4): 849–859.

Delaval, K.; Wagschal, A. & Feil, R. (2006). Epigenetic deregulation of imprinting in congenital diseases of aberrant growth. *Bioessays* Vol.28 (No.5): 453–459.

De Sousa, PA.; King, T.; Harkness, L.; Young, LE.; Walker, SK. & Wilmut, I. (2001). Evaluation of gestational deficiencies in cloned sheep fetuses and placentae. *Biology of reproduction* Vol.65 (No.1): 23–30.

Doherty, AS.; Mann, MR.; Tremblay, KD.; Bartolomei, MS. & Schultz, RM. (2000). Differential effects of culture on imprinted H19 expression in the preimplantation mouse embryo. *Biology of reproduction* Vol.62 (No.6): 1526–1535.

Dupont, C.; Armant, DR. & Brenner, CA. (2009). Epigenetics: definition, mechanisms and clinical perspective. *Seminars in reproductive medicine* Vol.27 (No.5): 351-357.

Eggan, K.; Akutsu, H.; Loring, J.; Jackson-Grusby, L.; Klemm, M.; Rideout 3rd, WM.; Yanagimachi, R. & Jaenisch, R. (2001). Hybrid vigor, fetal overgrowth, and viability of mice derived by nuclear cloning and tetraploid embryo complementation. *Proceedings of the National Academy of Sciences of the United States of America* Vol.98 No.11): 6209–6214.

Enklaar, T.; Zabel, BU. & Prawitt, D. (2006). Beckwith-Wiedemann syndrome: multiple molecular mechanisms. *Expert reviews in molecular medicine* Vol.8 (No.17): 1–19.

Everts, RE.; Chavatte-Palmer, P.; Razzak, A.; Hue, I.; Green, CA.; Oliveira, R.; Vignon, X.; Rodriguez-Zas, SL.; Tian, XC.; Yang, X.; Renard, JP. & Lewin, HA. (2008). Aberrant gene expression patterns in placentomes are associated with phenotypically normal and abnormal cattle cloned by somatic cell nuclear transfer. Aberrant gene expression patterns in placentomes are associated with phenotypically normal and abnormal cattle cloned by somatic cell nuclear transfer. *Physiological genomics* Vol.33 (No.1): 65-77.

Farin, CE.; Farin, PW.& Piedrahita, JA. (2004). Development of fetuses from in vitro-produced and cloned bovine embryos. *Journal of animal science*. Vol.82 E-Suppl: E53-62.

Farin, PW. & Farin, CE. (1995). Transfer of bovine embryos produced in vivo or in vitro : survival and fetal development. *Biology of reproduction* Vol.52 (No.3): 676-682.

Fauque, P.; Jouannet, P.; Lesaffre, C.; Ripoche, MA.; Dandolo, L.; Vaiman, D. & Jammes, H. (2007). Assisted Reproductive Technology affects developmental kinetics, H19 Imprinting Control Region methylation and H19 gene expression in individual mouse embryos. *BMC developmental biology* Vol.7: 116.

Fernández-Gonzalez, R.; Moreira, P.; Bilbao, A.; Jiménez, A.; Pérez-Crespo, M.; Ramírez, MA.; Rodríguez De Fonseca, F.; Pintado, B. & Gutiérrez-Adán, A. (2004). Long-term effect of in vitro culture of mouse embryos with serum on mRNA expression of imprinting genes, development, and behavior. *Proceedings of the National Academy of Sciences of the United States of America* Vol.101 (No.16); 5880–5885.

Fiona, JS.; Daniel, P.; Thomas, AB. & Elisabeth, AR. (2000). Methyl-specific DNAbinding by McrBC, amodification-dependent restriction enzyme. *Journal of molecular biology* Vol.298 (No.4): 611–622.

Fitzpatrick, GV.; Soloway, PD. & Higgins, MJ. (2002). Regional loss of imprinting and growth deficiency in mice with a targeted deletion of KvDMR1. *Nature genetics* Vol.32 (No.3), 426–431.

Fletcher, CJ.; Roberts, CT.; Hartwich, KM.; Walker, SK. & McMillen, IC. (2007). Somatic cell nuclear transfer in the sheep induces placental defects that likely precede fetal demise. *Reproduction* Vol.133 (No.1): 243–255.

Fowden, AL.; Sibley, C.; Reik, W. & Constancia, M. (2006). Imprinted genes placental development and fetal growth. *Hormone research* Vol.65 (Suppl. 3): 50–58.

Gardner, DK. & Lane, M. (2005). Ex vivo early embryo development and effects on gene expression and imprinting. *Reproduction, fertility, and development* Vol.17 (No.3): 361-370.

Goshen, R.; Rachmilewitz, J.; Schneider, T.; de-Groot, N.; Ariel, I.; Palti, Z. & Hochberg, AA. (1993). The expression of the H-19 and IGF-2 genes during human embryogenesis and placental development. *Molecular reproduction and development* Vol.34 (No.4): 374-379.

Guo, L.; Choufani, S.; Ferreira, J.; Smith, A.; Chitayat, D.; Shuman, C.; Uxa, R.; Keating, S.; Kingdom, J. & Weksberg, R. (2008). Altered gene expression and methylation of the human chromosome 11 imprinted region in small for gestational age (SGA) placentae. *Developmental biology* Vol.320 (No.1): 79-91.

Hashizume, K.; Ishiwata, H.; Kizaki, K.; Yamada, O.; Takahashi, T.; Imai, K.; Patel, OV.; Akagi, S.; Shimizu, M.; Takahashi, S.; Katsuma, S.; Shiojima, S.; Hirasawa, A.; Tsujimoto, G.; Todoroki, J. & Izaike, Y. (2002). Implantation and placental development in somatic cell clone recipient cows. *Cloning and stem cells* Vol.4 (No.3): 197–209.

Herath, CB.; Ishiwata, H.; Shiojima, S.; Kadowaki, T.; Katsuma, S.; Ushizawa, K.; Imai, K.; Takahashi, T.; Hirasawa, A.; Takahashi, S.; Izaike, Y.; Tsujimoto, G. & Hashizume, K. (2006). Developmental aberrations of liver gene expression in bovine fetuses derived from somatic cell nuclear transplantation. *Cloning Stem Cells* Vol.8 (No.1): 79–95.

Higashimoto, K.; Soejima, H.; Saito, T.; Okumura, K. & Mukai, T. (2006). Imprinting disruption of the KCNQ1OT1/CDKN1C domain: the molecular mechanisms causing Beckwith-Wiedemann syndrome and cancer. *Cytogenetic and genome research* Vol.113 (No.1-4): 306–312.

Hitchins, MP. & Moore, GE. (2002). Genomic imprinting in fetal growth and development. *Expert reviews in molecular medicine* Vol.4 (No.11): 1–19.

Hill, JR.; Burghardt, RC.; Jones, K.; Long, CR.; Looney, CR.; Shin, T.; Spencer, TE.; Thompson, JA.; Winger, QA. & Westhusin, ME. (2000). Evidence for placental abnormality as the major cause of mortality in first-trimester somatic cell cloned bovine fetuses. *Biology of reproduction* Vol.63 (No.6): 1787–1794.

Hill, JR.; Edwards, JF.; Sawyer, N.; Blackwell, C. & Cibelli, JB. (2001). Placental anomalies in a viable cloned calf. *Cloning* Vol.3 (No.2): 83–88.

Ho, SM.; Tang, WY.; Belmonte de Frausto, J. & Prins, GS. (2006). Developmental exposure to estradiol and bisphenol A increases susceptibility to prostate carcinogenesis and epigenetically regulates phosphodiesterase type 4 variant 4. *Cancer research* Vol.66 (No.11): 5624-5632.

Hochedlinger, K. & Jaenisch, R. (2006). Nuclear reprogramming and pluripotency. *Nature* Vol.441 (No.29): 1061–1067.

Hori N, Nagai M, Hirayama M, Hirai T, Matsuda K, Hayashi M, Tanaka T, Ozawa T, Horike S. (2010). Aberrant CpG methylation of the imprinting control region KvDMR1 detected in assisted reproductive technology-produced calves and pathogenesis of large offspring syndrome. *Animal reproduction science* Vol.122 (No.3-4): 303-312.

Horike, S.; Ferreira, JC.; Meguro-Horike, M.; Choufani, S.; Smith, AC.; Shuman, C.; Meschino, W.; Chitayat, D.; Zackai, E.; Scherer, SW. & Weksberg, R. (2009). Screening of DNA methylation at the H19 promoter or the distal region of its ICR1 ensures efficient detection of chromosome 11p15 epimutations in Russell-Silver syndrome. *American journal of medical genetics. Part A* Vol.A149 (No.11): 2415–2423.

Horike, S.; Mitsuya, K.; Meguro, M.; Kotobuki, N.; Kashiwagi, A.; Notsu, T.; Schulz, TC.; Shirayoshi, Y. & Oshimura, M. (2000). Targeted disruption of the human LIT1 locus defines a putative imprinting control element playing an essential role in Beckwith-Wiedemann syndrome. *Human Molecular Genetics* Vol.9 (No.14): 2075–2083.

Horsthemke, B. & Wagstaff, J. (2008). Mechanisms of imprinting of the Prader-Willi/Angelman region. *American journal of medical genetics. Part A* Vol.146A (No.16): 2041–2052.

Ideraabdullah, FY.; Vigneau, S. & Bartolomei, MS. (2008). Genomic imprinting mechanisms in mammals. *Mutation research* Vol.647 (No.1-2): 77–85.

Kerjean, A.; Couvert, P.; Heams, T.; Chalas, C.; Poirier, K.; Chelly, J.; Jouannet, P.; Paldi, A. & Poirot, C. (2003). In vitro follicular growth affects oocyte imprinting establishment in mice. *European journal of human genetics* Vol.11 (No.7): 493-496.

Khosla S, Dean W, Brown D, Reik W, Feil R. (2001). Culture of preimplantation mouse embryos affects fetal development and the expression of imprinted genes. *Biology of reproduction* Vol.64 (No.3): 918-926.

Khoueiry, R,; Ibala-Rhomdane, S,; Méry, L,; Blachère, T,; Guérin, JF,; Lornage, J. & Lefèvre,
     A. (2008). Dynamic CpG methylation of the *KCNQ1OT1* gene during maturation of
     human oocytes. *Journal of medical genetics* Vol.45 (No.9): 583-588.

Khosla S.; Dean, W.; Brown, D. Reik, W. & Feil, R. (2001). Culture of preimplantation mouse
     embryos affects fetal development and the expression of imprinted genes. *Biology of
     reproduction* Vol.64 (No.3): 918–926.

Lee, MP.; DeBaun, MR.; Mitsuya, K.; Galonek, HL.; Brandenburg, S.; Oshimura, M. &
     Feinberg, AP. (1999). Loss of imprinting of a paternally expressed transcript, with
     antisense orientation to KVLQT1, occurs frequently in Beckwith-Wiedemann
     syndrome and is independent of insulin-like growth factor II imprinting. Proc.
     Proceedings of the National Academy of Sciences of the United States of America Vol.96
     (No.9): 5203–5208.

Lee, RS.; Depree, KM. & Davey, HW. (2002). The sheep (Ovis aries) H19 gene: genomic
     structure and expression patterns, from the preimplantation embryo to adulthood.
     Gene Vol.301 (No.1-2): 67-77.

Li, T.; Vu, TH.; Ulaner, GA.; Littman, E.; Ling, JQ.; Chen, HL.; Hu, JF.; Behr, B.; Giudice, L. &
     Hoffman, AR. (2005). IVF results in de novo DNA methylation and histone
     methylation at an Igf2-H19 imprinting epigenetic switch. *Molecular human
     reproduction* Vol.11 (No.9): 631–640.

Loi, P.; Clinton, M.; Vackova, I.; Fulka, J Jr.; Feil, R.; Palmieri, C.; Della Salda, L. & Ptak, G.
     (2006). Placental abnormalities associated with post-natal mortality in sheep
     somatic cell clones. *Theriogenology* Vol.65 (No.6): 1110–1121.

Maher, ER.; Brueton, LA.; Bowdin, SC.; Luharia, A.; Cooper, W.; Cole, TR.; Macdonald, F.;
     Sampson, JR.; Barratt, CL.; Reik, W. & Hawkins, MM. (2003). Beckwith-Wiedemann
     syndrome and assisted reproduction technology (ART). *Journal of medical genetics*
     Vol.40 (No.1): 62–64.

Maher, ER. (2005). Imprinting and assisted reproductive technology. *Human Molecular
     Genetics* Vol.14: R133–R138.

Manipalviratn, S.; DeCherney, A. & Segars, J. (2009). Imprinting disorders and assisted
     reproductive technology. *Fertility and sterility* Vol.91 (No.2): 305–315.

Marques, CJ.; Carvalho, F.; Sousa, M.& Barros, A. (2004). Genomic imprinting in disruptive
     spermatogenesis. *Lancet* Vol.363 (No.9422): 1700-1702.)

Maxfield, EK.; Sinclair, KD.; Dolman, DF.; Staines, ME. & Maltin, CA. (1997). In vitro culture
     of sheep embryos increases weight, primary fibre size and secondary to primary
     fibre ratio in fetal muscle at day 61 of gestation. *Theriogenology* Vol.47: 376.

Mitsuya, K.; Meguro, M.; Lee, MP.; Katoh, M.; Schulz, TC.; Kugoh, H.; Yoshida, MA,;
     Niikawa, N,; Feinberg, AP,; Oshimura, M. (1999) LIT1, an imprinted antisense RNA
     in the human KvLQT1 locus identified by screening for differentially expressed
     transcripts using monochromosomal hybrids. *Human Molecular Genetics* Vol.8
     (No.7): 1209–1217.

Moore, K.; Kramer, JM.; Rodriguez-Sallaberry, CJ.; Yelich, JV. & Drost, M. (2007). Insulin-
     like growth factor (IGF) family genes are aberrantly expressed in bovine
     conceptuses produced in vitro or by nuclear transfer. *Theriogenology* Vol.68 (No.5):
     717–727.

Oishi, M.; Gohma, H.; Hashizume, K.; Taniguchi, Y.; Yasue, H.; Takahashi, S.; Yamada, T. & Sasaki, Y. (2006). Early embryonic death-associated changes in genome-wide gene expression profiles in the fetal placenta of the cow carrying somatic nuclear-derived cloned embryo. *Molecular reproduction and development* Vol.73 (No.4): 404–409.

Owen, CM. & Segars, JH Jr. (2009). Imprinting disorders and assisted reproductive technology. *Seminars in reproductive medicine* Vol.27 (No.5): 417-28.

Pachnis, V.; Belayew, A. & Tilghman, SM. (1984). Locus unlinked to alpha-fetoprotein under the control of the murine raf and Rif genes. Proceedings of the National Academy of Sciences of the United States of America Vol.81 (No.17): 5523-5527.

Panne, D.; Raleigh, EA. & Bickle, TA. (1999). The McrBC endonuclease translocates DNA in a reaction dependent on GTP hydrolysis. *Journal of molecular biology* Vol.290 (No.1): 49–60.

Pfister-Genskow, M.; Myers, C.; Childs, LA.; Lacson, JC.; Patterson, T.; Betthauser, JM.; Goueleke, PJ.; Koppang, RW.; Lange, G.; Fisher, P.; Watt, SR.; Forsberg, EJ.; Zheng, Y.; Leno, GH.; Schultz, RM.; Liu, B.; Chetia, C.; Yang, X.; Hoeschele, I. & Eilertsen, KJ.(2005). Identification of differentially expressed genes in individual bovine preimplantation embryos produced by nuclear transfer: improper reprogramming of genes required for development. *Biology of reproduction* Vol.72 (No.3): 546–555.

Reichenbach, HD.; Liebrich, J.; Berg, U. & Berm, G. (1992). Pregnancy rates and births after unilateral or bilateral transfer of bovine embryos produced in vitro. *Journal of reproduction and fertility* Vol.95 (No.2): 363-370.

Reik, W. & Walter, J. (2001). Genomic imprinting: parental influence on the genome. Nature reviews. Genetics No.1: 21-32.

Schmidt, M.; Greve, T.; Avery, B.; Beckers, JF.; Sulon, J. & Hansen, HB. (1996). Pregnancies, calves and calf viability after transfer of in vitro produced bovine embryos. *Theriogenology* Vol.46 (No.3): 527–539.

Shimada, M.; Kawano, N. & Terada, T. (2002). Delay of nuclear maturation and reduction in developmental competence of pig oocytes after mineral oil overlay of in vitro maturation media. *Reproduction* Vol.124 (No.4): 557-564.

Shiota, K. & Yamada, S. (2005). Assisted reproductive technologies and birth defects. *Congenital anomalies* Vol.45 (No.2): 39–43.

Shiota, K. & Yamada, S. (2009). Intrauterine environment-genome interaction and children's development (3): assisted reproductive technologies and developmental disorders. *The Journal of toxicological sciences* Vol.34 (Suppl.2), SP287–SP291.

Smith, AC.; Choufani, S.; Ferreira, JC. & Weksberg, R. (2007). Growth regulation, imprinted genes, and chromosome 11p15.5. *Pediatric research* Vol.61 (No.1): 43–47.

Smith, SL.; Everts, RE.; Sung, LY.; Du, F.; Page, RL.; Henderson, B.; Rodriguez-Zas, SL.; Nedambale, TL.; Renard, JP.; Lewin, HA.; Yang, X. & Tian, XC. (2009). Gene expression profiling of single bovine embryos uncovers significant effects of in vitro maturation, fertilization and culture. *Molecular Reproduction and Development* Vol.76 (No.1): 38-47.

Smith, SL.; Everts, RE.; Tian, XC.; Du, F.; Sung, L-Y.; Rodriguez-Zas, S.; Jeong, B-S.; Renard, JP.; Lewin, HA. & Yang, X. (2005). Global gene expression profiles reveal significant nuclear reprogramming by the blastocyst stage after cloning. *Proceedings of the*

National Academy of Sciences of the United States of America Vol.102 (No.49): 17582–17587.

Somers, J.; Smith, C.; Donnison, M.; Wells, DN.; Henderson, H.; McLeay, L. & Pfeffer, PL. (2006). Gene expression profiling of individual bovine nuclear transfer blastocysts. *Reproduction* Vol.131 (No.6): 1073–1084.

Thorvaldsen, JL.; Duran, KL. & Bartolomei, MS. (1998). Deletion of the H19 differentially methylated domain results in loss of imprinted expression of H19 and Igf2. *Genes & development* Vol.12 (no.23): 3693–3702.

Tveden-Nyborg, PY.; Alexopoulos, NI.; Cooney, MA.; French, AJ.; Tecirlioglu, RT.; Holland, MK.; Thomsen, PD. & D'Cruz, NT. (2008). Analysis of the expression of putatively imprinted genes in bovine peri-implantation embryos. *Theriogenology* Vol.70 (No.7): 1119-28.

Wakayama, T.; Perry, AC.; Zuccotti, M.;, Johnson, KR. & Yanagimachi, R. (1998). Full-term development of mice from enucleated oocytes injected with cumulus cell nuclei. *Nature* Vol.394 (No.6691): 369–374.

Walker, SK.; Hartwich, KM. & Seamark, RF. (1996). The production of unusually large offspring following embryo manipulation: concepts and challenges. *Theriogenology* Vol.45: 111–120.

Weksberg, R.; Shuman, C. & Smith, AC. (2005). Beckwith-Wiedemann syndrome. *American Journal of Medical Genetics Part C* Vol.137C: 12–23.

Weksberg, R.; Smith, AC.; Squire, J. & Sadowski, P. (2003). Beckwith-Wiedemann syndrome demonstrates a role for epigenetic control of normal development. *Human Molecular Genetics* Vol.12, R61–R68.

Wilmut, I.; Beaujean, N.; de Sousa, PA.; Dinnyes, A.; King, TJ.; Paterson, LA.; Wells, DN. & Young, LE. (2002). Somatic cell nuclear transfer. *Nature* Vol.419 (No.6907), 583–586.

Wilmut, I.; Schnieke, AE.; McWhir, J.; Kind, AJ. & Campbell, KH. (1997). Viable offspring derived from fetal and adult mammalian cells. *Nature* Vol.385 (No.6619): 810–813.

Willadsen, SM.; Janzen RE.; McAlister RJ.; Shea, BF.; Hamilton, G. & McDermand, D. (1991). The viability of late morular and blastocysts produced by nuclear transplantation in cattle. *Theriogenology* Vol.35 (No.1): 161-170.

Yamada, Y.; Watanabe, H.; Miura, F.; Soejima, H.; Uchiyama, M.; Iwasaka, T.; Mukai, T.; Sakaki, Y. & Ito, T. (2004). A comprehensive analysis of allelic methylation status of CpG islands on human chromosome 21q. *Genome research* Vol.14 (No.2): 247–266.

Yan, Y.; Frisén, J.; Lee, MH.; Massagué, J. & Barbacid, M. (1997). Ablation of the CDK inhibitor p57Kip2 results in increased apoptosis and delayed differentiation during mouse development. *Genes & development* Vol.11 (No.8): 973-983.

Yang, L.; Chavatte-Palmer, P.; Kubota, C.; O'Neill, M.; Hoagland, T.; Renard, JP.; Taneja, M.; Yang, X. & Tian, XC. (2005). Expression of imprinted genes is aberrant in deceased newborn cloned calves and relatively normal in surviving adult clones. *Molecular reproduction and development* Vol.71 (No.4): 431–438.

Young, LE.; Fairburn, HR. (2000). Improving the safety of embryo technologies: possible role of genomic imprinting. *Theriogenology* Vol.53 (No.2): 627-648.

Young, LE.; Schnieke, AE.; McCreath, KJ.; Wieckowski, S.; Konfortova, G.; Fernandes, K.; Ptak, G.; Kind, AJ.; Wilmut, I.; Loi, P. & Feil, R. (2003). Conservation of IGF2-H19 and IGF2R imprinting in sheep: effects of somatic cell nuclear transfer. *Mechanisms of development* Vol.120 (No.12): 1433–1442.

Young, LE.; Sinclair, KD. & Wilmut, I. (1998). Large offspring syndrome in cattle and sheep. *Reviews of Reproduction* Vol.3 (No.3): 155–163.

Zhang, S.; Kubota, C.; Yang, L.; Zhang, Y.; Page, R.; O'Neill, M.; Yang, X. & Tian, XC. (2004). Genomic imprinting of H19 in naturally reproduced and cloned cattle. *Biology of reproduction* Vol.71 (No.5): 1540–1544.

Part 3

Methylation Changes and Cancer

Investigating the Role DNA Methylations Plays in Developing Hepatocellular Carcinoma Associated with Tyrosinemia Type 1 Using the Comet Assay

Johannes F. Wentzel and Pieter J. Pretorius
Division for Biochemistry, North-West University,
South Africa

1. Introduction

Although our understanding of DNA methylation mechanisms and functions has vastly improved over the past two decades, the entire spectrum of influences of this potent mechanism is not fully grasped yet. Today, a wide variety of techniques are used to examine DNA methylation patterns but most of these methods are still relatively expensive and many are platform specific (Prokhortchouk and Defossez, 2008). The need has arisen to develop an economically viable and uncomplicated technique for DNA methylation analysis. The comet assay (single cell gel electrophoresis or SCGE) is a cost-effective, sensitive and simple technique which is traditionally used for analysing and quantifying DNA damage in individual cells (Azqueta, et al., 2011, Fairbairn, et al., 1995). By modifying this assay to be methylation sensitive, global DNA methylation can be routinely measured in cell cultures while simultaneously being able to deduce the integrity of the genetic material of the examined cells. The methylation sensitive comet assay was used to examine the effect of the accumulating metabolite, succinylacetone (SA), in hereditary Tyrosinemia type I (HT1), since hepatocellular carcinoma (HCC) frequently develops into this disease.

2. The role of DNA methylation in maintaining genome stability

DNA undergoes several modifications following each replication cycle of which DNA methylation is one. DNA methylation, as part of the epigenetic code, involves the addition of a methyl group to cytosine without altering the original DNA sequence. DNA methylation may take place on the number five carbon of the cytosine pyrimidine ring and this modification has been observed in every vertebrate examined (Herman, 2001, Waggoner, 2007). DNA methylation alters the biophysical characteristics of DNA which may either inhibit the recognition of certain DNA sequences by functional proteins or enable the binding of others (Prokhortchouk and Defossez, 2008).

DNA methylation is carried out by a group of enzymes called DNA methyltransferases (DNMTs) (Pogribny, et al., 2004). These enzymes not only determine the DNA methylation patterns during early development, but are also responsible for copying these patterns to the new strands formed during DNA replication. Based on their preferred DNA substrates, these enzymes can be divided into two different classes – maintenance enzymes (DNMT1) and *de novo* or pioneering enzymes (DNMT3A + B) (Sweatt, 2009).

The maintenance methyltransferase DNMT1 is responsible for copying existing methylation patterns to newly replicated, hemimethylated DNA (Kanai, et al., 2003, Sweatt, 2009). DNMT1 is also associated with other epigenetic processes such as silencing gene transcription and is known to interact with histone deacetylase 1 and 2 (HDAC1/2) (Kanai, et al., 2003). While some studies have indicated that DNMT1 has some *de novo* methylation activity *in vitro*, there is no substantial evidence indicating that this is the case *in vivo* (Okano, et al., 1999). Due to the relation of DNMT1 with HDAC1 and 2, it directly affects gene transcription and chromatin structure. Although the identification of hemimethylated sequences by DNMTs are still a point of discussion and uncertainty, there is some evidence suggesting that DNMT1 may be recruited by UHRF1 which can bind to partially methylated DNA (Prokhortchouk and Defossez, 2008). In short, DNMT1 is the key maintenance DNA methyltransferases which is responsible for copying the previously established methylation blueprint after every replication cycle.

The *de novo* methyltransferase DNMT3 typically establishes new DNA methylation patterns in specific regions, mainly in satellite repeats and retrotransposon (transposons via RNA intermediates) sequences (Gopalakrishnan, et al., 2008). The *dnmt3* gene family consists of *dnmt3a* and *dnmt3b* which seems to have related functions (Okano *et al*, 1999). Studies done on rodent embryos lacking the *dnmt3a* showed that these embryos developed normally but died a few weeks after birth (Bird, 2002, Mohn and Schubeler, 2009). On the other hand, mice lacking *dnmt3b* suffered from multiple developmental defects and were aborted (Gopalakrishnan, et al., 2008, Mohn and Schubeler, 2009). This is not only a clear indication of the importance of DNMT3 in *de novo* methylation but also illustrates the fundamental role of DNA methylation in early development. To summarize, DNMT3A and B are responsible for methylating previously unmethylated DNA and seem to play an imperative role in development and growth.

A fourth DNA methyltransferase, DNMT2, exists and is structurally similar to prokaryotic and eukaryotic methyltransferases. However, it shows weak or no DNA methyltransferase activity *in vitro* and targeted deletion of the DNMT2 gene in embryonic stem cells causes no detectable effect on global DNA methylation, suggesting that this enzyme has little involvement in establishing DNA methylation patterns (Okano *et al*, 1999; Klose, 2006). The methyltransferase DNMT3L is thought to be a co-factor for DNMT3a and DNMT3b and modulates their catalytic activity (Klose and Bird, 2006, Prokhortchouk and Defossez, 2008). S-adenosylmethionine (SAM) acts as a methyl donor and donates a methyl group to unmethylated cytosines within CpG islands of genomic DNA. The methyl group is transferred from SAM to the 5′ position from cytosine through the actions of DNMTs.

To summarize, DNMTs are at the core of the DNA methylation machinery and is not only responsible for establishing new methylation patterns in DNA but also to maintain these patterns after replication. It may even be an inseparable part of transcription. Elucidating the full extent of influence of the DNMTs will undoubtedly better our understanding of DNA methylation and will be indispensable for future medical endeavours to ultimately prevent disease.

Investigating the Role DNA Methylations Plays in Developing Hepatocellular Carcinoma Associated with
Tyrosinemia Type 1 Using the Comet Assay

213

Maintaining global genome stability is crucial for the development and normal functioning of any organism. In eukaryotic organisms, DNA methylation is crucial for maintaining genome stability, as well as playing an important role in routine gene expression (Bird, 2002, Nag and Smerdon, 2009). Apart from ensuring a stable genome in adults, DNA methylation also plays a central role in the ordered differentiation of mammalian cells during embryonic development by activating specific genes and silencing others (Mohn and Schubeler, 2009).

Two common patterns of cytosine methylation at CpG dinucleotides have been described in the eukaryotic genome - genome-wide CpG methylation density, and non-random regional CpG methylation (Cottrell, 2004, Ohgane, et al., 2008). DNA methylation in adult somatic tissues typically occurs in regions called CpG islands while non-CpG methylation (methylation outside of CpG islands) is most prevalent in embryonic stem cells (Haines, et al., 2001). CpG islands are characterized by high CpG density and tend to be unmethylated under normal conditions (Duffy, et al., 2009).

As part of the epigenetic code, DNA methylation ensures diverse cellular differentiation despite a primarily static genome. There exists a complex collaboration between epigenetic mechanisms (including DNA methylation, histone modification and micro RNA systems) to orchestrate a sophisticated and cell specific gene regulation (Guil and Esteller, 2009). These epigenetic processes connect the largely static genome and transcriptome by introducing complex and dynamic networking layers of gene control (Ohgane, et al., 2008). Due to this intensive involvement of DNA methylation in the epigenome, it comes as no surprise that the abnormal regulation of these mechanisms may cause altered gene expression and ultimately disease.

3. Abnormal DNA methylation patterns associated with cancer

Taken into account the crucial role epigenetics play in many cellular processes, it is logical that abnormal regulation of these mechanisms may lead to misinterpreted gene expression and eventually disease. This relationship between disease and the epigenome prompted scientists to perform comprehensive research into the role epigenetics play in the aetiology of these diseases. Today there is mounting evidence of irregular epigenetic regulation in many diseases (Gopalakrishnan, et al., 2008).

In the late 1980s abnormal DNA methylation patterns were observed in all cancer types investigated. There is ample evidence indicating that disruption of epigenetically regulated expression of genes play a direct, and maybe even a causative, role in the development and manifestation of many diseases including cancer (Hirst and Marra, 2009, Sawan, et al., 2008).

In cancer cells, methylated CpGs loose their DNA methylation status and unmethylated promoter regions may become densely methylated. This puzzling phenomenon prompted scientists to believe that both the loss and/or gain of DNA methylation are linked to cancer (Ehrlich, 2002, Gopalakrishnan, et al., 2008, Hirst and Marra, 2009). DNA methylation, as part of the epigenome, is critical in expanding and regulating the expression of the genome which in its turn determines the phenotype. The epigenome is subsequently influenced by internal cues as well as a large range of environmental factors (Lambert and Herceg, 2008). If these processes are not tightly regulated it can lead to changes in DNA methylation and histone modification patterns resulting in the disruption of important cellular processes including gene expression, DNA repair and tumour suppression - which may lead to cancer development. For instance, mutations in genes encoding for DNA methyltransferases lead to changes in their expression and to altered DNA methylation patterns (Brenner, et al., 2005). CpG islands are usually located in non-tissue specific promoter regions of genes and are normally unmethylated. Tumour cells exhibit global hypomethylation of the genome accompanied by region-specific hypermethylation (Szyf, 2006) (see figure 1).

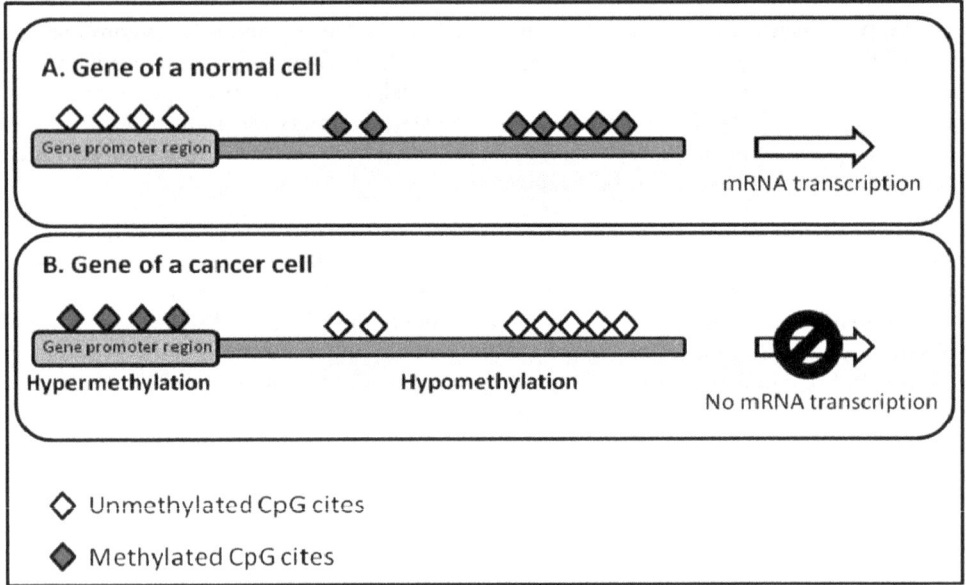

Fig. 1. Different methylation patterns in normal (A) and malignant (B) cells. Cancer cells exhibit global hypomethylation of the genome accompanied by region-specific hypermethylation.

DNA hypomethylation can be described as the demethylation of the typically methylated regions of genes. The three main mechanisms proposed to contribute to cancer development due to hypomethylation is: (1) an increase in genomic instability, (2) reactivation of transposable elements and (3) loss of imprinting (Hirst and Marra, 2009). This may lead to the weakening, or even lifting, of the transcriptional repression in silenced gene promoters which in turn may contribute to the unwanted or abnormal expression of genes (Wilson, et al., 2007). This demethylation pattern can be seen in several cancers including metastatic hepatocellular cancer, cervical cancer, prostate tumours, and brain cancers (Das and Singal, 2004). This abnormal demethylation can also cause chromosomal rearrangement and translocations which might negatively influence genome stability (Ehrlich, 2002).

Hypermethylation of DNA is far more common in tumour cells than hypomethylation and mainly occurs in the promoter regions of genes leading to their transcriptional silencing (Moss and Wallrath, 2007). This is apparent from the inactivation of the tumour suppressor genes, adhesion molecules and repair enzymes which contribute to cancer development (Gopalakrishnan, et al., 2008). It has been confirmed that one or more genes are hypermethylated in tumours. Examples of such studies are those concerning lung cancer (where over 40 genes were found to have altered methylation patterns) and leukaemia where many genes exhibit hypermethylation (Tsou, et al., 2002). CpG island hypermethylation also seems to increase in metastasis suppressor genes while increasing histone alterations (Lujambio and Esteller, 2007).

Over the past decade studies have shown that epigenetic alterations are present in many malignancies and in contrast to genetic alterations, epigenetic changes are theoretically reversible - making them opportune targets for the development of therapeutic and preventive drugs. Better insight into the mechanisms involved in DNA methylation-related diseases are of cardinal importance for the development of more effective treatments.

From this description of the crucial role DNA methylation plays in maintaining the integrity of the vertebrate genome, accurate measurement of the basic level of DNA methylation and deviations thereof become of paramount importance. Methods used to measure DNA methylation range from array technology to describe the methylome to evaluating the promoter assosiaded CpG islands of individual genes. The majority of the methods where global DNA methylation is investigated give a combined measure of DNA methylation of all the cells involved. The comet assay can be modified to measure DNA methylation of single cells. The advantages of this approach is that, because of the versatility of the comet assay, a spectrum of additional information pertaining to the integrity of the genome can be collected in one experiment e.g. DNA damage and repair.

4. Using the methylation sensitive comet assay to investigate the role DNA methylation plays in hepatocellular carcinoma associated with tyrosinemia type 1

4.1 Comet assay

Presently, a wide variety of techniques are used to examine DNA methylation patterns, but most of these methods are relatively expensive and many are platform specific (Prokhortchouk and Defossez, 2008). The need has arisen to develop alternative and more economically viable techniques for DNA methylation analysis.

The comet assay (or single cell gel electrophoresis assay) is a cost-effective, sensitive and simple technique, which is traditionally used for analysing and quantifying DNA damage in individual cells (Azqueta, et al., 2011, Fairbairn, et al., 1995). In 1984 Östling and Johanson developed the alkaline single cell gel electrophoresis assay as a novel approach for detecting DNA lesions. Today this method is regularly used in biomonitoring and mechanistic studies in a large range of *in vitro* and *in vivo* systems (Lovell and Omori, 2008). In this method, specific cells are harvested and encapsulated in a low melting point agarose gel. This gel solution is then applied to a glass slide covered in high melting point agarose.

The agarose gel acts to keep the cells separate from each other and to serve as staging point for all the treatments that follow. Encapsulation is followed by chemical (usually a high salt concentration solution) lysis. The supercoiled DNA attached to the nuclear matrix unwinds due to the high pH (~12.3) of the alkaline lysing solution. The encapsulated nucleoids are now exposed to an electric current causing the damaged fragments of the DNA to migrate towards the anode - forming the so called tail of the comet. After electrophoresis, Tris-HCl is used to neutralize the alkaline electrophoresis buffer followed by staining with ethidium bromide. Imaging software is then used to measure the fluorescence and to determine the extent of DNA damage (Collins, et al., 1997, Rojas, et al., 1999). The genomic integrity of the individual cell is then calculated by the tail migration by either measuring the tail length, tail moment or percentage DNA in the tail in (see figure 2 for an example of a comet) (Lovell and Omori, 2008).

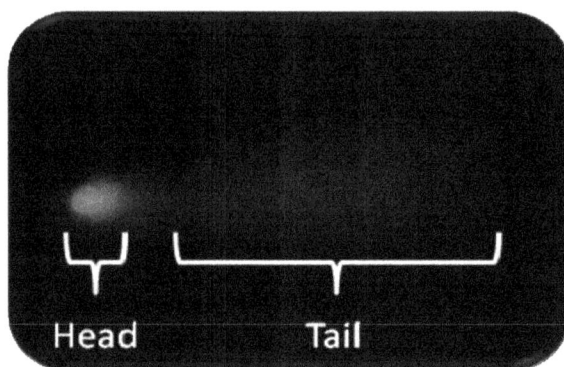

Fig. 2. The DNA fragments migrate from the nucleoid forming a comet like appearance. The general shape of a comet clearly indicating its head (nucleoid) and tail (migrated DNA) after electrophoresis and staining.

The comet assay is widely used for genotoxicity studies and determining DNA repair capacity and a variety of DNA lesions can be detected using the alkaline version of the single cell gel electrophoresis (comet) assay, including DNA double (DSB) and single strand breaks (SSB), as well as alkali-labile sites (ALS) (Collins and Gaivao, 2007).The flexibility of the comet assay is further expanded with the use of specific restriction endonucleases. Modifications to the comet assay have allowed the use of lesion specific restriction enzymes to detect specific base modifications as DNA single strand breaks (Collins and Gaivao, 2007, Epe, et al., 1993). For example, Fpg acts both as an AP-lyase and a N-glycosylase, allowing it to release modified purines from double stranded DNA (Tice, et al., 2000). It was also shown that Fpg strongly enhances the detection of mitomycin C (MMC) and ethylmethanesulphonate (EMS) induced DNA modifications (Andersson and B.E.Hellman, 2005). Damaged pyrimidines are removed in a similar manner from double stranded DNA by the endonuclease *Endo lll* (Speit, et al., 2009).

By using methylation sensitive restriction endonucleases, the traditional alkaline comet assay can be modified to be methylation sensitive. This method enables the routine measurement of global, as well as CpG island DNA methylation in a variety of cells while simultaneously determining the genetic integrity of examined cells.

4.2 Modification of the comet assay to be methylation sensitive

The methylation sensitive comet assay employs the isoschizomeric restriction enzymes HpaII and MspI. These enzymes recognize the same tetranucleotide sequence (5'-C C G G-

3') but display differential sensitivity to DNA methylation. HpaII is inactive when any of the two cytosines is methylated, but it digests the hemimethylated 5'-CCGG-3' at a lower rate compared with the unmethylated sequences. On the other hand, MspI digests 5'-C^{m}CGG-3' but not 5'-^{m}CCGG-3'. These restriction enzymes features have been used to assess the global DNA methylation status of DNA preparations in radio labeling experiments (Fujiwara and Ito, 2002, Pogribny, et al., 2004). Difference in methylation sensitivity of the isoschizomeric restriction endonucleases *Hpa*II and *Msp*I was also used to demonstrate the feasibility of the comet assay to measure global DNA methylation level in individual cells. Thus, when applied to the comet assay, one would expect that a higher level of DNA methylation of the CpG sequence in this site would result in a larger difference in the global amount of DNA in the comet tails of *Hpa*II digested versus *Msp* I digested DNA.

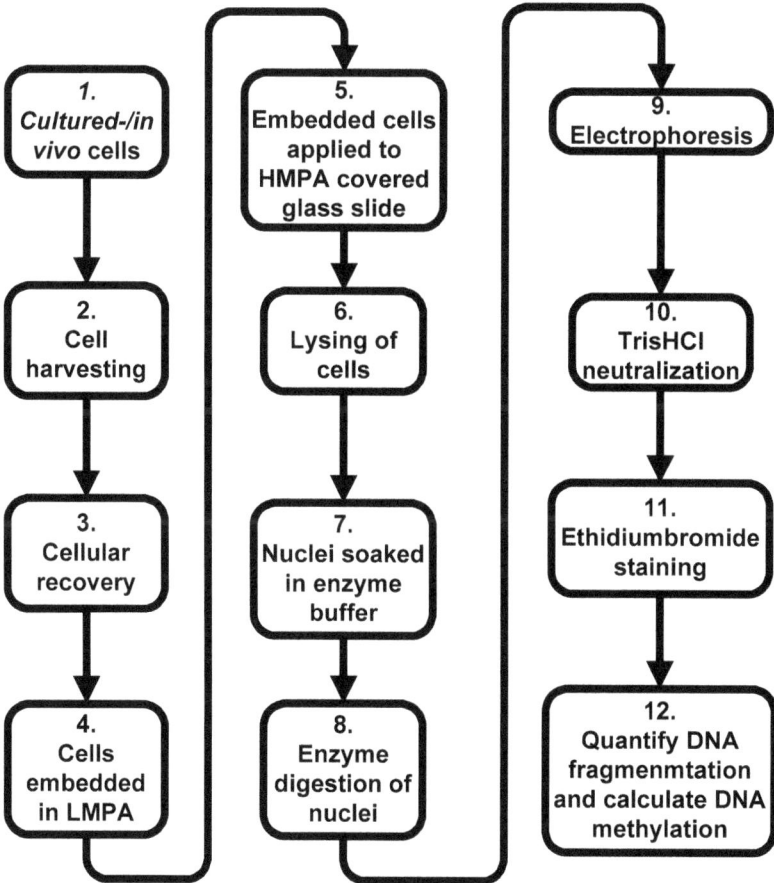

Fig. 3. A summary of the methylation sensitive comet assay method

4.2.1 Cell harvesting and repair

One of the advantages of the comet assay is the possibility to employ a wide variety of cells, ranging from cultured cells to cells harvested *in vivo*. Different cell lines require specific cellular harvesting techniques, for instance, cultured HepG2 is an adherent cell line and must be separated from their growth chambers, adjacent cells and other material in suspension, while leucocytes have to be separated from other cells and substances present in blood. Many of the cultured cell lines require a proteinase harvesting step to detach them from the growth chamber and adjacent cells (see section 4.2.4). During this harvesting process, cells are exposed to trypsin which negatively influences the cell's cellular integrity that may lead to DNA fragmentation. This makes a repair phase essential before the cells can be used in the comet assay. For recovery, cells were incubated in an orbital shaker (to prevent cells from adhering to the container) in DMEM nutrient medium for 2 hours at 37°C directly after trypsin harvesting. In order to undergo electrophoresis to quantify the extent of DNA damage and/or fragmentation, cells must first be encapsulated in agarose gel.

4.2.2 Cell encapsulation and lysis

After the recovery phase, cells are encapsulated in a low melting point agarose gel and applied to a glass slide covered with a thin layer of high melting point agarose. In order for the comet assay to illustrate the integrity of a cell's genetic material, the nucleoides must first be exposed. The encapsulated cells are submerged in lysis solution consisting of sodium chloride (NaCl) [2.5 - 5M], ethylenediaminetetraacetic acid (EDTA), 10% dimethyl sulfoxide (DMSO) and 1% Triton X-100 at 4°C. This solution is formulated to disrupt cellular material in the shortest amount of time, while leaving the nucleus unharmed. The high NaCl concentration draws water out of the cells by osmotic action. The exact NaCl concentration should be determined empirically for the specific type of cells used to prevent it from negatively influencing the cell's genomic integrity. EDTA is commonly used to forage metal ions in order to deactivate metal dependant enzymes that may cause DNA damage. DMSO is an excellent polar aprotic solvent and is especially effective solvent in reactions involving salts. It acts as a hydroxyl radical scavenger formed when iron is released from blood while also acting to inhibit the formation of secondary DNA structures (Chakrabarti and Schutt, 2001). Studies concluded that an 8-15% dilution of DMSO yields the best lysis results (Wentzel, unpublished M.Sc results). Finally, the non-ionic detergent Triton X-100 is used to solubilise cell proteins and membranes.

4.2.3 Treatment of exposed nucleoids with restriction enzymes HpaII and MspI

Nucleoids are treated with restriction enzymes *Hpa*II and *Msp*I just after the cell lysis step. In order to create favourable restriction conditions, the slides containing the nucleoids are soaked in enzyme reaction buffer for at least 10 minutes. The enzyme reaction buffer consist of 5×10^{-3} mol/l TrisHCl, 5×10^{-3} mol/l NaCl, 5×10^{-4} mol/l merkapto-ethanol and 1×10^{-3} mol/l EDTA which is a close recreation of the Tango buffer usually used with these enzymes. Experiments have indicated that this composition of the reaction buffer is absolutely crucial for normal enzyme function and care should be taken in its accurate preparation. After the slides have been soaked in reaction buffer (and the excess liquid removed to prevent any negative affect it may have on the enzyme reaction) the enzyme mixture is directly applied to the slides. Each enzyme mixture is composed of 1.5 unit of *Msp*I or *Hpa*II, 10 μl of Tango buffer (Fermentas) and filled to 100 μl with molecular grade H_2O. 100 μl of this enzyme mix is then carefully applied to each slide and covered with a cover slip. The slides are then placed in a damp plastic container lined with towel paper that was preheated to 37°C. After 5 minutes of incubation the slides are covered with towel paper soaked in reaction buffer to keep the slides from drying out while incubating for another 55 minutes. In some cases, experimental results showed little or no DNA digestion. This may be the result of tightly packed nuclei where the majority of the enzymes recognition sites are unavailable, thus leading to poor DNA digestion. This can be overcome by treating the cells directly with a 5% proteinase K solution just after lysis.

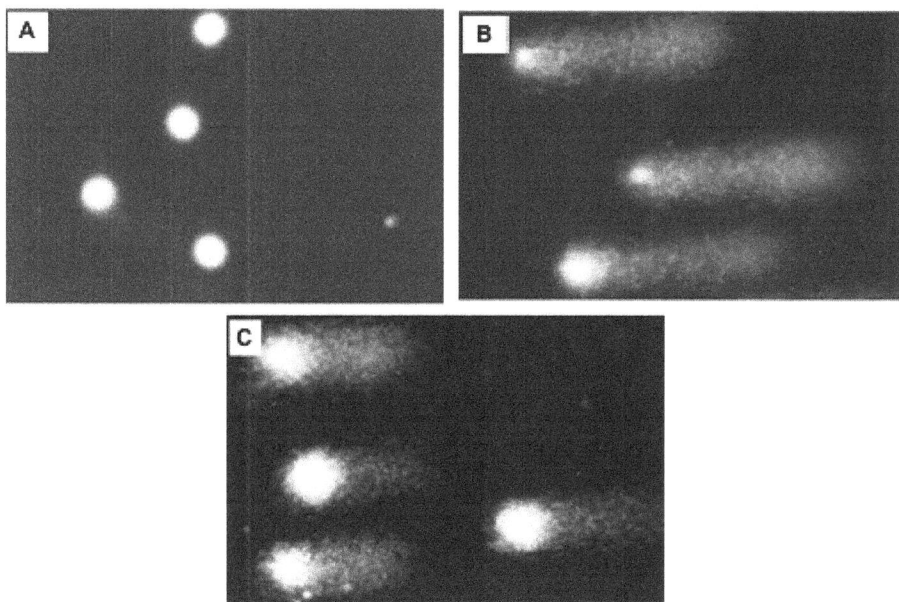

Fig. 4. Comets created by the treating nucleoids with the isoschizomeric enzymes MspI and HpaII: Nucleoids without enzyme treatment (A), with MspI treatment (B), and with HpaII treatment (C).

4.2.4 Proteinase K treatment of cells

Nucleoids prepared from cells in the later phases of the growth curve (or even cells with a high passage number) tend to be tightly packed, making enzyme digestion only partially effective. As a last resort, this problem can be overcome by soaking the slides in electrophoresis buffer for 30 minutes just before enzyme treatment. This alkaline solution assists with unwinding the nucleus but can cause DNA damage leading to unwanted fragmentation. An alternative is to treat the cells with proteinase K (Qiagen) (1.0 -1.5 mM) solution which is a subtilisin-type protease. This broad spectrum serine protease is commonly used to inactivate nucleases when isolating or purifying DNA (QIAGEN, 2005). Proteinase K acts on the nucleosomes causing the DNA to relax and unwind, making restriction enzyme recognition sites more accessible for *Msp*I and *Hpa*II (DNA-protein interactions that have survived the lysis step, are destroyed with this enzyme treatment). After proteinase K treatment there is a visible increase in the average tail percentage between the different enzyme digestions indicating the effectiveness of proteinase K in relaxing the nucleoids of HepG2tTS cells (see figure 5). Although the increase in DNA methylation was small, we observed these differences in several experiments And can also safely say that there was no DNase activity in the Proteinase K preparations The tail-DNA increase observed after restriction enzyme digestion in the Proteinase K treated cells can thus be ascribed to an increase in the number of available restriction sites due to differences in DNA methylation sensitivity.

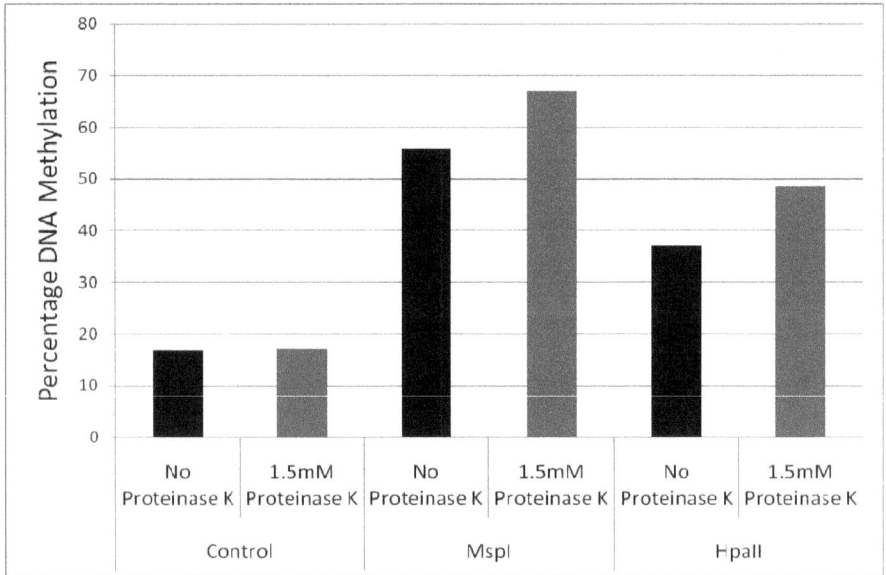

Fig. 5. Proteinase K treatment of the nucleoids. The nucleoids were treated with Proteinase K before treatment with restriction enzymes.

4.2.5 Investigation of DNA methylation within CpG islands using the methylation-sensitive endonuclease NotI

In adult somatic tissues DNA methylation typically occurs in regions called CpG islands. CpG islands are characterized by high CpG density and tend to be unmethylated under normal conditions (Duffy, et al., 2009). According to Gardiner-Garden sequence criteria, a CpG island is defined as a region greater than 200 bp with a G + C content greater than 50% (Gardiner-Garden and Frommer, 1987). Interestingly, over 40% of human genes contain CpG islands in their promoter regions and it is strongly believed that transcription activation and silencing of these genes are directly influenced by DNA methylation (Illingworth and Bird, 2009). To examine methylation in CpG islands, the methylation-sensitive enzyme NotI can be used, since it has a 5'-GC^GGCCGC-3' recognition site and acts mainly on target sites within CpG islands (Fazzari and Greally, 2004). Therefore, the

Investigating the Role DNA Methylations Plays in Developing Hepatocellular Carcinoma Associated with
Tyrosinemia Type 1 Using the Comet Assay

221

slides were soaked in restriction enzyme reaction buffer (50 mM Tris–HCl, 10 mM MgCl$_2$, 100 mM NaCl, and 0.1 mg/ml bovine serum albumin [BSA]) for 10 min. An enzyme mixture consisting of NotI (1.5 U/100 µl) in 1x Buffer O (Fermentas) was prepared, and 100 µl of this enzyme mixture was applied to the slides and covered with a coverslip. The comet assay was then performed as described above.

4.2.6 Electrophoresis, staining and quantification of nucleoids

After incubation, the slides are placed in an electrophoresis tank and covered with electrophoresis buffer (usually around 5 mol/l sodium hydroxide (NaOH) and 0.4 mol/l EDTA, pH 13). Electrophoresis is performed at a potential difference of 30V and a current of 300mA for 45 minutes while the buffer is maintained at 4°C. Experiments showed that electrophoresis at 4°C is extremely crucial for optimum DNA fragment migration. After electrophoresis the slides are placed in a Tris-HCl neutralisation buffer for approximately 15 minutes. This buffer acts to neutralize the alkaline electrophoresis and prevents it from causing unwanted DNA damage\fragmentation. The slides are then stained with 10µg/ml ethidiumbromide for at least an hour at 4°C and rinsed afterwards with distilled water.

Finally, the comet images are captured with a fluorescence microscope (x200 magnification) and scored using computer software (Comet IV from Perceptive Instruments Ltd). Due to intercellular variation, it is highly recommended that at least 200 comets are scored at random and that all experiments be done in duplicate. The data generated can then be analyzed and the results are expressed as percentage CpG methylation. This is calculated using the ratio between the average percentage tail DNA of HpaII- and MspI-digested DNA, that is, 100 – HpaII/MspI x 100, where HpaII and MspI are the average percentage tail DNA of HpaII- and MspI-digested nucleoids, respectively. To investigate the distribution properties of the percentage CpG methylation, the bootstrap analytical technique can be employed. The bootstrap is a technique that can estimate population parameters and distribution properties of statistics by substituting the population mechanism used to obtain the parameter with an empirical equivalent. These estimates can be obtained analytically, but they are obtained mostly through the use of resembling and Monte Carlo methods carried out by a computer (Wentzel, et al., 2010).

4.2.7 Validation of methylation sensitive comet assay results

To validate the results obtained with the comet assay, we performed the Cytosine Extension Assay (CEA) on the DNA isolated from the remaining cells of the same batch used for the comet assay. The CEA is also based on the selective use of the methylation-sensitive restriction enzymes HpaII and MspI, both of which leave a 5' guanine overhang after DNA cleavage followed by a single nucleotide extension with [3H] dCTP (Pogribny et al, 2004). The level of radioactive cytosine incorporation is then determined by scintillation counting, and the methylation percentage can subsequently be calculated.

The methylation-sensitive comet assay was applied to cultured cells that were treated with the metabolite succinylacetone (SA), which accumulates in hereditary tyrosinemia type I.

Tyrosinemia is a genetic disorder characterized by elevated blood levels of the amino acid tyrosine. Type I tyrosinemia (HT1), the most severe form of tyrosinemia, is caused by a defect of the enzyme fumarylacetoacetate hydrolase. Symptoms usually appear in the first few months of life and may include failure to thrive, diarrhea, vomiting, jaundice and increased tendency to bleed. Type I tyrosinemia can lead to liver and kidney failure, problems affecting the nervous system, and an increased risk of liver cancer (Al-Dhalimy, 2002; Orejuela, 2008). In a healthy liver, an enzyme mediated five-step process breaks down tyrosine into harmless molecules that are either excreted by the kidneys or used in energy the metabolism. Mutations in the *fah* gene cause a shortage of the enzyme Fumarylacetoacetate Hydrolase in this multistep pathway. This resulting enzyme deficiency leads to a toxic accumulation of tyrosine and its by-products, which can damage the liver, kidneys, nervous system and other organs and tissues. The exact cause of hepatocellular carcinoma associated with HT1 is unclear at the moment but it is speculated that some of the accumulating metabolites may have carcinogenic properties.

4.3 Application of the methylation sensitive comet assay

The HT1 metabolic defect leads to a toxic accumulation of tyrosine and its by-products, which can damage the liver, kidneys, nervous system, and other organs and tissues. The exact cause of hepatocellular carcinoma associated with HT1 is unclear at the moment but because there is some evidence indicating that HTI-accumulating metabolites may alter DNA methylation (Fazzari and Greally, 2004). There is some evidence suggesting that one of the accumulating metabolites of HT1, p-hydroxyphenylpyruvic acid or pHPPA, is responsible for the long-term impairment of the DNA repair, leading ultimately to cellular hepatocarcinoma (van Dyk and Pretorius, 2005).

We examined the effect of another accumulating metabolite of HT1, succinylacetone (SA), on cultured liver cells using the metyhelation sensitive comet assay. The cultured cells were exposed to 50 uM SA for a period of 24 h, and the comet assay was performed. It was found that in contrast to a notable decline in DNA methylation outside the CpG islands of approximately 20% after 24 h of exposure to SA, an increase of approximately 8% in CpG island methylation was detected after SA treatment (see figure 6). These methylation patterns are typical of those observed in cancer cells (global demethylation and gene specific hypermethylation) and indicates that SA may indeed cause aberrant DNA methylation patterns in HTI, which can contribute to the initiation of hepatocarcinoma associated with this disease (Wentzel, et al., 2010).

It is still not clear if the cell's exposure to SA directly alters the DNA methylation profile, subsequently leading to cancer or if the change in DNA methylation is only a result of important disrupted cellular processes such as DNA repair mechanisms and tumour suppression – caused by the malignant state of the cell.

5. Summary and conclusion

Despite the progress that has been made to better our understanding of DNA methylation, it seems we are only at the first chapter of this intricate book and many unanswered questions remain surrounding the extent in which DNA methylation influences normal as well as malignant cell activity. One thing that can be said with certainty is that the main methylation function involves the regulation of gene expression. This function is extended via other actions, e.g. chromatin compaction, x-chromosome inactivation, maintaining

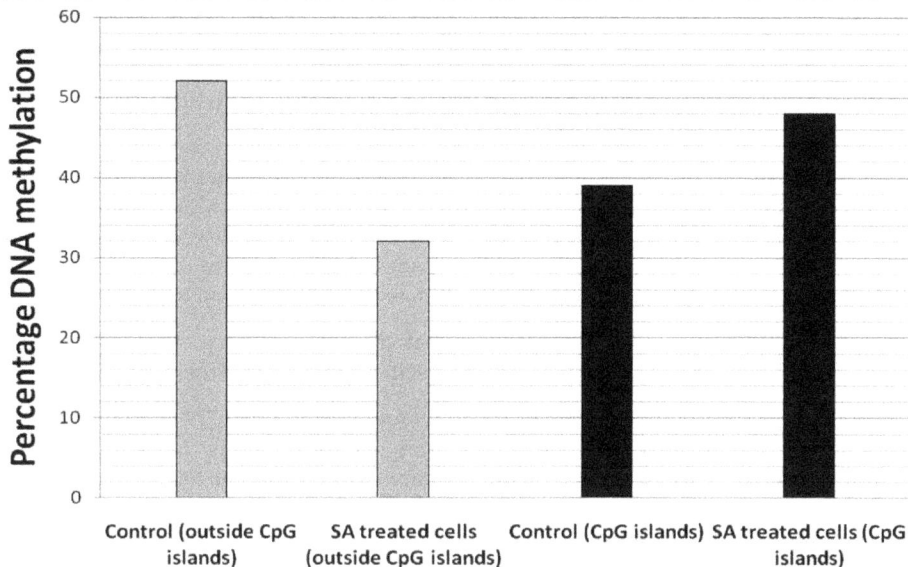

Fig. 6. DNA methylation of HepG2 cells treated with SA. Nucleoids digested with MspI and HpaII give an indication of DNA methylation outside the CpG islands, whereas NotI is used to determine DNA methylation in CpG islands.

genome stability and maintaining cellular identity. These functions are, to a great extent, dependent on the physical properties of cytosine methylation and properly established and maintained DNA methylation patterns seems to be essential for mammalian development and normal functioning of the adult organism.

According to the American Cancer Association nearly 13% of all deaths worldwide are cancer related. Currently, there are numerous publications suggesting that aberrant DNA methylation patterns play a direct, and likely causative, role in cancer initiation and development. Through better understanding of the role DNA methylation plays in cancer, we can develop more effective treatments for this devastating disease.

The comet assay (single cell gel electrophoresis or SCGE) is a cost-effective, sensitive and simple technique which is traditionally used for analyzing and quantifying DNA damage in individual cells. A variety of DNA lesions can also be detected using the alkaline version of comet assay, including DNA double and single strand breaks (SSB), as well as alkali-labile sites. Modifications to the comet assay have allowed the use of lesion specific endonucleases to detect specific base modifications as DNA single strand breaks.

Experiments with the methylation sensitive comet assay have shown that this technique can be successfully modified to determine changes in the level of global and regional DNA methylation of single cells. The study revealed that the hereditary tyrosinemia type I metabolite succinylacetone, may indeed cause DNA hypermethylation in HTI, which may contribute to the initiation of hepatocarcinoma associated with this disease.

Metabolic diseases can dramatically alter the metabolism of a cancer cell which can lead to dysfunctional electron transport chain, higher levels of reactive oxygen species and abnormal redox status. It is still unclear if metabolic alterations leads to epigenetic changes or vice versa, but it is apparent that the combination of these two modified states contribute to aberrant gene expression and cancer.

The advantage of the comet assay above other methylation detection techniques is its cost effectiveness, ability to investigate the DNA integrity and the DNA methylation status of cells simultaneously, while being able to measure global as well as gene specific DNA methylation. This modification of the comet assay further expands the versatility of the comet assay by increasing the variety of observations that can be made in one experiment.

One of the drawbacks of this technique is that the assay is very sensitive to external conditions and must be optimized to suite the specific laboratory setting.

Because DNA methylation was shown to be a tissue-specific event (Pogribny, et al., 2004), this modification of the comet assay provides the opportunity to study the DNA methylation status of single cells that are prepared from different tissues under various physiological conditions. By applying the methylation sensitive comet assay to other cancer cell types the technology can be extended to investigate DNA methylation patterns in a variety of cancers.

6. References

Andersson, M. A. and B.E.Hellman. (2005). Different roles of Fpg and Endo lll on catechol-induced DNA damage in extended-term cultures of human lymphocytes and L5178Y mouse lymphoma cells. *Toxicology in Vitro.* Vol.19, pp. 151-158

Azqueta, A., K. B. Gutzkow, G. Brunborg and A. R. Collins. (2011). Towards a more reliable comet assay: optimising agarose concentration, unwinding time and electrophoresis conditions. *Mutation Research.* Vol.724, No.1-2, (2011 May), pp. 41-45

Bird, A. (2002). DNA methylation patterns and epigenetic memory. *Genes and Development.* Vol.16, No.1, (January 2002), pp. 6-21, ISSN 0890-9369

Brenner, C., R. Deplus, C. Didelot, A. Loriot, E. Vire, C. De Smet, A. Gutierrez, D. Danovi, D. Bernard, T. Boon, P. G. Pelicci, B. Amati, T. Kouzarides, Y. de Launoit, L. Di Croce and F. Fuks. (2005). Myc represses transcription through recruitment of DNA methyltransferase corepressor. *Embo Journal.* Vol.24, No.2, (December 2004), pp. 336-346

Chakrabarti, R. and C. E. Schutt. (2001). The enhancement of PCR amplification by low molecular-weight sulfones. *Gene.* Vol.274, No.1-2, pp. 293-298.,

Collins, A., M. Dusinska, M. Franklin, M. Somorovska, H. Petrovska, S. Duthie, L. Fillion, M. Panayiotidis, K. Raslova and N. Vaughan. (1997). Comet assay in human biomonitoring studies: reliability, validation, and applications. *Environmental and Molecular Mutagenesis.* Vol.30, No.2, pp. 139-146.

Collins, A. R. and I. Gaivao. (2007). DNA base excision repair as a biomarker in molecular epidemiology studies. *Molecular Aspects in Medicine.* Vol.28, No.3-4, (June 2002), pp. 307-322

Cottrell, S. E. (2004). Molecular diagnostic applications of DNA methylation technology. *Clinical Biochemistry.* Vol.37, No.7, pp. 595-604.,

Das, P. M. and R. Singal. (2004). DNA methylation and cancer. *Journal of Clinical Oncology.* Vol.22, No.22, pp. 4632-4642

Duffy, M. J., R. Napieralski, J. W. Martens, P. N. Span, F. Spyratos, F. C. Sweep, N. Brunner, J. A. Foekens and M. Schmitt. (2009). Methylated genes as new cancer biomarkers. *European Journal of Cancer.* Vol.45, No.3, (January 2009), pp. 335-346

Ehrlich, M. (2002). DNA methylation in cancer: too much, but also too little. *Oncogene.* Vol.21, No.35, pp. 5400-5413

Epe, B., M. Pflaum, M. Haring, J. Hegler and H. Rudiger. (1993). Use of repair endonucleases to characterize DNA damage induced by reactive oxygen species in cellular and cell-free systems. *Toxicology Letters.* Vol.67, No.1-3, pp. 57-72

Fairbairn, D. W., P. L. Olive and K. L. O'Neill. (1995). The comet assay: a comprehensive review. *Mutation Research.* Vol.339, No.1, pp. 37-59

Fazzari, M. J. and J. M. Greally. (2004). Epigenomics: beyond CpG islands. *Nature Reviews Genetics.*Vol.5, No.6, pp. 446-455

Fujiwara, H. and M. Ito. (2002). Nonisotopic cytosine extension assay: a highly sensitive method to evaluate CpG island methylation in the whole genome. *Anallytical Biochemistry.* Vol.307, No.2, pp. 386-389

Gardiner-Garden, M. and M. Frommer. (1987). CpG islands in vertebrate genomes. *Journal of Molecular Biology.* Vol.196, No.2, pp. 261-282

Gopalakrishnan, S., B. O. Van Emburgh and K. D. Robertson. (2008). DNA methylation in development and human disease. *Mutation Research.* Vol.647, No.1-2, (Augustus 2008), pp. 30-38

Guil, S. and M. Esteller. (2009). DNA methylomes, histone codes and miRNAs: tying it all together. *International Journal of Biochemistry & Cell Biol.* Vol.41, No.1, (September 2008), pp. 87-95

Haines, T. R., D. I. Rodenhiser and P. J. Ainsworth. (2001). Allele-specific non-CpG methylation of the Nf1 gene during early mouse development. *Developmental Biology.* Vol.240, No.2, pp. 585-598

Hirst, M. and M. A. Marra. (2009). Epigenetics and human disease. *International Journal of Biochemistry & Cell Biol.*Vol.41, No.1, (September 2008) pp. 136-146

Kanai, Y., S. Ushijima, Y. Nakanishi, M. Sakamoto and S. Hirohashi. (2003). Mutation of the DNA methyltransferase (DNMT) 1 gene in human colorectal cancers. *Cancer Letters.* Vol.192, No.1, pp. 75-82

Klose, R. J. and A. P. Bird. (2006). Genomic DNA methylation: the mark and its mediators. *Trends Biochemical Scinces.* Vol.31, No.2, (January 2006), pp. 89-97

Lambert, M. P. and Z. Herceg. (2008). Epigenetics and cancer, 2nd IARC meeting, Lyon, France, 6 and 7 December 2007. *Mol Oncology.* Vol.2, No.1, (March 2008), pp. 33-40

Lovell, D. P. and T. Omori. (2008). Statistical issues in the use of the comet assay. *Mutagenesis.* Vol.23, No.3, (April 2008), pp. 171-182

Lujambio, A. and M. Esteller. (2007). CpG island hypermethylation of tumor suppressor microRNAs in human cancer. *Cell Cycle.* Vol.6, No.12, (May 2007), pp. 1455-1459

Mohn, F. and D. Schubeler. (2009). Genetics and epigenetics: stability and plasticity during cellular differentiation. *Trends in Genetics.* Vol.25, No.3, (January 2009), pp. 129-136

Moss, T. J. and L. L. Wallrath. (2007). Connections between epigenetic gene silencing and human disease. *Mutation Research.* Vol.618, No.1-2, pp. 163-174

Nag, R. and M. J. Smerdon. (2009). Altering the chromatin landscape for nucleotide excision repair. *Mutation Research*. Vol.682, No.1, (January 2009), pp. 13-20

Ohgane, J., S. Yagi and K. Shiota. (2008). Epigenetics: the DNA methylation profile of tissue-dependent and differentially methylated regions in cells. *Placenta*. Vol.29, No.Suppl A, (November 2007), pp. 29-35

Okano, M., D. W. Bell, D. A. Haber and E. Li. (1999). DNA methyltransferases Dnmt3a and Dnmt3b are essential for de novo methylation and mammalian development. *Cell*. Vol.99, No.3, pp. 247-257

Pogribny, I. P., S. J. James, S. Jernigan and M. Pogribna. (2004). Genomic hypomethylation is specific for preneoplastic liver in folate/methyl deficient rats and does not occur in non-target tissues. *Mutation Research*. Vol.548, No.1-2, pp. 53-59

Prokhortchouk, E. and P. A. Defossez. (2008). The cell biology of DNA methylation in mammals. *Biochimica Biophysica Acta*. Vol.1783, No.11, (July 2008), pp. 2167-2173

QIAGEN. 2005. Introducing QIAGEN Proteinase K- the robust proteinase. 04.11.2009 Available from
https://www.roche-applied-science.com/pack-insert/3115887a.pdf

Rojas, E., M. C. Lopez and M. Valverde. (1999). Single cell gel electrophoresis assay: methodology and applications. *Chromatography B: Biomedical Sciences*. Vol.722, No.1-2, pp. 225-254

Sawan, C., T. Vaissiere, R. Murr and Z. Herceg. (2008). Epigenetic drivers and genetic passengers on the road to cancer. *Mutation Research*. Vol.642, No.1-2, (March 2008), pp. 1-13

Speit, G., M. Vasquez and A. Hartmann. (2009). The comet assay as an indicator test for germ cell genotoxicity. *Mutation Research*. Vol.681, No.1, (March 2008), pp. 3-12

Sweatt, J. D. (2009). Experience-dependent epigenetic modifications in the central nervous system. *Biological Psychiatry*. Vol.65, No.3, (November 2008), pp. 191-197

Szyf, M. (2006). Targeting DNA methylation in cancer. *Bull Cancer*. Vol.93, No.9, pp. 961-972

Tice, R. R., E. Agurell, D. Anderson, B. Burlinson, A. Hartmann, H. Kobayashi, Y. Miyamae, E. Rojas, J. C. Ryu and Y. F. Sasaki. (2000). Single cell gel/comet assay: guidelines for in vitro and in vivo genetic toxicology testing. *Environmental and Molecular Mutagenesis*. Vol.35, No.3, pp. 206-221

Tost, J. (2009). DNA methylation: an introduction to the biology and the disease-associated changes of a promising biomarker. *Methods in Molecular Biology*. Vol.507, pp3-20

Tsou, J. A., J. A. Hagen, C. L. Carpenter and I. A. Laird-Offringa. (2002). DNA methylation analysis: a powerful new tool for lung cancer diagnosis. *Oncogene*. Vol.21, No.35, pp. 5450-5461

van Dyk, E. and P. J. Pretorius. (2005). DNA damage and repair in mammalian cells exposed to p-hydroxyphenylpyruvic acid. *Biochemical and Biophysical Research Communications*. Vol.338, No.2, (October 2005), pp. 815-819

Waggoner, D. (2007). Mechanisms of Disease: Epigenesis. *Seminars in Pediatric Neurology*. Vol.14, No.1, pp. 7-14

Wentzel, J. F., C. Gouws, C. Huysamen, E. Dyk, G. Koekemoer and P. J. Pretorius. (2010). Assessing the DNA methylation status of single cells with the comet assay. *Analytical Biochemistry* Vol.400, No.2, (February 2010) , pp. 190-194

Wilson, A. S., B. E. Power and P. L. Molloy. (2007). DNA hypomethylation and human diseases. *Biochimica Biophysica Acta*. Vol.1775, No.1, (September 2006), pp. 138-162

12

DNA Methylation and Histone Deacetylation: Interplay and Combined Therapy in Cancer

Yi Qiu, Daniel Shabashvili, Xuehui Li,
Priya K. Gopalan, Min Chen and Maria Zajac-Kaye
*Department of Anatomy and Cell Biology
and Department of Medicine,
College of Medicine, University of Florida,
Gainesville, Florida,
USA*

1. Introduction

In mammalian cells, DNA can be modified by methylation of cytosine residues in CpG dinucleotides, and the N-terminal tails of histone proteins are subject to a wide range of different modifications, including acetylation, methylation, phosphorylation and ubiquitylation. All of these chemical changes have a substantial influence on chromatin structure and gene expression. These epigenetic modification patterns can be regarded as heritable marks over many cell generations. Importantly, patterns and levels of DNA methylation and histone acetylation/deacetylation are profoundly altered in human cancers. Inhibitors of DNA methytransferases (DNMTs) and histone deacetylases (HDACs) have been shown to inhibit tumor growth by reactivating epigenetically silenced tumor suppressor genes. Although DNA methylation and histone deacetylation are carried out by different chemical reactions and require different sets of enzymes, it seems that there is a biological relationship between the two systems in modulating gene repression programming. Accumulating evidence also suggests that this epigenetic cross-talk may be involved in gene transcription and aberrant gene silencing in tumors. Thus, combined therapy with both DNMT and HDAC inhibitors can be a promising approach for cancer treatment.

2. Epigenetic gene silencing through DNA methylation and histone deacetylation

2.1 DNA methylation in regulating gene transcription

DNA methylation is a covalent chemical modification of DNA occurring at cytosine residues in CpG dinucleotides. Approximately 70–80% of cytosine in CpG dyads is methylated on both strands in human somatic cells(Chen & Riggs, 2011). DNA methylation is a stable epigenetic mark that is linked to the maintenance of chromatin in a silent state, therefore regulating chromatin structure and gene expression involved in processes such as X chromosome inactivation, genomic imprinting, embryogenesis, gametogenesis, and silencing of repetitive DNA elements. Deregulation of DNA methylation directly affects mammalian development and development of cancer (Gopalakrishnan et al., 2008).

The mammalian DNA methyltransferases (DNMTs) are enzymes that catalyze the transfer of a methyl group from S-adenosyl-L-methionine to cytosine. Among the three enzymatically active DNMTs, DNMT1 is thought to function as the major maintenance methyltransferase(Chen & Riggs, 2011). This enzyme maintains DNA methylation at hemi-methylated DNA after DNA replication (Pradhan et al., 1999), and it is responsible for copying pre-existing methylation patterns to the newly synthesized strand (Chen & Li, 2004). DNMT3A and DNMT3B are *de novo* methyltransferases active on unmethylated DNA. Both of them are responsible for establishing new DNA methylation patterns during early development (Okano et al., 1999) as well as maintaining these patterns during mitosis (Chen et al., 2003). DNMT3L is homologous to DNMT3A and DNMT3B within the N-terminal regulatory region and is highly expressed in germ cells. Although catalytically inactive, DNMT3L regulates DNMT3A and DNMT3B by stimulating their catalytic activity (Chen et al., 2011).

Our knowledge about the function of DNA methylation in mammals comes mainly from DNMTs transgenic mice studies(Chen & Li, 2004). Studies of the zygotic functions of DNMTs have shown that the establishment of embryonic methylation patterns requires both *de novo* and maintenance methyltransferase activities, and that the maintenance of DNA methylation above a threshold level is essential for embryonic development (Lei et al., 1996). Complete elimination of DNMT1 function results in embryonic lethality around E9.5, with extensive loss of global DNA methylation (Li et al., 1992). DNMT3B is also essential for embryogenesis. DNMT3B-deficient embryos show growth impairment and multiple developmental defects after E9.5 and die after E12.5. DNMT3A mutant mice die around 4 weeks of age. DNMT3A/DNMT3B-double knockout embryos die around E9.5, similar to DNMT1-null mutants (Okano et al., 1999). Loss of DNA methylation does not affect ES proliferation and viability, and the effect of demethylation only becomes apparent during or after gastrulation when the pluripotent embryonic cells begin to differentiate (Li et al., 1992,Okano et al., 1999). Conditional disruption of DNMT1 in mouse embryonic fibroblasts (MEFs) results in severe demethylation and cell death, and DNMT3B deficient MEFs show moderate demethylation, chromosomal instability, and abnormal proliferation (Jackson-Grusby et al., 2001,Farthing et al., 2008,Dodge et al., 2005). These findings suggest that DNA methylation is essential for cellular differentiation and normal functioning of differentiated cells.

Development: In mammals, DNA methylation patterns undergo dramatic changes during early development. During preimplantation development, both paternal and maternal genomes undergo a wave of demethylation, which erases the methylation marks inherited from the gametes. Within 3-6 hours after fertilization, the maternal genome is rapidly methylated through *de novo* DNA methylation, while the paternal genome is actively demethylated before the first replication event occurs (Mayer et al., 2000,Oswald et al., 2000). This is followed in the maternal genome by a loss of DNA methylation gradually over several DNA replication cycles (Mayer et al., 2000,Rougier et al., 1998). Replication-dependent demethylation is caused by the exclusion of maintenance methyltransferase DNMT1 from the nucleus of embryos (Carlson et al., 1992,Cardoso & Leonhardt, 1999). As a result, maternal and parental genomes become almost equally low in DNA methylation by the eight cell stage of development (Mayer et al., 2000). At morula stage, there is an increase in *de novo* methylation. This methylation occurs through the activity of *de novo* methyltransferases DNMT3A and 3B (Okano et al., 1999). Consistent with these data, DNMT1-deficient or DNMT3A/DNMT3B-double mutant ES cells show severe differentiation defects (Tucker et al., 1996).

Genomic imprinting: Another developmental stage that exhibits substantial *de novo* DNA methylation in mammals is gametogenesis. DNA methylation in both male and female germ cells plays a critical role in the establishment of genomic imprinting. Genomic imprinting is an epigenetic process that marks alleles according to their parental origin and results in monoallelic expression of a small subset of genes (Reik & Walter, 2001). Genetic studies demonstrate that DNMT3A and DNMT3L are essential for setting up DNA methylation imprints in germ cells (Kaneda et al., 2004,Bourc'his et al., 2001). Although DNMT3L has no enzymatic activity, it has been shown to interact with DNMT3A and stimulate its activity (Suetake et al., 2004). Active transcription across imprinting control regions also appears to be required for the establishment of DNA methylation imprints in female germ cells (Chotalia et al., 2009). It is possible that the removal of H3K4 methylation and active transcription at imprinted loci create or maintain a chromatin state that facilitates access of the DNMT3A-DNMT3L complex to these loci (Ooi et al., 2007).

Differentiation: Differentiation is an epigenetic process associated with selective temporal activation of lineage-specific genes and regulated silencing of pluripotency genes. Pluripotency genes are those that are important for maintaining the unrestricted developmental potential of the early embryo and ES cells. Undifferentiated ES cells do not show global differences in DNA methylation levels compared to somatic cells (Mohn et al., 2008). However, changes in DNA methylation at promoter regions were observed (Meissner et al., 2008). Among those promoter regions that gain DNA methylation upon differentiation of ES cells in culture are pluripotent genes and germ line-specific genes (Farthing et al., 2008,Mohn et al., 2008). *De novo* DNA methylation and silencing of the pluripotency genes contributes to loss of pluripotency in differentiated cells (Farthing et al., 2008). Oct4 and Nanog are two such genes which are essential for pluripotency and early development and become methylated after differentiation (Hattori et al., 2004,Hattori et al., 2007,You et al., 2011). Thus, *de novo* methylation at promoter regions during cellular differentiation may in part control the loss of pluripotency by silencing stem cell-specific genes.

Once cells have passed through the early embryonic stage, differentiation toward a specific lineage pathway occurs. These somatic tissues then contain specific gene expression patterns unique for that terminally differentiated cell type. Although DNMT1 is dispensable for ES cell maintenance, it is required for maintaining the somatic progenitor state through cell divisions. Depletion of the maintenance DNA methyltransferase DNMT1 in epidermal progenitors leads to premature differentiation. Genome-wide DNA methylation analysis showed that some epidermal differentiation gene promoters were methylated in self-renewing progenitor cells but were subsequently demethylated during differentiation (Sen et al., 2010). The correlation between gene expression and DNA methylation was investigated in diverse tissues and suggests that DNA methylation is critical for regulating the expression of some tissue specific genes (Shen et al., 2007,Illingworth et al., 2008). In the case of neuronal development, this methylation involves DNMT3A, which is expressed in postnatal neural stem cells and is required for neurogenesis (Mohn et al., 2008,Wu et al., 2010).

2.2 Histone deacetylation in regulating gene repression

Histone deacetylases (HDACs) catalyze the removal of acetyl groups from the ε-amino groups of lysine residues. The reversible acetylation of histones and non-histone proteins by histone acetyltransferases (HATs) and deacetylases plays a critical role in transcriptional regulation and many other cellular processes in eukaryotic cells. Acetylation of histones is commonly associated with the transcriptional activation of genes, and is thought to be responsible for the formation of a local "open chromatin" structure required for the binding of multiple transcription factors (Sterner & Berger, 2000). In contrast, the removal of acetyl groups by HDACs frequently accompanies the suppression of gene activity (Ng & Bird, 2000). However, non-histone protein lysine acetylation plays a diverse role in the regulation of all aspects of cellular processes (Glozak et al., 2005).

Classification: Mammalian HDACs are classified into four classes (I, II, III and IV) based on the sequence homology of the yeast histone deacetylases Rpd3 (reduced potassium dependency), Hda1 (histone deacetylase1), and Sir2 (silent information regulator 2), respectively. Class I HDACs include HDAC1, 2, 3 and 8. Class II HDACs contain HDAC4, 5, 6, 7, 9 and 10. Class III enzymes, however, require the coenzyme NAD+ as a cofactor. HDAC11 belongs to the class IV family (Reviewed in (Yang & Seto, 2008)). Although the precise cellular functions of the different HDAC enzymes are still poorly understood, evidence suggests that different members of the HDAC family have distinct functions (Cho et al., 2005,Foglietti et al., 2006).

Post-translational modifications: All class I HDACs can be phosphorylated. Phosphorylation of HDAC1, 2 and 3 increases deacetylase activities while phosphorylation of HDAC8 inhibits deacetylase activity (Sengupta & Seto, 2004). HDAC1 can also be acetylated at the catalytic core domain and the C-terminal region, resulting in dramatic reduction of enzymatic activity (Qiu et al., 2006). Other modifications of HDAC1 include sumoylation and ubiquitylation, which also influence deacetylase activity and protein stability (David et al., 2002,Oh et al., 2009). Class II HDACs can also be phosphorylated. Phosphorylation does not appear to regulate enzymatic activity of class II HDACs; instead, it modulates their subcellular localization (Yang & Gregoire, 2005). Class III deacetylases can also be phosphorylated. One report showed that phosphorylation of SIRT1 enhances deacetylation of p53 (Guo et al., 2010).

Function and regulation of Class I histone deacetylases: Class I HDACs are ubiquitously expressed nuclear proteins. Although they share a high level of sequence homology and common substrates, each of the enzymes has a unique role in cell function and cannot compensate for the other enzymes' functions, as deletion of each class I deacetylase leads to lethality (Reviewed in (Haberland et al., 2009)). HDACs can be recruited to genes by sequence-specific or non sequence-specific DNA-binding factors. Class I deacetylases are found to associate with a variety of proteins, such as transcription factors, coactivators, corepressors, chromatin remodeling proteins, etc, adding to the complexity of HDACs functions (Yang & Seto, 2007). Class I HDACs have been widely implicated in gene repression through the hypoacetylation of localized chromatin domains (Strahl & Allis, 2000). It is a traditionally held view that HDACs are associated with repressed promoters and are replaced by coactivators during gene activation (Berger, 2007). However, recent advancements in methodology allow studying the localization of HDACs or HATs on a genome-wide scale. The surprising result is that HATs and HDACs are positively correlated with gene expression. Levels of HDAC1,2 and 3 are high in active genes and absent from silenced genes (Wang et al., 2009). The emerging new model for the role of HDACs is that they can counteract histone acetyltransferase to maintain transcription activation within a certain level, and they can regulate gene activation as deacetylation resets chromatin for subsequent rounds of transcription (Perissi et al., 2010). This model is also supported by data that shows HDAC complexes in yeast, Rpd3s, can interact with active chromatin in order to prevent transcription initiation from cryptic sites (Li et al., 2007). In some cases, HDAC can even function as coactivator (Qiu et al., 2006). Besides regulation of gene transcription, class I HDACs have been shown to also be involved in many other cellular processes, especially events that are linked to oncogenesis, such as DNA repair, recombination, and replication, cell cycle check point control, and other signaling regulators (Spange et al., 2009).

Class I HDACs, except HDAC8, are often found in multiprotein complexes and their activity is regulated through associated complexes (reviewed in (Sengupta & Seto, 2004). HDAC1 and 2 are found to coexist in at least three evolutionarily conserved, distinct protein complexes: the Sin3, the CoREST and the NuRD/Mi2 complexes (Grozinger & Schreiber, 2002,Yang & Seto, 2008). All complexes are recruited to target genes through interactions with DNA binding transcription factors (Yang & Seto, 2007). Sin3 contains the conserved basic structure of multiple paired amphipathic helix (PAH) domains for protein-protein interaction. Mammals have two isoforms, Sin3A and Sin3B, which provide more diverse protein complexes for gene regulation. Sin3 does not bind to DNA and has no known enzymatic activity of its own. It is suggested that it functions through its ability to interact with other proteins. The CoREST complex is a multi-subunit complex containing the lysine demethylase LSD1, corepressors CtBP, BHC80, CoREST, HDAC1 and HDAC2 (Lee et al., 2005,Shi et al., 2003). Thus, this complex is capable of deacetylating as well as demethylating nucleosomes. Mi2 belongs to the CHD (chromo-helicase DNA-binding) protein family (Woodage et al., 1997). The NuRD complex includes the ATPase/helicase Mi-2, HDAC1 / HDAC2, MTA2 (Metastasis-associated) proteins, MBD3 (methyl CpG-binding domain-1), RbAp46/48 (Wade et al., 1998,Zhang et al., 1998). Interestingly, recent reports suggest that LSD1, a lysine specific demethylase, is also recruited to the NuRD complex through its interaction with MTA2 (Wang et al., 2009). Thus, the NuRD complex may regulate gene transcription through a combination of deacetylation, demethylation and remodeling. The NuRD complex is also part of a larger protein complex, MeCP1 (methyl CpG-binding protein 1) complex (Fatemi & Wade, 2006). The MeCP1 complex contains the additional component MBD2, a CpG methyl-binding protein. Therefore, NuRD/MeCP1 is targeted to methylated CpG sites leading to histone deacetylation and repression at the gene promoter (Nan et al., 1998).

HDAC3 forms a stable complex with nuclear receptor corepressor (N-CoR) and silencing mediator of retinoic and thyroid receptors (SMRT) (Wen et al., 2000,Guenther et al., 2001,Guenther et al., 2000). Both N-CoR and SMRT had been discovered as interacting partners of unliganded nuclear receptors, such as TR and RAR and mediators of their repressive functions. The interaction between HDAC3 and SMRT/N-CoR seems to be essential for HDAC3 deacetylase activity (Guenther et al., 2001).

HDAC8 was identified as a ubiquitously distributed nuclear enzyme (Hu et al., 2000). However, it is also found in the cytosol of smooth muscle cells where it associates with αactin cytoskeleton (Waltregny et al., 2004). A gene deletion study also shows that HDAC8 plays an important role in skull formation (Haberland et al., 2009). HDAC8 is phosphorylated by cyclic AMP-dependent protein kinase A (PKA) *in vitro* and *in vivo*. Induction of phosphorylation decreases HDAC8's enzymatic activity. Remarkably, inhibition of HDAC8 activity by hyperphosphorylation leads to hyperacetylation of histones H3 and H4, suggesting that PKA-mediated phosphorylation of HDAC8 plays a central role in the overall acetylation status of histones (Lee et al., 2004).

Functions of Class II deacetylases: class IIa mammalian HDACs mainly function as transcriptional corepressor through their deacetylase domain and other repression domains (Yang & Gregoire, 2005). Class IIa HDACs share conserved motifs, such as MEF-2 binding, 14-3-3 binding and nuclear localization signal at N-terminal region which are important for the function and regulation of class IIa members (Yang & Gregoire, 2005). These HDACs possess intrinsic nuclear import and export signals for nucleo-cytoplasmic trafficking (McKinsey et al., 2000). MEF2 and 14-3-3 are major HDAC4 binding partners. While MEF2 promotes nuclear localization of class IIa HDACs (Miska et al., 1999,Wang & Yang, 2001), 14-3-3 proteins stimulate cytoplasmic retention (Grozinger & Schreiber, 2000). Class IIa HDACs have low deacetylase activity on their own, it is suggested that HDAC3 is needed for class IIa HDACs to exert full deacetylase activity (Fischle et al., 2002).

A class IIb member, HDAC6 possesses two deacetylase domains and a zinc finger motif (Zn-UBP, ubiquitin carboxyl-terminal hydrolase-like zinc finger) (Seigneurin-Berny et al., 2001). HDAC6 deacetylates α-tubulin, Cortactin and HSP90 to regulate cell motility, cilium assembly, cell adhesion, the immune synapse, macropinocytosis, maturation of the glucocorticoid receptor (GR) and activation of some protein kinases (Yang & Seto, 2008) . Depending on its availability in the nucleus, HDAC6 is also able to deacetylate histones (Yoshida et al., 2004). In addition to its deacetylase domains, HDAC6 possesses a Zn-UBP finger that binds to ubiquitin and is involved in ubiquitin-dependent aggresome formation and cellular clearance of misfolded proteins (Rodriguez-Gonzalez et al., 2008,Yang & Seto, 2008). Both deacetylase and ubiquitin-binding activities of HDAC6 are required for these processes. Therefore, HDAC6 regulates various processes in the cytoplasm. Several lines of evidence suggest that HDAC6 also plays a role in the nucleus. It interacts with several nuclear proteins, including HDAC11 (Gao et al., 2002), sumoylated p300 (Girdwood et al., 2003), transcriptional corepressors such as ETO2 and L-CoR, and sequence specific transcription factors such as NF-kB and Runx2 (Yang & Gregoire, 2005). HDAC10 is also a class IIb histone deacetylase. The N-terminal half of HDAC10 is more similar to the first catalytic domain of HDAC6 than to the second (Guardiola & Yao, 2002). The C-terminal half of HDAC10 is leucine rich and shows limited sequence similarity to the second deacetylase domain of HDAC6 (Guardiola & Yao, 2002,Yang & Seto, 2008), but its function remains elusive.

Class IV deacetylase, HDAC11, is a highly conserved deacetylase. It shows sequence similarity to class I and II HDACs. Little is known about its function or regulation (Gao et al., 2002).

Class III deacetylases and function: Mammals have seven Sir2 homologs (sirtuins, SIRT1–7). These proteins have a highly conserved NAD-dependent sirtuin core domain, first identified in the founding yeast SIR2 protein. Mammalian sirtuins have diverse cellular locations, multiple substrates, and affect a broad range of cellular functions (Haigis & Guarente, 2006). Three mammalian sirtuins (SIRT1, SIRT6, and SIRT7) are localized to the nucleus. SIRT1 is most extensively studied, has more than a dozen known substrates (Haigis & Guarente, 2006). SIRT1 regulates histone acetylation levels (mainly K16-H4 and K9-H3 positions) (Vaquero et al., 2004,Pruitt et al., 2006), and the acetylation of transcription factors such as p53 (Vaziri et al., 2001), p300 histone acethyltransferase (Bouras et al., 2005), E2F1 (Wang et al., 2006), the DNA repair ku70 (Cohen et al., 2004), NF-KB (Yeung et al., 2004), and the androgen receptor (Fu et al., 2006). It is also responsible for tissue metabolism, cellular oxidative stress and DNA repair (Finkel et al., 2009, Uhl et al., 2010). SIRT6 and SIRT7 may also be important regulators of DNA damage and metabolism (Finkel et al., 2009). SIRT2 is a predominantly cytoplasmic protein. It colocalizes with tubulin, and can deacetylate a number of substrates *in vitro*, including α tubulin and histones (North et al., 2003). SIRT2 may be important in regulating mammalian cell cycle (Dryden et al., 2003). SIRT3-5 are localized at the mitochondria and are important for micochondria energy usage (Pereira et al., 2011).

3. Connections between DNA methylation and histone deacetylation for gene silencing

Alteration of DNA methylation affects histone acetylation, or vice versa: Both DNA methylation and hypoacetylation of histones H3 and H4 are frequently associated with silent genes(Dobosy & Selker, 2001). For example, in DNMT1 knockout cancer cells there is an increase in the amount of acetylated forms of histone H3 and a decrease in that of the methylated forms of histone H3. These changes are associated with a loss of interaction of HDACs and the heterochromatin protein HP1 with histone H3. These data strongly indicate that histone hyperacetylation is not always a result of a loss of HDAC activity, but that it could be due to a loss of HDACs targeted to specific DNA sequences. One possible explanation is that changes in DNA methylation also cause histone modification due to direct interactions between the enzymes regulating different epigenetic modifications (Espada et al., 2004). The evidence supporting the communication between DNA methylation and histone deacetylation has been demonstrated from inhibitor studies. HDAC inhibitors do not only change acetylation of histones, but also induce DNA demethylation. Using DNA demethylating agents, like 5-aza-2'-deoxycytidine, in combination with HDAC inhibitors changes the status of DNA methylation, reactivates gene transcription, and inhibits cancer cell growth (reviewed in (Dobosy & Selker, 2001,Hellebrekers et al., 2007)). Recent studies show that histone deacetylase inhibitor Trichostatin A (TSA) treatment reduces global DNA methylation and DNMT1 protein level, alters DNMT1 nuclear dynamics and interactions with chromatin. The mechanisms underlying these effects are apparently distinct from the mechanisms of action of the DNMT inhibitor 5-Azacytidine (Arzenani et al., 2011). Therefore, communication between histone deacetylation and DNA methylation is likely to be a dynamic process in the regulation of gene silencing.

Connections through methyl-CpG-binding proteins: Early studies show that DNA methylation can lead to transcriptional repression through interaction with methyl-CpG-binding proteins (Nan et al., 1997). Methyl-CpG-binding proteins are a group of proteins that bind to methylated CpG sites through their mentyl-CpG-binding domain (MBD). The founding member of the MBD family was MeCP2 (Lewis et al., 1992). It is a multidomain protein and is associated with and localized to densely methylated chromatin regions. This protein also contains a transcriptional repression domain (TRD). Subsequently, MBD1 to MBD4 were all discovered as EST clones with sequence similarity to the MBD motif of MeCP2 (Hendrich & Bird, 1998). All these proteins, except MBD3, contain both MBD domain and TRD domain, and recognize methylated DNA. The identification of methyl-CpG binding proteins (MBDs) leads to insights into the communication between DNA methylation and HDACs as all MBDs are components of co-repressor complexes that contain histone deacetylases (Fatemi & Wade, 2006).

Transcription repression mediated through MeCP2 requires both MBD and TRD domains. It was then demonstrated that MeCP2 can interact with transcription repressor SIN3. Furthermore, the region of interaction with Sin3 on MeCP2 significantly overlapped the TRD domain and tethering of the TRD resulted in repression that is sensitive to inhibitors of histone deacetylase. Thus, recruitment of the Sin3 repressive complex containing HDAC1 and HDAC2 to the methylation sites in the gene promoters results in a deacetylated repressive chromatin structure (Jones et al., 1998,Nan et al., 1997) . In neural progenitor/ES cells, genes are regulated by the REST corepressor complex which contain CoREST and histone deacetylases. MeCP2/Sin3 corepressor complexes can interact with REST repressor complex in the promoter region, resulting in gene inactivation (Ballas et al., 2005). It is noteworthy that MeCP2 can mediate transcriptional repression in both HDAC-dependent and –independent manners (Kaludov & Wolffe, 2000). In addition, a recent study suggests that MeCP2 can also function as an activator of gene transcription by recruiting transcriptional activator CREB1 to an activated gene promoter (Chahrour et al., 2008).

MeCP1 complex was purified using methylated DNA oligoes (Meehan et al., 1989). Later studies demonstrated that MeCP1 complex contains MBD2, HDAC1, HDAC2, and RbAp48 (Ng et al., 1999). Exogenous MBD2 represses transcription and this repression can be relieved by the deacetylase inhibitor trichostatin A (Ng et al., 1999). Interestingly, MBD2 is also capable of recruiting Mi-2/NuRD complex to methylated DNA through its heterodimerization with MBD3 (Feng & Zhang, 2001,Hendrich et al., 2001). This suggests an interplay of DNA methylation, nucleosome remodeling and histone deacetylation in gene silencing. In addition, activity of MBD2 is modulated through association with cofactors. A recent study indicates that methyl-CpG-binding protein 3-like-2 (MBD3L2) may function as a transcriptional modulator in MBD2-MeCP1-NuRD-mediated methylation silencing (Jin et al., 2005). MBD3L2 can convert inactivated genes to activated genes by displacing MBD2 present in the NuRD complex, which sequesters the MeCP1-NuRD complex away from methylated DNA. In addition, MBD*in* (Lembo et al., 2003) and transforming-acid-coiled-coil 3 (TACC3) can interact with MBD2 and reactivate MBD2-repressed genes through different regulation mechanisms. The recruitment of MBD*in* to MBD2 disrupts the association of MBD2 to its repressor complex (Lembo et al., 2003). Unlike MBD*in*, TACC3 can reactivate methylated genes by forming a complex with MBD2 and histone acetyltransferase PCAF (Angrisano et al., 2006).

MBD1 represses transcription of methylated reporter constructs. The repression requires both the TRD and MBD motifs and is sensitive to HDAC inhibitors (Ng et al., 2000). However, it remains undetermined which HDACs associate with the MBD1 complex.

Connections through DNMT-HDAC interaction: DNMTs, including DNMT1 and DNMT3s, can directly interact with HDAC1 and HDAC2 to repress gene transcription *in vitro* and *in vivo* (Geiman & Robertson, 2002). A study shows that repression of gene transcription by direct interaction between DNMT3A and HDAC1/2 can be abolished by sumoylation of DNMT3A, but not the other interaction partner, DNMT3B (Ling et al., 2004), which suggests a specific regulation of DNMT3A-mediated gene silencing. Notably, DNMT3L, which lacks the catalytic domain, has been shown to bind to HDAC1 and regulate DNA methylation independent of methylating activity (Deplus et al., 2002).

Another connection between DNA methyl transferase and deacetylase is that DNMT1 is stabilized by HDAC1 and the deubiquitinase HAUSP (herpes virus-associated ubiquitin-specific protease). In human colon cancers, the abundance of DNMT1 is correlated with that of HAUSP. HAUSP knockdown rendered colon cancer cells more sensitive to killing by HDAC inhibitors both in tissue culture and in tumor xenograft models (Du et al., 2010). HDACs may also be involved in regulating DNMT1 gene expression as HDAC inhibitor TSA decreased DNMT1 mRNA stability and reduced its transcript's half-life (Januchowski et al., 2007).

DNA methylation and Histone deacetylation, who comes first? Although it is well accepted that cross-talk between DNA methylation and histone deacetylation plays a pivotal role in gene silencing, it is still unclear which event is the dominant epigenetic determinant in this communication process. Some evidence suggests that DNA methylation may be the primary event to trigger histone deacetylation and lead to gene silencing (Irvine et al., 2002,Jones et al., 1998,Kaludov & Wolffe, 2000,Kudo, 1998,Nan et al., 1998,Rountree et al., 2000). Controversially, loss of histone acetylation may guide the local DNA hypermetylation and initiate transcriptional repression. Felsenfeld's group has shown that chicken β-globin insulator recruits HAT to establish a high level of histone acetylation which prevents promoter CpG methylation by blocking binding of the transcriptional repressor complex to the promoters (Mutskov et al., 2002). A global methylation study shows that histone hypoacetylation caused by TSA induces a significant decrease in global methylation (Ou et al., 2007). In addition, inhibition of HDAC1 by sodium butyrate induces promoter demethylation and reactivation of RARbeta2 in colon cancer cells (Spurling et al., 2008). A recent study in *Neurospora crassa* found that specific deacetylation of histone H2B and H3 is required for DNA methylation and heterochromatin formation (Smith et al., 2010). These results indicate that histone deacetylation status is crucial to sustaining DNA methylation of the promoters and gene silencing.

4. Aberrant DNA methylation and histone deacetylation in cancer

4.1 Aberrant DNA methylation and DNMTs recruitment in cancer

Altered methylation patterns are found in the majority of tumor types. Methylation changes generally occur early in cancer development, which supports the hypothesis that epigenetic changes precede cancer development (Linhart et al., 2007,Belinsky et al., 1998,Derks et al., 2006,Palmisano et al., 2000). A hallmark of many cancers is global hypomethylation and regional hypermethylation of CpG islands. This local hypermethylation in cancer is usually found on CpG islands associated with promoters of tumor suppressors or other genes involved in cell cycle regulation, leading to inactivation of these genes. Promoter hypermethylation also facilitates mutations in these regions as methylated cytosine residues are spontaneously deaminated to thymine residues, causing mutational silencing of the genes (Sulewska et al., 2007,Fryxell & Zuckerkandl, 2000,Rideout et al., 1990). This hypothesis explains the high incidence of CpG to TpG transition mutations observed in the promoters of tumor suppressors, for example, the p53 tumor suppressor gene (Rideout et al., 1990).

The effects of global hypomethylation are more varied and not well understood. In mouse models, hypomethylation has been shown to induce genomic instability and tumorigenesis (Gaudet et al., 2003,Eden et al., 2003). It has been suggested that global hypomethylation can induce reexpression of normally silenced genes, some of which may be oncogenic. Genes reactivated by global hypomethylation can include silenced oncogenes, imprinted genes, genes on the inactivated X-chromosome (Sharp et al., 2011), endogenous retroviruses and transposons (Yoder et al., 1997), as well as silenced drug resistance genes (Chekhun et al., 2007). For example, a cell-cell adhesion glycoprotein P-cadherin is often overexpressed in breast cancer, but not in normal breast tissue. The aberrant expression of P-cadherin in breast cancer is regulated by gene promoter hypomethylation. (Paredes et al., 2005). A similar mechanism regulates overexpression of cyclin D2 in gastric and ovarian cancer (Sakuma et al., 2007,Oshimo et al., 2003), and MAGE in melanomas (De Smet et al., 2004). Global hypomethylation also induces chromosomal instability by a mechanism that is not well understood; one possible cause is a large number of derepressed transposons and retroviruses created by this hypomethylated state (Florl et al., 1999,Howard et al., 2008).

Tumor heterogeneity is a major barrier to effective cancer diagnosis and treatment. Recent studies suggest that methylation patterns can be different in different cancer types and tumor stages (Wermann et al., 2010). Epigenetic analysis of large gene panels and genome-wide screening of DNA methylation levels discovered that overall methylation patterns can be used as biomarkers for cancer risk and/or tumor type (Kondo & Issa, 2010,Figueroa et al., 2010,Hawes et al., 2010,Worthley et al., 2010). Cancer-specific differentially methylated regions (cDMRs) were identified; methylation variation within these cDMRs distinguishes various cancers from normal tissue, with intermediate variation in adenomas (Hansen et al., 2011). Whole-genome bisulfite sequencing shows these variable cDMRs are related to loss of sharply delimited methylation boundaries at CpG islands. It suggests that the loss of epigenetic stability of well-defined genomic domains underlies increased methylation variability in cancer and may contribute to tumor heterogeneity. The distinct methylation patterns can be used not only to differentiate carcinoma from other tumor types, but also to predict tumor progression stage, with potential clinical applications in diagnosis and prognosis (Hernandez-Vargas et al., 2010).

Exact nature of the defect in the cellular methylation machinery in tumor cells remains unknown. It is proposed that inappropriate DNMT expression pattern or timing during the cell cycle could disrupt the regulation of DNA methylation patterns as DNMT1, 3A, and 3B are expressed differentially during the cell cycle (Robertson, 2001). Global hypomethylation in cancer cells may also be due to upregulation of DNA demethylase system (Rai et al., 2010). Increased expression of DNMTs can result in hypermethylation of CG islands in cancer cells and may play important roles in malignant progression of cancer, leading to aberrant methylation in many important tumor suppressor genes. In fact, it had been shown that DNMT overexpression is an early and significant event in urothelial, hepatic, gastric, pancreatic, lung, breast, and uterine cervix carcinogenesis (Daniel et al., 2011).

Although the DNMT1 and DNMT3 family of proteins have been considered either maintenance or *de novo* methyltransferases, respectively, it is likely that all three DNMTs possess both functions *in vivo*, particularly during carcinogenesis (Robertson, 2001). DNMT1 has been shown to be essential for the survival and proliferation of human cancer cells (Chen et al., 2007). Increased DNMT1 protein expression correlates significantly with frequent DNA hypermethylation of multiple CpG islands, poorer tumor differentiation and malignant progression (Etoh et al., 2004,Nakagawa et al., 2005,Saito et al., 2003). In bladder cancer, progressive increase of expression of DNMT1 protein occurs during the precancerous stages (Nakagawa et al., 2003). Depletion of DNMT1 resulted in lower cellular maintenance methyltransferase activity, global and gene-specific demethylation and re-expression of tumor-suppressor genes in human cancer cells. Specific depletion of DNMT1 but not DNMT3A or DNMT3B markedly potentiated the ability of 5-aza-2'-deoxycytidine to reactivate silenced tumor-suppressor genes, indicating that inhibition of DNMT1 function is the principal means by which 5-aza-2'-deoxycytidine reactivates genes. These results indicate that DNMT1 is necessary and sufficient to maintain global methylation and aberrant CpG island methylation in human cancer cells (Robert et al., 2003).

DNMT3B depletion reactivated methylation-silenced gene expression but did not induce global or juxtacentromeric satellite demethylation as did specific depletion of DNMT1, indicating that DNMT3B has significant site selectivity that is distinct from DNMT1 (Beaulieu et al., 2002). It is shown that DNMT3B1 but not DNMT3A1 efficiently methylates the same set of genes in tumors and in nontumor tissues, demonstrating that *de novo* methyltransferases can initiate methylation and silencing of specific genes in phenotypically normal cells. This suggests that DNA methylation patterns in cancer are the result of specific targeting of at least some tumor suppressor genes, such as Sfrp2, Sfrp4, and Sfrp5, rather than of random, stochastic methylation followed by clonal selection due to a proliferative advantage caused by tumor suppressor gene silencing (Linhart et al., 2007).

4.2 HDACs and cancer

One common theme in cancer cells is elevated HDAC expression and global hypoacetylation. Loss of acetylated Lys16 (K16-H4) and trimethylated Lys20 (K20-H4) of histone H4 may be a common event in human cancer (Fraga et al., 2005), while other studies also show that the decrease in histone acetylation is not only involved in tumorogenesis, but also in tumor invasion and metastasis (Yasui et al., 2003).

It has become increasingly clear that class I HDAC enzymes are clinically relevant to cancer therapy (Haberland et al., 2009,Ropero & Esteller, 2007). Increased HDAC1 expression levels have been reported to in a variety of cancers, such as gastric (Choi et al., 2001), prostate (Halkidou et al., 2004), colon (Wilson et al., 2006) and breast (Zhang et al., 2005) carcinomas. Overexpression of HDAC2 has been found in cervical (Huang et al., 2005), gastric (Song et al., 2005), and colorectal carcinoma (Ashktorab et al., 2009). Other studies have reported high levels of HDAC3 (Wilson et al., 2006) expression in colon cancer specimens. HDAC8 has been found to be associated with various types of leukemia (Balasubramanian et al., 2008). HDAC1, 6 and 8 could also be important for breast cancer invasion (Park et al., 2011). These observations suggest that transcriptional repression of tumor-suppressor genes by overexpression or aberrant recruitment of HDACs to their promoter regions could be a common phenomenon in tumor onset and progression. It is now known that HDACs have been associated with the deregulation of a number of well-characterized cellular oncogenes and tumor-suppressor genes. For example, Class I HDACs promote cell proliferation by inhibiting p21 and p27 promoter activity (Lagger et al., 2002,Wilson et al., 2006). In some tumors, p21(WAF1/cip1) is epigenetically inactivated by hypoacetylation of the promoter, and treatment with HDAC inhibitors leads to inhibition of tumor cell growth and an increase in both acetylation of the promoter and gene expression (Gui et al., 2004). The transcription factor Snail recruits HDAC1, HDAC2, and the corepressor complex mSin3A to the E-cadherin promoter to repress its expression (Peinado et al., 2004). Downregulation or loss of function of E-cadherin has been implicated in the acquisition of invasive potential by carcinomas (Hajra & Fearon, 2002), and so aberrant recruitment of HDACs to this promoter may have a crucial role in tumor invasion and metastasis. The role of HDACs in cancer is not restricted to their contribution to histone deacetylation, but also includes their role in deacetylation of non-histone proteins. For example, HDAC1 interacts with the tumor suppressor p53 and deacetylates it *in vivo* and *in vitro* (Juan et al., 2000,Luo et al., 2000). p53 is phosphorylated and acetylated under stress conditions. Since lysine residues acetylated in p53 overlap with those that are ubiquitinated, p53 acetylation serves to promote protein stability and activation, inducing checkpoints in the cell-division cycle, permanent cell-division arrest, and cell death. Aberrant recruitment of HDACs to specific promoters through the interaction with fusion proteins that result from chromosomal translocations in hematological malignancies has also been intensively studied. In acute promyelocytic leukemia, leukemic fusion between the PML (promyelocytic leukemia) gene and the retinoic acid receptor (RAR) gene suppresses transcription through recruitment of HDACs. Thus, cancer cells are unable to undergo differentiation, leading to excessive proliferation (Lin et al., 2001,He et al., 2001). Similar phenomena have been described for the RARα-PLZF (promyelocytic leukemia zinc finger protein) fusion, the AML1 (acute myelocytic leukemia protein1)-ETO fusion, and for the myc/Mad/Max signaling pathway involved in solid malignancies (Minucci et al., 2001,Ferrara et al., 2001,Kitamura et al., 2000,David et al., 1998).

Class II HDACs have also been shown to associate with various cancer types. Inhibition of class II HDACs induces p21 expression in breast cancer cell lines, suggesting that class II HDAC subfamily may exert specific roles in breast cancer progression (Duong et al., 2008). HDAC4 inhibits p21 gene expression through interaction with Sp1 at p21 proximal promoter. Induction of p21 mediated by silencing of HDAC4 arrested cancer cell growth *in vitro* and inhibited tumor growth in an *in vivo* human glioblastoma model (Mottet et al., 2009). HDAC4 also interacts with PLZF and represses PLZF-RARα fusion protein activity (Chauchereau et al., 2004,Yuki et al., 2004). In the prostate cancer model, HDAC4 is recruited to the nuclei of cancer cells, where it may exert an inhibitory effect on differentiation and contribute to the development of the aggressive phenotype during late stage of prostate cancer (Halkidou et al., 2004). HDAC5 and HDAC9 are significantly upregulated in high-risk medulloblastoma in comparison with low-risk medulloblastoma, and their expression is associated with poor survival (Milde et al., 2010). Higher expression of HDAC7 and HDAC9 is associated with pancreatic adenocarcinomas and poor prognosis in childhood ALL (Ouaissi et al., 2008, Moreno et al., 2010).

Class IIb deacetylase HDAC6 is linked to breast cancer. HDAC6 is expressed at significantly higher levels in breast cancer patients with small tumors and low histologic grade, and in estrogen receptor α- and progesterone receptor-positive tumors. Furthermore, patients with high levels of HDAC6 mRNA tended to be more responsive to endocrine treatment than those with low levels, indicating that HDAC6 may be an early prognosis marker (Zhang et al., 2004,Saji et al., 2005). Overexpression of HDAC6 in MCF-7 breast cancer cells increased cell motility, suggesting a role for HDAC6 in metastases (Saji et al., 2005). HDAC6 has additional functions in integrating signaling and cytoskeleton remodeling. It is shown that cortactin, a genuine substrate of HDAC6, is overexpressed in several carcinomas (Zhang et al., 2007,Luxton & Gundersen, 2007). Therefore, HDAC6 could be a viable target for cancer therapy. There is emerging evidence that inhibiting HDAC6 mediated aggresome pathway leads to the accumulation of misfolded proteins and apoptosis in tumor cells through autophagy (Rodriguez-Gonzalez et al., 2008).

Class III deacetylases, the NAD+ dependent SIRT proteins, are also connected to cancer. For instance, SIRT1 is upregulated in human lung cancer, prostate cancer and leukemia and has been found to be downregulated in colon tumors (Reviewed in (Ropero & Esteller, 2007)). SIRT1 is also responsible for the loss of the acetylation levels of K16-H4 and K9-H3, which is common in human cancer at early cancer development (Fraga et al., 2005). Upregulation of SIRT1 expression in human cancer can also induce deregulation of key proteins, such as p53 and E2F (Chen et al., 2005,Wang et al., 2006).

5. Epigenetic agents and combinatorial therapies in cancer treatment

5.1 DNA methylation inhibitors

The most common DNMT inhibitors in clinical use, 5-azacytidine (5-aza) and 5-aza-2'-deoxycytidine (decitabine), were synthesized almost 50 years ago (Sorm et al., 1964). 5-aza is a cytidine analog. It was originally developed as a nucleoside antimetabolite that could be incorporated into nucleic acids to induce chromosome breakage and mutations, and inhibit protein synthesis by interfering with tRNA and rRNA function (Viegas-Pequignot & Dutrillaux, 1976,Cihak, 1974,Karon & Benedict, 1972). Subsequently it was shown that 5-aza incorporates into nucleic acids and covalently binds to DNMTs, leading to a rapid loss of methylation as a result of DNMT depletion (reviewed in (Christman, 2002)). Decitabine is a very structurally similar compound that was suggested as a less toxic and more specific alternative to 5-aza, as it is not integrated into RNA (Vesely & Cihak, 1977,Bouchard & Momparler, 1983)(reviewed in (Christman, 2002). In the last twenty years, 5-aza and decitabine have been extensively tested in the clinic and approved by the FDA for treatment of myelodysplastic syndrome (MDS) and acute myelogenous leukemia (AML), respectively (section 5.3). However, both 5-aza and decitabine are toxic and highly unstable in aqueous solutions (reviewed in (Stresemann & Lyko, 2008). This makes them difficult to use in clinical settings, especially in solid tumors, so there is a need for other DNMT inhibitors with more favorable properties. Several new DNMT inhibitors have been designed in the last two decades, but most have minimal efficacy *in vivo*, with the exception of zebularine (reviewed in (Brueckner et al., 2007)). Zebularine is also a cytidine analog, but it is more stable and less toxic than 5-aza and decitabine. Zebularine was originally developed as a cytidine deaminase inhibitor, but was later shown to potently inhibit DNMT activity (Zhou et al., 2002) and cancer cell growth (Cheng et al., 2004,Balch et al., 2005) (reviewed in (Yoo et al., 2004). The mechanism of action is largely the same as 5-aza and decitabine, and although its IC50 is higher, lower toxicity and longer biological half-life makes it an attractive candidate for pre-clinical testing and future clinical trials.

5.2 Histone deacetylase inhibitors

HDAC inhibitors were discovered in the 1970s, when it was shown that treatment of cells with sodium butyrate led to hyperacetylation of histones (Candido et al., 1978). Sodium butyrate was found to be difficult to use clinically due to its poor pharmacological properties (Miller et al., 1987). In the following two decades, several more promising antitumor agents that inhibit HDACs were discovered. These include valproic acid (VPA), suberoylanilide hydroxamic acid (SAHA, trade name Vorinostat), Trichostatin A (TSA), and depsipeptide or FK228 (FR901228, trade name Romidepsin). All of them inhibit HDACs by binding to the active site (reviewed in (Marks & Dokmanovic, 2005), (Martinet & Bertrand, 2011)), resulting in release of epigenetic repression. HDAC inhibitors also synergize with DNA damaging treatments such as radiotherapy or chemotherapy with nucleoside analogs (Munshi et al., 2005,Chinnaiyan et al., 2005), probably because HDACs are also involved in DNA replication and DNA repair (section 2.2) (Spange et al., 2009). The three most popular HDAC inhibitors in widespread clinical use today are VPA, SAHA, and depsipeptide, while Trichostatin A is not used clinically due to toxic side effects. VPA has the longest clinical history as it has been used for treatment of epilepsy since the 1960s. It inhibits proliferation of cultured cancer cells at millimolar concentrations and shows synergistic effects in combination with decitabine or hydralazine (Table 1), and it is currently in clinical trials for multiple kinds of cancer (section 5.4). SAHA is more potent, as it was shown to induce growth inhibition, differentiation or apoptosis in cultured cancer cells (Richon et al., 1996,Butler et al., 2000) and inhibit cancer cell growth synergistically with decitabine or zebularine at micromolar concentrations (Table 1); however, it has a short biological half-life of about 2 hours (Kelly et al., 2005). Depsipeptide, discovered in 1994 (Ueda et al., 1994), is a cyclic tetrapeptide that preferentially targets class I HDACs (Furumai et al., 2002). It shows synergistic effects in combination with decitabine, zebularine, Trichostatin A, and 5-aza, and inhibits cancer cell proliferation at sub-micromolar concentrations (Table 1). Currently, SAHA and depsipeptide have been approved by FDA for treatment of cancer, while VPA is still in clinical trials (reviewed in section 5.4).

5.3 DNA methylation and HDAC inhibitors in combination in cancer cells and mouse models

DNMT inhibitors and HDAC inhibitors in combination show synergistic growth inhibition in cancer cell lines and in animal models of cancer. Table 1 summarizes combined treatments with DNMT and HDAC inhibitors in cancer cell lines, organized by cancer type, in alphabetical order. Combination treatments caused synergistic growth inhibition and/or differentiation in the majority of cancer cell lineages tested, including leukemia and lymphoma, small-cell and non-small cell lung cancer, esophageal, liver, breast, and pancreatic cancers (Table 1 for references). This growth inhibition is generally followed by apoptosis as drug dose is increased.

DNMT inhibitors and HDAC inhibitors synergistically affect chromatin state and lead to a more pronounced re-expression of epigenetically silenced tumor suppressor genes and cell cycle regulators (Cameron et al., 1999) (see Figure 1). For example, decitabine plus VPA synergistically induce NY-ESO-1 antigen in glioma cells (Oi et al., 2009), making them a target for immunotherapy. Epigenetic agents are more effective when key tumor suppressor

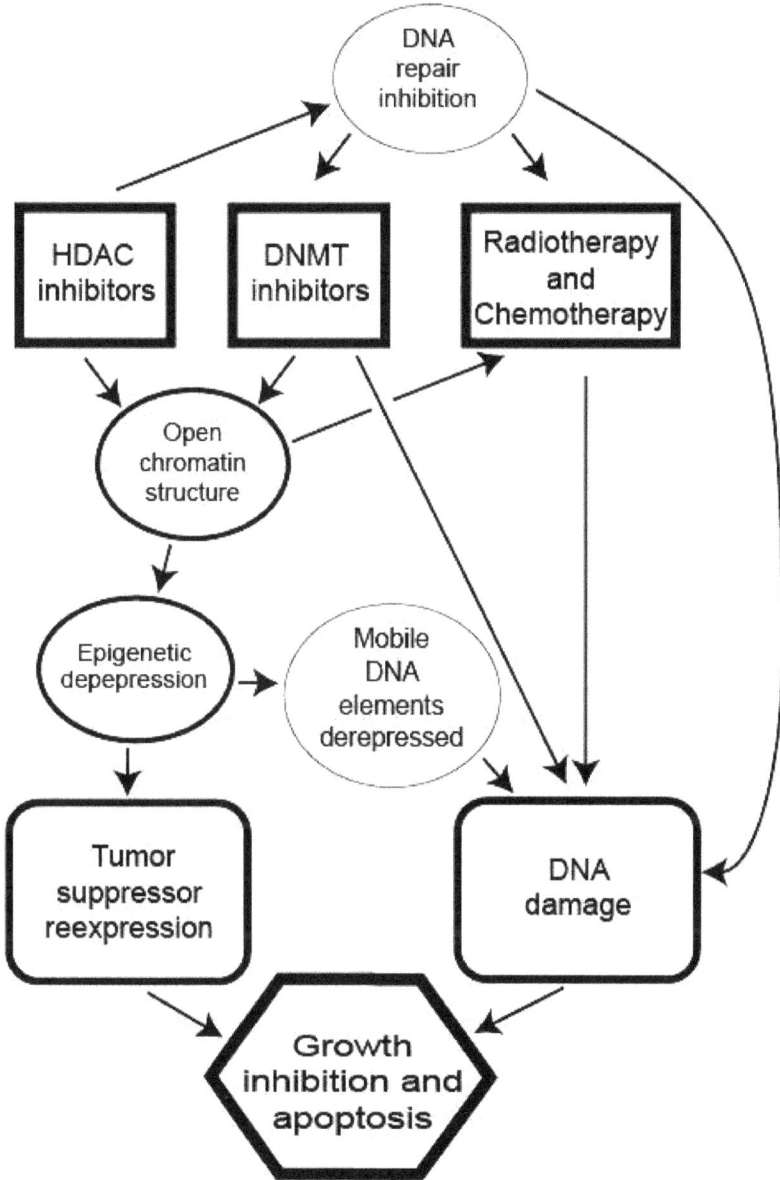

Fig. 1. HDAC and DNMT inhibitors synergistically affect chromatin state, leading to an open chromatin structure and derepression of tumor suppressors, resulting in growth inhibition and apoptosis. Open chromatin structure, as well as DNA repair inhibition induced by HDAC inhibitors, makes cells more susceptible to DNA damage induced by DNMT inhibitors, DNA damaging treatments such as radiotherapy, and epigenetic derepression of mobile DNA elements. DNA damage also leads to cell death.

genes, such as CDKN2A (p16) and p18, are epigenetically silenced (Table 1) (Gore et al., 2006), suggesting that the induction of silenced tumor suppressor genes may be critical for therapeutic effect. For example, the efficacy of DNMT inhibitors decitabine or zebularine combined with the HDAC inhibitor depsipeptide in lung and breast tumor cells with defined CDKN2A status was described recently (Chen et al., 2010). It was shown that non-small cell lung cancer cells with methylated CDKN2A are significantly more sensitive to methylation inhibitors than cell lines with deleted CDKN2A, and the combination of zebularine/depsipeptide results in a synergistic effect on cell growth inhibition that is also linked with the presence of epigenetically silenced CDKN2A (Chen et al., 2010). These data strongly support the importance of prospective pre-selection of patients in future clinical trials and suggest CDKN2A status as a key biomarker for DNMT/HDAC inhibition studies.

DNMT inhibitors and HDAC inhibitors do not necessarily act only on DNA methylation and histone acetylation, respectively, but have other functions as well (also see section 2). It was shown that HDAC inhibitors can affect DNA methylation (Sarkar et al., 2011), and DNMT inhibitors can affect histone methylation (Kondo et al., 2003). When these agents are used in combination, one inhibitor can affect epigenetic changes brought about by the other (Dobosy & Selker, 2001,Oi et al., 2009). DNMT inhibitors can also inhibit proliferation of cancer cells by causing DNA damage and chromosomal instability. Most DNMT inhibitors in widespread use are cytidine analogs (decitabine, 5-aza, zebularine, etc.) that can integrate into DNA and introduce perturbations in its normal structure leading to strand breaks (Karon & Benedict, 1972), or derepress endogenous transposons and retroviral sequences silenced by methylation (Groudine et al., 1981), which can lead to DNA damage and chromosomal instability as well (see Figure 1). These effects are likely augmented by HDAC inhibitors, which can introduce further changes into DNA structure, contribute to the activation of silenced transposons and retroviral sequences, or inhibit DNA damage repair (see Figure 1). For example, it was shown that depsipeptide can slow down removal of decitabine from the DNA (Chai et al., 2008) (possibly by inhibiting DNA repair mechanisms), suggesting an explanation for synergy between these compounds as decitabine staying on the DNA longer would lead to greater DNA demethylation and DNA damage. Also, DNA repair inhibition or chromatin structure changes may account for the fact that the interactions between DNMT inhibitors and HDAC inhibitors can depend on the order in which they are used. For example, depsipeptide treatment for 24 hours followed by 5-aza for 24 hours caused a synergistic induction of Gadd45, but this effect was not present when the compounds were used simultaneously or in the reverse order (Jiang et al., 2007).

HDAC inhibitors and DNMT inhibitors can also synergize with radiotherapy(Munshi et al., 2005,Chinnaiyan et al., 2005), CDK inhibitors (Almenara et al., 2002), TRAIL cytokine (Kaminskyy et al., 2011), and conventional chemotherapy agents such as cisplatin (Shang et al., 2008), paclitaxel, doxorubicin , and 5-fluorouracil (Mirza et al., 2010). Furthermore, they can induce a response in tumors which are resistant to chemotherapeutic agents (Plumb et al., 2000). In addition to the observation that HDAC and DNMT inhibitors can reactivate epigenetically silenced tumor suppressor genes and/or inhibit DNA repair, making conventional treatment regimens more effective, another possible mechanism behind this synergy is that the open chromatin structure induced by epigenetic agents makes DNA more accessible for chemotherapy drugs and radiation (Falk et al., 2008)(see Figure 1). These results show clinical potential for using combinations of epigenetic drugs and conventional agents, especially to overcome or prevent chemotherapy resistance (reviewed by (Gravina et al., 2010) and (Thurn et al., 2011)).

Several other conclusions can be drawn from the work summarized in Table 1. Cytidine analogs (decitabine, 5-aza, and zebularine) synergize well with HDAC inhibitors when acting on cultured cancer cells, making these combinations good candidates for future clinical trials. However, there are long-term cancer risks associated with using nucleoside analogs, as they induce DNA damage, increasing the probability of mutations. Among the HDAC inhibitors, depsipeptide is the most tested in culture over the recent years, because it preferentially inhibits class I HDACs (Furumai et al., 2002) that have been shown to be overexpressed and associated with poor prognosis in multiple kinds of cancer(Nakagawa et al., 2007,Weichert et al., 2008,Weichert et al., 2008).

The effects of epigenetic drugs on tumor growth in xenograft and in genetically engineered mouse models (GEMM) (Tables 2 and 3) are similar to the effects observed in cancer cell lines (Table 1).

Treatment of mice with DNMT1 and HDAC inhibitors causes induction of tumor suppressor genes and pro-apoptotic proteins leading to inhibition of tumor growth and apoptosis. Epigenetic agents were tested in xenografts (Table 2), and GEMMs, which more closely resemble human cancers (Becher & Holland, 2006)(Table 3). Consistent growth inhibition by epigenetic agents (Tables 2 and 3) further indicated that DMNTs and HDACs were potential targets for future clinical trials. The effects of epigenetic agents were similar using several distinct drugs and tumor models – e.g. p21 protein expression was induced with five different drugs (SAHA, MS-275, PXD101, LBH-589, decitabine) in xenograft models of breast, lung, and ovarian cancer (Table 2). CDKN1A/p21 was also induced by zebularine in a transgenic mouse model of breast cancer (Table 3). Other genes induced by epigenetic agents in mouse models included tumor suppressors such as p16 and MLH-1 and pro-apoptotic proteins Bax, caspase 3/7, and death receptor proteins 4 and 5 (Table 2). In contrast, expression of proto-oncogenes such as cyclin D1 and VEGF and anti-apoptotic proteins Bcl-2 and Bcl-XL was suppressed (Tables 2). These results are similar to what was observed in cancer cell lines, where epigenetic agents induced the expression of p16, p21, caspase 3 and MLH1, while downregulating Bcl-2 and other anti-apoptotic proteins (Table 1).

Most epigenetic agents that were tested in mouse models, e.g. decitabine, SAHA, and MS-275, are currently in clinical trials (see section 5.4). Several xenograft and GEMM studies focused on a novel DNMT inhibitor, zebularine, that is not yet approved for human use (Table 2 and 3). It has been shown that zebularine can be administered orally due to its longer half-life than other DNMT inhibitors and can inhibit tumor growth and induce expression of tumor suppressor genes (Cheng et al., 2003) and our unpublished observations (Chen et al, unpublished data). Subsequent xenograft studies showed that short-term treatment with zebularine (Dote et al., 2005) or even a single injection (Neureiter et al., 2007) can inhibit tumor growth. The GEMM studies focused on long-term therapy with an oral formulation of zebularine for intestinal adenomas (Yoo et al., 2008) and mammary tumors (Chen et al, unpublished data). In our unpublished study, high-dose zebularine treatment delayed tumor growth and reduced tumor burden in MMTV-PyMT mouse model. In the Yoo et al. study, continuous treatment of Min transgenic mice with low-dose zebularine prevented polyp formation in the majority of treated animals (while controls all developed intestinal polyps). This study makes low-dose oral zebularine an attractive target for future clinical trials.

Cancer	Drugs	Cell lines	Effects on growth	Genes affected	Ref
Brain cancer	Decitabine & VPA	U251, T98 glioma cells		Strong synergistic induction of NY-ESO-1 reexpression in both cell lines.	(Oi et al., 2009)
Breast	Decitabine & Depsipeptide	MDA-MB-231, MDA-MB-435, and MCF-7	Synergistic growth inhibition, enchanced apoptosis	Induced NY-ESO-1, Maspin, gelsolin.	{Weiser, 2001; Murakami, 2004} (Primeau et al., 2003)
	Hydralazine &VPA	MCF-7	Synergistic growth inhibition		(Chavez-Blanco et al., 2006)
	Zebularine & Depsipeptide	MDA-MB231, MDA-MB435	Synergistic growth inhibition in a cell line where CDKN2A was silenced by methylation (MDA-MB-435), but nct in MDA-MB-231 cells with deleted CDKN2A.		(Chen et al., 2010)
Cervical cancer	Hydralazine & VPA	Hela	Reduced resistance to standard chemotherapy agents such as cisplatin, adrimycin or gemcitabine		(Chavez-Blanco et al., 2006)
Colon Cancer	Hydralazine & VPA	SW480	Synergistic growth inhibition.	Synergistic effect on gene reexpression (352 genes induced by the combination vs 153 and 178 by hydralazine and VPA alone, respectively)	(Chavez-Blanco et al., 2006)
	Decitabine & TSA	HCT116, HCT15, SW48		Synergistic reexpression of p16 and p14.	(Magdinier & Wolffe, 2001)
Esophageal	Decitabine & depsipeptide	BE-3, SKGT-5, TE-1, -2, -3, -12, and -13	Enhanced apoptosis after sequential treatment.	Induced NY-ESO-1(CTA) reexpression.	(Weiser et al., 2001)

Cancer	Drugs	Cell lines	Effects on growth	Genes affected	Ref
Leukemia	Decitabine & TSA/ depsiptide	HL-60, KG1a	Both combinations produced much greater growth inhibition than single agents. Colony formation assay showed synergy.		(Shaker et al., 2003)
	Decitabine & VPA	HL-60 and MOLT4 cells	Synergistic growth inhibition and apoptosis induction	Induced expression of p21CIP1 and p57KIP2.	(Yang et al., 2005)
	Decitabine & depsipeptide	Patient cells and Kasumi-1, NB-4 and K562 cells	Synergistic reduction of cell viability in all cell lines	Very strong synergistic effect inducing IL-3 expression in Kasumi-1 cells.	(Klisovic et al., 2003)
	Decitabine & MS-275	MV4-11. Leukemic blood and bone marrow patient cells	Synergistic effects inhibiting the proliferation of both cultured cell lines and patient-derived cells.	Downregulated Mcl-1 and upregulated caspase-3 and p21^{wafl}. Effects of the combination were reduced when caspase cascade was inhibited.	(Nishioka et al., 2011)
Liver cancer	5-aza& Depsipeptide	HepG2 carcinoma cells	Synergistic growth inhibition	Synergistic induction of Gadd45.	(Jiang et al., 2007)
Lymphoma	Decitabine & LBH589	OCI-Ly1, OCI-Ly7 and Su-DHL6 germinal center DLBCL cell lines. OCI-Ly10, RIVA and Su-DHL2 activated B-cell DLBCL lines. Primary tumor cells from patients.	All drugs had an effect on cell proliferation. DAC and LBH589 combination showed clear synergy in inhibiting proliferation and inducing apoptosis in DLBCL cell lines and CD19-positive primary tumor cells from patients.	Synergistic induction of caspase 3 expression and histone H3 acetylation. Other genes afffected: GATA1, GATA4, SMAD ,DNMT3A, ABCD3, C20orf75, CD19, CEACAM5, CHKA, DIRAS3, FUBP1, GALT, HSPC138, HSPC268, METAP1, PGM2L1, TCEB1, VHL, WT1 and ZFP95. Skp2 was down-regulated, p27 was upregulated.	(Kalac et al., 2011)

Cancer	Drugs	Cell lines	Effects on growth	Genes affected	Ref
Lung cancer: Non-small cell	Decitabine & Depsipeptide	NSLC Calu-6, H-358, H-596, H-1299, H-1355, H-1650, H-2087, and H-2228 cells; mesothelioma H-2373 cells	Enhanced apoptosis after combination treatment.	Induced NY-ESO-1 expression	(Weiser et al., 2001)
	Decitabine & TSA/ Depsipeptide	A549 and H719, cells	Combinations decreased proliferation much more than single agents.		(Chai et al., 2008)
Lung Cancer: Small-cell	Decitabine& Depsipeptide	H69, H82, H209, H211, H719, H792, H841, N417 cells	Synergistic growth inhibition	Induced expression of $p19^{INK4d}$, $p16^{INK4a}$ and $p18^{INK4c}$	(Zhu et al., 2001)
	Decitabine & LBH589 / MGCD0103	H82, H146, H196, H526, DMS114, SW1271, H1688, H1048, and H2195 cells	Both combinations caused synergistic growth inhibition and DNA damage in multiple cell lines, accompanied by an increased number of necrotic cells.		(Luszczek et al., 2010)
	Decitabine & VPA/CI-994	H69, H82, H1417, H2171, and U1906 cells	Both combinations sensitized cells to chemotherapy drug TRAIL.	The combinations synergistically induced caspase-8 and Mcl-2.	(Kaminskyy et al., 2011)
	Zebularine /Decitabine & Depsipeptide	H719 and H865 cells	Combinations synergistically inhibited growth in H719 cell line where CDKN2A is silenced by methylation	In H719 cell line, induced CDKN2A derepression and p16 protein expression.	(Chen et al., 2010)
Melanoma	Decitabine & depsipeptide	F045, 586mel, 1300mel, and 1363mel cells	Enhanced apoptosis after sequential treatment compared to either agent alone.	Combination induced NY-ESO-1(CTA) reexpression.	(Weiser et al., 2001)

Cancer	Drugs	Cell lines	Effects on growth	Genes affected	Ref
Oral	Decitabine / Depsi peptide	HO-1-u-1, HSC2, HSC3, HSC4, SAS, KB, Hep2, Ca9-22, H-O-N-1, KOSC2, KOSC3, SCC25 cells		DAC and Depsipeptide synergistically induce the expression of maspin	(Murakami et al., 2004)
Ovarian	Decitabine & PXD101	A2780/cp7 - cisplatin-resistant A2780 cells		DAC & PXD101 synergistically increase MLH1 expression	(Steele et al., 2009)
	Decitabine & VPA	Hey, SKOv3 cells	Synergistic growth inhibition and induction of apoptosis	Synergistic induction of caspase 3/7 activity. Dose-dependent induction of ARHI and PEG3	(Chen et al., 2011)
Pancreatic Endocrine	Decitabine & TSA	CM (metastatic insulinoma), BON (metastatic carcinoid), QGP-1 (somatostatinoma) cells	Very strong synergistic effect in all three cell lines, inducing growth inhibition and then apoptosis as dose was increased.	In QGP-1 cell line TIMM50 was downregulated and DIABLO, Cathepsin B, SET were upregulated. In BON cell line, VDAC1, OP18, PHB, GRB2, DEC, TIMM50 were downregulated and Cathepsin D was upregulated. In CM cell line NPM1, LGALS1, S100A4, HSPA9B were downregulated.	(Cecconi et al., 2009)
Pancreatic Exocrine	Zebularine & SAHA	YAP C, DAN G and Panc-89 cells	Strong additive effect both in inducing apoptosis and inhibiting cell proliferation.	Pro-apoptotic bax / anti-apoptotic bcl-2 ratio increased after treatment.	(Neureiter et al., 2007)

Abbreviations: 5-FU – 5-fluorouracil, AML - acute myeloid leukemia, CTA – cancer testis antigen, DLBCL diffuse large B-cell lymphoma, LBH589 – trade name Panobinostat, PXD101 – trade name Belinostat

Table 1. Effects of HDAC and DNMT inhibitors on proliferation and gene expression in cancer cell lines

Combinations of DNMT and HDAC inhibitors were tested in three xenograft models and two GEMM's. Four models showed synergy between DMNT and HDAC inhibitors in the delay or reduction of tumor growth (Tables 2, 3). In addition, epigenetic agents in combination with chemotherapy drugs such as cisplatin, etoposide, daclizumab or bortezomib, enhanced the antitumor effect (Table 2). Animal studies paved the way for clinical trials using DNA methylation and HDAC inhibitors alone or in combination with chemotherapy (Section 5.4. and Tables 4, 5, 6, 7).

5.4 DNA methylation and HDAC inhibitors in clinical trials

The first epigenetic modulator in clinical use was a demethylating agent, 5-azacytidine. A phase I clinical trial was first reported in 1972, in which 30 patients with advanced solid tumors were treated with 5-azacytidine (Weiss 1972). Responses were seen in 7 of 11 patients with breast cancer, 2 of 5 patients with melanoma, and 2 of 6 patients with colon cancer. However, significant myelosuppression, nausea and diarrhea were observed. In a second study conducted in 1973, 37 children with acute leukemia were treated with 5-azacytidine (Karon 1973). Of the 14 patients with acute myelogenous leukemia (AML) enrolled in the trial, 5 achieved a complete remission (CR) lasting at least 2 months. Toxicities included myelosuppression, nausea, vomiting, diarrhea, and a transient, pruritic rash. Interest in the drug waned in the 1970s and 1980s. However, in 1993, the promising results of a single-arm phase I/II trial in patients with myelodysplastic syndrome (MDS, types refractory anemia with excess blasts (RAEB-t) and RAEB in transformation (RAEB) with 5-azacytidine were published (Silverman 1993). Out of 43 evaluable patients, the response rate was 32%, with 5 CRs (10%) and 11 partial remissions (PRs) (22%). An interesting finding was that the median time to initial response for several of these trials was 90 or more days, suggesting that longer assessments were necessary to see a significant effect on outcome. Eventually, a phase III trial was conducted, which led to its approval by the FDA for use in patients with MDS in 2004 (Table 3) (Silverman 2002). In this study, 191 patients with all French-American-British (FAB) subtypes of MDS were randomized to either best supportive care (BSC), or 5-azacytidine with BSC. The BSC-only patients could crossover to the 5-azacytidine arm if they had worsening disease after 4 months on the study. The response rate was 23% (7% CR, 16% PR), and median time to leukemic transformation or death was 21 months with 5-azacytidine, compared to 13 months for BSC (p=0.007). A landmark analysis to eliminate the confounding effect of early crossover to the 5-azacytidine arm showed a median survival of an additional 18 months for 5-azacytidine and 11 months for BSC (p=0.03). The most common adverse events were myelosuppression and infection. In a Quality of Life analysis, patients on the treatment arm, and those who crossed over, experienced a significant improvement in fatigue, dyspnea and physical functioning.

5-aza-2′-deoxycytidine (decitabine), another demethylating agent that has been extensively studied, was first used in clinical trials in patients with acute leukemia, in which 21 children and adults with acute lymphocytic leukemia (ALL) and AML were treated (Marmparler 1985). The response rate was 37%, with 6 CRs (22%) and 4 PRs (15%). There was a suggestion that the drug was more effective in AML than ALL, although the sample sizes in each group were too small. An analysis of DNA methylation by high-performance liquid

Mouse model	Drug	Effect	Molecular target	Cell line	Reference
Bladder cancer xenograft	Zebularine	Inhibited tumor growth.	Reactivated p16 expression	EJ6	(Cheng et al., 2003)
	Decitabine		Upregulated tumor suppressor RASSF1A, but also oncogenes uPA, HEPARANASE, CXCR4,SNCG, TGF-β and VEGF.	MCF-7	(Ateeq et al., 2008)
	SAHA	Inhibited tumor growth.	Induced p21 reexpression	MDA-231	(Pratap et al., 2010)
Breast cancer xenograft	SAHA plus TRAIL	Inhibited tumor growth, angiogenesis, and spread of metastases. Induced apoptosis.	Upregulated genes: DR4, DR5, Bak, Bax, Bim, Noxa, PUMA, p21CIP1, TIMP-1, and TIMP-2. Downregulated genes: cyclin D1, Bcl-2, Bcl-XL, VEGF, HIF-1α, IL-6, IL-8, MMP-2, and MMP-9	TRAIL-resistant MDA-MB-468	(Shankar et al., 2009)
	MS-275	Induced apoptosis, inhibited tumor cell proliferation, angiogenesis, spread of metastases. Reversed epithelial-to-mesenchymal transition.	Upregulated genes: DR4, DR5, Bax, Bak, and p21/CIP1. Downregulated genes: cyclin D1, Bcl-2, Bcl-XL, VEGF, HIF-1α, IL-8, MMP-2, and MMP-9.	MDA-MB-468	(Srivastava et al., 2010)
Breast cancer Orthotopic	Panobinostat plus MD5-1	Inhibited tumor growth.		4T1.2 or EO771	(Martin et al., 2011)

Mouse model	Drug	Effect	Molecular target	Cell line	Reference
Glioblastoma	Zebularine	Inhibited tumor growth.	Upregulated genes: RASSF1A, HIC-1, and 14-3-3r	U251	(Dote et al., 2005)
xenograft	SAHA	Inhibited tumor growth and increased survival.		U87, T98G and U118	(Yin et al., 2007)
	Decitabine & Zebularine	Increased survival.		L1210	(Lemaire et al., 2005)
Leukemia xenograft	Depsipeptide &Daclizumab	Enhanced antitumor effect and survival		MET-1	(Chen et al., 2009)
	MS-275 & RAD101	Inhibited tumor growth.		HL60	(Nishioka et al., 2008)
Myeloma xenograft	SAHA plus melphalan / bortezomib	Enhanced antitumor effect.		RPMI8226, U266, and MM1S	(Campbell et al., 2010)
Myeloma orthotopic	JNJ-26481585	Inhibited tumor growth and angiogenesis.		5TMM	(Deleu et al., 2009)
	JNJ-26481585 & bortezomib	Reduced osteoclasts		5TMM	(Deleu et al., 2009)
Lung cancer orthotopic rat model	Decitabine & MS-275	Reduced tumor burden.	Upregulated Bad, Bik, Bak, Bok and p21. Demethylated genes: p16, SFRP. Downregulated DNMT1	Calu-6	(Belinsky et al., 2003)
Lung cancer xenograft and orthotopic	Panobinostat & Etoposide	Enhanced antitumor effect.	Upregulated p21, caspase 3/7 and cleaved poly[ADP-ribose] polymerase. Decreased Bcl-2 and Bcl-XL protein expression	Murine TC-1, AE17, M30, A549, H69, BK-T, H526, RG1	(Crisanti et al., 2009)
Lymphoma Xenograft	Decitabine & Panobinostat	Induced apoptosis.		Ly1 line	(Kalac et al., 2011)

Mouse model	Drug	Effect	Molecular target	Cancer	Reference
Ovarian carcinoma xenograft	Decitabine & Belinostat	Reduced tumor growth.	Upregulated MLH1 and MAGE-A1. Demethylated MAGE-A1	A2780/cp70	(Steele et al., 2009)
	Decitabine & Cisplatin or Carboplatin or Epirubicin or Temozolomide	Enhanced antitumor effect.	Upregulated MLH1 gene	A2780/cp70	(Plumb et al., 2000)
	Belinostat		Upregulated genes CDKN1A, CTGF, DHRS2, DNAJB1, H1F0, MAP1LC3B, ODC, SAT, TACC1. Downregulated genes ABL1, CTPS, EIF4G2, KPNB1, CAD, RAN, TP53, TYMS, TD-60.	A2780	(Monks et al., 2009)
Pancreatic cancer xenograft	Zebularine	Inhibited tumor growth.		Panc-89	(Neureiter et al., 2007)
	Decitabine	Inhibited tumor growth and induced apoptosis.	Increased expression of ARHI protein.	PANC-1	(Yang et al., 2010)
Sarcoma xenograft	MS–275	Inhibited tumor growth.		Neuroblastoma, rhabdomyosarcoma, Ewing's sarcoma, retinoblastoma, medulloblastoma	(Jaboin et al., 2002)
	SAHA	Reduced tumor growth and induced apoptosis.		Sarcoma, osteosarcoma, and rhabdoid tumors	(Hrzenjak et al., 2010)
	Decitabine	Inhibited tumor growth.	Upregulated genes: GADD45A, HSPA9B, PAWR, PDCD5, NFKBIA, TNFAIP3, IGFBP6	U2OS	(Al-Romaih et al., 2007)

Table 2. Effects and molecular tragets of DNMT and HDAC inhibitors in xenograft or orthotopic mouse tumor models.

Mouse tumor model	Transgene or Knock out gene	Drug	Effect	Molecular target	Reference
Colorectal	APCmin	SB939	Inhibited adenoma formation		(Novotny-Diermayr et al., 2010)
Intestinal adenomas	Min	Zebularine	Prevented polyp formation in Min females	Demethylation of B1 repetitive element	(Yoo et al., 2008)
Lung	DNMT1 +/- (heterozygous)	Decitabine plus sodium phenylbutyrate	Inhibited tumor development		(Belinsky et al., 2003)
Medulloblastoma (MB) and rhabdomyosarcoma (RMS)	Ptchtm1Zim	Decitabine/ VPA	Combination prevented MB and RMS formation		(Ecke et al., 2009)
Mammary gland	MMTV-PyMT	Zebularine	Delayed onset of tumor growth; statistically significant reduction in total tumor burden.		(Chen et al., unpublished)

Table 3. Effects and molecular targets of DNMT and HDAC inhibitors in transgenic mouse tumor models.

chromatography (HPLC) on 2 patient samples showed that decitabine therapy produced > 70% inhibition in DNA methylation. Decitabine was subsequently studied in patients with solid tumors (Abele 1987), where only 1 response in 82 evaluable patients was observed (1%), but significant myelosuppression was seen in many patients. A phase III trial demonstrated its efficacy compared to BSC in patients with intermediate- (INT-1 and INT-2) and high-risk MDS (Table 3), which led to its approval by the FDA in 2006. The response rate was significantly better than BSC (30% versus 7%), but there was no significant difference in median time to leukemic transformation or death (12.1 versus 7.8 months, p = 0.16). Cytopenias were again the dominant adverse effect.

In 2010, Gurion et al. performed a meta-analysis of demethylating agents for the treatment of MDS (Gurion 2010). 952 patients with MDS enrolled in 4 randomized controlled clinical trials comparing demethylating agents with either BSC or chemotherapy. Treatment with a demethylating agent significantly improved overall survival (Hazard Ratio [HR] 0.72, 95% CI 0.60-0.85) and the median time to leukemic transformation or death (HR 0.69, 95% CI 0.58-0.82). In a subgroup analysis of the type of drug, these benefits were seen for 5-azacytidine but not for decitabine. Both agents improved response rates. However, a higher rate of grade 3/4 toxicities was also noted (RR 1.21, 95% CI 1.10 to 1.33). The data, however, were not all obtained in randomized trials, and it is possible that the patients who received 5-azacytidine or decitabine had a better prognosis despite a similar International Prognostic Scoring System (IPSS) score.

Valproic acid was one of the first HDAC inhibitors investigated in clinical trials in cancer patients. However, as an anticonvulsant, it was initially explored for seizure prophylaxis (Glantz 1996) and for neuropathic pain (Hardy 2001) in cancer patients. In 2005, three articles were published, detailing the results of phase II clinical trials of VPA in combination with all-trans retinoic acid (ATRA). The three trials included one in older (age $\geq$ 65) poor-risk patients with AML or MDS (RAEB) (6 of 11 had hematological improvements, with platelet and erythroid response) (Pailatrino 2005), one in poor-risk patients with AML (1 of 19 evaluable patients had a minor response and 2 patients had a PR and clearance of peripheral blasts (Bug 2005)) and one in elderly patients (age $\geq$ 70) with AML (1 of 11 patients had CR, 2 patients had CRi (morphological CR with incomplete blood count recovery and 2 patients had hematological improvement) (Raffoux 2005). The first clinical trial in solid tumors was conducted in 12 patients with cervical cancer (Chavez-Blanco 2005). No outcomes were reported, but hyperacetylation of both H3 and H4 in 7 of 12 patients were confirmed. The main adverse effects were sedation and fatigue. While valproic acid is FDA approved for seizure control, treatment of manic episodes and migraine prophylaxis, it is not yet approved for cancer.

Vorinostat (SAHA) is one of best-studied HDAC inhibitors in clinical trials. The first clinical trial was a phase I, dose-finding trial with an intravenous formulation, administered in 2 different schedules, in a total of 37 patients with advanced cancer (hematologic and solid tumors) (Kelly 2003). The maximum tolerated dose was determined (300 mg/m^2/day and 900 mg/m^2/day for 5 days for 3 weeks for hematologic and solid malignancies, respectively). Grade 3/4 toxicities in patients with hematologic malignancies included neutropenia (grade 4), anemia, thrombocytopenia, dyspnea (grade 4), thrombosis, diarrhea, constipation and fatigue. Grade 3/4 toxicities in patients with solid tumors included cardiac ischemia (grade 4), acute respiratory distress syndrome, thrombosis, constipation and abdominal pain. Four patients with solid tumors had an objective response. Post-therapy tumor biopsies confirmed the presence of acetylated histones. FDA approval for vorinostat was given in 2006. It was based on the results of two phase II trials (Duvic 2007, Olsen 2007) (Table 3). The pivotal trial was a phase IIB single-arm trial with 74 patients with persistent, progressive or recurrent cutaneous T-cell lymphoma (CTCL; mycosis fungoides or Sezary syndrome subtypes) who were treated with oral vorinostat (Olsen 2007). The response rate was approximately 30%. The most common adverse effects were diarrhea, fatigue, nausea and anorexia. Grade 3/4 toxicities included fatigue, pulmonary embolism, thrombocytopenia and nausea. A second trial in support of vorinostat was a three-arm, non-randomized, single-center trial in patients with refractory CTCL (Duvic 2007). Thirty-three patients were enrolled, and 8 achieved a PR. Again, the most common adverse effects were fatigue, thrombocytopenia, diarrhea and nausea, with grade 3/4 thrombocytopenia and dehydration.

Depsipeptide is another HDAC inhibitor that was recently approved by the FDA. The first clinical trial reported with this agent was a phase I trial in refractory or advanced solid tumors, in which 37 patients received escalating doses of depsipeptide intravenously on days 1 and 5 every 21 days (Sandor 2002). The dose-limiting toxicities included fatigue, nausea/vomiting, thrombocytopenia (grade 4) and cardiac arrhythmia (grade 4, atrial fibrillation). Based on preclinical data, there was a concern for cardiac toxicity, but no myocardial damage was evident, with only frequent, reversible ECG ST/T-wave changes noted. The maximum tolerated dose was determined to be 17.8 mg/m^2. One PR was observed in a patient with renal cell carcinoma, and 8 patients had stable disease. Two phase II trials led to FDA approval of depsipeptide in 2009, in patients with cutaneous T-cell lymphoma who had received at least one prior systemic therapy (Table 3). In the first trial, a multi-institutional, international, single-arm study, 96 patients were enrolled (Whittaker 2010). The response rate was 34%, and 6 patients (6%) achieved a CR. The median duration of response was 15 months. The most common adverse events were nausea, vomiting, fatigue, anorexia and infections, with serious adverse events including infection, sepsis and pyrexia. In the second trial, another multi-institutional single-arm study, a total of 71 patients were enrolled (Piekarz 2009). The response rate was 34%, with 4 CRs (6%). The median duration of response was 13.7 months. The main toxicities included nausea, vomiting, fatigue, and transient thrombocytopenia and neutropenia, anemia, and ECG T-wave changes, with serious adverse events including infection, supraventricular arrhythmia, ventricular arrhythmia, edema, pyrexia, nausea, leukopenia and thrombocytopenia. There are several other HDAC inhibitors currently in clinical trials as single agents, with similar promising results.

There are several compelling clinical reasons to use these epigenetic modalities in combination. As described above (Section 5.3), combining a DNA demethylating agent and an HDAC inhibitor appears to improve efficacy. In addition, the combination could decrease toxicity. In general, demethylating agents and HDAC inhibitors can cause significant toxicity, including grades 3 and 4 myelosuppression. The combination of the demethylating agent, decitabine, and the HDAC inhibitor, depsipeptide, was recently demonstrated to synergistically inhibit growth of lung and breast cancer cells, at 1000-fold lower doses of depsipeptide than what is used clinically (Chen 2010).

Clinical trials with demethylating agents and HDAC inhibitors in combination are still in the early stages of investigation, in phase I and II trials (Tables 4-6). At this time only eight trials have been published (Table 4). In a phase II trial with valproic acid in combination with hydralazine in 12 patients with refractory MDS, the CR/PR rate was 16%, with an overall RR of 50% (Candelaira 2011) (Table 4). The response rate includes a hematologic improvement (HI) rate of 34%, assessed per the International Working Group criteria (Cheson 2006). In comparison, a phase II trial with valproic acid in 43 MDS/AML patients showed a response in 18 patients (overall RR 24%), with no CR, 1 PR (2%) and 18 HI (22%) (Kuendgen 2005). It is important to note that the valproic acid dose in the Candelaira trial was fixed at 30 mg/kg/day, while valproic acid dose in the Kuendgen trial was titrated to valproic acid serum levels, which is more accurate. Therefore, the patients in the Candelaira trial might have been underdosed, and true outcomes in that trial may have been even better. However, while it appears that the addition of hydralazine to valproic acid might improve response, the study numbers for both trials were low, particularly the Candelaira trial (Candelaira 2011), and it would be premature to draw any conclusions about the superiority of this combination treatment.

Decitabine and valproic acid were combined in a phase I/II trial with 54 patients with MDS and AML (Garcia-Manero 2006) (Table 4). Six of the patients had MDS, of whom 3 responded (CR and PR) (50%). The number of CRs in this group was not reported. The sample size is too small to draw any conclusions about the efficacy of adding valproic acid to decitabine. Nevertheless, this combination warrants further investigation in an MDS-only population, as the response rate in a phase III trial with decitabine alone in 170 MDS patients was only 17% (Kantarjian 2006). The remaining 48 patients in the phase I/II trial had AML, among whom 9 patients experienced a response (19%). The number of CRs among AML patients was not reported. In a smaller phase I trial, 11 AML patients were also treated with decitabine and valproic acid (Blum 2007). Six patients (55%) responded, with 2 CRs (18%). As a comparison, in a phase II trial of decitabine monotherapy, limited to 55 previously-untreated older patients (age > 60), the overall response rate was 25%, with a 24% CR rate (Cashen 2010). The 19% RR for the AML patients in the phase I/II combination trial (Garcia-Manero 2006) is comparable to the 25% RR in the monotherapy trial (Cashen 2010), while the 55% RR observed in the small phase I trial (Blum 2007) is impressive. However, the interpretation of the small phase I trial in particular is limited by its small sample size (n=11).

In a study by Soriano et al. (2007), in which 5-azacytidine and valproic acid in combination with all-trans retinoic acid (ATRA) were administered to 53 AML and high-risk MDS patients, patients had impressive outcomes (Table 4). They had a RR of 42%, and a CR rate of 22%. However, another study of 5-azacytidine, valproic acid and ATRA by Raffoux et al. (Raffoux 2010) in 65 patients, also with AML and high-risk MDS, showed a markedly lower RR of 26%, with a similar CR rate of 21%. It is difficult to determine the reason for the discrepancies between these studies in response rates. They may be due to slightly different dosing schedules. In comparison to three clinical trials with valproic acid and ATRA in older and/or poor-risk patients with AML, with response rates of 11-55% (Pailatrino 2005, Bug 2005, Raffoux 2005), it does not appear that adding 5-azacytidine to valproic acid and ATRA adds to efficacy.

In solid tumors, combination therapy has not appeared to be effective. Stathis et al. did not observe any responses with decitabine and vorinostat (Stathis 2011) (Table 4). However, it was a phase I trial, and most of the 43 patients were not treated with the optimal doses (i.e.

recommended phase II doses) of the drugs. Nevertheless, the lack of response is similar to the lack of response observed with each individual agent in solid tumors. For example, Abele et al. (Abele 1987) observed only 1 response in 82 evaluable patients (1%) with advanced solid tumors treated with decitabine. Similarly, in a phase I trial in 73 patients with advanced solid tumors treated with oral vorinostat, only 1 CR (1%) and 3 PRs were noted (4%) (Kelly 2005). However, in another phase I trial with intravenous vorinostat, 17 advanced solid tumor patients were enrolled, of whom 4 had a response (23%, all PR) (Kelly 2003) (Table 4). It is unclear if this response is due to the use of intravenous vorinostat, but this method of delivery is no longer in development, and all current trials with vorinostat are using the oral form. Candelaria's 2007 trial with hydralazine, valproic acid and chemotherapy in 17 patients with solid tumors looks promising, with a 27% response rate (Table 4) (Candelaria 2007). It is especially impressive since valproic acid as monotherapy in solid tumors is not effective, as demonstrated in a phase I clinical trial in which none of 18 evaluable patients had a response (Atmaca 2007). However, it is difficult to assess the effect of chemotherapy itself. Since each patient received a different chemotherapy regimen, comparisons to historical controls with chemotherapy-only are difficult. In this trial, the authors noted that hydralazine and valproic administration permitted a lower chemotherapy dose intensity (Candelaria 2007), which might have led to lower rates of toxicity. This outcome warrants further clinical trials of chemotherapy administered sequentially after epigenetic agents.

One of the most impressive results with combination therapy appears to be the trial of hydralazine and valproic acid with chemotherapy (doxorubicin [Adriamycin]/ cyclophosphamide [AC]) as neoadjuvant therapy in 16 patients with locally-advanced breast cancer, where the response rate was 81%, with a clinical CR rate of 31% and pathologic CR rate of 6.6% (Arce-Salinas 2006) (Table 4). However, these outcomes are similar to those observed with AC chemotherapy alone in the neoadjuvant setting, with a response rate of 75%, clinical CR of 31%, and pathologic CR of 16%, albeit for patients with early breast cancer (Smith 2004). Currently, taxanes are often used with anthracyclines as neoadjuvant chemotherapy in the locally-advanced setting. The response rates range from 71 to 92%, with clinical CR rates from 17 to 31%, and pathologic CR rates 5-26% (Guarneri 2007), which are similar to the outcomes of AC with hydralazine and valproic acid (Arce-Salinas 2006), with potentially less hematologic toxicity.

To date, all of the studies with epigenetic agents in combination have been phase I or phase II trials, with sub-optimal historical controls with which we can compare response rates. Most have demonstrated transient hypomethylation or induction of histone acetylation (Table 4). Currently, there are many combination phase I and II trials in progress (Tables 6-7), using newer and potentially less toxic HDAC inhibitors, or studying the combination in non-hematologic malignancies (e.g. lung cancer). However, in order to truly assess the utility of combination epigenetic therapy, large, randomized, placebo-controlled phase III trials should be performed, directly comparing monotherapy with combination therapy, to assess overall survival, the most clinically meaningful endpoint. Furthermore, future clinical trial participation should be based on molecular biomarkers to predict which patients would respond best to the epigenetic agents. For example, it was demonstrated that a combination of a demethylating agent, zebularine, and HDAC inhibitor, depsipeptide, was only effective in tumor cells with silenced but not deleted CDKN2A (Chen 2010). A future clinical trial might focus on patients with silenced CDKN2A.

Cancer Type	Study Population	Agent	Phase	Schema	Number of Patients	Endpoints	Major Toxicities	Reference
Lymphoma	Refractory/relapsed CTCL (stage IB to IV) who failed 2 systemic therapies	Vorinostat	IIB	Vorinostat 400 mg daily	74	RR 30% (CR 1%, PR 29%); Median Time To Progression: 4.9 mo; Median duration of response: ≥185 days	Grade 3/4: Fatigue, pulmonary embolism, thrombocytopenia, nausea	(Olsen et al., 2007)
	CTCL (stage IIB to IV) - up to 2 prior cytotoxic therapies	Depsipeptide	II	Depsipeptide 14 mg/m2 on days 1, 8 and 15, every 28 days	71	RR 33% (CR 7%, PR 26%); Med. duration of response: 13.7 mo	Grade 3/4: Myelosuppression; Other toxicities: Nausea, vomiting, fatigue, anorexia	(Piekarz et al., 2009)
	Refractory/relapsed CTCL (stage IB to IVA)	Depsipeptide	II	Depsipeptide 14 mg/m2 on days 1, 8 and 15, every 28 days	96	RR 34% (CR 6%, PR 28%); Med. duration of response: 15 mo	GI disturbances, weakness	(Whittaker et al., 2010)
MDS	MDS (all subtypes, all IPSS risk groups)	5-aza	III	5-aza (75 mg/m2/d s.c. x 7 days every 28 days) vs BSC	191	5-aza vs BSC: med OS 20 mo vs 14 mo (p = NS); Time to leukemic transformation/death: 21 mo vs 13 mo (p = 0.007)	5-aza: myelosuppression, nausea, vomiting	(Silverman et al., 2002)
	MDS (all FAB subtypes, high-risk, INT-2, INT-1)	Decitabine	III	Decitabine (15 mg/m2 i.v. every 8 hrs for 3 days, every 6 wks) vs BSC	170	Decitabine vs BSC: med OS 14 mo vs 14.9 mo (p=NS); Time to leukemic transformation/death: 12.1 mo vs 7.8 mo, p=NS)	Decitabine: myelosuppression, pneumonia, hyperbilirubinemia, fever	(Kantarjian et al., 2006)

CTCL, cutaneous T-cell lymphoma; RR, response rate; CR, complete response; PR, partial response; mo, month; med, median; LFTs, liver function tests; GI, gastrointestinal; IPSS, International Prognostic Scoring System; BSC, best-supportive care; FAB, French-American-British; INT, intermediate-risk.

Table 4. Clinical trials leading to FDA approval of hypomethylating agents or HDAC inhibitors.

Cancer Type	Study Population	Agents	Phase	Schema	Number of Patients	Endpoints	Correlative Studies	Major Toxicities	Reference
Breast	Locally-advanced breast cancer	Hydralazine + VPA + chemo	-	Hydralazine (182 mg or 83 mg) + VPA (30 mg/kg) + chemo (AC every 21 days for 4 cycles beginning on day 7) THEN surgery	16	RR 81% (CR 31%, PR 50%); pCR 6.6% (1/15 operated pts)	Significant decrease in global methylation and HDAC activity	Grade 3/4: Myelo-suppression, tremor, somnolence, nausea, vomiting; Other toxicites: fatigue, headache, edema, diarrhea, anorexia	(Arce-Salinas et al., 2011)
Leukemia	Untreated or relapsed AML	Decitabine + VPA	I	Decitabine alone (dose-escalated; 14 pts), then Decitabine (at OBD) + VPA (dose-escalated) (11 pts)	25	RR 52% (21 pts) (CR 19%, PR 33%); Decitabine OBD 20 mg/m2/d on days 1-10; MTD VPA 20 mg/kg/d on days 5-21	Decitabine vs Decitabine+VPA: no significant difference in histone acetylation	Neutropenic fever, fatigue, infection, pneumonia, encephalopathy, differentiation syndrome	(Karpenko et al., 2008)
MDS	Refractory MDS	Hydralazine + VPA	II	Hydralazine (182 mg [fast acetylators] or 83 mg [slow acetylators]) + VPA (30 mg/kg t.i.d.)	12	RR 16% (CR 8%, PR 8%); Progression to leukemia 16.6%	NR	No grade 3/4 toxicities; Grade 1/2 toxicities: Somnolence, nausea, headache, edema, tremor	(Juergens et al., 2008)

Cancer Type	Study Population	Agents	Phase	Schema	Number of Patients	Endpoints	Correlative Studies	Major Toxicities	Reference
MDS or Leukemia	Phase I - acute or chronic leukemia or MDS; Phase II - AML or high-risk MDS	Decitabine + VPA	I/II	Decitabine (15 mg/m2/d i.v. x 10 days) + VPA (escalating doses p.o. daily x 10 days)	54 (22 in Phase I, of whom 8 were evaluable for phase II)	VPA MTD 50 mg/kg; RR 22% (CR 19%, PR 3%); med. OS 6 mo (15.3 mo in responders); med. remission duration 7.2 mo	Transient DNA hypo-methylation and induction of histone acetylation (associated with p15 reactivation)	Non-hematologic: fatigue, somnolence, nausea, vomiting, anorexia/weight loss, diarrhea, confusion/mental status changes, dizziness, mucositis	(Garcia-Manero et al., 2007)
	Refractory/relapsed AML or high-risk MDS	5-aza + VPA + ATRA	I/II	5-Aza (75 mg/m2/d s.c. x 7 days every 28 days) + VPA (dose-escalated p.o. daily x7 days) + ATRA (45 mg/m2/d p.o. x 5 days beginning on day 3)	53 (19 in phase I; 34 in phase II, including 10 pts in an expansion cohort treated at the MTD)	RR 42% (CR 22%, PR 18%); med. remission duration 26 wks	Transient decrease in global DNA methylation ($p<0.05$) and induction of histone acetylation	DLT: reversible neurotoxicity; Most common: somnolence, confusion, nausea/vomiting, fatigue, hepatic dysfunction, diarrhea	(Silverman et al., 2008)
	High-risk AML or MDS (untreated or refractory/relapsed)	5-aza + VPA + ATRA	II	5-Aza + VPA x 7 days, then ATRA x 21 days for 6 cycles	65	RR 26% (CR 21%, PR 5%); med OS 12.4 mo; No restoration of ATRA-induced differentiation with therapy	Demethylation of FZD9, ALOX12, HPN and CALCA genes associated with clinical response;	Infection (septicemia, pneumonia), confusion, fatigue, constipation, hemorrhage, somnolence, nausea/vomiting, injection site reaction, mucosal dryness	(Raffoux et al., 2010)

Cancer Type	Study Population	Agents	Phase	Schema	Number of Patients	Endpoints	Correlative Studies	Major Toxicities	Reference
Solid tumors	Refractory solid tumors (ovarian, cervix, breast, lung, testis)	Hydralazine + VPA + chemo	II	Hydralazine (182 mg [fast acetylators] or 83 mg [slow acetylators]) + VPA (40 mg/kg p.o. daily) + chemo (beginning 1 wk after hydralazine/VPA)	17 (15 evaluable for response)	27% (CR 0%, PR 27%), SD 53%; med OS 6.1 mo; med PFS 3.3 months	Reduction in global DNA methylation (in 6/8 evaluable pts), and HDAC activity (in 7/7 evaluable pts)	Grade 3/4: Myelosuppression, infection, hypoalbuminemia, somnolence, delirium, tremor	(Candelaria et al. 2007)
	Advanced solid tumors (25% colorectal, 9% NHL, 7% breast, 7% melanoma)	Decitabine + vorinostat	I	9 dose levels with sequential vs concurrent DAC/vorinostat	43	RP2D: Decitabine 10 mg/m2/d days 1-5 AND vorinostat 200 mg b.i.d. days 3-9; RR 0%; SD for 4+ cycles: 29%	NR	DLTs: Myelosuppression, fatigue, anorexia, dehydration, constipation; Other toxicities: Nausea, vomiting, diarrhea	(Stathis et al., 2011)

VPA, valproic acid; AC, Adriamycin/cyclophosphamide; RR, response rate; CR, complete response; PR, partial response; pCR, pathologic complete response; HDAC, histone deacetylase; AML, acute myelogenous leukemia; OBD, optimal biologic dose; MTD, maximum tolerated dose; NR, not reported; MDS, myelodysplastic syndrome; med, median; 5-aza, 5-azacytidine; ATRA, All-trans retinoic acid; DLT, dose-limiting toxicity; PFS, Progression-Free Survival; DAC, Decitabine; RP2D, recommended phase II dose

Table 5. Published trials of methylation and HDAC inhibitors in combination.

Cancer Type	Specific Population	Agents	Phase	Schema	Number of Patients	Endpoints	Correlative Studies	Major Toxicities	Reference
Breast	Operable and locally-advanced breast cancer	VPA + hydrala zine + chemo	II	VPA + hydralazine vs placebo for one week, then FAC x 4 cycles then paclitaxel weekly x 12 wks, then surgery	33	VPA/hydralazine vs placebo: pCR in 3/17 pts (30%) vs 7/16 (70%) (p=NS); Inoperable pts: 3 vs 1 (p=NS)	NR	NR	(Arce-Salinas et al., 2011)
Lung	Metastatic NSCLC	Decitabine + VPA	I	Decitabine (5-15 mg/m2 on days 1-10 of 28-day cycle) and VPA (10-20 mg/kg/d on days 5-21)	8	MTD Decitabine 5 mg/m2 + VPA 10 mg/kg/d; No responses; SD 1/8 pt	Decreased MGMT methylation in 1 pt; increase in HbF levels in 5/7 pts	DLTs: Neurologic, neutropenia; Gr 3 toxicities: neutropenia, muscle weakness, pleural effusion	(Karpenko et al., 2008)
	Relapsed advanced NSCLC	5-aza + Entinostat	I	Escalating doses of 5-aza (days 1-6 and 8-10 of 28-day cycle) + Entinostat (7 mg on days 3 and 10)	10	PR 1/10 pt, SD 2/10 pt	PK and PD analysis pending	DLTs: Neutropenia, thrombocytopenia; Other toxicities: injection site reactions, nausea/vomiting, constipation, fatigue	(Juergens et al., 2008)
MDS or AML	High-risk MDS (≥10% marrow blasts) or relapsed/ refractory AML	5-aza + Mocetino stat	I/II	Phase I: 5-aza 75 mg/m2 on days 1-7, Mocetinostat (escalating dose) 3X/ week starting on day 5 (28-day cycles)	24	MTD of Mocetinostat 110 mg; CR 3/14 pts, PR 4/14 pts	7/9 pts had reduction of whole cell HDAC activity	DLTs: Nausea, vomiting, anorexia,	(Garcia-Manero et al., 2007)

Cancer Type	Specific Population	Agents	Phase	Schema	Number of Patients	Endpoints	Correlative Studies	Major Toxicities	Reference
	MDS, AML	5-aza + vorinostat	I/II	Phase I: 5-aza (55 or 75 mg/m2/d on days 1-7) + Vorinostat (escalating doses)	Phase I: 20	CR 5/11 pts, PR 4/11 pts	NR	No Gr 3/4 toxicities; Gr 2 toxicities: anorexia, fatigue	(Silverman et al., 2008)
MDS, AML or PMF	Refractory/relapsed AML (53%), t-AML/MDS (18%), MDS/CMML (27%), PMF (2%)	5-aza + Belinostat	I	Part 1: 5-aza (75 mg/m2/d on days 1-5) + escalating doses of Belinostat; Part 2: 5-aza vs 5-aza/Bel (at MTD) for one cycle then all got 5-aza/Bel	56	MTD 1000 mg/m2 (max dose); CR 6/56 pts, PR 3/56	MDR1 transcripts increased 3.1-fold in 5-aza/Bel vs 5-aza alone (day 0 vs 5), no change in p21 or HIST1H3H transcript levels	No DLTs; Gr 3 myalgias	(Odenike et al., 2011)
Solid tumors	Advanced solid tumors	5-aza + VPA	I	VPA 75-100 mcg/ml + 5-aza x 10 days (dose escalated)	47	MTD of 5-aza 75 mg/m2; SD 34%	Transient DNA hypomethylation in all pts (median methylation 65% on day 0, 61% on day 10, 65% on day 0 of next cycle); histone acetylation in 53% of pts	DLTs: neutropenic fever, thrombocytopenia; Other toxicities: drowsiness, tremor, low Mg, anemia, vomiting	(Soriano et al., 2007)

VPA, valproic acid; FAC, 5-fluorouracil, doxorubicin, cyclophosphamide; pCR, pathologic complete response; pt(s), patient(s); NS, not significant; NR, not reported; NSCLC, non-small cell lung cancer; MTD, maximum tolerated dose; SD, stable disease; MGMT, O6-methylguanine–DNA methyltransferase; HgF, hemoglobin F; DLT, dose-limiting toxicity; Gr, grade; 5-aza, 5-azacytidine; PR, partial response; PK, pharmacokinetics; PD, pharmacodynamics; MDS, myelodysplastic syndrome; AML, acute myelogenous leukemia; CR, complete response; HDAC, histone deacetylase; t-AML, therapy-related AML; CMML, chronic myelomonocytic leukemia; PMF, primary myelofibrosis; Bel, Belinostat; Mg, magnesium

Table 6. Trials in progress with preliminary results

Cancer Type	Specific Population	Agents	Phase	Schema	Planned # of Patients	Endpoints	Clinical Trials.gov Identifier
ALL or lymphoblastic lymphoma	ALL or relapsed/refractory lymphoblastic lymphoma	Decitabine &vorinostat + chemo	II	Decitabine (15 mg/m2 on days 1-4) and vorinostat (230 mg/m2 on days 1-4) THEN induction chemo	40	Remission rate at day 33 (by BM) or day 42 (by RECIST criteria), toxicity, methylation status (blood, BM) before and after treatment	NCT0088206
AML	Relapsed/refractory AML age ≥ 50 (non-APL)	5-aza & Vorinostat & Gemtuzumab	I/II	Fixed-dose 5-aza + dose-escalated Vorinostat + fixed-dose Gemtuzumab	52	MTD, CR rate, incomplete count recovery rate, DFS, whether characteristics associated with gemtuzumab efficacy predict clinical benefit	NCT00895934
Breast cancer	Advanced breast cancer (triple negative or hormone-resistant and HER2 negative)	5-aza & Entinostat	II	5-aza + Entinostat	60	RR, toxicities, PFS, OS, clinical benefit rate	NCT01349959
Colorectal cancer	Recurrent metastatic colorectal cancer	5-aza & Entinostat	II	5-aza + Entinostat	41	RR; PFS; toxicity; gene expression by qRT-PCR; changes in gene methylation, HDAC activity and acetylation of H3 and H4 histones, and their correlation with response and time to progression	NCT01105377
Lung cancer	Recurrent, advanced NSCLC	5-aza & Entinostat	I/II	5-aza (days 1-6 and 8-10 of each 28-day cycle) + Entinostat (days 3 and 10)	76	MTD, RR, toxicity, PK, PFS (1-year), OS (1-year), DNA methylation, histone acetylation, gene re-expression (blood, sputum, tissue biopsies)	NCT00387465

Cancer Type	Specific Population	Agents	Phase	Schema	Planned # of Patients	Endpoints	ClinicalTrials.gov Identifier
Lung cancer	Resected stage I NSCLC	5-aza & Entinostat	II	Randomized 5-aza + Entinostat versus BSC	258	PFS (3-year), toxicity, OS, DNA methylation, gene re-expression, predict clinical outcome in patients by GWAS on pre-treatment tumor, PFS of patients with methylated vs unmethylated N2 lymph nodes	NCT01207726
	Previously-treated advanced lung cancer	Hydralazine & VPA	I	Hydralazine (25 mg days 1-28) + VPA (titrated to serum level)	30	MTD, DLT, RR, clinical benefit, time to tumor response, TTP, OS	NCT00996060
Lymphoma	Relapsed/refractory DLBCL	5-aza & Vorinostat	I/II	Dose-escalated 5-aza + Vorinostat	32	MTD, RR	NCT01120834
MDS or AML	Patients age ≥ 60 with IPSS Int-2 or high-risk MDS, or AML	Decitabine & Panobinostat	I/II	Fixed dose Decitabine (20 mg/m2/day on days 1-5) and dose-escalated Panobinostat	66	MTD, DLT, CR rate, cytogenic CR rate with incomplete count recovery, RR, QOL, time to response, response duration, PFS, EFS, OS, toxicity	NCT00691938
MDS or AML	Newly-diagnosed MDS (IPSS Int-1, Int-2, high-risk) or AML (non-favorable cytogenetics)	5-aza & Vorinostat	II	5-aza (75 mg/m2 on days 1-5) + Vorinostat (200 mg p.o. TID on days 1-5)	30	60-day survival rate, RR, toxicity	NCT00948064

Cancer Type	Specific Population	Agents	Phase	Schema	Planned Number of Patients	Endpoints	ClinicalTrials.gov Identifier
MDS, AML or CMML	MDS (IPSS INT-2 or high-risk), CMML, AML with multi-lineage dysplasia and ≤ 30% marrow blasts	5-aza & Panobinostat	Ib (Dose-findin g and Expansion phases)	Dose-finding phase: 5-aza (75 mg/m2 on days 1-7) + Panobinostat (escalating doses)	At least 9 for dose-finding phase, then 11 additional pts in expansion phase	MTD, toxicities, RR, gene expression, methylation status	NCT00946647, Ottmann et al. ASCO 2010
	MDS (any IPSS), CMML (dysplastic type), AML with multilineage dysplasia (formerly RAEB-t)	5-aza & Entinostat	II	Randomized: 5-aza (days 1-10) or 5-aza (days 1-10) + Entinostat (on days 3 and 10)	196	RR (CR, PR, trilineage response), toxicity, methylation by methylation-specific RT-PCR	NCT00313586
Melanoma	Metastatic melanoma	Decitabine & Panobino stat & temozolo mide	Ib/II	Fixed dose temozolomide & dose-escalated Decitabine & dose-escalated Panobinostat	70	Safety/tolerability, TTP (compared to DTIC historical controls), MTD, DLT, OS, evaluate response with FDG-PET and compare with conventional imaging	NCT00925132

ALL, Acute Lymphblastic Leukemia; BM, bone marrow; RECIST, Response Evaluation Criteria In Solid Tumors; PFS, Progression-Free Survival; †, trial in progress; AML, acute myelogenous leukemia; APL, acute promyelocytic leukemia; 5-aza, 5-azacytidine; MTD, maximum tolerated dose; CR, complete response; DFS, disease-free survival; HER2, Human Epidermal Growth Factor Receptor 2; RR, response rate; PFS, Progression-free survival; OS, overall survival; qRT-PCR, quantitative real-time polymerase chain reaction; HDAC, histone deacetylase; NSCLC, non-small cell lung cancer; PK, pharmokinetics; GWAS, genome-wide association study; VPA, valproic acid; DLT, dose-limiting toxicity; TTP, time-to-progression; DLBCL, diffuse large B-cell lymphoma; MDS, myelodysplastic syndrome; IPSS, International Prognostic Scoring System; Int, intermediate-risk; QOL, quality of life; EFS, event-free survival; TID, three times a day; CMML, chronic myelomonocytic leukemia; RAEB-t, refractory anemia with excess blasts in transformation; PR, partial response; DTIC, dacarbazine; FDG-PET, fluorodeoxyglucose positron emission tomography.

Table 7. Trials in progress.

6. References

Al-Romaih, K.;Somers, G. R.;Bayani, J.;Hughes, S.;Prasad, M.;Cutz, J. C.;Xue, H.;Zielenska, M.;Wang, Y. & Squire, J. A. (2007) Modulation by decitabine of gene expression and growth of osteosarcoma U2OS cells in vitro and in xenografts: identification of apoptotic genes as targets for demethylation. *Cancer Cell Int* Vol.7, pp.14

Almenara, J.;Rosato, R. & Grant, S. (2002) Synergistic induction of mitochondrial damage and apoptosis in human leukemia cells by flavopiridol and the histone deacetylase inhibitor suberoylanilide hydroxamic acid (SAHA). *Leukemia* Vol.16, No.7, pp.1331-1343

Angrisano, T.;Lembo, F.;Pero, R.;Natale, F.;Fusco, A.;Avvedimento, V. E.;Bruni, C. B. & Chiariotti, L. (2006) TACC3 mediates the association of MBD2 with histone acetyltransferases and relieves transcriptional repression of methylated promoters. *Nucleic Acids Res* Vol.34, No.1, pp.364-372

Arce-Salinas, C.;Dueñas-González, A.;Mohar, A.;Lara-Medina, F. U.;Perez-Sanchez, V. & Bargallo, E. (2011) Randomized, double blind, phase II trial: Epigenetic therapy (valproic acid and hydralazine) plus chemotherapy versus placebo plus neoadjuvant chemotherapy in operable and locally advanced breast cancer. *Journal of Clinical Oncology* Vol.29, No.18S, pp.e13550

Arzenani, M. K.;Zade, A. E.;Ming, Y.;Vijverberg, S. J.;Zhang, Z.;Khan, Z.;Sadique, S.;Kallenbach, L.;Hu, L.;Vukojevic, V. & Ekstrom, T. J. (2011) Genomic DNA hypomethylation by HDAC inhibition implicates DNMT1 nuclear dynamics. *Mol Cell Biol*

Ashktorab, H.;Belgrave, K.;Hosseinkhah, F.;Brim, H.;Nouraie, M.;Takkikto, M.;Hewitt, S.;Lee, E. L.;Dashwood, R. H. & Smoot, D. (2009) Global histone H4 acetylation and HDAC2 expression in colon adenoma and carcinoma. *Dig Dis Sci* Vol.54, No.10, pp.2109-2117

Ateeq, B.;Unterberger, A.;Szyf, M. & Rabbani, S. A. (2008) Pharmacological inhibition of DNA methylation induces proinvasive and prometastatic genes in vitro and in vivo. *Neoplasia* Vol.10, No.3, pp.266-278

Balasubramanian, S.;Ramos, J.;Luo, W.;Sirisawad, M.;Verner, E. & Buggy, J. J. (2008) A novel histone deacetylase 8 (HDAC8)-specific inhibitor PCI-34051 induces apoptosis in T-cell lymphomas. *Leukemia* Vol.22, No.5, pp.1026-1034

Balch, C.;Yan, P.;Craft, T.;Young, S.;Skalnik, D. G.;Huang, T. H. & Nephew, K. P. (2005) Antimitogenic and chemosensitizing effects of the methylation inhibitor zebularine in ovarian cancer. *Mol Cancer Ther* Vol.4, No.10, pp.1505-1514

Ballas, N.;Grunseich, C.;Lu, D. D.;Speh, J. C. & Mandel, G. (2005) REST and its corepressors mediate plasticity of neuronal gene chromatin throughout neurogenesis. *Cell* Vol.121, No.4, pp.645-657

Beaulieu, N.;Morin, S.;Chute, I. C.;Robert, M. F.;Nguyen, H. & MacLeod, A. R. (2002) An essential role for DNA methyltransferase DNMT3B in cancer cell survival. *J Biol Chem* Vol.277, No.31, pp.28176-28181

Becher, O. J. & Holland, E. C. (2006) Genetically engineered models have advantages over xenografts for preclinical studies. *Cancer Res* Vol.66, No.7, pp.3355-3358, discussion 3358-3359

Belinsky, S. A.;Nikula, K. J.;Palmisano, W. A.;Michels, R.;Saccomanno, G.;Gabrielson, E.;Baylin, S. B. & Herman, J. G. (1998) Aberrant methylation of p16(INK4a) is an early event in lung cancer and a potential biomarker for early diagnosis. *Proc Natl Acad Sci U S A* Vol.95, No.20, pp.11891-11896

Belinsky, S. A.;Klinge, D. M.;Stidley, C. A.;Issa, J. P.;Herman, J. G.;March, T. H. & Baylin, S. B. (2003) Inhibition of DNA methylation and histone deacetylation prevents murine lung cancer. *Cancer Res* Vol.63, No.21, pp.7089-7093

Berger, S. L. (2007) The complex language of chromatin regulation during transcription. *Nature* Vol.447, No.7143, pp.407-412

Bouchard, J. & Momparler, R. L. (1983) Incorporation of 5-Aza-2'-deoxycytidine-5'-triphosphate into DNA. Interactions with mammalian DNA polymerase alpha and DNA methylase. *Mol Pharmacol* Vol.24, No.1, pp.109-114

Bouras, T.;Fu, M.;Sauve, A. A.;Wang, F.;Quong, A. A.;Perkins, N. D.;Hay, R. T.;Gu, W. & Pestell, R. G. (2005) SIRT1 deacetylation and repression of p300 involves lysine residues 1020/1024 within the cell cycle regulatory domain 1. *J Biol Chem* Vol.280, No.11, pp.10264-10276

Bourc'his, D.;Xu, G. L.;Lin, C. S.;Bollman, B. & Bestor, T. H. (2001) Dnmt3L and the establishment of maternal genomic imprints. *Science* Vol.294, No.5551, pp.2536-2539

Brueckner, B.;Kuck, D. & Lyko, F. (2007) DNA methyltransferase inhibitors for cancer therapy. *Cancer J* Vol.13, No.1, pp.17-22

Butler, L. M.;Agus, D. B.;Scher, H. I.;Higgins, B.;Rose, A.;Cordon-Cardo, C.;Thaler, H. T.;Rifkind, R. A.;Marks, P. A. & Richon, V. M. (2000) Suberoylanilide hydroxamic acid, an inhibitor of histone deacetylase, suppresses the growth of prostate cancer cells in vitro and in vivo. *Cancer Res* Vol.60, No.18, pp.5165-5170

Cameron, E. E.;Bachman, K. E.;Myohanen, S.;Herman, J. G. & Baylin, S. B. (1999) Synergy of demethylation and histone deacetylase inhibition in the re-expression of genes silenced in cancer. *Nat Genet* Vol.21, No.1, pp.103-107

Campbell, R. A.;Sanchez, E.;Steinberg, J.;Shalitin, D.;Li, Z. W.;Chen, H. & Berenson, J. R. (2010) Vorinostat enhances the antimyeloma effects of melphalan and bortezomib. *Eur J Haematol* Vol.84, No.3, pp.201-211

Candelaria, M.;Gallardo-Rincón, D.;Arce, C.;Cetina, L.;Aguilar-Ponce, J.;Arrieta, Ó.;González-Fierro, A.;Chávez-Blanco, A.;de la Cruz-Hernández, E.;Camargo, M.;Trejo-Becerril, C.;Pérez-Cárdenas, E.;Pérez-Plasencia, C.;Taja-Chayeb, L.;Wegman-Ostrosky, T.;Revilla-Vazquez, A. & Dueñas-González, A. (2007) A phase II study of epigenetic therapy with hydralazine and magnesium valproate to overcome chemotherapy resistance in refractory solid tumors. *Annals of Oncology* Vol.18, No.9, pp.1529-1538

Candido, E. P.;Reeves, R. & Davie, J. R. (1978) Sodium butyrate inhibits histone deacetylation in cultured cells. *Cell* Vol.14, No.1, pp.105-113

Cardoso, M. C. & Leonhardt, H. (1999) DNA methyltransferase is actively retained in the cytoplasm during early development. *J Cell Biol* Vol.147, No.1, pp.25-32

Carlson, L. L.;Page, A. W. & Bestor, T. H. (1992) Properties and localization of DNA methyltransferase in preimplantation mouse embryos: implications for genomic imprinting. *Genes Dev* Vol.6, No.12B, pp.2536-2541

Cecconi, D.;Donadelli, M.;Dalla Pozza, E.;Rinalducci, S.;Zolla, L.;Scupoli, M. T.;Righetti, P. G.;Scarpa, A. & Palmieri, M. (2009) Synergistic effect of trichostatin A and 5-aza-2'-

deoxycytidine on growth inhibition of pancreatic endocrine tumour cell lines: a proteomic study. *Proteomics* Vol.9, No.7, pp.1952-1966

Chahrour, M.;Jung, S. Y.;Shaw, C.;Zhou, X.;Wong, S. T.;Qin, J. & Zoghbi, H. Y. (2008) MeCP2, a key contributor to neurological disease, activates and represses transcription. *Science* Vol.320, No.5880, pp.1224-1229

Chai, G.;Li, L.;Zhou, W.;Wu, L.;Zhao, Y.;Wang, D.;Lu, S.;Yu, Y.;Wang, H.;McNutt, M. A.;Hu, Y. G.;Chen, Y.;Yang, Y.;Wu, X.;Otterson, G. A. & Zhu, W. G. (2008) HDAC inhibitors act with 5-aza-2'-deoxycytidine to inhibit cell proliferation by suppressing removal of incorporated abases in lung cancer cells. *PLoS One* Vol.3, No.6, pp.e2445

Chauchereau, A.;Mathieu, M.;de Saintignon, J.;Ferreira, R.;Pritchard, L. L.;Mishal, Z.;Dejean, A. & Harel-Bellan, A. (2004) HDAC4 mediates transcriptional repression by the acute promyelocytic leukaemia-associated protein PLZF. *Oncogene* Vol.23, No.54, pp.8777-8784

Chavez-Blanco, A.;Perez-Plasencia, C.;Perez-Cardenas, E.;Carrasco-Legleu, C.;Rangel-Lopez, E.;Segura-Pacheco, B.;Taja-Chayeb, L.;Trejo-Becerril, C.;Gonzalez-Fierro, A.;Candelaria, M.;Cabrera, G. & Duenas-Gonzalez, A. (2006) Antineoplastic effects of the DNA methylation inhibitor hydralazine and the histone deacetylase inhibitor valproic acid in cancer cell lines. *Cancer Cell Int* Vol.6, pp.2

Chekhun, V. F.;Lukyanova, N. Y.;Kovalchuk, O.;Tryndyak, V. P. & Pogribny, I. P. (2007) Epigenetic profiling of multidrug-resistant human MCF-7 breast adenocarcinoma cells reveals novel hyper- and hypomethylated targets. *Mol Cancer Ther* Vol.6, No.3, pp.1089-1098

Chen, J.;Zhang, M.;Ju, W. & Waldmann, T. A. (2009) Effective treatment of a murine model of adult T-cell leukemia using depsipeptide and its combination with unmodified daclizumab directed toward CD25. *Blood* Vol.113, No.6, pp.1287-1293

Chen, M.;Voeller, D.;Marquez, V. E.;Kaye, F. J.;Steeg, P. S.;Giaccone, G. & Zajac-Kaye, M. (2010) Enhanced growth inhibition by combined DNA methylation/HDAC inhibitors in lung tumor cells with silenced CDKN2A. *Int J Oncol* Vol.37, No.4, pp.963-971

Chen, M. Y.;Liao, W. S.;Lu, Z.;Bornmann, W. G.;Hennessey, V.;Washington, M. N.;Rosner, G. L.;Yu, Y.;Ahmed, A. A. & Bast, R. C., Jr. (2011) Decitabine and suberoylanilide hydroxamic acid (SAHA) inhibit growth of ovarian cancer cell lines and xenografts while inducing expression of imprinted tumor suppressor genes, apoptosis, G2/M arrest, and autophagy. *Cancer*

Chen M.Y; Shabashvili D.E.;, Nawab A; Yang S. X.; Hollingshead M.; Hunter K. W.; Kaye F. J.; Hochwald S. N.; Marquez V; Steeg P & Zajac-Kaye M (2011) Induction of apoptosis and delay of tumor growth by DNA methyltransferase inhibitor in genetically engineered mouse model of breast cancer. Unpublished

Chen, T.;Ueda, Y.;Dodge, J. E.;Wang, Z. & Li, E. (2003) Establishment and maintenance of genomic methylation patterns in mouse embryonic stem cells by Dnmt3a and Dnmt3b. *Mol Cell Biol* Vol.23, No.16, pp.5594-5605

Chen, T. & Li, E. (2004) Structure and function of eukaryotic DNA methyltransferases. *Curr Top Dev Biol* Vol.60, pp.55-89

Chen, T.;Hevi, S.;Gay, F.;Tsujimoto, N.;He, T.;Zhang, B.;Ueda, Y. & Li, E. (2007) Complete inactivation of DNMT1 leads to mitotic catastrophe in human cancer cells. *Nat Genet* Vol.39, No.3, pp.391-396

Chen, W. Y.;Wang, D. H.;Yen, R. C.;Luo, J.;Gu, W. & Baylin, S. B. (2005) Tumor suppressor HIC1 directly regulates SIRT1 to modulate p53-dependent DNA-damage responses. *Cell* Vol.123, No.3, pp.437-448

Chen, Z. X. & Riggs, A. D. (2011) DNA methylation and demethylation in mammals. *J Biol Chem* Vol.286, No.21, pp.18347-18353

Cheng, J. C.;Matsen, C. B.;Gonzales, F. A.;Ye, W.;Greer, S.;Marquez, V. E.;Jones, P. A. & Selker, E. U. (2003) Inhibition of DNA methylation and reactivation of silenced genes by zebularine. *J Natl Cancer Inst* Vol.95, No.5, pp.399-409

Cheng, J. C.;Weisenberger, D. J.;Gonzales, F. A.;Liang, G.;Xu, G. L.;Hu, Y. G.;Marquez, V. E. & Jones, P. A. (2004) Continuous zebularine treatment effectively sustains demethylation in human bladder cancer cells. *Mol Cell Biol* Vol.24, No.3, pp.1270-1278

Chinnaiyan, P.;Vallabhaneni, G.;Armstrong, E.;Huang, S. M. & Harari, P. M. (2005) Modulation of radiation response by histone deacetylase inhibition. *Int J Radiat Oncol Biol Phys* Vol.62, No.1, pp.223-229

Cho, Y.;Griswold, A.;Campbell, C. & Min, K. T. (2005) Individual histone deacetylases in Drosophila modulate transcription of distinct genes. *Genomics* Vol.86, No.5, pp.606-617

Choi, J. H.;Kwon, H. J.;Yoon, B. I.;Kim, J. H.;Han, S. U.;Joo, H. J. & Kim, D. Y. (2001) Expression profile of histone deacetylase 1 in gastric cancer tissues. *Jpn J Cancer Res* Vol.92, No.12, pp.1300-1304

Chotalia, M.;Smallwood, S. A.;Ruf, N.;Dawson, C.;Lucifero, D.;Frontera, M.;James, K.;Dean, W. & Kelsey, G. (2009) Transcription is required for establishment of germline methylation marks at imprinted genes. *Genes Dev* Vol.23, No.1, pp.105-117

Christman, J. K. (2002) 5-Azacytidine and 5-aza-2'-deoxycytidine as inhibitors of DNA methylation: mechanistic studies and their implications for cancer therapy. *Oncogene* Vol.21, No.35, pp.5483-5495

Cihak, A. (1974) Biological effects of 5-azacytidine in eukaryotes. *Oncology* Vol.30, No.5, pp.405-422

Cohen, H. Y.;Miller, C.;Bitterman, K. J.;Wall, N. R.;Hekking, B.;Kessler, B.;Howitz, K. T.;Gorospe, M.;de Cabo, R. & Sinclair, D. A. (2004) Calorie restriction promotes mammalian cell survival by inducing the SIRT1 deacetylase. *Science* Vol.305, No.5682, pp.390-392

Crisanti, M. C.;Wallace, A. F.;Kapoor, V.;Vandermeers, F.;Dowling, M. L.;Pereira, L. P.;Coleman, K.;Campling, B. G.;Fridlender, Z. G.;Kao, G. D. & Albelda, S. M. (2009) The HDAC inhibitor panobinostat (LBH589) inhibits mesothelioma and lung cancer cells in vitro and in vivo with particular efficacy for small cell lung cancer. *Mol Cancer Ther* Vol.8, No.8, pp.2221-2231

Daniel, F. I.;Cherubini, K.;Yurgel, L. S.;de Figueiredo, M. A. & Salum, F. G. (2011) The role of epigenetic transcription repression and DNA methyltransferases in cancer. *Cancer* Vol.117, No.4, pp.677-687

David, G.;Alland, L.;Hong, S. H.;Wong, C. W.;DePinho, R. A. & Dejean, A. (1998) Histone deacetylase associated with mSin3A mediates repression by the acute promyelocytic leukemia-associated PLZF protein. *Oncogene* Vol.16, No.19, pp.2549-2556

David, G.;Neptune, M. A. & DePinho, R. A. (2002) SUMO-1 modification of histone deacetylase 1 (HDAC1) modulates its biological activities. *J Biol Chem* Vol.277, No.26, pp.23658-23663

De Smet, C.;Loriot, A. & Boon, T. (2004) Promoter-dependent mechanism leading to selective hypomethylation within the 5' region of gene MAGE-A1 in tumor cells. *Mol Cell Biol* Vol.24, No.11, pp.4781-4790

Deleu, S.;Lemaire, M.;Arts, J.;Menu, E.;Van Valckenborgh, E.;Vande Broek, I.;De Raeve, H.;Coulton, L.;Van Camp, B.;Croucher, P. & Vanderkerken, K. (2009) Bortezomib alone or in combination with the histone deacetylase inhibitor JNJ-26481585: effect on myeloma bone disease in the 5T2MM murine model of myeloma. *Cancer Res* Vol.69, No.13, pp.5307-5311

Deplus, R.;Brenner, C.;Burgers, W. A.;Putmans, P.;Kouzarides, T.;de Launoit, Y. & Fuks, F. (2002) Dnmt3L is a transcriptional repressor that recruits histone deacetylase. *Nucleic Acids Res* Vol.30, No.17, pp.3831-3838

Derks, S.;Postma, C.;Moerkerk, P. T.;van den Bosch, S. M.;Carvalho, B.;Hermsen, M. A.;Giaretti, W.;Herman, J. G.;Weijenberg, M. P.;de Bruine, A. P.;Meijer, G. A. & van Engeland, M. (2006) Promoter methylation precedes chromosomal alterations in colorectal cancer development. *Cell Oncol* Vol.28, No.5-6, pp.247-257

Dobosy, J. R. & Selker, E. U. (2001) Emerging connections between DNA methylation and histone acetylation. *Cell Mol Life Sci* Vol.58, No.5-6, pp.721-727

Dodge, J. E.;Okano, M.;Dick, F.;Tsujimoto, N.;Chen, T.;Wang, S.;Ueda, Y.;Dyson, N. & Li, E. (2005) Inactivation of Dnmt3b in mouse embryonic fibroblasts results in DNA hypomethylation, chromosomal instability, and spontaneous immortalization. *J Biol Chem* Vol.280, No.18, pp.17986-17991

Dote, H.;Cerna, D.;Burgan, W. E.;Carter, D. J.;Cerra, M. A.;Hollingshead, M. G.;Camphausen, K. & Tofilon, P. J. (2005) Enhancement of in vitro and in vivo tumor cell radiosensitivity by the DNA methylation inhibitor zebularine. *Clin Cancer Res* Vol.11, No.12, pp.4571-4579

Dryden, S. C.;Nahhas, F. A.;Nowak, J. E.;Goustin, A. S. & Tainsky, M. A. (2003) Role for human SIRT2 NAD-dependent deacetylase activity in control of mitotic exit in the cell cycle. *Mol Cell Biol* Vol.23, No.9, pp.3173-3185

Du, Z.;Song, J.;Wang, Y.;Zhao, Y.;Guda, K.;Yang, S.;Kao, H. Y.;Xu, Y.;Willis, J.;Markowitz, S. D.;Sedwick, D.;Ewing, R. M. & Wang, Z. (2010) DNMT1 stability is regulated by proteins coordinating deubiquitination and acetylation-driven ubiquitination. *Sci Signal* Vol.3, No.146, pp.ra80

Duong, V.;Bret, C.;Altucci, L.;Mai, A.;Duraffourd, C.;Loubersac, J.;Harmand, P. O.;Bonnet, S.;Valente, S.;Maudelonde, T.;Cavailles, V. & Boulle, N. (2008) Specific activity of class II histone deacetylases in human breast cancer cells. *Mol Cancer Res* Vol.6, No.12, pp.1908-1919

Ecke, I.;Petry, F.;Rosenberger, A.;Tauber, S.;Monkemeyer, S.;Hess, I.;Dullin, C.;Kimmina, S.;Pirngruber, J.;Johnsen, S. A.;Uhmann, A.;Nitzki, F.;Wojnowski, L.;Schulz-Schaeffer, W.;Witt, O. & Hahn, H. (2009) Antitumor effects of a combined 5-aza-

2'deoxycytidine and valproic acid treatment on rhabdomyosarcoma and medulloblastoma in Ptch mutant mice. *Cancer Res* Vol.69, No.3, pp.887-895

Eden, A.;Gaudet, F.;Waghmare, A. & Jaenisch, R. (2003) Chromosomal instability and tumors promoted by DNA hypomethylation. *Science* Vol.300, No.5618, pp.455

Espada, J.;Ballestar, E.;Fraga, M. F.;Villar-Garea, A.;Juarranz, A.;Stockert, J. C.;Robertson, K. D.;Fuks, F. & Esteller, M. (2004) Human DNA methyltransferase 1 is required for maintenance of the histone H3 modification pattern. *J Biol Chem* Vol.279, No.35, pp.37175-37184

Etoh, T.;Kanai, Y.;Ushijima, S.;Nakagawa, T.;Nakanishi, Y.;Sasako, M.;Kitano, S. & Hirohashi, S. (2004) Increased DNA methyltransferase 1 (DNMT1) protein expression correlates significantly with poorer tumor differentiation and frequent DNA hypermethylation of multiple CpG islands in gastric cancers. *Am J Pathol* Vol.164, No.2, pp.689-699

Falk, M.;Lukasova, E. & Kozubek, S. (2008) Chromatin structure influences the sensitivity of DNA to gamma-radiation. *Biochim Biophys Acta* Vol.1783, No.12, pp.2398-2414

Farthing, C. R.;Ficz, G.;Ng, R. K.;Chan, C. F.;Andrews, S.;Dean, W.;Hemberger, M. & Reik, W. (2008) Global mapping of DNA methylation in mouse promoters reveals epigenetic reprogramming of pluripotency genes. *PLoS Genet* Vol.4, No.6, pp.e1000116

Fatemi, M. & Wade, P. A. (2006) MBD family proteins: reading the epigenetic code. *J Cell Sci* Vol.119, No.Pt 15, pp.3033-3037

Feng, Q. & Zhang, Y. (2001) The MeCP1 complex represses transcription through preferential binding, remodeling, and deacetylating methylated nucleosomes. *Genes Dev* Vol.15, No.7, pp.827-832

Ferrara, F. F.;Fazi, F.;Bianchini, A.;Padula, F.;Gelmetti, V.;Minucci, S.;Mancini, M.;Pelicci, P. G.;Lo Coco, F. & Nervi, C. (2001) Histone deacetylase-targeted treatment restores retinoic acid signaling and differentiation in acute myeloid leukemia. *Cancer Res* Vol.61, No.1, pp.2-7

Figueroa, M. E.;Lugthart, S.;Li, Y.;Erpelinck-Verschueren, C.;Deng, X.;Christos, P. J.;Schifano, E.;Booth, J.;van Putten, W.;Skrabanek, L.;Campagne, F.;Mazumdar, M.;Greally, J. M.;Valk, P. J.;Lowenberg, B.;Delwel, R. & Melnick, A. (2010) DNA methylation signatures identify biologically distinct subtypes in acute myeloid leukemia. *Cancer Cell* Vol.17, No.1, pp.13-27

Finkel, T.;Deng, C. X. & Mostoslavsky, R. (2009) Recent progress in the biology and physiology of sirtuins. *Nature* Vol.460, No.7255, pp.587-591

Fischle, W.;Dequiedt, F.;Hendzel, M. J.;Guenther, M. G.;Lazar, M. A.;Voelter, W. & Verdin, E. (2002) Enzymatic activity associated with class II HDACs is dependent on a multiprotein complex containing HDAC3 and SMRT/N-CoR. *Mol Cell* Vol.9, No.1, pp.45-57

Florl, A. R.;Lower, R.;Schmitz-Drager, B. J. & Schulz, W. A. (1999) DNA methylation and expression of LINE-1 and HERV-K provirus sequences in urothelial and renal cell carcinomas. *Br J Cancer* Vol.80, No.9, pp.1312-1321

Foglietti, C.;Filocamo, G.;Cundari, E.;De Rinaldis, E.;Lahm, A.;Cortese, R. & Steinkuhler, C. (2006) Dissecting the biological functions of Drosophila histone deacetylases by RNA interference and transcriptional profiling. *J Biol Chem* Vol.281, No.26, pp.17968-17976

Fraga, M. F.;Ballestar, E.;Villar-Garea, A.;Boix-Chornet, M.;Espada, J.;Schotta, G.;Bonaldi, T.;Haydon, C.;Ropero, S.;Petrie, K.;Iyer, N. G.;Perez-Rosado, A.;Calvo, E.;Lopez, J. A.;Cano, A.;Calasanz, M. J.;Colomer, D.;Piris, M. A.;Ahn, N.;Imhof, A.;Caldas, C.;Jenuwein, T. & Esteller, M. (2005) Loss of acetylation at Lys16 and trimethylation at Lys20 of histone H4 is a common hallmark of human cancer. *Nat Genet* Vol.37, No.4, pp.391-400

Fryxell, K. J. & Zuckerkandl, E. (2000) Cytosine deamination plays a primary role in the evolution of mammalian isochores. *Mol Biol Evol* Vol.17, No.9, pp.1371-1383

Fu, M.;Liu, M.;Sauve, A. A.;Jiao, X.;Zhang, X.;Wu, X.;Powell, M. J.;Yang, T.;Gu, W.;Avantaggiati, M. L.;Pattabiraman, N.;Pestell, T. G.;Wang, F.;Quong, A. A.;Wang, C. & Pestell, R. G. (2006) Hormonal control of androgen receptor function through SIRT1. *Mol Cell Biol* Vol.26, No.21, pp.8122-8135

Furumai, R.;Matsuyama, A.;Kobashi, N.;Lee, K. H.;Nishiyama, M.;Nakajima, H.;Tanaka, A.;Komatsu, Y.;Nishino, N.;Yoshida, M. & Horinouchi, S. (2002) FK228 (depsipeptide) as a natural prodrug that inhibits class I histone deacetylases. *Cancer Res* Vol.62, No.17, pp.4916-4921

Gao, L.;Cueto, M. A.;Asselbergs, F. & Atadja, P. (2002) Cloning and functional characterization of HDAC11, a novel member of the human histone deacetylase family. *J Biol Chem* Vol.277, No.28, pp.25748-25755

Garcia-Manero, G.;Yang, A. S.;Klimek, V.;Luger, S.;Newsome, W. M.;Berman, N.;Patterson, T.;Maroun, C.;Li, Z.;Ward, R. & Martell, R. (2007) Phase I/II study of a novel oral isotype-selectie histone deacetylase (HDAC) inhibitor MGCD01013 in combination with azacitidine in patients (pts) with high-risk myelodysplastic syndrome (MDS) or acute myelogenous leukemia (AML). *Journal of Clinical Oncology* Vol.25, No.18S, pp.7062

Gaudet, F.;Hodgson, J. G.;Eden, A.;Jackson-Grusby, L.;Dausman, J.;Gray, J. W.;Leonhardt, H. & Jaenisch, R. (2003) Induction of tumors in mice by genomic hypomethylation. *Science* Vol.300, No.5618, pp.489-492

Geiman, T. M. & Robertson, K. D. (2002) Chromatin remodeling, histone modifications, and DNA methylation-how does it all fit together? *J Cell Biochem* Vol.87, No.2, pp.117-125

Girdwood, D.;Bumpass, D.;Vaughan, O. A.;Thain, A.;Anderson, L. A.;Snowden, A. W.;Garcia-Wilson, E.;Perkins, N. D. & Hay, R. T. (2003) P300 transcriptional repression is mediated by SUMO modification. *Mol Cell* Vol.11, No.4, pp.1043-1054

Glozak, M. A.;Sengupta, N.;Zhang, X. & Seto, E. (2005) Acetylation and deacetylation of non-histone proteins. *Gene* Vol.363, pp.15-23

Gopalakrishnan, S.;Van Emburgh, B. O. & Robertson, K. D. (2008) DNA methylation in development and human disease. *Mutat Res* Vol.647, No.1-2, pp.30-38

Gore, S. D.;Baylin, S.;Sugar, E.;Carraway, H.;Miller, C. B.;Carducci, M.;Grever, M.;Galm, O.;Dauses, T.;Karp, J. E.;Rudek, M. A.;Zhao, M.;Smith, B. D.;Manning, J.;Jiemjit, A.;Dover, G.;Mays, A.;Zwiebel, J.;Murgo, A.;Weng, L. J. & Herman, J. G. (2006) Combined DNA methyltransferase and histone deacetylase inhibition in the treatment of myeloid neoplasms. *Cancer Res* Vol.66, No.12, pp.6361-6369

Gravina, G. L.;Festuccia, C.;Marampon, F.;Popov, V. M.;Pestell, R. G.;Zani, B. M. & Tombolini, V. (2010) Biological rationale for the use of DNA methyltransferase inhibitors as new strategy for modulation of tumor response to chemotherapy and radiation. *Mol Cancer* Vol.9, pp.305

Groudine, M.;Eisenman, R. & Weintraub, H. (1981) Chromatin structure of endogenous retroviral genes and activation by an inhibitor of DNA methylation. *Nature* Vol.292, No.5821, pp.311-317

Grozinger, C. M. & Schreiber, S. L. (2000) Regulation of histone deacetylase 4 and 5 and transcriptional activity by 14-3-3-dependent cellular localization. *Proc Natl Acad Sci U S A* Vol.97, No.14, pp.7835-7840

Grozinger, C. M. & Schreiber, S. L. (2002) Deacetylase enzymes: biological functions and the use of small-molecule inhibitors. *Chem Biol* Vol.9, No.1, pp.3-16

Guardiola, A. R. & Yao, T. P. (2002) Molecular cloning and characterization of a novel histone deacetylase HDAC10. *J Biol Chem* Vol.277, No.5, pp.3350-3356

Guenther, M. G.;Lane, W. S.;Fischle, W.;Verdin, E.;Lazar, M. A. & Shiekhattar, R. (2000) A core SMRT corepressor complex containing HDAC3 and TBL1, a WD40-repeat protein linked to deafness. *Genes Dev* Vol.14, No.9, pp.1048-1057

Guenther, M. G.;Barak, O. & Lazar, M. A. (2001) The SMRT and N-CoR corepressors are activating cofactors for histone deacetylase 3. *Mol Cell Biol* Vol.21, No.18, pp.6091-6101

Gui, C. Y.;Ngo, L.;Xu, W. S.;Richon, V. M. & Marks, P. A. (2004) Histone deacetylase (HDAC) inhibitor activation of p21WAF1 involves changes in promoter-associated proteins, including HDAC1. *Proc Natl Acad Sci U S A* Vol.101, No.5, pp.1241-1246

Guo, X.;Williams, J. G.;Schug, T. T. & Li, X. (2010) DYRK1A and DYRK3 promote cell survival through phosphorylation and activation of SIRT1. *J Biol Chem* Vol.285, No.17, pp.13223-13232

Haberland, M.;Johnson, A.;Mokalled, M. H.;Montgomery, R. L. & Olson, E. N. (2009) Genetic dissection of histone deacetylase requirement in tumor cells. *Proc Natl Acad Sci U S A* Vol.106, No.19, pp.7751-7755

Haberland, M.;Mokalled, M. H.;Montgomery, R. L. & Olson, E. N. (2009) Epigenetic control of skull morphogenesis by histone deacetylase 8. *Genes Dev* Vol.23, No.14, pp.1625-1630

Haberland, M.;Montgomery, R. L. & Olson, E. N. (2009) The many roles of histone deacetylases in development and physiology: implications for disease and therapy. *Nat Rev Genet* Vol.10, No.1, pp.32-42

Haigis, M. C. & Guarente, L. P. (2006) Mammalian sirtuins--emerging roles in physiology, aging, and calorie restriction. *Genes Dev* Vol.20, No.21, pp.2913-2921

Hajra, K. M. & Fearon, E. R. (2002) Cadherin and catenin alterations in human cancer. *Genes Chromosomes Cancer* Vol.34, No.3, pp.255-268

Halkidou, K.;Gaughan, L.;Cook, S.;Leung, H. Y.;Neal, D. E. & Robson, C. N. (2004) Upregulation and nuclear recruitment of HDAC1 in hormone refractory prostate cancer. *Prostate* Vol.59, No.2, pp.177-189

Hansen, K. D.;Timp, W.;Bravo, H. C.;Sabunciyan, S.;Langmead, B.;McDonald, O. G.;Wen, B.;Wu, H.;Liu, Y.;Diep, D.;Briem, E.;Zhang, K.;Irizarry, R. A. & Feinberg, A. P. (2011) Increased methylation variation in epigenetic domains across cancer types. *Nat Genet* Vol.43, No.8, pp.768-775

Hattori, N.;Nishino, K.;Ko, Y. G.;Hattori, N.;Ohgane, J.;Tanaka, S. & Shiota, K. (2004) Epigenetic control of mouse Oct-4 gene expression in embryonic stem cells and trophoblast stem cells. *J Biol Chem* Vol.279, No.17, pp.17063-17069

Hattori, N.;Imao, Y.;Nishino, K.;Hattori, N.;Ohgane, J.;Yagi, S.;Tanaka, S. & Shiota, K. (2007) Epigenetic regulation of Nanog gene in embryonic stem and trophoblast stem cells. *Genes Cells* Vol.12, No.3, pp.387-396

Hawes, S. E.;Stern, J. E.;Feng, Q.;Wiens, L. W.;Rasey, J. S.;Lu, H.;Kiviat, N. B. & Vesselle, H. (2010) DNA hypermethylation of tumors from non-small cell lung cancer (NSCLC) patients is associated with gender and histologic type. *Lung Cancer* Vol.69, No.2, pp.172-179

He, L. Z.;Tolentino, T.;Grayson, P.;Zhong, S.;Warrell, R. P., Jr.;Rifkind, R. A.;Marks, P. A.;Richon, V. M. & Pandolfi, P. P. (2001) Histone deacetylase inhibitors induce remission in transgenic models of therapy-resistant acute promyelocytic leukemia. *J Clin Invest* Vol.108, No.9, pp.1321-1330

Hellebrekers, D. M.;Griffioen, A. W. & van Engeland, M. (2007) Dual targeting of epigenetic therapy in cancer. *Biochim Biophys Acta* Vol.1775, No.1, pp.76-91

Hendrich, B. & Bird, A. (1998) Identification and characterization of a family of mammalian methyl-CpG binding proteins. *Mol Cell Biol* Vol.18, No.11, pp.6538-6547

Hendrich, B.;Guy, J.;Ramsahoye, B.;Wilson, V. A. & Bird, A. (2001) Closely related proteins MBD2 and MBD3 play distinctive but interacting roles in mouse development. *Genes Dev* Vol.15, No.6, pp.710-723

Hernandez-Vargas, H.;Lambert, M. P.;Le Calvez-Kelm, F.;Gouysse, G.;McKay-Chopin, S.;Tavtigian, S. V.;Scoazec, J. Y. & Herceg, Z. (2010) Hepatocellular carcinoma displays distinct DNA methylation signatures with potential as clinical predictors. *PLoS One* Vol.5, No.3, pp.e9749

Howard, G.; Eiges, R.; Gaudet, F.; Jaenisch, R. & Eden, A. (2008) Activation and transposition of endogenous retroviral elements in hypomethylation induced tumors in mice. *Oncogene* Vol.27, No.3, pp.404-408

Hrzenjak, A.;Moinfar, F.;Kremser, M. L.;Strohmeier, B.;Petru, E.;Zatloukal, K. & Denk, H. (2010) Histone deacetylase inhibitor vorinostat suppresses the growth of uterine sarcomas in vitro and in vivo. *Mol Cancer* Vol.9, pp.49

Hu, E.;Chen, Z.;Fredrickson, T.;Zhu, Y.;Kirkpatrick, R.;Zhang, G. F.;Johanson, K.;Sung, C. M.;Liu, R. & Winkler, J. (2000) Cloning and characterization of a novel human class I histone deacetylase that functions as a transcription repressor. *J Biol Chem* Vol.275, No.20, pp.15254-15264

Huang, B. H.;Laban, M.;Leung, C. H.;Lee, L.;Lee, C. K.;Salto-Tellez, M.;Raju, G. C. & Hooi, S. C. (2005) Inhibition of histone deacetylase 2 increases apoptosis and p21Cip1/WAF1 expression, independent of histone deacetylase 1. *Cell Death Differ* Vol.12, No.4, pp.395-404

Illingworth, R.;Kerr, A.;Desousa, D.;Jorgensen, H.;Ellis, P.;Stalker, J.;Jackson, D.;Clee, C.;Plumb, R.;Rogers, J.;Humphray, S.;Cox, T.;Langford, C. & Bird, A. (2008) A novel CpG island set identifies tissue-specific methylation at developmental gene loci. *PLoS Biol* Vol.6, No.1, pp.e22

Irvine, R. A.;Lin, I. G. & Hsieh, C. L. (2002) DNA methylation has a local effect on transcription and histone acetylation. *Mol Cell Biol* Vol.22, No.19, pp.6689-6696

Jaboin, J.;Wild, J.;Hamidi, H.;Khanna, C.;Kim, C. J.;Robey, R.;Bates, S. E. & Thiele, C. J. (2002) MS-27-275, an inhibitor of histone deacetylase, has marked in vitro and in vivo antitumor activity against pediatric solid tumors. *Cancer Res* Vol.62, No.21, pp.6108-6115

Jackson-Grusby, L.;Beard, C.;Possemato, R.;Tudor, M.;Fambrough, D.;Csankovszki, G.;Dausman, J.;Lee, P.;Wilson, C.;Lander, E. & Jaenisch, R. (2001) Loss of genomic methylation causes p53-dependent apoptosis and epigenetic deregulation. *Nat Genet* Vol.27, No.1, pp.31-39

Januchowski, R.;Dabrowski, M.;Ofori, H. & Jagodzinski, P. P. (2007) Trichostatin A down-regulate DNA methyltransferase 1 in Jurkat T cells. *Cancer Lett* Vol.246, No.1-2, pp.313-317

Jiang, C.;Zhou, B.;Fan, K.;Heung, E.;Xue, L.;Liu, X.;Kirschbaum, M. & Yen, Y. (2007) A sequential treatment of depsipeptide followed by 5-azacytidine enhances Gadd45beta expression in hepatocellular carcinoma cells. *Anticancer Res* Vol.27, No.6B, pp.3783-3789

Jin, S. G.;Jiang, C. L.;Rauch, T.;Li, H. & Pfeifer, G. P. (2005) MBD3L2 interacts with MBD3 and components of the NuRD complex and can oppose MBD2-MeCP1-mediated methylation silencing. *J Biol Chem* Vol.280, No.13, pp.12700-12709

Jones, P. L.;Veenstra, G. J.;Wade, P. A.;Vermaak, D.;Kass, S. U.;Landsberger, N.;Strouboulis, J. & Wolffe, A. P. (1998) Methylated DNA and MeCP2 recruit histone deacetylase to repress transcription. *Nat Genet* Vol.19, No.2, pp.187-191

Juan, L. J.;Shia, W. J.;Chen, M. H.;Yang, W. M.;Seto, E.;Lin, Y. S. & Wu, C. W. (2000) Histone deacetylases specifically down-regulate p53-dependent gene activation. *J Biol Chem* Vol.275, No.27, pp.20436-20443

Juergens, R. A.;Vendetti, F.;Coleman, B.;Sebree, R. S.;Rudek, M. A.;Belinsky, S. A.;Brock, M. V.;Herman, J. G.;Baylin, S. B. & Rudin, C. M. (2008) Phase I trial of 5-azacitidine (5AC) and SNDX-275 in advanced lung cancer (NSCLC). *Journal of Clinical Oncology* Vol.26, No.15S, pp.19036

Kalac, M.;Scotto, L.;Marchi, E.;Amengual, J.;Seshan, V. E.;Bhagat, G.;Ulahannan, N.;Leshchenko, V. V.;Temkin, A. M.;Parekh, S.;Tycko, B. & O'Connor, O. A. (2011) HDAC inhibitors and decitabine are highly synergistic and associated with unique gene expression and epigenetic profiles in models of DLBCL. *Blood*

Kaludov, N. K. & Wolffe, A. P. (2000) MeCP2 driven transcriptional repression in vitro: selectivity for methylated DNA, action at a distance and contacts with the basal transcription machinery. *Nucleic Acids Res* Vol.28, No.9, pp.1921-1928

Kaminskyy, V.;Surova, O.;Vaculova, A. & Zhivotovsky, B. (2011) Combined inhibition of DNA methyltransferase and histone deacetylase restores caspase-8 expression and sensitizes SCLC cells to TRAIL. *Carcinogenesis*

Kaneda, M.;Okano, M.;Hata, K.;Sado, T.;Tsujimoto, N.;Li, E. & Sasaki, H. (2004) Essential role for de novo DNA methyltransferase Dnmt3a in paternal and maternal imprinting. *Nature* Vol.429, No.6994, pp.900-903

Kantarjian, H.;Issa, J.-P. J.;Rosenfeld, C. S.;Bennett, J. M.;Albitar, M.;DiPersio, J.;Klimek, V.;Slack, J.;de Castro, C.;Ravandi, F.;Helmer, R.;Shen, L.;Nimer, S. D.;Leavitt, R.;Raza, A. & Saba, H. (2006) Decitabine improves patient outcomes in myelodysplastic syndromes. *Cancer* Vol.106, No.8, pp.1794-1803

Karon, M. & Benedict, W. F. (1972) Chromatid breakage: differential effect of inhibitors of DNA synthesis during G 2 phase. *Science* Vol.178, No.56, pp.62

Karpenko, M. J.;Liu, Z.;Aimiuwu, J.;Wang, L.;Wu, X.;Villalona-Calero, M. A.;Young, D.;Chan, K. K.;Grever, M. R. & Otterson, G. A. (2008) Phase I study of 5-aza-2'-deoxycytidine in combination with valproic acid in patients with NSCLC. *Journal of Clinical Oncology* Vol.26, No.15S, pp.3502

Kelly, W. K.;O'Connor, O. A.;Krug, L. M.;Chiao, J. H.;Heaney, M.;Curley, T.;MacGregore-Cortelli, B.;Tong, W.;Secrist, J. P.;Schwartz, L.;Richardson, S.;Chu, E.;Olgac, S.;Marks, P. A.;Scher, H. & Richon, V. M. (2005) Phase I study of an oral histone deacetylase inhibitor, suberoylanilide hydroxamic acid, in patients with advanced cancer. *J Clin Oncol* Vol.23, No.17, pp.3923-3931

Kitamura, K.;Hoshi, S.;Koike, M.;Kiyoi, H.;Saito, H. & Naoe, T. (2000) Histone deacetylase inhibitor but not arsenic trioxide differentiates acute promyelocytic leukaemia cells with t(11;17) in combination with all-trans retinoic acid. *Br J Haematol* Vol.108, No.4, pp.696-702

Klisovic, M. I.;Maghraby, E. A.;Parthun, M. R.;Guimond, M.;Sklenar, A. R.;Whitman, S. P.;Chan, K. K.;Murphy, T.;Anon, J.;Archer, K. J.;Rush, L. J.;Plass, C.;Grever, M. R.;Byrd, J. C. & Marcucci, G. (2003) Depsipeptide (FR 901228) promotes histone acetylation, gene transcription, apoptosis and its activity is enhanced by DNA methyltransferase inhibitors in AML1/ETO-positive leukemic cells. *Leukemia* Vol.17, No.2, pp.350-358

Kondo, Y.;Shen, L. & Issa, J. P. (2003) Critical role of histone methylation in tumor suppressor gene silencing in colorectal cancer. *Mol Cell Biol* Vol.23, No.1, pp.206-215

Kondo, Y. & Issa, J. P. (2010) DNA methylation profiling in cancer. *Expert Rev Mol Med* Vol.12, pp.e23

Kudo, S. (1998) Methyl-CpG-binding protein MeCP2 represses Sp1-activated transcription of the human leukosialin gene when the promoter is methylated. *Mol Cell Biol* Vol.18, No.9, pp.5492-5499

Lagger, G.;O'Carroll, D.;Rembold, M.;Khier, H.;Tischler, J.;Weitzer, G.;Schuettengruber, B.;Hauser, C.;Brunmeir, R.;Jenuwein, T. & Seiser, C. (2002) Essential function of histone deacetylase 1 in proliferation control and CDK inhibitor repression. *Embo J* Vol.21, No.11, pp.2672-2681

Lee, H.;Rezai-Zadeh, N. & Seto, E. (2004) Negative regulation of histone deacetylase 8 activity by cyclic AMP-dependent protein kinase A. *Mol Cell Biol* Vol.24, No.2, pp.765-773

Lee, M. G.;Wynder, C.;Cooch, N. & Shiekhattar, R. (2005) An essential role for CoREST in nucleosomal histone 3 lysine 4 demethylation. *Nature* Vol.437, No.7057, pp.432-435

Lei, H.;Oh, S. P.;Okano, M.;Juttermann, R.;Goss, K. A.;Jaenisch, R. & Li, E. (1996) De novo DNA cytosine methyltransferase activities in mouse embryonic stem cells. *Development* Vol.122, No.10, pp.3195-3205

Lemaire, M.;Momparler, L. F.;Bernstein, M. L.;Marquez, V. E. & Momparler, R. L. (2005) Enhancement of antineoplastic action of 5-aza-2'-deoxycytidine by zebularine on L1210 leukemia. *Anticancer Drugs* Vol.16, No.3, pp.301-308

Lembo, F.;Pero, R.;Angrisano, T.;Vitiello, C.;Iuliano, R.;Bruni, C. B. & Chiariotti, L. (2003) MBDin, a novel MBD2-interacting protein, relieves MBD2 repression potential and reactivates transcription from methylated promoters. *Mol Cell Biol* Vol.23, No.5, pp.1656-1665

Lewis, J. D.;Meehan, R. R.;Henzel, W. J.;Maurer-Fogy, I.;Jeppesen, P.;Klein, F. & Bird, A. (1992) Purification, sequence, and cellular localization of a novel chromosomal protein that binds to methylated DNA. *Cell* Vol.69, No.6, pp.905-914

Li, B.;Gogol, M.;Carey, M.;Lee, D.;Seidel, C. & Workman, J. L. (2007) Combined action of PHD and chromo domains directs the Rpd3S HDAC to transcribed chromatin. *Science* Vol.316, No.5827, pp.1050-1054

Li, E.;Bestor, T. H. & Jaenisch, R. (1992) Targeted mutation of the DNA methyltransferase gene results in embryonic lethality. *Cell* Vol.69, No.6, pp.915-926

Lin, R. J.;Sternsdorf, T.;Tini, M. & Evans, R. M. (2001) Transcriptional regulation in acute promyelocytic leukemia. *Oncogene* Vol.20, No.49, pp.7204-7215

Ling, Y.;Sankpal, U. T.;Robertson, A. K.;McNally, J. G.;Karpova, T. & Robertson, K. D. (2004) Modification of de novo DNA methyltransferase 3a (Dnmt3a) by SUMO-1 modulates its interaction with histone deacetylases (HDACs) and its capacity to repress transcription. *Nucleic Acids Res* Vol.32, No.2, pp.598-610

Linhart, H. G.;Lin, H.;Yamada, Y.;Moran, E.;Steine, E. J.;Gokhale, S.;Lo, G.;Cantu, E.;Ehrich, M.;He, T.;Meissner, A. & Jaenisch, R. (2007) Dnmt3b promotes tumorigenesis in vivo by gene-specific de novo methylation and transcriptional silencing. *Genes Dev* Vol.21, No.23, pp.3110-3122

Luo, J.;Su, F.;Chen, D.;Shiloh, A. & Gu, W. (2000) Deacetylation of p53 modulates its effect on cell growth and apoptosis. *Nature* Vol.408, No.6810, pp.377-381

Luszczek, W.;Cheriyath, V.;Mekhail, T. M. & Borden, E. C. (2010) Combinations of DNA methyltransferase and histone deacetylase inhibitors induce DNA damage in small cell lung cancer cells: correlation of resistance with IFN-stimulated gene expression. *Mol Cancer Ther* Vol.9, No.8, pp.2309-2321

Luxton, G. W. & Gundersen, G. G. (2007) HDAC6-pack: cortactin acetylation joins the brew. *Dev Cell* Vol.13, No.2, pp.161-162

Magdinier, F. & Wolffe, A. P. (2001) Selective association of the methyl-CpG binding protein MBD2 with the silent p14/p16 locus in human neoplasia. *Proc Natl Acad Sci U S A* Vol.98, No.9, pp.4990-4995

Marks, P. A. & Dokmanovic, M. (2005) Histone deacetylase inhibitors: discovery and development as anticancer agents. *Expert Opin Investig Drugs* Vol.14, No.12, pp.1497-1511

Martin, B. P.;Frew, A. J.;Bots, M.;Fox, S.;Long, F.;Takeda, K.;Yagita, H.;Atadja, P.;Smyth, M. J. & Johnstone, R. W. (2011) Antitumor activities and on-target toxicities mediated by a TRAIL receptor agonist following cotreatment with panobinostat. *Int J Cancer* Vol.128, No.11, pp.2735-2747

Martinet, N. & Bertrand, P. (2011) Interpreting clinical assays for histone deacetylase inhibitors. *Cancer Manag Res* Vol.3, pp.117-141

Mayer, W.;Fundele, R. & Haaf, T. (2000) Spatial separation of parental genomes during mouse interspecific (Mus musculus x M. spretus) spermiogenesis. *Chromosome Res* Vol.8, No.6, pp.555-558

Mayer, W.;Niveleau, A.;Walter, J.;Fundele, R. & Haaf, T. (2000) Demethylation of the zygotic paternal genome. *Nature* Vol.403, No.6769, pp.501-502

McKinsey, T. A.;Zhang, C. L.;Lu, J. & Olson, E. N. (2000) Signal-dependent nuclear export of a histone deacetylase regulates muscle differentiation. *Nature* Vol.408, No.6808, pp.106-111

Meehan, R. R.;Lewis, J. D.;McKay, S.;Kleiner, E. L. & Bird, A. P. (1989) Identification of a mammalian protein that binds specifically to DNA containing methylated CpGs. *Cell* Vol.58, No.3, pp.499-507

Meissner, A.;Mikkelsen, T. S.;Gu, H.;Wernig, M.;Hanna, J.;Sivachenko, A.;Zhang, X.;Bernstein, B. E.;Nusbaum, C.;Jaffe, D. B.;Gnirke, A.;Jaenisch, R. & Lander, E. S. (2008) Genome-scale DNA methylation maps of pluripotent and differentiated cells. *Nature* Vol.454, No.7205, pp.766-770

Milde, T.;Oehme, I.;Korshunov, A.;Kopp-Schneider, A.;Remke, M.;Northcott, P.;Deubzer, H. E.;Lodrini, M.;Taylor, M. D.;von Deimling, A.;Pfister, S. & Witt, O. (2010) HDAC5 and HDAC9 in medulloblastoma: novel markers for risk stratification and role in tumor cell growth. *Clin Cancer Res* Vol.16, No.12, pp.3240-3252

Miller, A. A.;Kurschel, E.;Osieka, R. & Schmidt, C. G. (1987) Clinical pharmacology of sodium butyrate in patients with acute leukemia. *Eur J Cancer Clin Oncol* Vol.23, No.9, pp.1283-1287

Minucci, S.;Nervi, C.;Lo Coco, F. & Pelicci, P. G. (2001) Histone deacetylases: a common molecular target for differentiation treatment of acute myeloid leukemias? *Oncogene* Vol.20, No.24, pp.3110-3115

Mirza, S.;Sharma, G.;Pandya, P. & Ralhan, R. (2010) Demethylating agent 5-aza-2-deoxycytidine enhances susceptibility of breast cancer cells to anticancer agents. *Mol Cell Biochem* Vol.342, No.1-2, pp.101-109

Miska, E. A.;Karlsson, C.;Langley, E.;Nielsen, S. J.;Pines, J. & Kouzarides, T. (1999) HDAC4 deacetylase associates with and represses the MEF2 transcription factor. *Embo J* Vol.18, No.18, pp.5099-5107

Mohn, F.;Weber, M.;Rebhan, M.;Roloff, T. C.;Richter, J.;Stadler, M. B.;Bibel, M. & Schubeler, D. (2008) Lineage-specific polycomb targets and de novo DNA methylation define restriction and potential of neuronal progenitors. *Mol Cell* Vol.30, No.6, pp.755-766

Monks, A.;Hose, C. D.;Pezzoli, P.;Kondapaka, S.;Vansant, G.;Petersen, K. D.;Sehested, M.;Monforte, J. & Shoemaker, R. H. (2009) Gene expression-signature of belinostat in cell lines is specific for histone deacetylase inhibitor treatment, with a corresponding signature in xenografts. *Anticancer Drugs* Vol.20, No.8, pp.682-692

Moreno, D. A.;Scrideli, C. A.;Cortez, M. A.;de Paula Queiroz, R.;Valera, E. T.;da Silva Silveira, V.;Yunes, J. A.;Brandalise, S. R. & Tone, L. G. (2010) Differential expression of HDAC3, HDAC7 and HDAC9 is associated with prognosis and survival in childhood acute lymphoblastic leukaemia. *Br J Haematol* Vol.150, No.6, pp.665-673

Mottet, D.;Pirotte, S.;Lamour, V.;Hagedorn, M.;Javerzat, S.;Bikfalvi, A.;Bellahcene, A.;Verdin, E. & Castronovo, V. (2009) HDAC4 represses p21(WAF1/Cip1) expression in human cancer cells through a Sp1-dependent, p53-independent mechanism. *Oncogene* Vol.28, No.2, pp.243-256

Munshi, A.;Kurland, J. F.;Nishikawa, T.;Tanaka, T.;Hobbs, M. L.;Tucker, S. L.;Ismail, S.;Stevens, C. & Meyn, R. E. (2005) Histone deacetylase inhibitors radiosensitize human melanoma cells by suppressing DNA repair activity. *Clin Cancer Res* Vol.11, No.13, pp.4912-4922

Murakami, J.;Asaumi, J.;Maki, Y.;Tsujigiwa, H.;Kuroda, M.;Nagai, N.;Yanagi, Y.;Inoue, T.;Kawasaki, S.;Tanaka, N.;Matsubara, N. & Kishi, K. (2004) Effects of demethylating agent 5-aza-2(')-deoxycytidine and histone deacetylase inhibitor FR901228 on maspin gene expression in oral cancer cell lines. *Oral Oncol* Vol.40, No.6, pp.597-603

Mutskov, V. J.;Farrell, C. M.;Wade, P. A.;Wolffe, A. P. & Felsenfeld, G. (2002) The barrier function of an insulator couples high histone acetylation levels with specific protection of promoter DNA from methylation. *Genes Dev* Vol.16, No.12, pp.1540-1554

Nakagawa, M.;Oda, Y.;Eguchi, T.;Aishima, S.;Yao, T.;Hosoi, F.;Basaki, Y.;Ono, M.;Kuwano, M.;Tanaka, M. & Tsuneyoshi, M. (2007) Expression profile of class I histone deacetylases in human cancer tissues. *Oncol Rep* Vol.18, No.4, pp.769-774

Nakagawa, T.;Kanai, Y.;Saito, Y.;Kitamura, T.;Kakizoe, T. & Hirohashi, S. (2003) Increased DNA methyltransferase 1 protein expression in human transitional cell carcinoma of the bladder. *J Urol* Vol.170, No.6 Pt 1, pp.2463-2466

Nakagawa, T.;Kanai, Y.;Ushijima, S.;Kitamura, T.;Kakizoe, T. & Hirohashi, S. (2005) DNA hypermethylation on multiple CpG islands associated with increased DNA methyltransferase DNMT1 protein expression during multistage urothelial carcinogenesis. *J Urol* Vol.173, No.5, pp.1767-1771

Nan, X.;Campoy, F. J. & Bird, A. (1997) MeCP2 is a transcriptional repressor with abundant binding sites in genomic chromatin. *Cell* Vol.88, No.4, pp.471-481

Nan, X.;Ng, H. H.;Johnson, C. A.;Laherty, C. D.;Turner, B. M.;Eisenman, R. N. & Bird, A. (1998) Transcriptional repression by the methyl-CpG-binding protein MeCP2 involves a histone deacetylase complex. *Nature* Vol.393, No.6683, pp.386-389

Neureiter, D.;Zopf, S.;Leu, T.;Dietze, O.;Hauser-Kronberger, C.;Hahn, E. G.;Herold, C. & Ocker, M. (2007) Apoptosis, proliferation and differentiation patterns are influenced by Zebularine and SAHA in pancreatic cancer models. *Scand J Gastroenterol* Vol.42, No.1, pp.103-116

Ng, H. H.;Zhang, Y.;Hendrich, B.;Johnson, C. A.;Turner, B. M.;Erdjument-Bromage, H.;Tempst, P.;Reinberg, D. & Bird, A. (1999) MBD2 is a transcriptional repressor belonging to the MeCP1 histone deacetylase complex. *Nat Genet* Vol.23, No.1, pp.58-61

Ng, H. H. & Bird, A. (2000) Histone deacetylases: silencers for hire. *Trends Biochem Sci* Vol.25, No.3, pp.121-126

Ng, H. H.;Jeppesen, P. & Bird, A. (2000) Active repression of methylated genes by the chromosomal protein MBD1. *Mol Cell Biol* Vol.20, No.4, pp.1394-1406

Nishioka, C.;Ikezoe, T.;Yang, J.;Koeffler, H. P. & Yokoyama, A. (2008) Blockade of mTOR signaling potentiates the ability of histone deacetylase inhibitor to induce growth arrest and differentiation of acute myelogenous leukemia cells. *Leukemia* Vol.22, No.12, pp.2159-2168

Nishioka, C.;Ikezoe, T.;Yang, J.;Udaka, K. & Yokoyama, A. (2011) Simultaneous inhibition of DNA methyltransferase and histone deacetylase induces p53-independent apoptosis via down-regulation of Mcl-1 in acute myelogenous leukemia cells. *Leuk Res* Vol.35, No.7, pp.932-939

North, B. J.;Marshall, B. L.;Borra, M. T.;Denu, J. M. & Verdin, E. (2003) The human Sir2 ortholog, SIRT2, is an NAD+-dependent tubulin deacetylase. *Mol Cell* Vol.11, No.2, pp.437-444

Novotny-Diermayr, V.;Sangthongpitag, K.;Hu, C. Y.;Wu, X.;Sausgruber, N.;Yeo, P.;Greicius, G.;Pettersson, S.;Liang, A. L.;Loh, Y. K.;Bonday, Z.;Goh, K. C.;Hentze, H.;Hart, S.;Wang, H.;Ethirajulu, K. & Wood, J. M. (2010) SB939, a novel potent and orally active histone deacetylase inhibitor with high tumor exposure and efficacy in mouse models of colorectal cancer. *Mol Cancer Ther* Vol.9, No.3, pp.642-652

Odenike, O.;Godley, L. A.;Madzo, J.;Karrison, T.;Green, M.;Artz, A. S.;Mattison, R. J.;Yee, K. W. L.;Bennett, M.;Fulton, N.;Koval, G.;Malnassy, G.;Larson, R. A.;Ratain, M. J. & Stock, W. (2011) A phase I and pharmacodynamic (PD) study of the histone deacetylase (HDAC) inhibitor belinostat (BEL) plus azacitidine (AZC) in advanced myeloid malignancies. *Journal of Clinical Oncology* Vol.29, No.18S, pp.6521

Oh, Y. M.;Kwon, Y. E.;Kim, J. M.;Bae, S. J.;Lee, B. K.;Yoo, S. J.;Chung, C. H.;Deshaies, R. J. & Seol, J. H. (2009) Chfr is linked to tumour metastasis through the downregulation of HDAC1. *Nat Cell Biol* Vol.11, No.3, pp.295-302

Oi, S.;Natsume, A.;Ito, M.;Kondo, Y.;Shimato, S.;Maeda, Y.;Saito, K. & Wakabayashi, T. (2009) Synergistic induction of NY-ESO-1 antigen expression by a novel histone deacetylase inhibitor, valproic acid, with 5-aza-2'-deoxycytidine in glioma cells. *J Neurooncol* Vol.92, No.1, pp.15-22

Okano, M.;Bell, D. W.;Haber, D. A. & Li, E. (1999) DNA methyltransferases Dnmt3a and Dnmt3b are essential for de novo methylation and mammalian development. *Cell* Vol.99, No.3, pp.247-257

Olsen, E. A.;Kim, Y. H.;Kuzel, T. M.;Pacheco, T. R.;Foss, F. M.;Parker, S.;Frankel, S. R.;Chen, C.;Ricker, J. L.;Arduino, J. M. & Duvic, M. (2007) Phase IIB Multicenter Trial of Vorinostat in Patients With Persistent, Progressive, or Treatment Refractory Cutaneous T-Cell Lymphoma. *Journal of Clinical Oncology* Vol.25, No.21, pp.3109-3115

Ooi, S. K.;Qiu, C.;Bernstein, E.;Li, K.;Jia, D.;Yang, Z.;Erdjument-Bromage, H.;Tempst, P.;Lin, S. P.;Allis, C. D.;Cheng, X. & Bestor, T. H. (2007) DNMT3L connects unmethylated lysine 4 of histone H3 to de novo methylation of DNA. *Nature* Vol.448, No.7154, pp.714-717

Oshimo, Y.;Nakayama, H.;Ito, R.;Kitadai, Y.;Yoshida, K.;Chayama, K. & Yasui, W. (2003) Promoter methylation of cyclin D2 gene in gastric carcinoma. *Int J Oncol* Vol.23, No.6, pp.1663-1670

Oswald, J.;Engemann, S.;Lane, N.;Mayer, W.;Olek, A.;Fundele, R.;Dean, W.;Reik, W. & Walter, J. (2000) Active demethylation of the paternal genome in the mouse zygote. *Curr Biol* Vol.10, No.8, pp.475-478

Ou, J. N.;Torrisani, J.;Unterberger, A.;Provencal, N.;Shikimi, K.;Karimi, M.;Ekstrom, T. J. & Szyf, M. (2007) Histone deacetylase inhibitor Trichostatin A induces global and gene-specific DNA demethylation in human cancer cell lines. *Biochem Pharmacol* Vol.73, No.9, pp.1297-1307

Ouaissi, M.;Sielezneff, I.;Silvestre, R.;Sastre, B.;Bernard, J. P.;Lafontaine, J. S.;Payan, M. J.;Dahan, L.;Pirro, N.;Seitz, J. F.;Mas, E.;Lombardo, D. & Ouaissi, A. (2008) High histone deacetylase 7 (HDAC7) expression is significantly associated with adenocarcinomas of the pancreas. *Ann Surg Oncol* Vol.15, No.8, pp.2318-2328

Palmisano, W. A.;Divine, K. K.;Saccomanno, G.;Gilliland, F. D.;Baylin, S. B.;Herman, J. G. & Belinsky, S. A. (2000) Predicting lung cancer by detecting aberrant promoter methylation in sputum. *Cancer Res* Vol.60, No.21, pp.5954-5958

Paredes, J.;Albergaria, A.;Oliveira, J. T.;Jeronimo, C.;Milanezi, F. & Schmitt, F. C. (2005) P-cadherin overexpression is an indicator of clinical outcome in invasive breast carcinomas and is associated with CDH3 promoter hypomethylation. *Clin Cancer Res* Vol.11, No.16, pp.5869-5877

Park, S. Y.;Jun, J. A.;Jeong, K. J.;Heo, H. J.;Sohn, J. S.;Lee, H. Y.;Park, C. G. & Kang, J. (2011) Histone deacetylases 1, 6 and 8 are critical for invasion in breast cancer. *Oncol Rep* Vol.25, No.6, pp.1677-1681

Peinado, H.;Ballestar, E.;Esteller, M. & Cano, A. (2004) Snail mediates E-cadherin repression by the recruitment of the Sin3A/histone deacetylase 1 (HDAC1)/HDAC2 complex. *Mol Cell Biol* Vol.24, No.1, pp.306-319

Pereira, C. V.;Lebiedzinsk, M.;Wieckowski, M. R. & Oliveira, P. J. (2011) Regulation and protection of mitochondrial physiology by sirtuins. *Mitochondrion*

Perissi, V.;Jepsen, K.;Glass, C. K. & Rosenfeld, M. G. (2010) Deconstructing repression: evolving models of co-repressor action. *Nat Rev Genet* Vol.11, No.2, pp.109-123

Piekarz, R. L.;Frye, R.;Turner, M.;Wright, J. J.;Allen, S. L.;Kirschbaum, M. H.;Zain, J.;Prince, H. M.;Leonard, J. P.;Geskin, L. J.;Reeder, C.;Joske, D.;Figg, W. D.;Gardner, E. R.;Steinberg, S. M.;Jaffe, E. S.;Stetler-Stevenson, M.;Lade, S.;Fojo, A. T. & Bates, S. E. (2009) Phase II multi-institutional trial of the histone deacetylase inhibitor romidepsin as monotherapy for patients with cutaneous T-cell lymphoma. *J Clin Oncol* Vol.27, No.32, pp.5410-5417

Plumb, J. A.;Strathdee, G.;Sludden, J.;Kaye, S. B. & Brown, R. (2000) Reversal of drug resistance in human tumor xenografts by 2'-deoxy-5-azacytidine-induced demethylation of the hMLH1 gene promoter. *Cancer Res* Vol.60, No.21, pp.6039-6044

Pradhan, S.;Bacolla, A.;Wells, R. D. & Roberts, R. J. (1999) Recombinant human DNA (cytosine-5) methyltransferase. I. Expression, purification, and comparison of de novo and maintenance methylation. *J Biol Chem* Vol.274, No.46, pp.33002-33010

Pratap, J.;Akech, J.;Wixted, J. J.;Szabo, G.;Hussain, S.;McGee-Lawrence, M. E.;Li, X.;Bedard, K.;Dhillon, R. J.;van Wijnen, A. J.;Stein, J. L.;Stein, G. S.;Westendorf, J. J. & Lian, J. B. (2010) The histone deacetylase inhibitor, vorinostat, reduces tumor growth at the metastatic bone site and associated osteolysis, but promotes normal bone loss. *Mol Cancer Ther* Vol.9, No.12, pp.3210-3220

Primeau, M.;Gagnon, J. & Momparler, R. L. (2003) Synergistic antineoplastic action of DNA methylation inhibitor 5-AZA-2'-deoxycytidine and histone deacetylase inhibitor depsipeptide on human breast carcinoma cells. *Int J Cancer* Vol.103, No.2, pp.177-184

Pruitt, K.;Zinn, R. L.;Ohm, J. E.;McGarvey, K. M.;Kang, S. H.;Watkins, D. N.;Herman, J. G. & Baylin, S. B. (2006) Inhibition of SIRT1 reactivates silenced cancer genes without loss of promoter DNA hypermethylation. *PLoS Genet* Vol.2, No.3, pp.e40

Qiu, Y.;Zhao, Y.;Becker, M.;John, S.;Parekh, B. S.;Huang, S.;Hendarwanto, A.;Martinez, E. D.;Chen, Y.;Lu, H.;Adkins, N. L.;Stavreva, D. A.;Wiench, M.;Georgel, P. T.;Schiltz,

R. L. & Hager, G. L. (2006) HDAC1 acetylation is linked to progressive modulation of steroid receptor-induced gene transcription. *Mol Cell* Vol.22, No.5, pp.669-679

Raffoux, E.;Cras, A.;Recher, C.;Boëlle, P.-Y.;de Labarthe, A.;Turlure, P.;Marolleau, J.-P.;Reman, O.;Gardin, C.;Victor, M.;Maury, S.;Rousselot, P.;Malfuson, J.-V.;Maarek, O.;Daniel, M.-T.;Fenaux, P.;Degos, L.;Chomienne, C.;Chevret, S. & Dombret, H. (2010) Phase 2 clinical trial of 5-azacitidine, valproic acid, and all-trans retinoic acid in patients with high-risk acute myeloid leukemia or myelodysplastic syndrome. *Oncotarget* Vol.1, No.1, pp.34-42

Rai, K.;Sarkar, S.;Broadbent, T. J.;Voas, M.;Grossmann, K. F.;Nadauld, L. D.;Dehghanizadeh, S.;Hagos, F. T.;Li, Y.;Toth, R. K.;Chidester, S.;Bahr, T. M.;Johnson, W. E.;Sklow, B.;Burt, R.;Cairns, B. R. & Jones, D. A. (2010) DNA demethylase activity maintains intestinal cells in an undifferentiated state following loss of APC. *Cell* Vol.142, No.6, pp.930-942

Reik, W. & Walter, J. (2001) Genomic imprinting: parental influence on the genome. *Nat Rev Genet* Vol.2, No.1, pp.21-32

Richon, V. M.;Webb, Y.;Merger, R.;Sheppard, T.;Jursic, B.;Ngo, L.;Civoli, F.;Breslow, R.;Rifkind, R. A. & Marks, P. A. (1996) Second generation hybrid polar compounds are potent inducers of transformed cell differentiation. *Proc Natl Acad Sci U S A* Vol.93, No.12, pp.5705-5708

Rideout, W. M., 3rd;Coetzee, G. A.;Olumi, A. F. & Jones, P. A. (1990) 5-Methylcytosine as an endogenous mutagen in the human LDL receptor and p53 genes. *Science* Vol.249, No.4974, pp.1288-1290

Robert, M. F.;Morin, S.;Beaulieu, N.;Gauthier, F.;Chute, I. C.;Barsalou, A. & MacLeod, A. R. (2003) DNMT1 is required to maintain CpG methylation and aberrant gene silencing in human cancer cells. *Nat Genet* Vol.33, No.1, pp.61-65

Robertson, K. D. (2001) DNA methylation, methyltransferases, and cancer. *Oncogene* Vol.20, No.24, pp.3139-3155

Rodriguez-Gonzalez, A.;Lin, T.;Ikeda, A. K.;Simms-Waldrip, T.;Fu, C. & Sakamoto, K. M. (2008) Role of the aggresome pathway in cancer: targeting histone deacetylase 6-dependent protein degradation. *Cancer Res* Vol.68, No.8, pp.2557-2560

Ropero, S. & Esteller, M. (2007) The role of histone deacetylases (HDACs) in human cancer. *Mol Oncol* Vol.1, No.1, pp.19-25

Rougier, N.;Bourc'his, D.;Gomes, D. M.;Niveleau, A.;Plachot, M.;Paldi, A. & Viegas-Pequignot, E. (1998) Chromosome methylation patterns during mammalian preimplantation development. *Genes Dev* Vol.12, No.14, pp.2108-2113

Rountree, M. R.;Bachman, K. E. & Baylin, S. B. (2000) DNMT1 binds HDAC2 and a new co-repressor, DMAP1, to form a complex at replication foci. *Nat Genet* Vol.25, No.3, pp.269-277

Saito, Y.;Kanai, Y.;Nakagawa, T.;Sakamoto, M.;Saito, H.;Ishii, H. & Hirohashi, S. (2003) Increased protein expression of DNA methyltransferase (DNMT) 1 is significantly correlated with the malignant potential and poor prognosis of human hepatocellular carcinomas. *Int J Cancer* Vol.105, No.4, pp.527-532

Saji, S.;Kawakami, M.;Hayashi, S.;Yoshida, N.;Hirose, M.;Horiguchi, S.;Itoh, A.;Funata, N.;Schreiber, S. L.;Yoshida, M. & Toi, M. (2005) Significance of HDAC6 regulation via estrogen signaling for cell motility and prognosis in estrogen receptor-positive breast cancer. *Oncogene* Vol.24, No.28, pp.4531-4539

Sakuma, M.;Akahira, J.;Ito, K.;Niikura, H.;Moriya, T.;Okamura, K.;Sasano, H. & Yaegashi, N. (2007) Promoter methylation status of the Cyclin D2 gene is associated with poor prognosis in human epithelial ovarian cancer. *Cancer Sci* Vol.98, No.3, pp.380-386

Sarkar, S.;Abujamra, A. L.;Loew, J. E.;Forman, L. W.;Perrine, S. P. & Faller, D. V. (2011) Histone Deacetylase Inhibitors Reverse CpG Methylation by Regulating DNMT1 through ERK Signaling. *Anticancer Res* Vol.31, No.9, pp.2723-2732

Seigneurin-Berny, D.;Verdel, A.;Curtet, S.;Lemercier, C.;Garin, J.;Rousseaux, S. & Khochbin, S. (2001) Identification of components of the murine histone deacetylase 6 complex: link between acetylation and ubiquitination signaling pathways. *Mol Cell Biol* Vol.21, No.23, pp.8035-8044

Sen, G. L.;Reuter, J. A.;Webster, D. E.;Zhu, L. & Khavari, P. A. (2010) DNMT1 maintains progenitor function in self-renewing somatic tissue. *Nature* Vol.463, No.7280, pp.563-567

Sengupta, N. & Seto, E. (2004) Regulation of histone deacetylase activities. *J Cell Biochem* Vol.93, No.1, pp.57-67

Shaker, S.;Bernstein, M.;Momparler, L. F. & Momparler, R. L. (2003) Preclinical evaluation of antineoplastic activity of inhibitors of DNA methylation (5-aza-2'-deoxycytidine) and histone deacetylation (trichostatin A, depsipeptide) in combination against myeloid leukemic cells. *Leuk Res* Vol.27, No.5, pp.437-444

Shang, D.;Liu, Y.;Matsui, Y.;Ito, N.;Nishiyama, H.;Kamoto, T. & Ogawa, O. (2008) Demethylating agent 5-aza-2'-deoxycytidine enhances susceptibility of bladder transitional cell carcinoma to Cisplatin. *Urology* Vol.71, No.6, pp.1220-1225

Shankar, S.;Davis, R.;Singh, K. P.;Kurzrock, R.;Ross, D. D. & Srivastava, R. K. (2009) Suberoylanilide hydroxamic acid (Zolinza/vorinostat) sensitizes TRAIL-resistant breast cancer cells orthotopically implanted in BALB/c nude mice. *Mol Cancer Ther* Vol.8, No.6, pp.1596-1605

Sharp, A. J.;Stathaki, E.;Migliavacca, E.;Brahmachary, M.;Montgomery, S. B.;Dupre, Y. & Antonarakis, S. E. (2011) DNA methylation profiles of human active and inactive X chromosomes. *Genome Res*

Shen, L.;Kondo, Y.;Guo, Y.;Zhang, J.;Zhang, L.;Ahmed, S.;Shu, J.;Chen, X.;Waterland, R. A. & Issa, J. P. (2007) Genome-wide profiling of DNA methylation reveals a class of normally methylated CpG island promoters. *PLoS Genet* Vol.3, No.10, pp.2023-2036

Shi, Y.;Sawada, J.;Sui, G.;Affar el, B.;Whetstine, J. R.;Lan, F.;Ogawa, H.;Luke, M. P.;Nakatani, Y. & Shi, Y. (2003) Coordinated histone modifications mediated by a CtBP co-repressor complex. *Nature* Vol.422, No.6933, pp.735-738

Silverman, L. R.;Demakos, E. P.;Peterson, B. L.;Kornblith, A. B.;Holland, J. C.;Odchimar-Reissig, R.;Stone, R. M.;Nelson, D.;Powell, B. L.;DeCastro, C. M.;Ellerton, J.;Larson, R. A.;Schiffer, C. A. & Holland, J. F. (2002) Randomized Controlled Trial of Azacitidine in Patients With the Myelodysplastic Syndrome: A Study of the Cancer and Leukemia Group B. *Journal of Clinical Oncology* Vol.20, No.10, pp.2429-2440

Silverman, L. R.;Verma, A.;Odchimar-Reissig, R.;Cozza, A.;Najfeld, V.;Licht, J. D. & Zwiebel, J. A. (2008) A phase I/II study of vorinostat, an oral histone deacetylase inhibitor, in combination with azacitidine in patients with the myelodysplastic syndrome (MDS) and acute myeloid leukemia (AML). Initial results of the phase I trial: A

New York Cnacer Consortium. *Journal of Clinical Oncology* Vol.2008, No.15S, pp.7000

Smith, K. M.;Dobosy, J. R.;Reifsnyder, J. E.;Rountree, M. R.;Anderson, D. C.;Green, G. R. & Selker, E. U. (2010) H2B- and H3-specific histone deacetylases are required for DNA methylation in Neurospora crassa. *Genetics* Vol.186, No.4, pp.1207-1216

Song, J.;Noh, J. H.;Lee, J. H.;Eun, J. W.;Ahn, Y. M.;Kim, S. Y.;Lee, S. H.;Park, W. S.;Yoo, N. J.;Lee, J. Y. & Nam, S. W. (2005) Increased expression of histone deacetylase 2 is found in human gastric cancer. *Apmis* Vol.113, No.4, pp.264-268

Soriano, A. O.;Braiteh, F.;Garcia-Manero, G.;Camacho, L. H.;Hong, D.;Moulder, S.;Ng, C. & Kurzrock, R. (2007) Combination of 5-azacytidine (5-AZA) and valproic acid (VPA) in advanced solid cancers: A phase I study. *Journal of Clinical Oncology* Vol.25, No.18S, pp.3547

Sorm, F.;Piskala, A.;Cihak, A. & Vesely, J. (1964) 5-Azacytidine, a new, highly effective cancerostatic. *Experientia* Vol.20, No.4, pp.202-203

Spange, S.;Wagner, T.;Heinzel, T. & Kramer, O. H. (2009) Acetylation of non-histone proteins modulates cellular signalling at multiple levels. *Int J Biochem Cell Biol* Vol.41, No.1, pp.185-198

Spurling, C. C.;Suhl, J. A.;Boucher, N.;Nelson, C. E.;Rosenberg, D. W. & Giardina, C. (2008) The short chain fatty acid butyrate induces promoter demethylation and reactivation of RARbeta2 in colon cancer cells. *Nutr Cancer* Vol.60, No.5, pp.692-702

Srivastava, R. K.;Kurzrock, R. & Shankar, S. (2010) MS-275 sensitizes TRAIL-resistant breast cancer cells, inhibits angiogenesis and metastasis, and reverses epithelial-mesenchymal transition in vivo. *Mol Cancer Ther* Vol.9, No.12, pp.3254-3266

Stathis, A.;Hotte, S. J.;Chen, E. X.;Hirte, H. W.;Oza, A. M.;Moretto, P.;Webster, S.;Laughlin, A.;Stayner, L.-A.;McGill, S.;Wang, L.;Zhang, W.-j.;Espinoza-Delgado, I.;Holleran, J. L.;Egorin, M. J. & Siu, L. L. (2011) Phase I Study of Decitabine in Combination with Vorinostat in Patients with Advanced Solid Tumors and Non-Hodgkin's Lymphomas. *Clinical Cancer Research* Vol.17, No.6, pp.1582-1590

Steele, N.;Finn, P.;Brown, R. & Plumb, J. A. (2009) Combined inhibition of DNA methylation and histone acetylation enhances gene re-expression and drug sensitivity in vivo. *Br J Cancer* Vol.100, No.5, pp.758-763

Sterner, D. E. & Berger, S. L. (2000) Acetylation of histones and transcription-related factors. *Microbiol Mol Biol Rev* Vol.64, No.2, pp.435-459

Strahl, B. D. & Allis, C. D. (2000) The language of covalent histone modifications. *Nature* Vol.403, No.6765, pp.41-45

Stresemann, C. & Lyko, F. (2008) Modes of action of the DNA methyltransferase inhibitors azacytidine and decitabine. *Int J Cancer* Vol.123, No.1, pp.8-13

Suetake, I.;Shinozaki, F.;Miyagawa, J.;Takeshima, H. & Tajima, S. (2004) DNMT3L stimulates the DNA methylation activity of Dnmt3a and Dnmt3b through a direct interaction. *J Biol Chem* Vol.279, No.26, pp.27816-27823

Sulewska, A.;Niklinska, W.;Kozlowski, M.;Minarowski, L.;Naumnik, W.;Niklinski, J.;Dabrowska, K. & Chyczewski, L. (2007) DNA methylation in states of cell physiology and pathology. *Folia Histochem Cytobiol* Vol.45, No.3, pp.149-158

Thurn, K. T.;Thomas, S.;Moore, A. & Munster, P. N. (2011) Rational therapeutic combinations with histone deacetylase inhibitors for the treatment of cancer. *Future Oncol* Vol.7, No.2, pp.263-283

Tucker, K. L.;Talbot, D.;Lee, M. A.;Leonhardt, H. & Jaenisch, R. (1996) Complementation of methylation deficiency in embryonic stem cells by a DNA methyltransferase minigene. *Proc Natl Acad Sci U S A* Vol.93, No.23, pp.12920-12925

Ueda, H.;Nakajima, H.;Hori, Y.;Fujita, T.;Nishimura, M.;Goto, T. & Okuhara, M. (1994) FR901228, a novel antitumor bicyclic depsipeptide produced by Chromobacterium violaceum No. 968. I. Taxonomy, fermentation, isolation, physico-chemical and biological properties, and antitumor activity. *J Antibiot (Tokyo)* Vol.47, No.3, pp.301-310

Uhl, M.;Csernok, A.;Aydin, S.;Kreienberg, R.;Wiesmuller, L. & Gatz, S. A. (2010) Role of SIRT1 in homologous recombination. *DNA Repair (Amst)* Vol.9, No.4, pp.383-393

Vaquero, A.;Scher, M.;Lee, D.;Erdjument-Bromage, H.;Tempst, P. & Reinberg, D. (2004) Human SirT1 interacts with histone H1 and promotes formation of facultative heterochromatin. *Mol Cell* Vol.16, No.1, pp.93-105

Vaziri, H.;Dessain, S. K.;Ng Eaton, E.;Imai, S. I.;Frye, R. A.;Pandita, T. K.;Guarente, L. & Weinberg, R. A. (2001) hSIR2(SIRT1) functions as an NAD-dependent p53 deacetylase. *Cell* Vol.107, No.2, pp.149-159

Vesely, J. & Cihak, A. (1977) Incorporation of a potent antileukemic agent, 5-aza-2'-deoxycytidine, into DNA of cells from leukemic mice. *Cancer Res* Vol.37, No.10, pp.3684-3689

Viegas-Pequignot, E. & Dutrillaux, B. (1976) Segmentation of human chromosomes induced by 5-ACR (5-azacytidine). *Hum Genet* Vol.34, No.3, pp.247-254

Wade, P. A.;Jones, P. L.;Vermaak, D. & Wolffe, A. P. (1998) A multiple subunit Mi-2 histone deacetylase from Xenopus laevis cofractionates with an associated Snf2 superfamily ATPase. *Curr Biol* Vol.8, No.14, pp.843-846

Waltregny, D.;De Leval, L.;Glenisson, W.;Ly Tran, S.;North, B. J.;Bellahcene, A.;Weidle, U.;Verdin, E. & Castronovo, V. (2004) Expression of histone deacetylase 8, a class I histone deacetylase, is restricted to cells showing smooth muscle differentiation in normal human tissues. *Am J Pathol* Vol.165, No.2, pp.553-564

Wang, A. H. & Yang, X. J. (2001) Histone deacetylase 4 possesses intrinsic nuclear import and export signals. *Mol Cell Biol* Vol.21, No.17, pp.5992-6005

Wang, C.;Chen, L.;Hou, X.;Li, Z.;Kabra, N.;Ma, Y.;Nemoto, S.;Finkel, T.;Gu, W.;Cress, W. D. & Chen, J. (2006) Interactions between E2F1 and SirT1 regulate apoptotic response to DNA damage. *Nat Cell Biol* Vol.8, No.9, pp.1025-1031

Wang, Y.;Zhang, H.;Chen, Y.;Sun, Y.;Yang, F.;Yu, W.;Liang, J.;Sun, L.;Yang, X.;Shi, L.;Li, R.;Li, Y.;Zhang, Y.;Li, Q.;Yi, X. & Shang, Y. (2009) LSD1 is a subunit of the NuRD complex and targets the metastasis programs in breast cancer. *Cell* Vol.138, No.4, pp.660-672

Wang, Z.;Zang, C.;Cui, K.;Schones, D. E.;Barski, A.;Peng, W. & Zhao, K. (2009) Genome-wide mapping of HATs and HDACs reveals distinct functions in active and inactive genes. *Cell* Vol.138, No.5, pp.1019-1031

Weichert, W.;Roske, A.;Gekeler, V.;Beckers, T.;Ebert, M. P.;Pross, M.;Dietel, M.;Denkert, C. & Rocken, C. (2008) Association of patterns of class I histone deacetylase expression with patient prognosis in gastric cancer: a retrospective analysis. *Lancet Oncol* Vol.9, No.2, pp.139-148

Weichert, W.;Roske, A.;Niesporek, S.;Noske, A.;Buckendahl, A. C.;Dietel, M.;Gekeler, V.;Boehm, M.;Beckers, T. & Denkert, C. (2008) Class I histone deacetylase expression has independent prognostic impact in human colorectal cancer: specific role of class I histone deacetylases in vitro and in vivo. *Clin Cancer Res* Vol.14, No.6, pp.1669-1677

Weiser, T. S.;Guo, Z. S.;Ohnmacht, G. A.;Parkhurst, M. L.;Tong-On, P.;Marincola, F. M.;Fischette, M. R.;Yu, X.;Chen, G. A.;Hong, J. A.;Stewart, J. H.;Nguyen, D. M.;Rosenberg, S. A. & Schrump, D. S. (2001) Sequential 5-Aza-2 deoxycytidine-depsipeptide FR901228 treatment induces apoptosis preferentially in cancer cells and facilitates their recognition by cytolytic T lymphocytes specific for NY-ESO-1. *J Immunother* Vol.24, No.2, pp.151-161

Wen, Y. D.;Perissi, V.;Staszewski, L. M.;Yang, W. M.;Krones, A.;Glass, C. K.;Rosenfeld, M. G. & Seto, E. (2000) The histone deacetylase-3 complex contains nuclear receptor corepressors. *Proc Natl Acad Sci U S A* Vol.97, No.13, pp.7202-7207

Wermann, H.;Stoop, H.;Gillis, A. J.;Honecker, F.;van Gurp, R. J.;Ammerpohl, O.;Richter, J.;Oosterhuis, J. W.;Bokemeyer, C. & Looijenga, L. H. (2010) Global DNA methylation in fetal human germ cells and germ cell tumours: association with differentiation and cisplatin resistance. *J Pathol* Vol.221, No.4, pp.433-442

Whittaker, S. J.;Demierre, M.-F.;Kim, E. J.;Rook, A. H.;Lerner, A.;Duvic, M.;Scarisbrick, J.;Reddy, S.;Robak, T.;Becker, J. C.;Samtsov, A.;McCulloch, W. & Kim, Y. H. (2010) Final Results From a Multicenter, International, Pivotal Study of Romidepsin in Refractory Cutaneous T-Cell Lymphoma. *Journal of Clinical Oncology* Vol.28, No.29, pp.4485-4491

Wilson, A. J.;Byun, D. S.;Popova, N.;Murray, L. B.;L'Italien, K.;Sowa, Y.;Arango, D.;Velcich, A.;Augenlicht, L. H. & Mariadason, J. M. (2006) Histone deacetylase 3 (HDAC3) and other class I HDACs regulate colon cell maturation and p21 expression and are deregulated in human colon cancer. *J Biol Chem* Vol.281, No.19, pp.13548-13558

Woodage, T.;Basrai, M. A.;Baxevanis, A. D.;Hieter, P. & Collins, F. S. (1997) Characterization of the CHD family of proteins. *Proc Natl Acad Sci U S A* Vol.94, No.21, pp.11472-11477

Worthley, D. L.;Whitehall, V. L.;Buttenshaw, R. L.;Irahara, N.;Greco, S. A.;Ramsnes, I.;Mallitt, K. A.;Le Leu, R. K.;Winter, J.;Hu, Y.;Ogino, S.;Young, G. P. & Leggett, B. A. (2010) DNA methylation within the normal colorectal mucosa is associated with pathway-specific predisposition to cancer. *Oncogene* Vol.29, No.11, pp.1653-1662

Wu, H.;Coskun, V.;Tao, J.;Xie, W.;Ge, W.;Yoshikawa, K.;Li, E.;Zhang, Y. & Sun, Y. E. (2010) Dnmt3a-dependent nonpromoter DNA methylation facilitates transcription of neurogenic genes. *Science* Vol.329, No.5990, pp.444-448

Yang, H.;Hoshino, K.;Sanchez-Gonzalez, B.;Kantarjian, H. & Garcia-Manero, G. (2005) Antileukemia activity of the combination of 5-aza-2'-deoxycytidine with valproic acid. *Leuk Res* Vol.29, No.7, pp.739-748

Yang, H.;Lu, X.;Qian, J.;Xu, F.;Hu, Y.;Yu, Y.;Bast, R. C. & Li, J. (2010) Imprinted tumor suppressor gene ARHI induces apoptosis correlated with changes in DNA methylation in pancreatic cancer cells. *Mol Med Report* Vol.3, No.4, pp.581-587

Yang, X. J. & Gregoire, S. (2005) Class II histone deacetylases: from sequence to function, regulation, and clinical implication. *Mol Cell Biol* Vol.25, No.8, pp.2873-2884

Yang, X. J. & Seto, E. (2007) HATs and HDACs: from structure, function and regulation to novel strategies for therapy and prevention. *Oncogene* Vol.26, No.37, pp.5310-5318

Yang, X. J. & Seto, E. (2008) The Rpd3/Hda1 family of lysine deacetylases: from bacteria and yeast to mice and men. *Nat Rev Mol Cell Biol* Vol.9, No.3, pp.206-218

Yasui, W.;Oue, N.;Ono, S.;Mitani, Y.;Ito, R. & Nakayama, H. (2003) Histone acetylation and gastrointestinal carcinogenesis. *Ann N Y Acad Sci* Vol.983, pp.220-231

Yeung, F.;Hoberg, J. E.;Ramsey, C. S.;Keller, M. D.;Jones, D. R.;Frye, R. A. & Mayo, M. W. (2004) Modulation of NF-kappaB-dependent transcription and cell survival by the SIRT1 deacetylase. *Embo J* Vol.23, No.12, pp.2369-2380

Yin, D.;Ong, J. M.;Hu, J.;Desmond, J. C.;Kawamata, N.;Konda, B. M.;Black, K. L. & Koeffler, H. P. (2007) Suberoylanilide hydroxamic acid, a histone deacetylase inhibitor: effects on gene expression and growth of glioma cells in vitro and in vivo. *Clin Cancer Res* Vol.13, No.3, pp.1045-1052

Yoder, J. A.;Walsh, C. P. & Bestor, T. H. (1997) Cytosine methylation and the ecology of intragenomic parasites. *Trends Genet* Vol.13, No.8, pp.335-340

Yoo, C. B.;Cheng, J. C. & Jones, P. A. (2004) Zebularine: a new drug for epigenetic therapy. *Biochem Soc Trans* Vol.32, No.Pt 6, pp.910-912

Yoo, C. B.;Chuang, J. C.;Byun, H. M.;Egger, G.;Yang, A. S.;Dubeau, L.;Long, T.;Laird, P. W.;Marquez, V. E. & Jones, P. A. (2008) Long-term epigenetic therapy with oral zebularine has minimal side effects and prevents intestinal tumors in mice. *Cancer Prev Res (Phila)* Vol.1, No.4, pp.233-240

Yoshida, N.;Omoto, Y.;Inoue, A.;Eguchi, H.;Kobayashi, Y.;Kurosumi, M.;Saji, S.;Suemasu, K.;Okazaki, T.;Nakachi, K.;Fujita, T. & Hayashi, S. (2004) Prediction of prognosis of estrogen receptor-positive breast cancer with combination of selected estrogen-regulated genes. *Cancer Sci* Vol.95, No.6, pp.496-502

You, J. S.;Kelly, T. K.;De Carvalho, D. D.;Taberlay, P. C.;Liang, G. & Jones, P. A. (2011) OCT4 establishes and maintains nucleosome-depleted regions that provide additional layers of epigenetic regulation of its target genes. *Proc Natl Acad Sci U S A* Vol.108, No.35, pp.14497-14502

Yuki, Y.;Imoto, I.;Imaizumi, M.;Hibi, S.;Kaneko, Y.;Amagasa, T. & Inazawa, J. (2004) Identification of a novel fusion gene in a pre-B acute lymphoblastic leukemia with t(1;19)(q23;p13). *Cancer Sci* Vol.95, No.6, pp.503-507

Zhang, X.;Yuan, Z.;Zhang, Y.;Yong, S.;Salas-Burgos, A.;Koomen, J.;Olashaw, N.;Parsons, J. T.;Yang, X. J.;Dent, S. R.;Yao, T. P.;Lane, W. S. & Seto, E. (2007) HDAC6 modulates cell motility by altering the acetylation level of cortactin. *Mol Cell* Vol.27, No.2, pp.197-213

Zhang, Y.;LeRoy, G.;Seelig, H. P.;Lane, W. S. & Reinberg, D. (1998) The dermatomyositis-specific autoantigen Mi2 is a component of a complex containing histone deacetylase and nucleosome remodeling activities. *Cell* Vol.95, No.2, pp.279-289

Zhang, Z.;Yamashita, H.;Toyama, T.;Sugiura, H.;Omoto, Y.;Ando, Y.;Mita, K.;Hamaguchi, M.;Hayashi, S. & Iwase, H. (2004) HDAC6 expression is correlated with better survival in breast cancer. *Clin Cancer Res* Vol.10, No.20, pp.6962-6968

Zhang, Z.;Yamashita, H.;Toyama, T.;Sugiura, H.;Ando, Y.;Mita, K.;Hamaguchi, M.;Hara, Y.;Kobayashi, S. & Iwase, H. (2005) Quantitation of HDAC1 mRNA expression in invasive carcinoma of the breast*. *Breast Cancer Res Treat* Vol.94, No.1, pp.11-16

Zhou, L.;Cheng, X.;Connolly, B. A.;Dickman, M. J.;Hurd, P. J. & Hornby, D. P. (2002) Zebularine: a novel DNA methylation inhibitor that forms a covalent complex with DNA methyltransferases. *J Mol Biol* Vol.321, No.4, pp.591-599

Zhu, W. G.;Dai, Z.;Ding, H.;Srinivasan, K.;Hall, J.;Duan, W.;Villalona-Calero, M. A.;Plass, C. & Otterson, G. A. (2001) Increased expression of unmethylated CDKN2D by 5-aza-2'-deoxycytidine in human lung cancer cells. *Oncogene* Vol.20, No.53, pp.7787-7796

13

DNA Methylation in Acute Leukemia

Kristen H. Taylor and Michael X. Wang
University of Missouri-Columbia,
USA

1. Introduction

After birth, all blood cells are produced in the bone marrow by a process known as hematopoiesis. The basic biological process of hematopoiesis is the differentiation of hematopoietic stem cells (HSCs) to generate different types of mature blood cells (Kawamoto et al., 2010). In an adult, approximately 10^{12} blood cells are produced daily and released into the circulating peripheral blood. As new cells are released, some old cells undergo apoptosis in tissues or cleaned by spleen in order to maintain a homeostatic level of blood cells. Hematopoiesis is highly regulated through the interaction between hematopoietic cells and the bone marrow microenvironment. Many cytokines or growth factors and extracellular matrix molecules that are secreted either from stromal cells or from hematopoietic cells, as well as nutrients and vitamins, provide a favorable microenvironment for hematopoiesis to occur (Metcalf, 2008). Under the influence of specific growth factors such as c-KIT ligand and FLT-3 ligand, after rounds of asymmetric divisions of hematopoietic stem cells (HSCs), some daughter cells participate in lineage-commitment differentiation to become either lymphoid or myeloid progenitors. After several rounds of proliferation, these progenitor cells undergo terminal differentiation becoming mature lymphocytes (B-cells, T-cells and NK-cells) or myeloid cells (granulocytes, monocytes, red blood cells, mast cells and platelets). Generation of lymphocytes and myeloid cells in bone marrow during hematopoiesis is termed as lymphopoiesis and myelopoiesis, respectively. Disruption of these normal physiological processes at the stages of HSCs and/or progenitors may initiate leukemogenesis, a neoplastic transformation, and result in leukemia. Leukemia is classified as acute or chronic leukemia based on the clinical presentation and pathophysiological features. Acute leukemia is classified as acute lymphoblastic leukemia (ALL) and acute myeloid leukemia (AML) based on the cell lineages. Therefore, ALL may be B-cell, T-cell, or NK-cell in origin, while AML may be a granulocytic, monocytic, erythroid, or megakaryocytic subtype (Swerdlow et al., 2008; Vardiman et al., 2009).

2. Clinical aspects of acute leukemia

ALL and AML are the most common leukemias in children and adults, respectively (Siegel et al., 2011). Both diseases are characterized by acute onset and rapid accumulation of immature leukemic cells (blasts) in the bone marrow and blood (>20% of nucleated cells). Leukemic blasts are abnormal because they remain immature and do not function like mature white blood cells. ALL occurs mainly in children with peak prevalence between the ages of 2 and 5 years. Approximately 3,200 new ALL cases in childhood are diagnosed in the United States each year and two thirds of these cases are the B-cell subtype (Pui et al. 2004). Worldwide, ALL occurs approximately 5 cases per 100,000 populations per year. In contrast, AML occurs mainly in adults aged 65 years or older with a median age of 60 years (Löwenberg et al., 1999; Estey & Döhner, 2006). Approximately 13,400 new AML cases are diagnosed in the United States each year. Incidence of AML is about twice that of ALL worldwide. The prognosis is much poorer for AML than ALL, especially in elderly patients (Estey & Döhner, 2006).

Both ALL and AML are clonal disorders of hematopoietic stem cells (HSCs) or progenitor cells characterized by loss of normal maturation and gain of capacity of uncontrolled proliferation (Pui et al. 2004; Estey & Döhner, 2006; Becker et al. 2010). Leukemic blasts accumulate in blood and bone marrow and replace normal hematopoietic elements in the bone marrow. Furthermore, the blasts can infiltrate other organs and tissues including spleen, liver, lymph nodes and even skin. Most patients with acute leukemia present with the consequences of bone marrow failure presenting as anemia (decreased red blood cells), neutropenia (decreased neutrophils), and thrombocytopenia (decreased platelets). Pancytopenia (decreased all types of blood cells) with circulating blasts is the strong evidence for the diagnosis of acute leukemia. After diagnosis of a specific subtype of acute leukemia with advanced and integrated approaches including clinical information, laboratory data, morphology, immunophenotype (flow cytometry or immunohistochemistry), and genetic tests (cytogenetic karyotype and molecular genetic analysis), an effective chemotherapy regimen must be immediately initiated. The goal of treatment is to eliminate the leukemic blasts, preserve and restore normal hematopoiesis, and to prevent relapse (Bassan & Hoelzer, 2011; Burnett et al., 2011). To avoid the toxicity of chemotherapy, the patients are stratified based on the biological features of leukemia blasts and risk factors of the individual patient (host). In addition to clinical factors (age, gender, leukocyte count, etc.), specific genetic abnormalities, such as chromosomal translocations and/or gene mutations, are the most important factors in determining risk stratification in modern treatment of the leukemia patients (Pui et al., 2011). Although the treatment regimens are different in ALL and AML, a standard protocol typically consists of three phases: a remission-induction phase, an intensification (or consolidation) phase, and a continuation (maintenance) phase (Bassan & Hoelzer, 2011; Burnett et al., 2011). The goal of remission-induction treatment is to eradicate more than 99% of the initial leukemic cell burden and to restore normal hematopoiesis. The intensification treatment is generally aimed to eradicate drug-resistant residual leukemic cells including leukemia stem cells (Becker & Jordan, 2010). The last phase is maintenance chemotherapy for an additional 2.0-2.5 years to reduce the risk of relapse. The regimens usually include several drugs that have different pharmacological mechanisms of action to have maximal efficacy. In addition to routine chemotherapy, allogeneic hematopoietic stem-cell transplantation is the most intensive form of treatment for high risk acute leukemia (Gupta et al., 2010). Compared to various solid tumors, outcomes are excellent for treatment of acute leukemia, especially in young patients. In most large clinical trials, the cure rates are more than 80% and 60% in ALL and AML, respectively (Pui et al., 2011).

3. Genetic alterations

Like all other malignancies, acute leukemia is a genetic disease. Specific genetic alterations including chromosomal translocation, deletion, addition, and gene mutations (point mutation, copy number change) have been identified in all ALL and AML cases. These genetic alterations are widely utilized for diagnosis, risk stratification, prediction of response to chemotherapeutic reagents, prognosis and detection of minimal residual disease (Estey & Döhner, 2006; Swerdlow et al., 2008; Pui et al., 2008).

3.1 Genetic alterations in ALL

At the genetic level, ALL is a group of heterogeneous diseases. Standard cytogenetic analyses can detect primary chromosomal abnormalities in more than 70% of ALL cases (Mrózek et al., 2009). Using higher resolution and/or high throughput molecular methods, genetic alterations can be identified in virtually all cases of ALL tested (Mullighan et al., 2007). These alterations include gene rearrangements, gene copy number changes (deletions or duplications) and genomic sequence point mutations. Some of these changes are directly linked with leukemogenesis and affect important cellular pathways in cell differentiation, cell cycle regulation, tumor suppression, and apoptosis. For instance, PAX5, a B-cell specific transcription factor important for B-cell differentiation, was found frequently deleted or mutated in B-ALL (Mullighan et al., 2007). Other changes may be merely "passengers" and irrelevant to the biological properties and leukemogenesis of ALL.

The WHO classification of tumors of hematopoietic and lymphoid tissues (Swerdlow et al., 2008) designates ALL as B- or T-lymphoblastic leukemia/lymphoma based on cell lineages and percentage of the blasts in bone marrow. Each lineage group is further categorized as ALL with recurrent genetic abnormalities and not otherwise specified (NOS) based on identifiable chromosomal abnormalities by routine cytogenetic analysis (Figure 1). B-ALLs with recurrent genetic abnormalities include B-ALL with t(9;22)(q34;q11.2); *BCR-ABL1*, B-ALL with t(v;11q23); *MLL* rearranged, B-ALL with t(12;21)(p13;q22) *TEL-AML1(ETV6-RUNX1)*, B-ALL with hyperdiploidy, B-ALL with hypodiploidy, B-ALL with t(5;14)(q31;q32) *IL3-IGH*, B-ALL with t(1;19)(q23;p13.3); *E2A-PBX1 (TCF3-PBX1)*. Each designation contains the chromosomes involved, chromosomal loci, and the genes as well as alternative gene names. This subgroup accounts for 60% to 80% of B-ALL cases with distinct biologic and pharmacologic features that are important in diagnosis and risk stratification. The remaining B-ALL cases with no identifiable chromosomal abnormalities are characterized on the basis of morphologic and immunophenotypic features Figure 1A. Fifty to seventy percent of T-ALL patients demonstrate abnormal karyotypes. The most common recurrent abnormalities are translocations that involve the alpha and delta T-cell receptor loci at 14q11.2, the beta locus at 7q35, or the gamma locus at 7p14-15, and many partner genes (Pui et al., 2004; Giroux et al., 2006; Swerdlow et al., 2008). The pathological significance of these abnormalities is not as clearly understood as those that are associated with B-ALL.

3.2 Genetic alterations in AML

More than 90% of AML cases have at least one known genomic alteration as demonstrated by current routine cytogenetic or molecular analysis (Löwenberg et al., 1999) Figure 1B.

Using high resolution and high throughput methods, virtually all AML cases are identified as having distinct genetic mutations (Estey & Döhner, 2006; Godley et al., 2011).

A.

B.

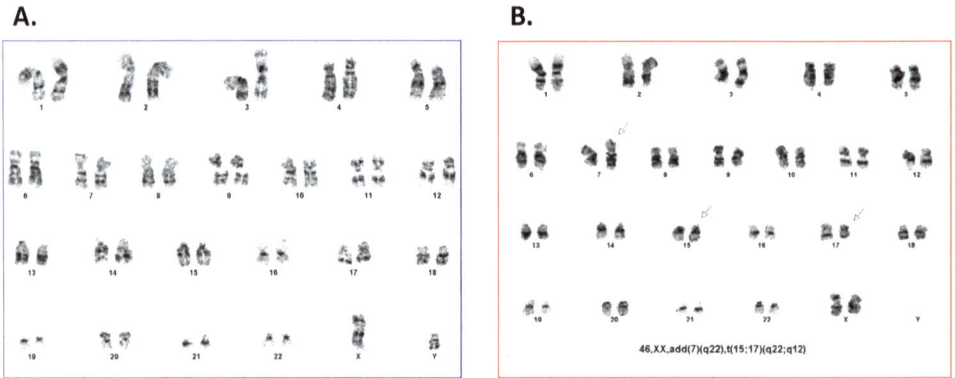

Fig. 1. Genetic alterations in ALL and AML. Genetic alterations can occur at either chromosomal level or DNA sequence level. Although some acute leukemia cases have a normal karyotype (A, an ALL patient), there are many mutations detectible at the DNA level. Most acute leukemia cases demonstrate specific chromosomal abnormalities that defined a distinct subtype of ALL or AML with important diagnostic and prognostic applications. The right panel (B) shows a case of acute promyelocytic leukemia (APL) with t(15;17)(q22;q12); *PML-RARA* (with addition at chromosome 7 in this case). APL represents a cytogenetically defined AML that can be treated with less toxic all-*trans* retinoid acid, a metabolite of vitamin A, with a favorable prognosis.

AML is a group of extremely heterogeneous diseases at the genetic level. Based on cytogenetic features and cell lineages, the WHO classification of tumors of hematopoietic and lymphoid tissues classifies AML into seven cytogenetic types with specific genetic abnormalities and nine not otherwise specified (NOS) types based on distinct morphology, cytochemistry and immunophenotype (Figure 1B). Furthermore, two provisional subtypes have been proposed based on point mutations in *NPM1* and *CEBPA* genes (Swerdlow et al., 2008, Vardiman et al., 2009; Döhner, 2010). AML with recurrent genetic abnormalities include AML with t(8;21)(q22;q22); *RUNX1-RUNX1T1*, AML with inv(16)(p13.1q22) or t(16;16)(p13.1;q22); *CBFB-MYH11*, APL with t(15;17)(q22;q12); *PML-RARA* (Figure 1B), AML with t(9;11)(p22;q23); *MLLT3-MLL*, AML with t(6;9)(p23;q34); *DEK-NUP214*, AML with inv(3)(q21q26.2) or t(3;3)(q21;q26.2); *RPN1-EVI1*, AML (megakaryoblastic) with t(1;22)(p13;q13); *RBM15-MKL1*. The chromosomes, loci and the genes are indicated in each specific designation. Many of these chromosomal translocations result in fusion proteins with altered functions of transcription factors critical for normal hematopoiesis and myeloid differentiation (Sternberg & Gilliland, 2004). These cytogenetic abnormalities also have important diagnostic and therapeutic implications. In contrast, the cases with mutations in nucleophosmin (*NPM1*), CCAAT/enhancer binding protein α (*CEBPA*), Fms-like tyrosine kinase (*FLIT3*) and *KIT* genes have been associated with prognosis. More than half of these cases with the gene mutations have normal karyotypes (Marcucci et al., 2005; Foran, 2010).

New approaches such as whole-genome sequencing, gene expression and microRNA profiling, proteomics and genome-wide DNA methylation profiling will not only identify all known mutations and epigenetic alterations, but can also potentially identify novel mutations and epigenetic lesions in biological pathways at the individual level (Boehm & Hahn, 2011). Deciphering the interactions between genetic and epigenetic alterations and expression profile will further provide a comprehensive and high resolution blueprint of leukemogenesis. That information, in turn, will be applied in clinical management of AML patients in the future.

Over the past few years, high-throughput next generation DNA sequencing (NGS) technologies have revolutionized the field of cancer genomics including leukemia (Metzker, 2010). Recently, Link and colleagues performed whole-genome sequencing on AML leukemic cells from a 37 year old woman with suspected cancer susceptibility syndrome (Link et al., 2011). The patient developed a therapy-related acute myeloid leukemia (t-AML) after chemotherapy for her breast cancer and ovarian cancer. Whole-genome sequencing revealed a novel, heterozygous 3-kilobase deletion removing exons 7-9 of *TP53* gene in germline DNA of patient normal skin cells. The deletion became homozygous in the leukemia DNA as a result of uniparental disomic recombination. Additionally, a total of 28 somatic single-nucleotide variations in coding regions, 8 somatic structural variants, and 12 somatic copy number alterations were identified in the patient's leukemia genome. Using a similar approach, Welch and colleagues performed whole-genome sequencing on a 39-year-old woman with a diagnosis of acute promyelocytic leukemia (APL) with unusual genetic lesion (Welch et al., 2011). The sequencing identified a novel cryptic insertion of *PML* gene into chromosome 17 that produced a classic pathogenic *PML/RARA* gene fusion figure 1B. This type of genetic event could not have been identified with routine cytogenetic and FISH techniques. The results led to a change in therapy using a less toxic targeted reagent, retinoic acid, rather than using the high risk procedure of allogeneic bone marrow transplantation.

These two cases represent excellent examples that whole-genome sequencing in leukemia not only detects novel genetic mutations in a cancer genome, but also directly benefits patient care. Although next generation sequencing has not been used for routine clinical diagnosis due to its high cost and long turnaround time (7 weeks in the latter case), it does provide insights for understanding the molecular mechanisms of leukemogenesis at the DNA sequence level in individual patients. Most likely, it will be used in clinical diagnosis with a "one-size-fits-all" feature in the near future.

4. Epigenetics

The completion of the human genome in 2003 held great promise for uncovering the cancer-causing genetic mutations that would allow for the development of targeted therapies and the eradication of cancer. Unfortunately, this has not come to fruition due, in large part, to the role of epigenetic modifications in the development of cancer. An individual's epigenetic makeup is much more complex than their genetic makeup. Human DNA sequence in all somatic tissue cells (except lymphocytes) is identical in a given individual. However, epigenetics presents in a developmental and tissue-specific manor. Epigenetic modifications are largely responsible for the differences in all somatic tissue cells such as brain, liver, skin cells, hair cells and the cells that make up the human eye. As a matter of fact, epigenetic modifications are responsible for regulating the expression of genes that establish each of the tissue types in the body.

Epigenetics is defined as the change of gene expression caused other than DNA sequence change (Bird, 1980; Robertson & Jones, 2000). There are 3 key epigenetic players which work in concert to control/alter gene expression. These include DNA methylation, histone modifications and non-coding RNAs inference (Figure 2). DNA cytosine methylation occurs at the 5th position of cytosine in CpG dinucleotide. Methyl binding proteins are recruited to methylated DNA regions and can block transcription factors in the promoter of a gene. The inaccessibility of transcription factors to the DNA results in a silenced gene. There are also numerous histone modifications such as acetylation, methylation, phosphorylation, ubiquitylation and sumoylation which can be present alone or in combination and each has an impact on chromatin structure. For example, some histone modifications (histone code) are associated with open chromatin (euchromatin) and some are associated with condensed chromatin (heterochromatin). If the chromatin is open, genes are accessible and can be transcribed. Alternately if the chromatin is closed, the genes are inaccessible to transcription

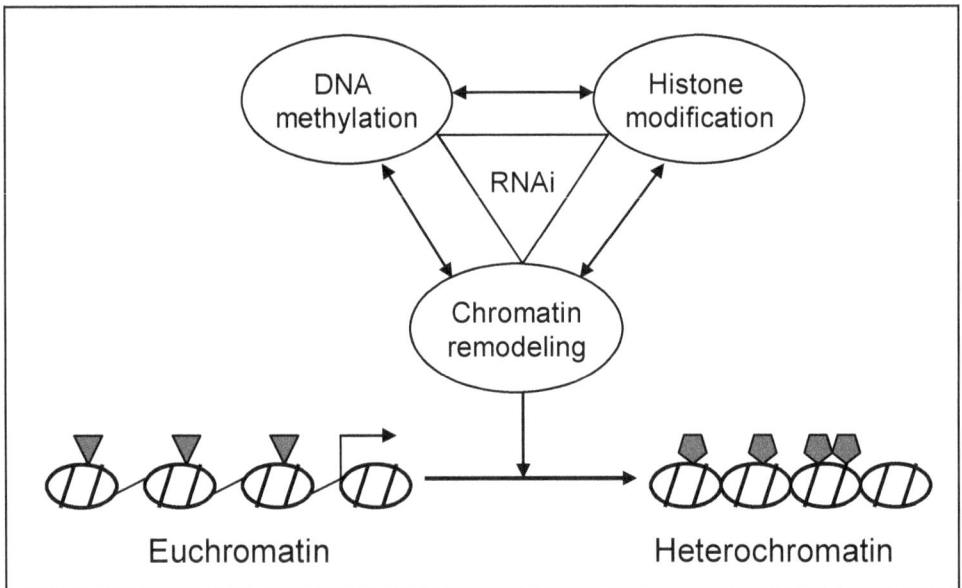

Fig. 2. Epigenetic network. Cell epigenetic network consists of DNA methylation, histone methylations (pentagon), histone acetylation (triangle), chromatin (nucleosome) remodeling, and RNA interference (RNAi) induced by microRNAs and short interfering RNAs (siRNAs). Interaction between these epigenetic components results in transition of euchromatin to heterochromatin. The transcription is inactivated (silenced) permanently when the focal open chromatin becomes a closed heterochromatin configuration (Adapted from Wang MX and Shi HD: Basics of Molecular Biology, Genetic Engineering and Metabolic Engineering. In: Fu, PC, Latterich, M, and Panke, S (Eds): Systems Biology and Synthetic Biology, John Wiley & Sons, Inc., pp.36, 2009, with permission)

factors and therefore cannot be transcribed. Finally, non-coding RNAs play a role in gene expression by binding and degrading messenger RNA and inhibiting protein assembly (Chen J et al, 2010, Melo and Esteller, 2011). While each of these modifications plays an important role in gene expression, DNA methylation has been the most widely studied epigenetic modification that is associated with the development of acute leukemia and other types of cancer.

DNA methylation occurs globally in the normal genome and is estimated to affect between 70 and 80% of all CpG dinucleotides in human cells (Bird, 1980; Robertson & Jones, 2000). These dinucleotides are not uniformly distributed across the genome but occur in clusters such as large repetitive sequences or in CG-rich DNA stretches known as CpG islands (CGIs). Normally, the majority of the CpG dinucleotides which are found in intragenic regions, including repetitive sequences such as satellite sequences and centromeric repeats, contains methylated CpG dinucleotides; while CGIs which are found preferentially in the promoter regions of genes typically contain unmethylated CpG dinucleotide (Craig & Bickmore, 1994). Some exceptions to this rule include those CGIs located on the inactive X chromosome in females (Goto & Monk,1998) and those associated with imprinted genes (genes for which only the paternally- or maternally-inherited allele is expressed) which are methylated in the normal state (Li et al., 1993; Razin & Cedar, 1994).

In leukemia, the normal pattern of methylation gets reversed. Those regions of the genome that are normally methylated (and inactivated) such as repetitive sequences lose methylation and result in genomic instability. Those regions of the genome which are typically unmethylated such as the promoters of tumor suppressor genes become methylated and gene expression is lost. A typical gene comprises a promoter region, transcriptional start site (TSS), 5′ untranslated region (UTR), exons, introns and a 3′UTR. CGIs may encompass each of these genetic regions. Typically methylation present within the gene body (exons, introns) is associated with gene expression and methylation present in gene promoters, 5′UTRs and the first exon is associated with gene silencing. The majority of genome-wide methylation studies to date have focused on CGI at the promoter regions of genes via microarray methylation analysis. Recently, it has been suggested that not only are the CGI themselves important but that the flanking genomic regions termed CpG shores are more highly correlated with gene silencing (Irizarry et al., 2009).

5. Emerging epigenetic technologies

DNA methylation plays an important role in acute leukemia. This is evidenced by the growing number of studies that have recently been published describing methylation profiles for many types and subtypes of leukemia. A number of strategies have been used to examine DNA methylation and over time these methods have progressed from small-scale candidate gene analysis to the ability to construct whole-genome methylation profiles (Laird, 2010). These can broadly be divided into three classes: those that require bisulfite conversion, those that are affinity based and those that require the use of restriction enzymes. Restriction enzyme-based methods such as DMH (Huang et al., 1999) and MCA (Huang et al., 1999; Toyota & Issa, 2002) rely on methylation-sensitive enzymes and are limited by the fact that restriction sites are not present in all possible CpG rich regions of interest. In affinity based approaches such as methylated DNA immunoprecipitation (MeDIP) (Weber et al., 2005) and methylated CpG island recovery assay (MIRA) (Rauch &

Pfeifer, 2005), fragmented DNA is immunoprecipitated using a monoclonal antibody to 5-methylcytosine or by methylated DNA binding proteins respectively. Methylated DNA is enriched in affinity based methods, however, these methods are limited by the sensitivity of the antibody or protein being used and are typically biased towards regions of the genome with a high density of methylation. Sodium bisulfite converts unmethylated cytosines to uracils in DNA strands. Once the conversion has taken place the sequence can be determined after PCR because the cytosines that were originally unmethylated will be converted to thymines and those that are methylated will remain cytosines. The net effect of the conversion process is that the complexity of the genome is reduced from 4 bases to only 3 bases. Bias may be introduced due to incomplete bisulfite conversion, destruction of DNA strands and also by the efficiency of bisulfite PCR. After the samples are prepared, they can be hybridized to microarrays or used to make sequencing libraries for next generation sequencing. The recent development of high throughput technologies holds the promise of providing biological insight and new avenues for translational research and clinical applications.

5.1 Arrays

The extent of genome coverage is primarily determined by the resolution of the utilized microarray platform which has progressed rapidly from arrays that examine the DNA methylation present in select CGI to those that examine all CGIs and/or promoters and even to those that examine the entire genome (i.e. Agilent and Nimblegen). Microarray analysis is dependent upon the sample preparation used. Those that use restriction enzymes typically involve a co-hybridization of a test (i.e. leukemia) sample and a reference sample (i.e. bone-marrow cells from a healthy donor). Differential methylation is determined by comparing the intensity of the test sample to the intensity of the reference sample at each locus represented on the array. If the DNA is prepared using an affinity method, the reference sample is usually an aliquot of the original sample DNA while the test sample is the immunoprecipitated portion derived from the original sample. Methylated sequences are detected by comparing the fluorescence signal for each probe corresponding to known genomic sequences for reference and the test samples. The loci that are enriched in the test sample are potential methylation candidates. There are also arrays that require DNA to be bisulfite converted (i.e. Illumina). These arrays are also known as SNP chips and methylated cytosines are identified by the presence of a cytosine at a particular locus as opposed to a thymine. These later arrays give site-specific methylation profiles whereas the oligonucleotide microarrays give region specific methylation profiles.

5.2 Next Generation Sequencing (NGS)

Next generation sequencing has revolutionized the ability to perform genomic research since it was introduced in 2007 (Metzker, 2010 for a review). More recently, this technology has been adapted to epigenomic research including DNA methylation analyses. NGS technologies can generate millions of sequencing reads in parallel and has led to a dramatic increase in the number of genomic and epigenomic sequences encompassing normal and diseased tissues. There are multiple sequencing platforms available and the choice of platform and methodology is dependent on the scientific application and the capacity for extensive bioinformatics analysis.

Whole genome bisulfite sequencing (WGBS) provides coverage at a single base pair resolution and is the most comprehensive NGS technology in DNA methylation analysis. WGBS provides unbiased coverage of the genome allowing for interrogation of whole regions of the genome that are often missed by other methodologies. This method requires the most extensive bioinformatics and is the most expensive because more sequences are required to cover the entire genome. Therefore, this option may not be suitable for laboratories without bioinformatics support or with a small budget.

Reduced representation bisulfite sequencing (RRBS) is a less expensive alternative to WGBS. As the name of the method implies, the genome is reduced in size using an enzyme which enriches for regions of the genome that contain CpG dinucleotides. To be more specific, coverage includes approximately 12% of all CpG dinucleotides and 84% of all CpG islands in the human genome (Smith et al., 2009). In total, about 1% of the genome is covered which greatly reduces the cost of sequencing and the resources needed for alignments while providing data for those regions of the genome that are enriched for CpG dinucleotides. RRBS is an excellent alternative to WGBS with the caveat that important genomic regions that are not enriched for CpG dinucleotides are not assayed. As an example of the consequence of this limitation, it was recently published that much of the methylation present in colon cancer occurred in CGI shores and that this methylation was highly correlated with gene expression (Irizarry et al., 2009). RRBS covers less than 50% of the CGI shores so in this particular region of the genome, important data may be lost (Gu et al., 2011). RRBS may become an optional platform for targeted DNA methylation analysis for clinical diagnostic purposes.

Affinity sequencing utilizes either an antibody against 5-methylcytosine or proteins with methyl binding domains to immunoprecipate (or enrich for) methylated DNA. This method does not require bisulfite conversion and produces sequences for regions of the genome that are methylated. However, it does not provide data for the methylation status of individual CpG dinucleotides. Bioinformatics analysis requires genome alignment and must take into account the density of CpG dinucleotides in a given region because the efficiency of anti-5' methylcytosine and methyl binding proteins is dependent on the number of methylated cytosines in a given region (Weber et al., 2005).

6. DNA methylation in acute leukemia

DNA methylation is an important epigenetic mechanism to control gene expression. In many cases, tumor suppressor genes (TSGs) are inactivated by somatic mutations (point mutations or deletions) as the "first hit", DNA methylation silences the gene expression of other allele as the "second hit", or in a reversed sequence. In this regards, aberrant DNA methylation plays a crucial role in leukemogenesis (Herman & Baylin, 2003; Galm et al., 2006). Using various technologies described above, many aberrant DNA methylation loci have been identified in both ALL and AML

6.1 DNA methylation in ALL

Genomic DNA methylation profile in a given cell is defined as methylome. There are numerous recurrent chromosomal and genetic abnormalities in ALL. However, these abnormalities are neither sufficient nor necessary in the development of ALL. Therefore it is likely that epigenetic modifications also contribute to the leukemogenesis in ALL. Numerous studies have been published that have focused on a single candidate gene or groups of genes involved in important cellular processes such as cell signaling pathways, apoptosis, regulation of transcription and cell cycle control in pediatric ALL (Agirre et al., 2006; Canalli et al., 2005; Cheng et al., 2006; Corn et al., 1999; Garcia-Manero et al., 2003; Gutierrez et al., 2003; Iravani et al., 1997; Paixao et al., 2006; Roman-Gomez et al., 2002; Roman-Gomez et al., 2004; Roman-Gomez et al., 2006, 2007; Sahu & Das, 2005; Scholz et al., 2005; Stam et al., 2006; Tsellou et al., 2005; Yang et al., 2006; Zheng et al., 2004), and adult ALL (Batova et al., 1997; Chim et al., 2001; Garcia-Manero et al., 2002a; Garcia-Manero et al., 2002b; Hutter et al., 2011; Jimenez-Velasco et al., 2005; Martin et al., 2008; Roman et al., 2001; Roman-Gomez et al., 2004; Scott et al., 2004; Shteper et al., 2001; Taniguchi et al., 2008; Yang et al., 2006). A search of PubMeth (www.pubmeth.org) provides a list of 46 methylated genes that have been associated with acute lymphoblastic leukemia including *ABCB1, ABL1, ADAMTS1, ADAMTS5, AHR, APAF1, BNIP3, CDH1, CDH13, CDKN1A, CDKN1C, CDKN2A, CDKN2B, CHFR, DAPK1, DIABLO, DKK3, ESR1, EXT1, FHIT, HCK, KLK10, LATS1, LATS2, LMNA, MGMT, MME, MYOD1, NNAT, NR0B2, PARK2, PGR, PPP1R13B, PTEN, PYCARD, RARB, SFRP1, SFRP2, SFRP4, SFRP5, SYK, THBS1, THBS2, TP73, WIF1* and *WRN*.

Recent studies have increased our knowledge of aberrantly methylated loci in ALL by utilizing genome-wide technologies to construct genome-scale methylomes. These studies have shown that methylation profiles can be used in diagnosis (Davidsson et al., 2009; Dunwell et al., 2010; Milani et al., 2010; Stumpel et al., 2009; Taylor et al., 2007a; Taylor et al., 2007b; Vilas-Zornoza et al., 2011), prognosis (Davidsson et al., 2009; Hogan et al., 2011; Kuang et al., 2008; Milani et al., 2010), and in the treatment (Hogan et al., 2011; Vilas-Zornoza et al., 2011) of individuals with ALL.

As an example of the breadth and depth provided by using genome-scale technologies, Davidsson and colleagues used bacterial artificial chromosome arrays and genome-wide methylation arrays in two of the most common subtypes of pediatric ALL, t(12;21)(p13;q22) and high hyperdiploidy (Davidsson et al., 2009). The methylation microarray used covers all of the UCSC-annotated CGI and promoter regions of all RefSeq genes and contains a total of 385,000 probes. A total of 8,662 genes were identified with significant methylation present within the promoter and the 10 individuals with hyperdiploid ALL had approximately twice as many hypermethylated genes as the 10 individuals with the t(12;21) abnormality. Of particular importance is that none of the 30 genes with the highest methylation peaks in the ALL patients have previously been shown to be methylated in ALL or any other neoplasia. An additional study by Hogan and colleagues (Hogan et al., 2011) used the Infinium Human Methylation27 BeadChip to create methylation profiles for paired diagnostic/relapse samples from 33 pediatric ALL patients. This study identified over 900 genes that were preferentially methylated in relapse samples when compared to samples at diagnosis. Further combinatorial expression and copy number variation analysis identified important biological pathways such as the WNT/beta-catenin pathway and the MAPK pathway which may be implicated in the relapse of pediatric ALL.

6.2 DNA methylation in AML

Like in the case of ALL, numerous studies have been published using single-gene based methods such as methylation-specific PCR (MSP), combined bisulfite restriction analysis

(COBRA) or array to identify a single or a group of genes with abnormal DNA methylations in AML. The examples include *CDH1, ESR1, IGSF4, FHIT, p15INK4B, p21CIP1/WAF1, MEG3, SNRPN, p73, SOCS1, CALC1, HIC-1, CTNNA1, CEBPA, MLH1, MGMT, CNAML, HOXA1, MYOD, KRT13, NR2F2, PITX2, RBP1, CEBPA, BAHCC1, EVI1,* and *DAPK* genes (Ekmekci et al., 2004; Agrawal et al., 2007; Desmond et al., 2007; Glasow et al., 2008; Rosu-Myles & Wolff, 2008; Melnick, 2010; Oki & Issa, 2010; Lugthart et al., 2011; Lin et al., 2011). Often hypermethylation in these gene CpG-island-promoters results in transcriptional silencing and loss of the function in important biological pathways (Calvanese et al, 2011).

High throughput technologies such as mass spectrometry, microarray and next generation sequencing (NGS) have been used to study altered DNA methylation at the genome-wide scale. Bullinger and colleagues used a combination of base-specific cleavage biochemistry and mass spectrometry (MALDI-TOF-MS) to quantify DNA methylation in 92 selected genomic regions of 256 AML patient samples (Bullinger et al., 2010). Distinct DNA methylation patterns were identified in abnormal cytogenetic subgroups and the DNA methylation levels (CpG units) could provide independent prognostic information. Alvarez and colleagues used a bead array-based methylation assay to examine the methylation status of 1,505 CpG-sites from 807 genes on 116 de novo AML patients (Alvarez et al., 2010). They confirmed that the DNA methylation signatures were associated with the specific cytogenetic status. In addition, aberrant DNA methylation of the promoter of *DBC1* could predict the disease-free and overall survival time in normal karyotype cases. Interestingly, the aberrant DNA methylation pattern could be induced by genetic transduction of *MLL* rearrangement fusion genes in normal human hematopoietic stem/progenitor cells (HSPC), but cannot be induced by *AML1/ETO* or *CBFB/MYH11* fusion gene. This is the direct evidence of interaction between genetic and epigenetic alteration in AML. Using the same platform, Wilop and colleagues found a gain of overall methylation in 32 AML patient samples (Wilop et al., 2011). The methylation pattern was maintained at relapse with increased density and extended to addition genes, consistent with the previous studies (Agrawal et al., 2007; Kroeger et al., 2008). These observations provided a strong scientific basis for DNA methylation to be used as a biomarker for diagnosis, minimal residual disease detection and clinical follow-up in AML patients.

A more comprehensive study was conducted by Figueroa and colleagues using HpaII tiny fragment enrichment by ligation-mediated PCR (HELP), linked with a microarray platform measuring the methylation abundance of 50,000 cytosines distributed among 14,000 gene promoters in 344 AML patient samples (Figueroa et al., 2010). Based on the DNA methylation signatures, these patients could be classified into 16 epigenetically unique subtypes. Although the DNA methylation patters were different among subtypes, none of the AML subtypes were similar to any of the stages of normal myeloid maturation indicating a distinct difference between leukemia and normal myeloid cell methylomes. Furthermore, they found a set of 45 genes to be aberrantly methylated common in all AML cases, but not in normal myeloid cells. Patients with a *CEPBA* signature have markedly poor clinical outcomes. Functionally, the hypermethylated genes were down-regulated and associated with biologically relevant pathways in leukemogenesis. These genes include zinc finger transcription factors, components in retinoic acid, STAT, p53 signal pathways, DNA-damage repair, immune response and tumor suppressors (Sternberg et al, 2004). It is anticipated that a complete AML methylome with single base pair resolution by next generation sequencing such as whole genome bisulfite sequencing (WGBS) will be published soon.

7. Interaction between genome and epigenome in leukemia

Data from cytogenetic karyotyping, conventional sequencing, microarray and whole genome sequencing indicates that there are extensive and distinct genetic and epigenetic alterations in acute leukemia genomes. Recent studies showed that genetic and epigenetic alterations are not independent events. Specific DNA methylation patterns are identified in specific cytogenetic subgroups of ALL (Davidsson et al., 2009) and AML (Figueroa et al., 2010). Chromosomal translocations result in fusion oncoproteins that can recruit components for DNA methylation, histone deacetylation and transcriptional repressor complexes (Croce LD, 2005; Chen et al., 2010). In contrast, DNA hypomethylation may lead to abnormal microRNA expression and chromosomal instability that in turn may result in chromosomal translocations (Eden et al. 2003; Calvanese et al, 2010; Popp & Bohlander et al., 2010; Toyota et al. 2010; Melo and Esteller, 2011).

AML with t(8;21)(q22;q22) translocation results in the fusion genes of *RUNX1-RUNX1T* (or *AML1- ETO*). The oncoprotein represses the transcription of wild-type *AML1* target genes by recruiting co-repressor complexes (Ferrara & Del Vecchio, 2002). AML1 (also known as RUNX1 or CBFA) is a transcription factor containing the DNA binding domain of the α-subunit of core binding factor (CBF). Another subtype of AML with inv(16)(p13.1q22) or t(16;16)(p13.1;q22) results in *CBFB-MYH11* fusion gene that contains a β subunit of CBF. Together, α-subunit and β subunit forms a heterodimeric core binding transcription factor and plays an important role in normal hematopoiesis and myeloid differentiation (Link et al., 2010). However, fusion forms of these truncated subunits lost the ability to form core binding factor and no longer induce myeloid differentiation (Paschka, 2008). Even more, the fusion proteins actively repress the transcription of normal AML1 target genes by either recruiting histone deacetylase (HDAC) or DNA methyltransferase 1 (DNMT1) or by cooperating nuclear receptor co-repressor 1 (NCOR1), NCOR2 and sIN3A15 to form a repressor complex (Liu et al., 2005). As a result, normal myeloid differentiation mediated by CBF is disrupted and hematopoietic stem cells (HSCs) and myeloid progenitor cells cannot achieve the next mature stages, and result in accumulation of leukemia blasts in bone marrow and blood. Since AML with t(8;21) and inv(16) or t(16;16) are involved in a common pathway at the molecular level and show specific clinical features, these two genetic subtypes are called   core binding factor leukemia (Ferrara & Del Vecchio, 2002). This leukemia responds well to high doses of cytorabine (HiDAC) and has a better prognosis (Dombret et al., 2009; Solis, 2011).

Acute promyelocytic leukemia (APL) is another example in which blockage of myeloid differentiation by the fusion oncoproteins is mediated by an interaction between genetic and epigenetic mechanisms. All patients with APL have the t(15;17) translocation or one of its variants t(11;17), t(5;17) (Warrell et al., 1993). Translocation t(15;17) results in a fusion protein PML-RARA typically comprised of variable portions of PML protein and all but the first 30 amino acids of retinoic acid receptor-α (RARA). Wild-type RARA protein is a transcriptional activator crucial for normal hematopoiesis and myeloid differentiation. Many RARA target genes including specific transcription factors such as *PU.1 (SPI1)* and *C/EBPβ (CEBPB)* have been identified to have RARA biding sites. The fusion protein PML-RARA functions as a transcriptional repressor, but not an activator, by binding to promoter region of RARA target genes and recruiting proteins including HDAC, NCOR1 and NCOR2

(N-CoR) complex, DNMT1, DNMT3A, repressive histone methyltransferases and polycomb group proteins (Licht et al., 2006). DNA methyltransferases (DNMT1 and DNMT3A) induce DNA methylation. Histone deacetylase (HDAC) removes acetyl group from histones. Together with other repressive proteins, the focal chromatin structure at the promoter regions of target genes are converted to a closed configuration and the transcription initiation is abolished. Promyelocytes of APL lack key transcription factors for further maturation. Accumulation of abnormal promyelocytes in bone marrow is a diagnostic feature of APL. A high dose of all-trans-retinoic acid (ATRA) relieves this repression by allowing the release of the N-CoR complex and the recruitment of a co-activator complex and it has become the cornerstone in treatment of APL by molecular targeting (Wang & Chen, 2008).

8. Molecular mechanisms of leukemogenesis

Among more than 100 types of cancer with different tissue origins, acute leukemia is a unique form that is originated from hematopoietic stem cells (HSC) or hematopoietic progenitor cells in bone marrow. In order to acquire a malignant phenotype, leukemic cells must have all the malignant biological properties including self-sufficiency in growth signals, insensitivity to growth-inhibitory signals, altered cellular metabolism, evasion of programmed cell death (apoptosis) and immunological destruction, limitless replicative proliferation and tissue invasion and metastasis (Hanahan & Weinberg, 2000; 2011). At the molecular level, the phenotype of leukemic cells represents a global change of gene expression due to irreversible genetic and epigenetic alterations. These changes affect biological pathways of cell differentiation, cell cycle regulation, tumor suppression, drug responsiveness, and apoptosis. Identification of the molecular signature of leukemia as well as the genetic background of the host individual will provide a unique biological road map for each patient that will become the foundation for personalized therapy in the future (Godley et al., 2011).

The etiology of acute leukemia is not completely clear. Some environmental risk factors for ALL including parental occupation, parental tobacco or alcohol use, prenatal vitamin use, diet, exposure to pesticides or solvents, infectious pathogens and exposure to ionizing radiation or the highest levels of residential power-frequency magnetic fields have been reported (Belson et al., 2007; Milne et al, 2010; Bailey et al., 2011). Environmental risk factors for AML include exposure to ionizing radiation and benzene (Bowen, 2006; Smith et al., 2011). The cytotoxic chemotherapy (alkylating agents and topoisomerase-II inhibitors) and/or radiotherapy for other solid tumors and pre-leukemic conditions myelodysplastic syndromes in the elderly are proven risk factors for AML (Löwenberg et al., 1999; Garcia-Manero et al., 2011).

ALL occurs exclusively in childhood although adult ALL exists. Screening of neonatal cord-blood samples has revealed several specific leukemic chromosomal translocations. One particular clone with the *TEL-AML1* fusion gene derived from chromosomal translocation t(12;21)(p13;q22) is found in 1% of newborn babies. The prevalence of B-ALL with this fusion gene is 100 times higher than those who do not have the fusion gene (Cobaleda et al., 2009; Lausten-Thomsen et al., 2011). Similarly, some leukemic translocations such as t(8;21)(q22;q22) resulting in *AML1-ETO* fusion gene can be detected in neonatal blood samples from the teenagers diagnosed with AML (Mori et al., 2002). In addition, ALL and AML occurs in approximately 10% of identical twins with these or other karyotypes (Mori et al., 2002; Greaves et al., 2003). These observations support the hypothesis that these specific genetic alterations at the fetal stage increases the frequency of ALL and AML, but additional postnatal events, either genetic or epigenetic, are required for full leukemic transformation (Greaves & Wiemels, 2003; McHale et al., 2004; Wiemels et al., 2009).

Recent studies suggest that the original leukemic clone is most likely raised from hematopoietic stem cells (HSC) or lineage committed precursor cells (Clarke et al. 1987; Lapidot et al., 1994; Cox et al., 2004, 2007; Jamieson et al., 2004). Under the influence of genetic and the environmental risk factors described above, normal HSC or precursor cells undergo malignant transformation and become leukemia stem cells (LSCs) (Passegué et al, 2003). LSCs have the distinct properties with partial normal HSC and partial leukemia cell features. These cells are characterized by self-renewal, over proliferation and the capacity to develop an entire leukemic blast population (Huntly & Gilliland, 2005; Becker & Jordan, 2010). Identification of LSCs by specific biomarkers and development of specific agents to target LSCs has significant clinical implication since eradication of LSCs will prevent the relapse and cure the leukemia (Jan et al., 2011).

At the molecular level, based on the facts that chromosomal translocations and point mutations can be found in the majority of AML patients, Kelly and colleagues suggested a two-hit model that AML leukemogenesis driven by two types of gene mutations (Kelly et al., 2002). The class 1 mutations result in constitutive activation of cell-surface receptors, such as receptor tyrosine kinases, FLT3 and KIT. Through various downstream signaling pathways, constitutive activation confers proliferation and survival advantage leading to clonal expansion of the affected hematopoietic stem cell or progenitors. The class 2 mutations, exemplified by formation of fusion genes from the t(8;21) or inv(16) chromosomal translocations or overexpression of HOX genes, block myeloid differentiation. Either class 1 or class 2 lesions alone does not cause leukemia in mouse models (Downing, 2003). AML develops only when both classes of lesions are present.

This model, however, provides a less cogent explanation for AML derived from myelodysplastic syndrome and therapy-related AML (t-AML) in elderly. These AML are frequently associated with chromosomal deletion or addition (Godley & Larson, 2008). Furthermore, this model also does not fully explain the AML containing normal karyotype with multiple point mutations in *FLIT3*, *NPM1*, and *CEBPA* genes (Foran, 2010). The class 1 mutations in ALL have not fully established. Epigenetic factors, especially DNA hypermethylation that can inactivate various putative tumor suppressor genes, DNA-repair, cell cycle, apoptosis related genes appear to play important roles in leukemogenesis (Issa et al., 1997; Esteller, 2008; Kulis & Esteller, 2010; Deaton & Bird, 2011). An integrated model combining genetic and epigenetic factors at the individual, cellular and molecular levels for acute leukemia is proposed (Figure 3).

9. Clinical applications

Genetic and epigenetic studies from basic science have been applied to many aspects in the clinical management of acute leukemia patients. The current WHO classification of tumors of hematopoietic and lymphoid tissues has included an increasing number of clinicopathologic entities defined by chromosomal abnormalities as well as gene mutations.

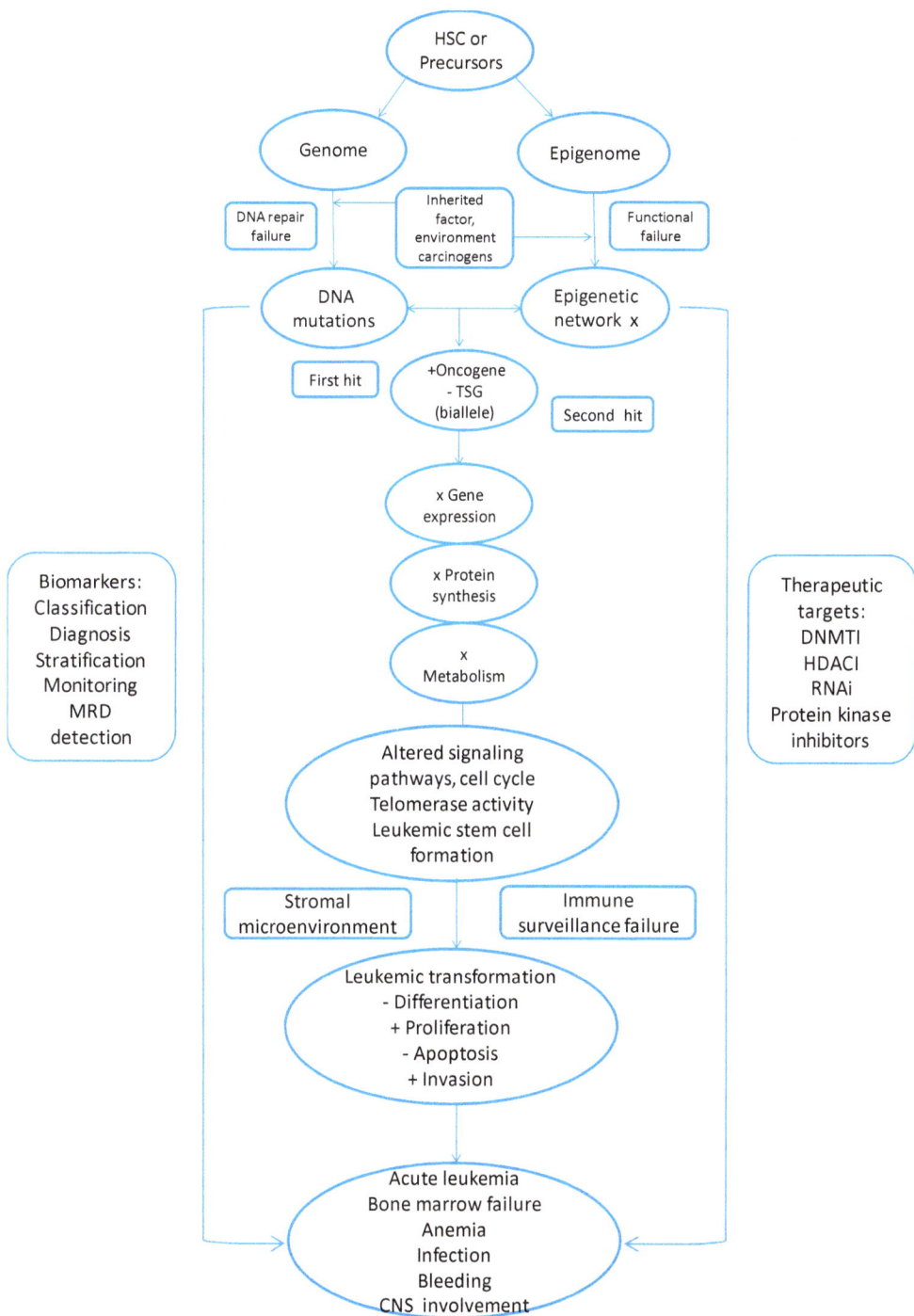

Fig. 3. A new model of leukemogenesis integrated genetic and epigenetic mechanisms and their clinical implications. Although the inherited factors in leukemogenesis of acute leukemia is not apparent, the genetic alterations including chromosomal translocations and numerical changes such as trisomy 21 have been found at prenatal stage. The changes may be related to maternal factors such as carcinogens exposure, nutrients (including folate) and aging in pregnancy. The incidence of acute leukemia is dramatically increased (~100 times higher), but not all children will have the leukemia when carrying the specific chromosomal abnormalities at the prenatal stage. It indicates the second hit, either genetic mutations or epigenetic alterations, is required for a full leukemic transformation. With an interaction between genetic and epigenetic networks, the gene expression profile is globally changed in hematopoietic stem cell s or precursors. Corresponding functional changes including cell signalings and cell cycle control result in a malignant leukemia phenotype. These leukemia cells escape from immune surveillance and accumulate in bone marrow and blood, thus acute leukemia is developed. Clinically, genetic abnormalities have been used as biomarker for disease classification and diagnosis, while aberrant epigenetic alterations have become therapeutic targets. Note: HSC: hematopoietic stem cell; TSG: tumor suppressor gene; DNMTI; HDACI; RNAi; Epigenetic network: DNA methylation, histone modifications and microRNA. siRNAs. +: increase; -: decrease; x: disruption.

These subtypes of AML or ALL often have a distinct morphology, immunophenotype and clinical course. Some of these patients with specific genetic or epigenetic alterations may respond to specific chemotherapeutic reagents or epigenetic modifiers. Mutation status of *NPM1*, *CEBPA* and *FLT3* genes has been used in risk assessment, prognostic evaluation and guidance of therapy (Foran, 2010). Detection of specific fusion RNA levels using quantitative RT-PCR molecular tests in patient blood has been used routinely for therapeutic monitoring and minimal residual disease detection (Gulley et al., 2010).

Because of the genetic heterogeneity and the limited number of meaningful genetic biomarkers identified in acute leukemia, the use of aberrant epigenetic alterations, especially DNA methylation and microRNA as biomarkers, is being studied at the single gene as well as genome-wide level. Agrawal and colleagues reported that the methylation of *ERa* and *p15INK4B* genes occurred frequently and specifically in acute leukemia but not in healthy controls or in nonmalignant hematologic diseases (Agrawal et al., 2007). Aberrant DNA methylation of these two genes was detectable in >20% of leukemia patients during clinical remission. The presence of detectable methylation was correlated to minimal residual disease (MRD) and associated with subsequent relapse (Agrawal et al., 2007). Wang and colleagues demonstrated that the aberrant DNA methylation of *DLC-1*, *PCDHGA12* and *RPIB9* genes can be identified in over 80% of ALL patients (Wang et al., 2010). Using a single gene *DLC-1*, we could trace clinical B-ALL cases up to 10 years retrospectively and the *DLC-1* methylation is correlated with patient clinical status. Importantly, these specific DNA methylation loci are retained in leukemia cells and can be detected in relapse. Compared with primary leukemia at diagnosis, relapsed leukemia maintains the original methylation loci, yet extents methylation in addition genes (Kroeger et al., 2008; Figueroa et al., 2010). These studies indicated that the DNA methylation is a biologically stable marker that can be used for MRD detection and patient follow up in acute leukemia.

In terms of therapy, there are two groups of epigenetic agents currently in clinical use, DNA methyltransferase inhibitor (DNMTI) and histone deacetylase inhibitor (HDACI) (Peters & Schwaller, 2011). The prototypic nucleoside analogue DNMT inhibitors include 5-azacytidine (5-Aza or azacitidine) and 5-aza-2'deoxycytidine (decitabine). They exert a demethylating effect by incorporating into DNA (5-Aza is also incorporated into RNA) and form a covalent complex with the DNMT enzymes. The enzymes are trapped and eventually degraded and the newly synthesized DNA strand will not be methylated (Schoofs & Müller-Tidow, 2011). These two agents are active in a broad range of myeloid neoplasms including AML and myelodysplastic syndrome (MDS). Because of its excellent efficacy (~50% response rate) in clinical trials, both agents have been approved by the US FDA for the treatment of MDS (Silverman & Mufti, 2005). The use of these reagents in treatment of AML has been actively investigated and showed promising utility especially in elderly patients (Musolino, 2010). The second group of epigenetic therapeutic agents is histone deacetylase inhibitor (HDACI). This group consists of heterogenic compounds that may reactivate the genes that have been turned off by histone deacetylation. Particularly, HDACI has demonstrated some efficacy in treat of core binding factor (CBF) leukemia. Clinical trials have been conducted using HDACI alone or in combination with DNMTI in CBF and other subtypes of leukemia patients (Quintás-Cardama et al., 2011).

10. Conclusion

Acute leukemia (ALL and AML), like all other cancer types, is a genetic disease. DNA sequence examination in the specific loci as well as at the genome-wide level has confirmed this original hypothesis. Epigenetic alterations including DNA methylation, histone modifications and microRNA play a functional role in leukemogenesis. Interaction between genetic and epigenetic elements changes the global landscape of gene expression, protein synthesis and metabolism in hematopoietic stem cells and/or committed precursor cells which results in leukemic transformation. Systemic study at the genome level in DNA sequence and DNA methylation, gene and microRNA expression profile, proteome and metabolism not only provides the insight for understanding leukemogenesis, but also identifies biomarkers for leukemia stem cell, leukemia classification, diagnosis, risk assessment, therapy selection, response prediction, prognosis, minimal residual disease detection and other aspects of clinical decision-making and applications. Toward this end, current advanced high throughput technologies including next generation sequencing, microarray, proteomics, targeted molecular testing and bioinformatics have provided powerful tools. Well-designed clinical trials will make a clinical connection with new scientific discoveries in leukemia genome and epigenome. Assembly and synthesis of the massive amounts of new information by systems biology will generate a high resolution picture of leukemogenesis of acute leukemia. With combined efforts from bench and bedside, the ultimate goal is to eradicate all leukemic blasts including leukemic stem cells in the patients by less toxic reagents to completely cure leukemia in the future.

11. Acknowledgments

We thank Ms. Marie Schultz for her editorial assistance.

12. References

Agirre, X., Roman-Gomez, J., Vazquez, I., Jimenez-Velasco, A., Garate, L., Montiel-Duarte, C., Artieda, P., Cordeu, L., Lahortiga, I., Calasanz, M. J., Heiniger, A., Torres, A., Minna, J. D., & Prosper, F. (2006). Abnormal methylation of the common PARK2 and PACRG promoter is associated with downregulation of gene expression in acute lymphoblastic leukemia and chronic myeloid leukemia. Int.J.Cancer, Vol.118, No.8, pp. 1945-1953.

Agrawal S, Unterberg M, Koschmieder S, zur Stadt U et al (2007). DNA methylation of tumor suppressor genes in clinical remission predicts the relapse risk in acute myeloid leukemia. Cancer Res. Vol. 67, No.3, pp.1370-1377.

Alvarez S, Suela J, Valencia A, Fernández A et al. (2010). DNA methylation profiles and their relationship with cytogenetic status in adult acute myeloid leukemia. PLoS One. Vol.5, No.8, pp. e12197.

Bailey HD, Armstrong BK, de Klerk NH, Fritschi L, Attia J, Scott RJ, Smibert E, Milne E; Aus-ALL Consortium. (2011). Exposure to professional pest control treatments and the risk of childhood acute lymphoblastic leukemia. Int J Cancer. Vol.129, No.7, pp.1678-188.

Bassan R & Hoelzer D. Modern therapy of acute lymphoblastic leukemia. J Clin Oncol. Vol.29, No.5, pp.532-543.

Batova, A., Diccianni, M. B., Yu, J. C., Nobori, T., Link, M. P., Pullen, J., & Yu, A. L. (1997). Frequent and selective methylation of p15 and deletion of both p15 and p16 in T-cell acute lymphoblastic leukemia. Cancer Res., Vol.57, No.5, pp. 832-836.

Becker MW & Jordan CT. (2011). Leukemia stem cells in 2010: current understanding and future directions. Blood Rev. Vol.25, No.2, pp.75-81.

Bell O, Tiwari VK, Thomä NH, Schübeler D. (2011). Determinants and dynamics of genome accessibility. Nat Rev Genet. Vol.12, No.8, pp.554-564.

Belson M, Kingsley B, Holmes A. (20070. Risk factors for acute leukemia in children: a review. Environ Health Perspect. Vol.115, No.1, pp.138-145.

Bird, A. P. (1980). DNA methylation and the frequency of CpG in animal DNA. Nucleic Acids Res, Vol.8, No.7, pp.1499-1504.

Boehm JS & Hahn WC. (2011). Towards systematic functional characterization of cancer genomes. Nat Rev Genet. Vol.12, No.7, pp.487-498.

Bowen DT. (2006). Etiology of acute myeloid leukemia in the elderly. Semin Hematol. Vol.43, No.2, pp.82-88.

Bullinger L, Ehrich M, Döhner K, Schlenk RF, Döhner H, Nelson MR, van den Boom D. (2010). Quantitative DNA methylation predicts survival in adult acute myeloid leukemia. Blood. Vol.115, No.3, pp.636-642.

Burnett A, Wetzler M, Löwenberg B. (2011). Therapeutic advances in acute myeloid leukemia. J Clin Oncol. Vol.29, No.5, pp.48794.

Calvanese V, Fernández AF, Urdinguio RG, Suárez-Alvarez B et al. (2011). A promoter DNA demethylation landscape of human hematopoietic differentiation. Nucleic Acids Res. 2011 Sep 12.

Canalli, A. A., Yang, H., Jeha, S., Hoshino, K., Sanchez-Gonzalez, B., Brandt, M., Pierce, S., Kantarjian, H., Issa, J. P., & Garcia-Manero, G. (2005). Aberrant DNA methylation of a cell cycle regulatory pathway composed of P73, P15 and P57KIP2 is a rare event in children with acute lymphocytic leukemia. Leuk.Res., Vol.29, No.8, pp. 881-885.

Chen J, Odenike O, Rowley JD. (2010). Leukaemogenesis: more than mutant genes. Nat Rev Cancer. Vol.10, No.1, pp.23-36.

Cheng, Q., Cheng, C., Crews, K. R., Ribeiro, R. C., Pui, C. H., Relling, M. V., & Evans, W. E. (2006). Epigenetic regulation of human gamma-glutamyl hydrolase activity in acute lymphoblastic leukemia cells. Am.J.Hum.Genet., Vol.79, No.2, pp. 264-274.

Chim, C. S., Tam, C. Y., Liang, R., & Kwong, Y. L. (2001). Methylation of p15 and p16 genes in adult acute leukemia: lack of prognostic significance. Cancer, Vol.91, No.12, pp. 2222-2229.

Clarke BJ, Liao SK, Leeds C, Soamboonsrup P, Neame PB. (1987). Distribution of a hematopoietic-specific differentiation antigen of K562 cells in the human myeloid and lymphoid cell lineages. Cancer Res. Vol.47, No.16, pp.4254-4259.

Cobaleda C & Sánchez-García I. (20090. B-cell acute lymphoblastic leukaemia: towards understanding its cellular origin. Bioessays. Vol.31, No.6, pp.600-609.

Corn, P. G., Kuerbitz, S. J., van Noesel, M. M., Esteller, M., Compitello, N., Baylin, S. B., & Herman, J. G. (1999). Transcriptional silencing of the p73 gene in acute lymphoblastic leukemia and Burkitt's lymphoma is associated with 5' CpG island methylation. Cancer Res., Vol.59, No.14, pp.3352-3356.

Cox CV, Evely RS, Oakhill A, Pamphilon DH, Goulden NJ, Blair A. (2004). Characterization of acute lymphoblastic leukemia progenitor cells. Blood. Vol.104, pp.2919-2925.

Cox CV, Martin HM, Kearns PR, Virgo P, Evely RS, Blair A. (2007).Characterization of a progenitor cell population in childhood T-cell acute lymphoblastic leukemia. Blood. Vol.109, pp.674-682.

Craig, J. M. & Bickmore, W. A. (1994). The distribution of CpG islands in mammalian chromosomes. Nat Genet, Vol.7, No.3, pp. 376-382.

Davidsson, J., Lilljebjorn, H., Andersson, A., Veerla, S., Heldrup, J., Behrendtz, M., Fioretos, T., & Johansson, B. (2009). The DNA methylome of pediatric acute lymphoblastic leukemia. Hum.Mol.Genet., Vol.18, No.21, pp.4054-4065.

Deaton AM & Bird A. (20110. CpG islands and the regulation of transcription. Genes Dev. Vol.25, No.10, pp.1010-1022.

Desmond JC, Raynaud S, Tung E, Hofmann WK, Haferlach T, Koeffler HP. (2007). Discovery of epigenetically silenced genes in acute myeloid leukemias. Leukemia. Vol.21, No.5, pp.1026-1034.

Di Croce L. (2005). Chromatin modifying activity of leukaemia associated fusion proteins. Hum Mol Genet. Vol.14 Spec No 1:R77-84.

Döhner H, Estey EH, Amadori S, Appelbaum FR, Büchner T, Burnett AK, Dombret H, Fenaux P, Grimwade D, Larson RA, Lo-Coco F, Naoe T, Niederwieser D, Ossenkoppele GJ, Sanz MA, Sierra J, Tallman MS, Löwenberg B, Bloomfield CD; European LeukemiaNet. (2010). Diagnosis and management of acute myeloid leukemia in adults: recommendations from an international expert panel, on behalf of the European LeukemiaNet. Blood. Vol.115, No.3, pp.453-474.

Dombret H, Preudhomme C, Boissel N. (2009). Core binding factor acute myeloid leukemia (CBF-AML): is high-dose Ara-C (HDAC) consolidation as effective as you think? Curr Opin Hematol. Vol.16, No.2, pp92-97.

Downing JR. (2003). The core-binding factor leukemias: lessons learned from murine models. Curr Opin Genet Dev. Vol.13, pp.48–54.

Dunwell, T., Hesson, L., Rauch, T. A., Wang, L., Clark, R. E., Dallol, A., Gentle, D., Catchpoole, D., Maher, E. R., Pfeifer, G. P., & Latif, F. (2010). A genome-wide screen identifies frequently methylated genes in haematological and epithelial cancers. Mol.Cancer, Vol.9, pp.44.

Eden A, Gaudet F, Waghmare A, Jaenisch R. (2003). Chromosomal instability and tumors promoted by DNA hypomethylation. Science. Vol.300, No.5618, pp.455.

Ekmekci CG, Gutiérrez MI, Siraj AK, Ozbek U, Bhatia K. (2004). Aberrant methylation of multiple tumor suppressor genes in acute myeloid leukemia. Am J Hematol. Vol.77, No.3, pp.233-340.

Esteller M. (2008). Epigenetics in cancer. N Engl J Med. Vol.358, No.11, pp.1148-1159.

Estey E & Döhner H. (2006). Acute myeloid leukaemia. Lancet. Vol.368, No.9550, pp.1894-1907.

Ferrara F & Del Vecchio L. (2002). Acute myeloid leukemia with t(8;21)/AML1/ETO: a distinct biological and clinical entity. Haematologica. Vol.87, No.3, pp.306-319.

Figueroa ME, Lugthart S, Li Y, Erpelinck-Verschueren C et al. (2010). DNA methylation signatures identify biologically distinct subtypes in acute myeloid leukemia. Cancer Cell. Vol.17, No.1, pp13-27.

Foran JM. (2010). New prognostic markers in acute myeloid leukemia: perspective from the clinic. Hematology Am Soc Hematol Educ Program. pp.47-55.

Garcia-Manero G. (20110. Myelodysplastic syndromes: 2011 update on diagnosis, risk-stratification, and management. Am J Hematol. Vol.86, No.6, pp.490-498.

Garcia-Manero, G., Bueso-Ramos, C., Daniel, J., Williamson, J., Kantarjian, H. M., & Issa, J. P. (2002a). DNA methylation patterns at relapse in adult acute lymphocytic leukemia. Clin.Cancer Res., Vol.8, No.6, pp.1897-1903.

Garcia-Manero, G., Daniel, J., Smith, T. L., Kornblau, S. M., Lee, M. S., Kantarjian, H. M., & Issa, J. P. (2002b). DNA methylation of multiple promoter-associated CpG islands in adult acute lymphocytic leukemia. Clin.Cancer Res., Vol.8, No.7, pp.2217-2224.

Garcia-Manero, G., Jeha, S., Daniel, J., Williamson, J., Albitar, M., Kantarjian, H. M., & Issa, J. P. (2003). Aberrant DNA methylation in pediatric patients with acute lymphocytic leukemia. Cancer, Vol.97, No.3, pp.695-702.

Glasow A, Barrett A, Petrie K, Gupta R, Boix-Chornet M, Zhou DC, Grimwade D, Gallagher R, von Lindern M, Waxman S, Enver T, Hildebrandt G, Zelent A. (2088). DNA methylation-independent loss of RARA gene expression in acute myeloid leukemia. Blood. Vol.111, No.4, pp.2374-2377.

Godley LA, Cunningham J, Dolan ME, Huang RS et al. (2011). An integrated genomic approach to the assessment and treatment of acute myeloid leukemia. Semin Oncol. Vol.38, No.2, pp.215-224.

Godley LA & Larson RA. (2008). Therapy-related myeloid leukemia. Semin Oncol. Vol.35, No.4, 418-429.

Goto, T. & Monk, M. (1998). Regulation of X-Chromosome Inactivation in Development in Mice and Humans. Microbiol.Mol.Biol.Rev., Vol.62, No.2, pp. 362-378.

Graux, C., Cools, J., Michaux, L., Vandenberghe, P., & Hagemeijer, A. (2006). Cytogenetics and molecular genetics of T-cell acute lymphoblastic leukemia: from thymocyte to lymphoblast. Leukemia, Vol.20, No.9, pp.1496-1510.

Greaves MF, Maia AT, Wiemels JL, Ford AM. (2003). Leukemia in twins: lessons in natural history. Blood. Vol.102, No.7, pp.2321-2333.

Greaves MF & Wiemels J. (2003). Origins of chromosome translocations in childhood leukaemia. Nat Rev Cancer. Vol.3, No.9, pp.639-649.

Gu, H., Smith, Z. D., Bock, C., Boyle, P., Gnirke, A., & Meissner, A. (2011). Preparation of reduced representation bisulfite sequencing libraries for genome-scale DNA methylation profiling. Nat.Protoc., Vol.6, No.4, pp.468-481.

Gulley ML, Shea TC, Fedoriw Y. (2010). Genetic tests to evaluate prognosis and predict therapeutic response in acute myeloid leukemia. J Mol Diagn. Vol.12, No.1, pp.3-16.

Gupta V, Tallman MS, Weisdorf DJ. (2011). Allogeneic hematopoietic cell transplantation for adults with acute myeloid leukemia: myths, controversies, and unknowns. Blood. Vol.117, No.8, 2307-2318.

Gutierrez, M. I., Siraj, A. K., Bhargava, M., Ozbek, U., Banavali, S., Chaudhary, M. A., El, S. H., & Bhatia, K. (2003). Concurrent methylation of multiple genes in childhood ALL: Correlation with phenotype and molecular subgroup. Leukemia, Vol.17, No.9, pp.1845-1850

Hanahan D & Weinberg RA. (2011)Hallmarks of cancer: the next generation. Cell. Vol.144, No.5, pp.646-674.

Hanahan, D. & Weinberg, R.A. (2000). The hallmarks of cancer. Cell. Vol.100, pp.57–70.

Hatziapostolou M & Iliopoulos D. (2011). Epigenetic aberrations during oncogenesis. Cell Mol Life Sci. Vol.68, No.10, 1681-1702.

Herman JG & Baylin SB. (2003). Gene silencing in cancer in association with promoter hypermethylation. N Engl J Med. Vol.349, No.21, pp.2042-2054.

Hogan, L. E., Meyer, J. A., Yang, J., Wang, J., Wong, N., Yang, W., Condos, G., Hunger, S. P., Raetz, E., Saffery, R., Relling, M. V., Bhojwani, D., Morrison, D. J., & Carroll, W. L. (2011). Integrated genomic analysis of relapsed childhood acute lymphoblastic leukemia reveals therapeutic strategies. Blood.

Huang, T. H., Perry, M. R., & Laux, D. E. (1999). Methylation profiling of CpG islands in human breast cancer cells. Hum.Mol.Genet., Vol.8, No.3, pp.459-470.

Huntly BJ & Gilliland DG. (2005). Leukaemia stem cells and the evolution of cancer-stem-cell research. Nat Rev Cancer. Vol.5, No.4, pp.311-321.

Hutter, G., Kaiser, M., Neumann, M., Mossner, M., Nowak, D., Baldus, C. D., Gokbuget, N., Hoelzer, D., Thiel, E., & Hofmann, W. K. (2011). Epigenetic regulation of PAX5 expression in acute T-cell lymphoblastic leukemia. Leuk.Res., Vol.35, No.5, pp.614-619.

Iravani, M., Dhat, R., & Price, C. M. (1997). Methylation of the multi tumor suppressor gene-2 (MTS2, CDKN1, p15INK4B) in childhood acute lymphoblastic leukemia. Oncogene, Vol.15, No.21, pp. 2609-2614.

Irizarry, R. A., Ladd-Acosta, C., Wen, B., Wu, Z., Montano, C., Onyango, P., Cui, H., Gabo, K., Rongione, M., Webster, M., Ji, H., Potash, J. B., Sabunciyan, S., & Feinberg, A. P. (2009). The human colon cancer methylome shows similar hypo- and hypermethylation at conserved tissue-specific CpG island shores. Nat.Genet., Vol.41, No.2, pp.178-186.

Issa JP, Baylin SB, Herman JG. (1997). DNA methylation changes in hematologic malignancies: biologic and clinical implications. Leukemia. Vol.11 Suppl, No.1:S7-11.

Jamieson CH, Ailles LE, Dylla SJ, Muijtjens M, Jones C, Zehnder JL, Gotlib J, Li K, Manz MG, Keating A, Sawyers CL, Weissman IL. (2004). Granulocyte-macrophage progenitors as candidate leukemic stem cells in blast-crisis CML. N Engl J Med. Vol.351, No.7, pp.657-667.

Jan M, Chao MP, Cha AC, Alizadeh AA, Gentles AJ, Weissman IL, Majeti R. (2011). Prospective separation of normal and leukemic stem cells based on differential expression of TIM3, a human acute myeloid leukemia stem cell marker. Proc Natl Acad Sci U S A. Vol.108, No.12, pp.5009-5014.

Jimenez-Velasco, A., Roman-Gomez, J., Agirre, X., Barrios, M., Navarro, G., Vazquez, I., Prosper, F., Torres, A., & Heiniger, A. (2005). Downregulation of the large tumor suppressor 2 (LATS2/KPM) gene is associated with poor prognosis in acute lymphoblastic leukemia. Leukemia, Vol.19, No.12, pp.2347-2350.

Kawamoto H, Wada H, Katsura Y. (2010). A revised scheme for developmental pathways of hematopoietic cells: the myeloid-based model. Int Immunol. Vol.22, No.2, pp.65-70.

Kelly LM & Gilliland DG. (2002). Genetics of myeloid leukemias. Annu Rev Genomics Hum Genet. Vol.3, pp.179-98.

Kroeger H, Jelinek J, Estécio MR, He R et al. (2008). Aberrant CpG island methylation in acute myeloid leukemia is accentuated at relapse. Blood. Vol.112, No.4, pp.1366-1373.

Kuang, S. Q., Tong, W. G., Yang, H., Lin, W., Lee, M. K., Fang, Z. H., Wei, Y., Jelinek, J., Issa, J. P., & Garcia-Manero, G. (2008). Genome-wide identification of aberrantly methylated promoter associated CpG islands in acute lymphocytic leukemia. Leukemia, Vol.22, No.8, pp.1529-1538.

Kulis M & Esteller M. (2010). DNA methylation and cancer. Adv Genet. Vol. 70, pp.27-56.

Laird, P. W. (2010). Principles and challenges of genome-wide DNA methylation analysis. Nat.Rev.Genet., Vol.11, No.3, pp.191-203.

Lapidot T, Sirard C, Vormoor J, Murdoch B, Hoang T, Caceres-Cortes J, Minden M, Paterson B, Caligiuri MA, Dick JE. (1994). A cell initiating human acute myeloid leukaemia after transplantation into SCID mice. Nature. Vol.367, pp.645-648.

Lausten-Thomsen U, Madsen HO, Vestergaard TR, Hjalgrim H, Nersting J, Schmiegelow K. (2011). Prevalence of t(12;21)[ETV6-RUNX1]-positive cells in healthy neonates. Blood, Vol.117, No.1, pp.186-189.

Lechner M, Boshoff C, Beck S. (2010). Cancer epigenome. Adv Genet. Vol.70, pp.247-276.

Ley TJ, Ding L, Walter MJ, McLellan MD et al. (2010). DNMT3A mutations in acute myeloid leukemia. N Engl J Med. Vol.363, No.25, pp.2424-2433.

Li, E., Beard, C., & Jaenisch, R. (1993). Role for DNA methylation in genomic imprinting. Nature, Vol.366, No.6453, pp.362-365.

Licht, J. D. (2006). Reconstructing a disease: What essential features of the retinoic acid receptor fusion oncoproteins generate acute promyelocytic leukaemia? Cancer Cell, Vol.9, pp.73–74.

Lin TC, Hou HA, Chou WC, Ou DL, Yu SL, Tien HF, Lin LI. (2011). CEBPA methylation as a prognostic biomarker in patients with de novo acute myeloid leukemia. Leukemia. Vol.25, No.1, pp.32-40.

Link DC, Schuettpelz LG, Shen D, Wang J et al. (2011). Identification of a novel TP53 cancer susceptibility mutation through whole-genome sequencing of a patient with therapy-related AML. JAMA. Vol.305, No.15, pp.1568-1576.

Link KA, Chou FS, Mulloy JC. (2010). Core binding factor at the crossroads: determining the fate of the HSC. J Cell Physiol. Vol.222, No.1, pp.50-56.

Liu S, Shen T, Huynh L, Klisovic MI, Rush LJ et al. (2005). Interplay of RUNX1/MTG8 and DNA methyltransferase 1 in acute myeloid leukaemia. Cancer Res. Vol.65, pp.1277–1284.

Löwenberg B, Downing JR, Burnett A. (1999). Acute myeloid leukemia. N Engl J Med. Vol.341, No.14, pp.1051-1062.

Lugthart S, Figueroa ME, Bindels E, Skrabanek L, Valk PJ, Li Y, Meyer S, Erpelinck-Verschueren C, Greally J, Löwenberg B, Melnick A, Delwel R. (2011). Aberrant DNA hypermethylation signature in acute myeloid leukemia directed by EVI1. Blood. Vol.117, No.1, pp.234-241.

Marcucci G, Mrózek K, Bloomfield CD. (2005). Molecular heterogeneity and prognostic biomarkers in adults with acute myeloid leukemia and normal cytogenetics. Curr Opin Hematol. Vol.12, No.1, pp.68-75.

Mardis ER, Ding L, Dooling DJ, Larson DE et al. (2009). Recurring mutations found by sequencing an acute myeloid leukemia genome. N Engl J Med. Vol.361, No.11, pp.1058-1066.

Martin, V., Agirre, X., Jimenez-Velasco, A., Jose-Eneriz, E. S., Cordeu, L., Garate, L., Vilas-Zornoza, A., Castillejo, J. A., Heiniger, A., Prosper, F., Torres, A., & Roman-Gomez, J. (2008). Methylation status of Wnt signaling pathway genes affects the clinical outcome of Philadelphia-positive acute lymphoblastic leukemia. Cancer Sci., Vol.99, No.9, pp.1865-1868.

Martín-Subero JI & Esteller M. (2011). Profiling epigenetic alterations in disease. Adv Exp Med Biol. Vol.711, pp.162-77.

McHale CM & Smith MT. (2004). Prenatal origin of chromosomal translocations in acute childhood leukemia: implications and future directions. Am J Hematol. Vol.75, No.4, pp.254-257.

Melnick AM. (2010). Epigenetics in AML. Best Pract Res Clin Haematol. Vol.23, No.4, pp.463-468.

Metcalf D. (2008). Hematopoietic cytokines. Blood. Vol.111, No.2, 485-491.

Metzker ML. (2010). Sequencing technologies - the next generation. Nat Rev Genet. Vol.11, No.1, pp.31-46.

Milani, L., Lundmark, A., Kiialainen, A., Nordlund, J., Flaegstad, T., Forestier, E., Heyman, M., Jonmundsson, G., Kanerva, J., Schmiegelow, K., Soderhall, S., Gustafsson, M. G., Lonnerholm, G., & Syvanen, A. C. (2010). DNA methylation for subtype classification and prediction of treatment outcome in patients with childhood acute lymphoblastic leukemia. Blood, Vol.115, No.6, pp.1214-1225.

Milne E, Royle JA, Miller M, Bower C, de Klerk NH, Bailey HD, van Bockxmeer F, Attia J, Scott RJ, Norris MD, Haber M, Thompson JR, Fritschi L, Marshall GM, Armstrong BK. (2010). Maternal folate and other vitamin supplementation during pregnancy and risk of acute lymphoblastic leukemia in the offspring. Int J Cancer. Vol.126, No.11, pp.2690-2699.

Mori H, Colman SM, Xiao Z, Ford AM, Healy LE, Donaldson C, Hows JM, Navarrete C, Greaves M. (2002). Chromosome translocations and covert leukemic clones are generated during normal fetal development. Proc Natl Acad Sci U S A. Vol.99, No.12, pp.8242-8247.

Mori H, Colman SM, Xiao Z, Ford AM. (2002). Chromosome translocations and covert leukemic clones are generated during normal fetal development. Proc Natl Acad Sci U S A. Vol.99, No.12, pp.8242-8247.

Mrózek K, Harper DP, Aplan PD. (2009). Cytogenetics and molecular genetics of acute lymphoblastic leukemia. Hematol Oncol Clin North Am. Vol.23,, No.5, pp.991-1010

Mullighan CG, Goorha S, Radtke I, et al. (2007). Genome-wide analysis of genetic alterations in acute lymphoblastic leukaemia. Nature vol.446, pp.758-764.

Musolino C, Sant'antonio E, Penna G, Alonci A, Russo S, Granata A, Allegra A. (2010). Epigenetic therapy in myelodysplastic syndromes. Eur J Haematol. Vol.84, No.6, pp.463-473.

Odenike O, Thirman MJ, Artz AS, Godley LA, Larson RA, Stock W. (2011). Gene mutations, epigenetic dysregulation, and personalized therapy in myeloid neoplasia: are we there yet? Semin Oncol. Vol.38, No.2, pp.196-214.

Oki Y & Issa JP. (2010). Epigenetic mechanisms in AML - a target for therapy. Cancer Treat Res. Vol.145, pp.19-40.

Paixao, V. A., Vidal, D. O., Caballero, O. L., Vettore, A. L., Tone, L. G., Ribeiro, K. B., & Lopes, L. F. (2006). Hypermethylation of CpG island in the promoter region of CALCA in acute lymphoblastic leukemia with central nervous system (CNS) infiltration correlates with poorer prognosis. Leuk.Res. Vol.30, No.7, pp.891-894.

Paschka P. (2008). Core binding factor acute myeloid leukemia. Semin Oncol. Vol.35, No.4, pp.410-417.

Passegué E, Jamieson CH, Ailles LE, Weissman IL. (2003). Normal and leukemic hematopoiesis: are leukemias a stem cell disorder or a reacquisition of stem cell characteristics? Proc Natl Acad Sci U S A. Vol.100 Suppl, No.1, pp.11842-11849.

Peters AH & Schwaller J. (2011). Epigenetic mechanisms in acute myeloid leukemia. Prog Drug Res. Vol.67, pp.197-219.

Popp HD & Bohlander SK. (2010). Genetic instability in inherited and sporadic leukemias. Genes Chromosomes Cancer. Vol.49, No.12, pp.1071-1081.

Pui CH, Carroll WL, Meshinchi S, Arceci RJ. (2011). Biology, risk stratification, and therapy of pediatric acute leukemias: an update. J Clin Oncol. Vol.29, No.5, pp.551-565.

Pui C-H, Relling MV, and Downing JR. (2004). Mechanisms of Disease: Acute Lymphoblastic Leukemia. N Engl J Med Vol.350, pp.1535-1548

Pui CH, Robison LL, Look AT. (2008). Acute lymphoblastic leukaemia. Lancet, Vol.371, pp.1030-1043.

Pui, C. H., Relling, M. V., & Downing, J. R. (2004). Acute lymphoblastic leukemia. N.Engl.J.Med., Vol.350, No.15, pp. 1535-1548.

Quintás-Cardama A, Santos FP, Garcia-Manero G. (2011). Histone deacetylase inhibitors for the treatment of myelodysplastic syndrome and acute myeloid leukemia. Leukemia. Vol.25, No.2, pp.226-235.

Rauch, T. & Pfeifer, G. P. (2005). Methylated-CpG island recovery assay: a new technique for the rapid detection of methylated-CpG islands in cancer. Lab Invest, Vol.85, No.9, pp.1172-1180.

Razin, A. & Cedar, H. (1994). DNA methylation and genomic imprinting. Cell, Vol.77, No.4, pp.473-476.

Robertson, K. D. & Jones, P. (2000). DNA methylation: past, present and future directions. Carcinogenesis, Vol.21, No.3, pp.461-467.

Roman, J., Castillejo, J. A., Jimenez, A., Bornstein, R., Gonzalez, M. G., del Carmen, R. M., Barrios, M., Maldonado, J., & Torres, A. (2001). Hypermethylation of the calcitonin gene in acute lymphoblastic leukaemia is associated with unfavourable clinical outcome. Br.J.Haematol., Vol.113, No.2, pp.329-338.

Roman-Gomez, J., Castillejo, J. A., Jimenez, A., Gonzalez, M. G., Moreno, F., Rodriguez, M. C., Barrios, M., Maldonado, J., & Torres, A. (2002). 5' CpG island hypermethylation is associated with transcriptional silencing of the p21(CIP1/WAF1/SDI1) gene and confers poor prognosis in acute lymphoblastic leukemia. Blood, Vol.99, No.7, pp.2291-2296.

Roman-Gomez, J., Jimenez-Velasco, A., Agirre, X., Castillejo, J. A., Navarro, G., Barrios, M., Andreu, E. J., Prosper, F., Heiniger, A., & Torres, A. (2004). Transcriptional silencing of the Dickkopfs-3 (Dkk-3) gene by CpG hypermethylation in acute lymphoblastic leukaemia. Br.J.Cancer, Vol.91, No.4, pp. 707-713.

Roman-Gomez, J., Jimenez-Velasco, A., Agirre, X., Castillejo, J. A., Navarro, G., Calasanz, M. J., Garate, L., San Jose-Eneriz, E., Cordeu, L., Prosper, F., Heiniger, A., & Torres, A. (2006). CpG island methylator phenotype redefines the prognostic effect of t(12;21) in childhood acute lymphoblastic leukemia. Clin.Cancer Res., Vol.12, No.16, pp.4845-4850.

Roman-Gomez, J., Jimenez-Velasco, A., Cordeu, L., Vilas-Zornoza, A., San Jose-Eneriz, E., Garate, L., Castillejo, J. A., Martin, V., Prosper, F., Heiniger, A., Torres, A., & Agirre, X. (2007). WNT5A, a putative tumour suppressor of lymphoid malignancies, is inactivated by aberrant methylation in acute lymphoblastic leukaemia. Eur.J.Cancer, Vol.43, No.18, pp.2736-2746

Rosu-Myles M & Wolff L. (2008). p15Ink4b: dual function in myelopoiesis and inactivation in myeloid disease. Blood Cells Mol Dis. Vol.40, No.3, pp.406-409.

Sahu, G. R. & Das, B. R. (2005). Alteration of p73 in pediatric de novo acute lymphoblastic leukemia. Biochem.Biophys.Res.Commun., Vol.327, No.3, pp.750-755.

Scholz, C., Nimmrich, I., Burger, M., Becker, E., Dorken, B., Ludwig, W. D., & Maier, S. (2005). Distinction of acute lymphoblastic leukemia from acute myeloid leukemia through microarray-based DNA methylation analysis. Ann.Hematol., Vol.84, No.4, pp.236-244.

Schoofs T & Müller-Tidow C. (2011). DNA methylation as a pathogenic event and as a therapeutic target in AML. Cancer Treat Rev. Vol.37 Suppl No.1, pp.S13-8.

Scott, S. A., Kimura, T., Dong, W. F., Ichinohasama, R., Bergen, S., Kerviche, A., Sheridan, D., & DeCoteau, J. F. (2004). Methylation status of cyclin-dependent kinase inhibitor genes within the transforming growth factor beta pathway in human T-cell lymphoblastic lymphoma/leukemia. Leuk.Res., Vol.28, No.12, pp.1293-1301.

Shteper, P. J., Siegfried, Z., Asimakopoulos, F. A., Palumbo, G. A., Rachmilewitz, E. A., Ben-Neriah, Y., & Ben-Yehuda, D. (2001). ABL1 methylation in Ph-positive ALL is exclusively associated with the P210 form of BCR-ABL. Leukemia, Vol.15, No.4, pp.575-582.

Siegel R, Ward E, Brawley O, Jemal A. (2011). Cancer statistics 2011, The impact of eliminating socioeconomic and racial disparities on premature cancer deaths. CA Cancer J Clin. Vol.61, pp.212-236.

Silverman LR & Mufti GJ. (2005). Methylation inhibitor therapy in the treatment of myelodysplastic syndrome. Nat Clin Pract Oncol. Vol.2 Suppl No.1, pp.S12-23.

Smith MT, Zhang L, McHale CM, Skibola CF, Rappaport SM. (2011). Benzene, the exposome and future investigations of leukemia etiology. Chem Biol Interact. Vol.192, No.1-2, pp.155-159.

Smith, Z. D., Gu, H., Bock, C., Gnirke, A., & Meissner, A. (2009). High-throughput bisulfite sequencing in mammalian genomes. Methods, Vol.48, No.3, pp.226-232.

Solis EC. (2011). Treatment strategies in patients with core-binding factor acute myeloid leukemia. Curr Oncol Rep. Vol.13, No.5, pp.359-360.

Stam, R. W., den Boer, M. L., Passier, M. M., Janka-Schaub, G. E., Sallan, S. E., Armstrong, S. A., & Pieters, R. (2006). Silencing of the tumor suppressor gene FHIT is highly characteristic for MLL gene rearranged infant acute lymphoblastic leukemia. Leukemia, Vol.20, No.2, pp.264-271.

Sternberg DW & Gilliland DG. (2004). The role of signal transducer and activator of transcription factors in leukemogenesis. J Clin Oncol. Vol.22, No.2, pp.361-371.

Stumpel, D. J., Schneider, P., van Roon, E. H., Boer, J. M., de, L. P., Valsecchi, M. G., de Menezes, R. X., Pieters, R., & Stam, R. W. (2009). Specific promoter methylation identifies different subgroups of MLL-rearranged infant acute lymphoblastic leukemia, influences clinical outcome, and provides therapeutic options. Blood, Vol.114, No.27, pp.5490-5498.

Swerdlow SH, Campo E, Harris NL, et al, eds. (2008). WHO Classification of Tumours of Haematopoietic and Lymphoid Tissues. Lyon, France: IARC.

Taberlay PC & Jones PA. (2011). DNA methylation and cancer. Prog Drug Res. Vol.67, pp.1-23

Taniguchi, A., Nemoto, Y., Yokoyama, A., Kotani, N., Imai, S., Shuin, T., & Daibata, M. (2008). Promoter methylation of the bone morphogenetic protein-6 gene in association with adult T-cell leukemia. Int.J.Cancer, Vol.123, No.8, pp.1824-1831.

Taylor, K. H., Kramer, R. S., Davis, J. W., Guo, J., Duff, D. J., Xu, D., Caldwell, C. W., & Shi, H. (2007a). Ultradeep bisulfite sequencing analysis of DNA methylation patterns in multiple gene promoters by 454 sequencing. Cancer Res, Vol.67, No.18, pp.8511-8518.

Taylor, K. H., Pena-Hernandez, K. E., Davis, J. W., Arthur, G. L., Duff, D. J., Shi, H., Rahmatpanah, F. B., Sjahputera, O., & Caldwell, C. W. (2007b). Large-Scale CpG Methylation Analysis Identifies Novel Candidate Genes and Reveals Methylation Hotspots in Acute Lymphoblastic Leukemia. Cancer Res, Vol.67, No.6, pp.2617-2625.

Toyota M & Suzuki H. (2010). Epigenetic drivers of genetic alterations. Adv Genet. Vol.70, pp.309-323.

Toyota, M. & Issa, J. P. (2002). Methylated CpG island amplification for methylation analysis and cloning differentially methylated sequences. Methods Mol.Biol., Vol.200, pp.101-110.

Tsellou, E., Troungos, C., Moschovi, M., Athanasiadou-Piperopoulou, F., Polychronopoulou, S., Kosmidis, H., Kalmanti, M., Hatzakis, A., Dessypris, N., Kalofoutis, A., & Petridou, E. (2005). Hypermethylation of CpG islands in the promoter region of the p15INK4B gene in childhood acute leukaemia. Eur.J.Cancer, Vol.41, No.4, pp.584-589.

Vardiman JW, Thiele J, Arber DA, Brunning RD, Borowitz MJ, Porwit A, Harris NL, Le Beau MM, Hellström-Lindberg E, Tefferi A, Bloomfield CD. (2009). The 2008 revision of the World Health Organization (WHO) classification of myeloid neoplasms and acute leukemia: rationale and important changes. Blood. Vol.114, No.5, pp.937-951.

Vilas-Zornoza, A., Agirre, X., Martin-Palanco, V., Martin-Subero, J. I., San Jose-Eneriz, E., Garate, L., Alvarez, S., Miranda, E., Rodriguez-Otero, P., Rifon, J., Torres, A., Calasanz, M. J., Cruz, C. J., Roman-Gomez, J., & Prosper, F. (2011). Frequent and simultaneous epigenetic inactivation of TP53 pathway genes in acute lymphoblastic leukemia. PLoS.One., Vol.6, No.2, p. (e17012),ISSN

Wang MX, Wang HY, Zhao X, Srilatha N et al. (2010). Molecular detection of B-cell neoplasms by specific DNA methylation biomarkers. Int J Clin Exp Pathol. Vol.3, No.3, pp.265-279.

Wang ZY & Chen Z. (2008). Acute promyelocytic leukemia: from highly fatal to highly curable. Blood. Vol.111, No.5, pp.2505-2515.

Warrell RP Jr, de Thé H, Wang ZY, Degos L. (1993). Acute promyelocytic leukemia. N Engl J Med. Vol.329, No.3, pp.177-189.

Weber, M., Davies, J. J., Wittig, D., Oakeley, E. J., Haase, M., Lam, W. L., & Schubeler, D. (2005). Chromosome-wide and promoter-specific analyses identify sites of differential DNA methylation in normal and transformed human cells. Nat Genet, Vol.37, No.8, pp.853-862.

Welch JS, Westervelt P, Ding L, Larson DE et al. (2011). Use of whole-genome sequencing to diagnose a cryptic fusion oncogene. JAMA. Vol.305, No.15, pp.1577-1584.

Wiemels J, Kang M, Greaves M. (2009). Backtracking of leukemic clones to birth. Methods Mol Biol. Vol.538, pp.7-27.

Wilop S, Fernandez AF, Jost E, Herman JG, Brümmendorf TH, Esteller M, Galm O. (2011). Array-based DNA methylation profiling in acute myeloid leukaemia. Br J Haematol. Vol.155, No.1, pp.65-72.

Wu G, Yi N, Absher D, Zhi D. (2011). Statistical quantification of methylation levels by next-generation sequencing. PLoS One. Vol.6, No.6, pp.e21034.

Yan XJ, Xu J, Gu ZH, Pan CM et al. (2011). Exome sequencing identifies somatic mutations of DNA methyltransferase gene DNMT3A in acute monocytic leukemia. Nat Genet. Vol.43, No.4, pp309-315.

Yang, Y., Takeuchi, S., Hofmann, W. K., Ikezoe, T., van Dongen, J. J., Szczepanski, T., Bartram, C. R., Yoshino, N., Taguchi, H., & Koeffler, H. P. (2006). Aberrant methylation in promoter-associated CpG islands of multiple genes in acute lymphoblastic leukemia. Leuk.Res., Vol.30, No.1, pp.98-102.

Zheng, S., Ma, X., Zhang, L., Gunn, L., Smith, M. T., Wiemels, J. L., Leung, K., Buffler, P. A., & Wiencke, J. K. (2004). Hypermethylation of the 5' CpG island of the FHIT gene is associated with hyperdiploid and translocation-negative subtypes of pediatric leukemia. Cancer Res., Vol.64, No.6, pp.2000-2006.

14

The Importance of Aberrant DNA Methylation in Cancer

Koraljka Gall Trošelj, Renata Novak Kujundžić
and Ivana Grbeša
*Rudjer Boskovic Institute,
Division of Molecular Medicine, Zagreb,
Croatia*

1. Introduction

Cancer has long been considered primarily a genetic disease, caused by different mutations throughout the genome. In 1983., Feinberg and Vogelstein discovered that the level of DNA methylation significantly varies between primary human malignant tumors and their normal counterparts (Feinberg & Vogelstein, 1983). Before this publication, there had been a paper describing changes in DNA methylation in cancer cell cultures, including the influence of N-methyl-N-nitrosourea on the level of DNA methylation in Raji cells (Boehm & Drahovsky, 1981). Currently, we are presented with much experimental data showing multilevel changes in cancer cells. In this context, two major areas of epigenomic research - DNA methylation and histone modifications appear most promising in understanding the multistep nature of carcinogenesis. Additionally, they seem to have the potential for being cancer biomarkers, useful in early detection, in predicting the biological behavior of tumors and for therapy monitoring, as recently reviewed (Rodriguez-Paredes & Esteller, 2011; Baylin & Jones, 2011). Finally, epigenetic changes are well-recognized targets for cancer therapy, alone, or in combination with various cytostatics (Ren et al, 2011). Epigenetic changes are also of the greatest importance in chemoprevention, as there is increasing data relating to possibly reversing epigenetic changes in the earliest phase of carcinogenesis, when genetic changes have yet to develop (reviewed, Huang et al., 2011). It is not easy to understand the particular rules applicable to epigenomic processes. If one specific epigenetic change, relating to a specific gene/its promoter, exists in a majority of tumors of a specific type, it does not necessarily mean that the same change exists in another type of tumor. This is consequential, and represents one reason for obvious differences in responses to epigenomic therapy. Recently, we wrote a review article on some aspects of epigenomic changes in which we used the term „epigenetic networking". If we imagine each of our living cells as an orchestra performing the symphony of life, then each player (a gene) of the orchestra needs to play in concert with 30,000 other players. The communication that produces the network of our epi-genome is established at many levels: transcriptionally, post-transcriptionally, through protein translation, at the level of post-translational protein modifications, through their orchestrated interactions and, finally, their interaction with the DNA that can be modified in order to accept or reject the protein partner. This is the way of controlling gene activity and why we, in the previously mentioned review paper, considered epigenomic changes as „A Bird's Eye Perspective on the Genome" (Gall Trošelj & Novak-Kujundžić, 2010).

2. Cancer, DNA methylation and factors beyond our control

A disturbed DNA methylation pattern represents, most probably, the best and most commonly studied epi-change. This change has been extensively documented, especially after the introduction of genome-wide analytical methods which clearly confirmed, on a very broad scale, that both gain and loss in DNA methylation are very frequent events in cancer (reviewed, Ndlovu et al., 2011). It has been known for some time that *de novo* DNA methylation of promoter CpG island (CpGI) does not occur accidentally (Choi et al., 2010). An arbitrary border at -1kB upstream and +0.5 kb downstream from the transcription start site shows that some 60% of human genes associate with CpG islands.

Until 2007, the definition of the CpG island was primarily related to the primary structure of DNA molecule. In 1987, Gardiner-Garden and Frommer established the generaly accepted definition (200 bp long stretch of DNA with a GC content >50% and an observed CpG/expected CpG ≥0.6) that, as realized later, did not make a stringent distinction between bona fide CpG islands and *Alu* repates (Gardiner-Garden & Frommer, 1987). In 2002, this obstacle was properly addressed and a new definition of the CpG island, commonly used in the field of cancer research, was offered: It is the DNA region longer than 500 bp, with a GC content ≥55% and observed CpG/expected CpG of 0.65. (Takai & Jones, 2002). The percentage number for defining „the promoter rich in CpG" varies; the most commonly used number is usually 55% (Espada & Esteller, 2007). However, it became obvious that these definitions lack a clear biological justification and do need improvement as they, although sufficiently sensitive in detecting majority of bona fide CpG islands in the humane genome, lack of specificity leading to a considerably high number of false positive results. In 2007, the computational modeling was used to estimate the "CpG Island Strength", based on predicted epigenetic state and chromatine structure, on non-repetitive parts of chromosomes 21 and 22. The "combined epigenetic score" that was based on available "open and transcriptionally competent chromatine structure" epigenomic data (including H3K4 di- and trimethylation, H3K9/14 acetylation, DNAse I hypersensitivity and Sp1 transcription factor binding), allowed for meaningful interpretation. Between the scores of "0" and "1" that related to a particular CpGI function (where "0" represented silenced, inactive and inaccessible island, and "1" represented unmethylated, highly accessible CpGI with prominent promoter activity), the value of 0.5 turned out to be equally likely to correspond to both, *bona fide* CpGI and the region of DNA that is not a CpGI. Hence, the 0.5 value was recommended as a threshold for majority (although not all) future applications. This approach has profiled 21,631 CpG islands on the tested chromosomes and, for high specificity mapping of CpGI, the map of predicted CpGIs based on the combined epigenetic score was suggested (Bock et al., 2007).

The CpG islands are rarely methylated in normal tissue (except for X-chromosome inactivated and imprinted genes). However, in cancer, the picture changes dramatically. Aging also represents a process relating to a linear increase of the level of DNA methylation in CpG rich gene promoters. On the other hand, paradoxically, the global level of methylation in older cells/tissues seems to be decreased. This clearly mimics the methylation status of a cancer cell. However, one should be critical when trying to understand what really happens in the living cell: very similar cell types derive from different stem cell niches and their epigenomes may differ significantly (Kim at al., 2005). In addition, as recently discussed (Ehrlich, 2009), the major problem in quantitative DNA methylation studies dealing with native clinical samples is the presence of cells that are non-neoplastic. When dealing with a tumor tissue that was taken in a surgical theater and immersed in liquid nitrogen immediately after extirpation, one can be more than convinced that non-tumorous cells are present in a sample. The percentage of „contaminating", non-tumorous cells varies from sample to sample. Even if we deal with very similar, relatively „clean", native tumors, we must be aware that every cell divides with its own dynamics. Hence, not all cells are in the same phase of the cell cycle. So, the whole cellular content of the tumor represents, in a percentage that varies, a mixture of very heterogeneous, cell-cycle related, methylomes. Accordingly, what we measure when using the methods that are not *in situ*, is a mixture of signals and we (usually) focus on the most prominent ones. But it does not mean that the signals that are less prominent are less important for the tumor *in toto*.

The problem becomes even more prominent in comparative analyses, when tumor tissue needs to be compared to non-tumorous, adjacent tissue. Our group was not the only one that has shown, unexpectedly, the change in imprinting status of *IGF2* in a tissue adjacent to laryngeal cancer (Grbeša et al., 2008). It seemed „normal" to the surgeon, and, back then in 2005, the simplest thought was that we mixed up tumorous and non-tumorous samples. Even at that time, we were quite careful with tumor samples, as years of experience taught us to take only a small portion of tissue for analyses, leaving at least one small piece of tissue in our tumor bank (Spaventi et al. 1994). After obtaining confusing results, this residual piece of tissue was given to a very experienced pathologist who needed to answer our question: „Is this the tumor"? Morphologically, it was not the tumor. Epigenetically, it showed loss of *IGF2* imprinting. Based on that finding – it did not appear as „normal tissue". We still think that, especially in smoking-related cancer, this specific change may be the first sign of "abnormality".

In addition to obvious problems relating to exploring the DNA methylation status in native tumors, there are also very specific problems when using cell cultures. As shown by Asada, who used several different rat liver cell lines (including a primary cell line), methylation level increases significantly after 10 passages. Hence, the authors concluded that, at least in their experimental model, „a cautious approach is required when cell lines are utilized to study methylation-related carcinogenesis, metastatic or tumoricidal mechanisms" (Asada et al., 2006).

Based on this brief but, hopefully, informative data relating to objective limitations of the system based on factors beyond our control, we enter the field of cancer epigenomics.

3. Cancer and DNA methyltransferases

DNA methyltransferases (DNMTs) are the only enzymes which have been shown to mediate the transfer of a methyl group from S-adenosylmethionine (SAM) to the C-5 position of cytosine, mainly in CpG nucleotides, in mammalian genomes. Although detected, cytosine methylation is very rare in the outside of CpG sequences, at least in differentiated cells. For example, 99.98% of all methylation in mature fibroblasts occur at CpG dimers. This number is significantly reduced in both embryonic and induced stem cells (Lister et al., 2009).

In mammals, there are four DNMTs: DNMT1, DNMT2, DNMT3A, DNMT3B. While DNMT1 has the highest importance in maintaining post-replicative DNA methylation patterns, DNMT3A and -3B are considered critical players in establishing *de novo* methylation patterns. They also assume maintenance activity. DNMT3L, discovered in 2000 (Aapola et al., 2000), is a regulatory factor for *de novo* methylation. Its amino acid sequence is very similar to that of DNMT3a and DNMT3b but lacks the residues required for DNA methyltransferase activity in the C-terminal domain.

It was shown that the fidelity in replicating methylation patterns in human non-cancerous, dividing cells reaches 99.85-99.92% per site, per generation in CGs reach promoters and 99.56-99.83% in CGs outside the promoters (Ushijima et al., 2003). Human cancer gastric cell lines showed decreased fidelity in maintaining the methylation pattern which manifested through an increased level of *de novo* DNA methylation in promoter regions of five tested cancer-related genes and 4- to 8-fold higher expression of *de novo* DNMT3B. This increase was highest in two cell lines that showed the highest level of decreased fidelity (Ushijama et al., 2005). The question remains: was the increased level of DNMT3B alone sufficient to induce so prominent change at the promoters of these genes?

3.1 DNA Methyltransferase 1 (DNMT1)

Homozygous knockout of DNMT1 is lethal to the embryo in mammals. On the other hand, studies on DNMT1-overexpression in embryonic stem cells also resulted in lethality of the embryo, suggesting that accurate expression of DNMT1 is a key factor in maintaining embryonic development (Biniszkiewicz et al., 2002).

For maintaining the methylation pattern during cell division, the cellular machinery uses DNMT1. After replication, 5-mC is present only on one parental DNA strand and the methylation of cytosines on the newly synthesized strand takes place on the cytosine that lies diagonally opposite to 5-mC in the parent DNA strand. Keeping the methylation pattern as inheritable modification that needed to be preserved during cell division was originally published in 1975 by three independent researchers/research groups (Holliday & Pugh 1975; Riggs, 1975). Since then, our knowledge has been significantly advancing, especially as a result of fast developing molecular techniques. However, it does not matter how rapidly our research progresses, the importance of the discovery published in 1975 remains astonishing, even from the most sophisticated molecular perspective.

During the S-phase of the cell cycle, DNMT1 was found to be localized to DNA replication foci through its interaction with proliferating cell nuclear antigen (PCNA). The precise cell-cycle-dependent localization of DNMT1 depends on the protein UHRF1, also known as ICBP and NP95. This protein shows strong preferential binding to hemimethylated CG sites through its methyl DNA binding domain, and tethers DNMT1 to replication fork (Bostick et al., 2007; Sharif at al., 2007). The DNMT1 also interacts with histone deacetylases resulting in repressing gene expression or forming heterochromatin structure.

3.1.2 DNMT1 and post-translational modifications

Little is known about post-translational modification of DNMT1, that may, possibly, change its functioning, especially in cancer. There are several *in vitro* studies pointing out the protein kinases involved in DNMT1 phosphorylation. From the perspective of cancer research, the AKT and PKC certainly are very promising candidates that may help us to better understand the functioning of DNMT1. Both kinases were shown to phosphorylate recombinant DNMT1 at Ser127. AKT additionally phosphorylates Ser143. This modification decreases the ability of DNMT1 to interact with PCNA and UHRF1. As a consequence, DNMT1 shows increased cell-cycle-dependent stability (Esteve et al., 2011).

3.2 DNMT2

DNMT2 is expressed in most human and mouse adult tissues (Goll & Bestor, 2005) and its role seems to be related to methylation of cytosine 38 (C38) of RNA^{Asp} (Goll et al., 2006). There are no strong evidence on DNA methylation activity of DNMT2. The Dnmt2 defficient mouse embryonic stem cells do not show measurable alteration of genomic DNA methylation pattern (Okano et al., 1998). Additionally, in contrast to exclusive nuclear localization of Dnmt1 and Dnmt3, Dnmt2 is primarily localized in the cytoplasm of transfected mouse 3T3 fibroblasts (Goll et al., 2006).

The level of DNMT2 expression in human cancer cell line is quite variable: high in K562 (leukemia) and MCF-7 (breast cancer) and very low, almost undetectable in A549 (lung cancer) and HepG2 cells (liver cancer) (Schaefer et al., 2009). It has been shown that the treatment with 5-azacytidine inhibits C38 methylation at RNA^{Asp}. These findings open the possibility that DNMT2 may contribute to neoplastic process through a novel pathways, related to RNA methylation. Clearly, much research should be performed in this area in order to understand all possible roles of DNMT2.

3.3 DNMT3 family

In mature cells which divide, DNMT1 is predominant DNA methyltransferase. However, there are two other DNMTs, DNMT3a and DNMT3b, which cannot differentiate between unmethylated and hemimethylated CpG sites. Their role is primarily *de novo* DNA methylation. Accordingly, they are highly expressed in early embryonic cells when programmed waves of *de novo* methylation occur. Their level is considerably lower after differentiation and in adult somatic tissues, but it significantly increases in cancer cells. Both enzymes contain large N-terminal parts which interact with other proteins. The C-terminal domain represents the catalytic center (Gowher & Jeltsch, 2002). In 1999, mice with targeted disruption of the Dnmt3a and Dnmt3b genes was an excellent model for exploring the activity of these two enzymes. Experiments showed lack of *de novo* methylation in embryonic stem cells and early embryos but without any effect on the maintenance of imprinted methylation patterns (Okano et al., 1999).

3.3.1 DNMT3B

The significance of this enzyme in cancer has been well recognized. The most recent research publications present its role in silencing tumor suppressor genes, through methylation of their promoters. In a study of hepatocellular carcinoma, DNMT3B overexpression was correlated to the level of promoter methylation and expression of *MTSS*1 (Metastasis Suppressor 1). There was negative correlation with DNMT3B expression and *MTSS*1 expression, but not with its promoter methylation. The DNMT3B was found to be directly bound to the 5'-flanking *MSS1* region that was sparsely methylated and methylation inhibitors failed to recover the *MSS1* expression. Based on these findings, the conclusion was that DNMT3B may repress *MTSS1* through a DNA methylation-independent mechanism (Fan et al., 2011). This should not be surprising, keeping in mind that different protein complexes include the DNMT3B. The thought is that DNMT3B, through a partnership with a transcription repressive complex, inhibits gene expression without necessarily exhibiting its genuine *de novo* methyltransferase function.

3.3.2 DNMTs in cancer

As shown in a comprehensive review in 2011, incorporating most available data relating to the level of DNMTs in cancer, these enzymes are increased in all tested cancer clinical specimens and cancer cell lines (Daniel et al., 2011). The methods used for the quantification were primarily Real-Time PCR and immunochistochemical methods, or both. When we perform these experiments, we must question: Does the amount of mRNA reflect the amount of the protein? Can we reach any conclusion without measuring protein activity? On all these, the answer is, or should be, a resounding „no". However, it has happened all too many times that we do not see clearly and that we reach our conclusion prematurely. If premature – then it is, unfortunately often – wrong. The consequence of all too many examples of this kind of unfortunate mistake is an enormous waste of time, as it takes years to get back on the right track. Many recently retracted papers, including those published in journals with the highest impact factors, are extremely consequential. Many researchers who are initially on the right path, change their hypotheses after reading what they had been led to believe, mistakenly, to be a break-through article. This mistaken action took them straight into the disaster zone of irreproducible results. It takes years for an article to be retracted. Meanwhile, many scientific careers are damaged in an effort to reproduce a result that cannot be reproduced.

4. Cancer and global DNA hypomethylation

The reasons for global DNA hypomethylation combined with hypermethylation at many 5' gene or promoter regions in cancer is not understood. In prior years, research related to this phenomenon was performed on several models: prostate carcinomas, Wilms's tumors and gastric cancer (Ehrlich et al., 2002; Kaneda et al., 2004; Santourlidis et al., 1999). In order to clarify this phenomenon, Ehrlich and co-workers analyzed the relationship of cancer-linked hypermethylation and hipomethylation at 55 gene loci (mostly CpG islands overlapping the 5' promoter regions), three classes of repetitive elements and global hypomethylation profile in epithelial ovarian malignant tumors (19 ovarian carcinomas, 20 LMP (low malignant potential) tumors and 21 cystadenomas) (Ehrlich et al., 2006). They proved that promoter 5' gene hypermethylation and both satellite or global DNA hypomethylation occur independently. This was shown in a multivariate regression analysis where, in a final model, hypermethylation variables and hypomethylation variables independently predicted the degree of malignancy in ovarian tumors as follows: *LTB4*R (P<0.005), *MTHFR1* (P=0.006), *CDH1* (P=0.005) and Satα (P=0.005). After making an adjustment for multiple comparisons, the *LTB4*R and *MTHFR1* showed an association of DNA methylation with DNMT1 mRNA levels (P<0.01), in carcinomas. However, this association was not seen when combining them with LMP tumors.

These examples lead to the following thought: if the total amount of the enzyme is increased, and the system is (globally) hypomethylated, then something has to impact its function. The focus should be on protein interactions because, as shown very recently, the interaction between DNMT1, PCNA and UHFR1 may be disrupted in human and mice astrocytes and glial progenitor cells. This specific change was shown to be an oncogenic event (Hervouet et al., 2010). The same paper shows that gliomagenesis relates to the decrease of 5-mC, the expression level of Dnmt1 remains stabile, but the catalytic activity of the enzyme decreases. This knowledge was applied to analyses measuring maintaining DNMT1 activity in 45 glioma patients who were divided into two groups: those with low (N=23), and those with high level of methyltransferase activity (N=22). Very significant differences in survival time was found between these two groups (p=0.0019), indicating that the level of DNMT1 activity, rather than the absolute amount, could be used as a survival prognostic factor. However, this conclusion must be taken with caution because of the limited number of patients. The results also clearly show that DNMT1/PCNA/UHRF1 interactions inversely correlate with the level of DNMT1 phosphorylation, reflecting, as proved in the cited paper, that the DNMT1 phosphorylation represents the hallmark of DNMT1/PCNA/UHRF1 interaction. The loss of this interaction represents a milestone for chromosomal instability induced by hypomethylated DNA repeat elements and also mediates overexpression of several very potent oncogenes such H-*ras* and *survivin*. However, it has to be noted, once again, that the number of patients was rather small and more research, based on a larger number of patients, must be performed in order to convert very strong indications into conclusions relating to DNMT1 phosphorylation as a prognostic cancer marker.

5. Loss of Imprinting (LOI) and cancer

Genomic imprinting is an epigenetic phenomenon that ensures monoallelic gene expression in a parent-of-origin dependent manner. Accordingly, imprinted genes are expressed only from a paternal or maternal allele. If we consider the biallelic expression as a full activity of a certain gene, then the imprinted gene gives "half" of the information which makes it very vulnerable to pathogenetic processes. If the gene is biallelically expressed, then any kind of damage affecting one allele still leaves 50% of overall function. As is the case with tumor suppressor genes, this may be sufficient for normal functioning. If the same happens with the active copy of the imprinted gene, the other allele, silenced through established imprinting marks, cannot add to the functioning. Hence, there is a haplo-insufficiency related to imprinted genes that makes them more "vulnerable".

5.1 Regulation of genomic imprinting

Estimation of the total number of imprinted genes in the human genome varies according to the methodology used. There are ~100 imprinted genes in the mammalian genome and ~70 imprinted genes have been experimentally verified and catalogued (Morison et al., 2001). These genes are not randomly scattered throughout the genome. They are clustered in the domains containing regulatory DNA elements - imprinting control regions, ICRs. These *cis*-regulatory elements are methylated only on one allele and that is the reason for calling them differentially methylated regions/domains, DMRs/DMDs. DNMT1 has the most important role in DNA methylation maintenance at ICRs. In addition to DNA methylation, other epigenetic modifications (post-translational histone tail modifications, binding of polycomb proteins, non-coding RNAs) play an important role in regulating ICRs.

5.2 IGF2/H19 imprinting

IGF2, coding for IGF2 mitogenic peptide and *H19,* a protein non-coding gene, at the human chromosome 11p15.5, are reciprocally imprinted, in most tissues studied to date. This is controlled by the *IGF2/H19* ICR which lies upstream of *H19* and is methylated only on the paternal allele (Tremblay et al., 1997). Accordingly, *H19* promoter is also methylated on the paternal allele and *H19* is silent (Zhang et al, 1993).

The *insulator model* (Figure 1) describes, roughly, how *IGF2/H19* ICR regulates monoallelic expression of *IGF2* and *H19*. The insulators are DNA sequences which block contact between promoters and nearby enhancers/silencers. The *IGF2/H19* ICR is positioned between *IGF2* and *H19*, ~100 kb downstream of the *IGF2*. The downstream enhancers are shared by *IGF2* and *H19* (Leighton et al., 1995). On the maternal allele, the CCCTC binding factor (CTCF) binds to unmethylated *IGF2/H19* ICR and insulates *Igf2* promoters from the enhancers (Bell & Felsenfeld, 2000; Hark et al., 2000). The human *IGF2/H19* ICR region has seven CTCF binding sites, but only the methylation of the sixth one acts as a key regulatory domain (Takai et al., 2001) through abolishing the CTCF binding to the paternal *IGF2/H19* ICR, leading to *IGF2* expression (Bell & Felsenfeld, 2000; Hark et al., 2000). In humans, the CTCF binding to both *IGF2/H19* ICR and the *IGF2* promoters P2-P4, and insulation of the *IGF2* promoters from enhancers on the maternal allele, involves long-range intrachromosomal interactions (Vu et al., 2010).

a) Paternal *IGF2/H19* allele

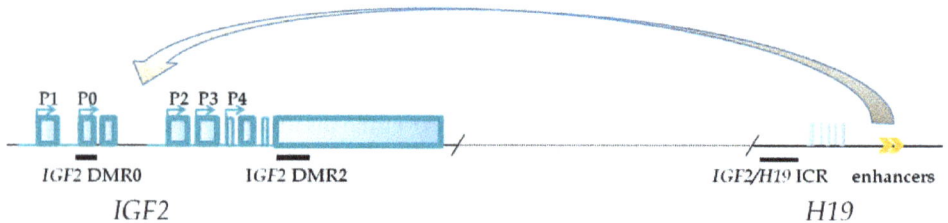

b) Maternal *IGF2/H19* allele

Fig. 1. The insulator model. Arrows: Five *IGF2* promoters and a *H19* promoter; shaded boxes: nine *IGF2* (lined blue) and five *H19* (lined pink) exons. Black filled lines: DMR0, DMR2 and *IGF2/H19* ICR, shown below the exons as (methylated) on the paternal allele, and without any fill (unmethylated) on the maternal allele. Orange arrowheads: enhancers. Yellow sun: CTCF.

In humans, there are still some missing links relating to *IGF2/H19* imprinting mechanisms. One of them includes two additional differentially methylated regions (Figure 1): DMR0, surrounding *IGF2* promoter P0 and methylated on the paternal allele (Murrell et al., 2008) and DMR2, in *IGF2* exon 9 (Murrell et al., 2008). DMR0 does not bind CTCF (Vu et al., 2010). The existence of two additional imprinted genes in this region *IGF2AS* (Okutsu et al., 2000), and *PIHit (paternally expressed Igf2/H19 intergenic transcript)* (Court et al., 2011) adds even more complexity to the whole chromosomal locus; together with the CTCF paralogue, BORIS/CTCFL (Brother of Regulator of Imprinted Sites/CTCF-like) (Loukinov et al., 2002).

Both proteins, CTCF (Uniprot: P49711) and BORIS (Uniprot: Q8NI51), share 74% homology in their 11-Zn-finger, DNA binding domains. Their N' and C' termini have less than a 10% sequence homology. This implies that they can bind to the same DNA sequences (for example, *IGF2/H19* ICR) but each of them interacts with different protein partners and has different function. For example, BORIS is involved in the *establishment* of *IGF2/H19* imprinting marks in the male germline (Jelinic et al., 2006). CTCF is involved in the *interpretation* of these imprinted marks in somatic cells (Hore et al., 2008).

5.3 Loss of imprinting in cancer

Loss of imprinting (LOI) in cancer is manifested as either activation of normally epigenetically silenced allele resulting in biallelic expression, or as silencing of normally active allele (Table 1; *IGF2* and *H19* not shown).

Imprinted gene	Cancer type	Reference
	Biallelic expression	
IGF2AS	Wilms' tumor	(Vu et al., 2003)
PEG1/MEST	Invasive breast cancer	(Pedersen et al., 1999)
	Lung cancer	(Kohda et al., 2001)
LIT1	Colorectal cancer	(Nakano et al., 2006)
IPW	Testicular germ cell tumor	(Rachmilewitz et al., 1996)
	Loss of expression	
PEG3	Glioma	(Maegawa et al., 2001)
	Endometrial, cervical and ovarian cancer cell lines	(Dowdy et al., 2005)
PLAGL1/ZAC1	Ovarian cancer	(Abdollahi et al., 2003)
	Breast cancer	(Abdollahi et al., 2003)
ARHI/DIRAS3/NOEY2	Breast cancer	(Yuan et al., 2003)
ITUP1	Glioma cell lines	(Maegawa et al., 2004)
CDKN1C	Bladder cancer	(Hoffmann et al., 2005)
	Lung cancer	(Kondo et al., 1996)
MEG3	Meningioma	(Zhang, X. et al., 2010)

Table 1. Imprinted genes - Loss of imprinting in cancer

Theoretically, in the case of imprinted tumor suppressor genes, loss of the expression from only one functional allele could contributes to tumorigenesis by mimicking "the second hit", according to Knudson's Two Hits Hypothesis (Knudson, 1971). The same effect on the cell dividing potential and growth has the biallelic expression of the imprinted, growth promoting gene; mice with *Igf2* LOI in intestines have less differentiated intestines and develop twice as many adenomas, in comparison with control (Sakatani et al., 2005). This is, of course, simplified presentation which may not be realistic at all due to existence of many different regulating processes and signaling molecules included in the process of carcinogenesis.

5.3.1 Loss of IGF2 and H19 imprinting in cancer

IGF2 and *H19* LOI has been demonstrated in many different types of cancer (reviewed in Jelinic & Shaw, 2007). We were the first to analyze the *IGF2/H19* imprinting status in laryngeal squamous cell carcinoma samples (LSCCs) (Grbesa et al., 2008) where we detected *IGF2* LOI in 33% of LSCCs and 28% of adjacent non-tumorous laryngeal tissues. This finding was already discussed. At that time, the *IGF2* LOI in normal tissues has been detected only in colon mucosa of patients with colorectal cancer (Cui et al., 2003), and it was also known that *IGF2* LOI exists in peripheral blood lymphocytes, in 10% of normal population (Sakatani et al., 2001). It is difficult to imagine the *IGF2* LOI as a cancer biomarker as, presently, we detect LOI at the level of easily degraded mRNA (that needs to be entirely free of DNA). Additional problem presents restriction of analysis to the polymorphic sites (SNP), that are not necessarily informative for certain markers, in certain populations (Kaaks et al., 2009). We have also detected *H19* LOI in 23% of LSCCs, in line with el-Naggar's results (el-Naggar et al., 1999). Increased *H19* expression in LSCCs was reconfirmed recently (Mirisola et al., 2011).

In a non-tumorous cell, based on the insulator model, one expects existence of MOI (maintenance of imprinting) through monoallelic *IGF2* and monoallelic *H19* expression. If there is a LOI, the insulator model proposes biallelic expression of *IGF2* or *H19*, and no expression of the partner (for example, *IGF2* biallelically expressed, *H19* silenced). But, this is not the case. We have detected, in a small subset of samples (2/10) informative for both *IGF2* and *H19* imprinting analyses, biallelic *IGF2* expression (*IGF2* LOI) joined with *H19* MOI. In remaining eight samples, the imprinting was maintained. This was also observed in a broad group of head and neck cancers (among them, 14 LSCCs) (el-Naggar et al., 1999). Again, we are here facing the challenge related to the presence of "contaminating", non-tumorous cell with an open possibility that they contributed to the "mixed" result. The solution could be the usage of methods which enable analysis of *IGF2/H19* imprinting in the individual cells (for example RNA fluorescent *in situ* hybridization).

We have also analyzed the methylation of the 6th CTCF-binding site (CBS6) within the ICR by methylation restriction PCR, MR-PCR (Ulaner et al., 2003a). In the samples with *IGF2* and *H19* MOI, the CBS6 was hemimethylated, while its methylation appeared aberrant in the tissue samples with *IGF2* or *H19* LOI (Grbesa et al., 2008). The analysis of CTCF binding to the CBS6 by chromatin immunoprecipitation (ChIP) was not performed due to the well known problem in this part of molecular oncology: the limited amount of tissue. However, based on ours and other groups results, it seems that *IGF2* and *H19* LOI cannot be explained solely on the basis of the level of CBS6 methylation (Cui et al., 2002) and its occupancy by the CTCF (Ulaner et al., 2003b). In cancer cell lines with IGF2 LOI, the whole 3-D structure of the IGF2/H19 locus is dramatically changed, in comparison with normal cells and cancer cells with IGF2 MOI (Vu et al., 2010). As the prototype of the method that should be used for this kind of analysis (The Chromosome Conformation Capture Original Copy Assay (3C-OC) coupled to QPCR and 3D software analysis) does not represent the standard technique applicable for clinical tissue specimen analyses, it will take some time before implementing this type of research on native clinical material.

5.3.2 New player on the scene – BORIS

The BORIS is involved in establishment of methylation marks at paternal *IGF2/H19* ICR during spermatogenesis (Jelinic et al., 2006). Its expression has been detected in various cancers and cancer cell lines (D'Arcy et al., 2008; Hoffmann et al., 2006; Hong et al., 2005; Jones et al., 2011; Kholmanskikh et al., 2008; Looijenga et al., 2006; Renaud et al., 2007; Risinger et al., 2007; Smith et al., 2009; Ulaner et al., 2003a; Woloszynska-Read et al., 2007).

In our LSSC samples, BORIS was expressed with both maintained *IGF2* and *H19* imprinting but also in the samples with IGF2/H19 LOI (Grbesa I, unpublished results). Recently, 23 *BORIS* transcript variants that may potentially produce 17 BORIS polypeptides were discovered (Pugacheva et al., 2010). In human tissues, polypeptides that correspond to calculated molecular weight of some of the BORIS isoforms, have been detected with polyclonal anti-BORIS antibody (Jones et al., 2011) but the role of different BORIS isoforms in establishment and maintenance of IGF2/H19 imprinting remains to be elucidated.

6. Poly(ADP-ribosyl)ation in regulation of DNA methylation

Since the seminal work by Feinberg and Vogelstein pointing to the global hypomethylation in tumor cells (Feinberg & Vogelstein, 1983), many reports followed documenting aberrant aquisition of methylation marks at discrete loci in the genome, most notably those comprising genes involved in cell cycle control. Such methylation pattern is opposite to the bimodal methylation pattern, characterized by global DNA methylation and hypomethylation of CpG islands, that is physiologically aquired at the time of embryo implantation and faithfully maintained throughout life (Brandeis et al., 1993). The search for *cis*- and *trans*-acting factors that orchestrate such bimodal methylation pattern has since been the focus of scientific interest.

Twenty years ago, linker histone H1 has been identified to have inhibitory effect on DNA methylation (Caiafa et al., 1991). Shortly thereafter, the difference between H1 histone isoforms, at that time termed as "tightly-bound" and "loosely-bound," in regulation of DNA methylation were observed. In contrast to "typical" loosly-bound histone H1, tightly-bound histone H1 has been shown to facilitate methylation of linker DNA (Santoro et al., 1993). The histone H1, which is able to bind CpG-rich DNA sequences and inhibit double-stranded DNA methylation, has later been identified as variant H1e (Santoro et al., 1995) that promotes chromatin condensation or, upon poly(ADP-ribosyl)ation (pARylation), chromatin decondensation (D'Erme et al., 1996). Appart from the importance of this histone variant and its pARylation in chromatin decondensation, which allows recruitment of transcription factors, the same group has demonstrated the mandatory role of pARylation in the maintenance of hypomethylated state of CpG islands in mouse fibroblasts (Zardo & Caifa, 1998).

Inhibition of pARylation by competitive PARP inhibitor, 3-aminobenzamide (3-AB), enhances DNA metylation (Zardo & Caifa, 1998). pARylation is process catalyzed by poly(ADP-ribose) polymerases (PARP), which use NAD as a substrate to build up polymers of ADP-ribose on acceptor proteins, including PARP-1 (D'Amours et al., 1999). It is the founding member of this enzyme family, accounting for more than 90% of cellular pARylating capacity. It is able to form long linear or branched ADP-ribose polymers composed of several up to 200 ADP-ribose units. There are at least 28 sites in PARP-1 automodification domain upon which long and branched ADP-ribose polymers bind (Juarez-Salinas et al., 1982). The negative charge of ADP-ribose polymers makes them resemble nucleic acids and compete with them for binding different protein partners. ADP-ribose polymers, either covalently linked to acceptor proteins or protein-free, are also able to non-covalently bind proteins (Malanga & Althaus, 2005). The binding of ADP-ribose polymers to acceptor proteins is dependent on the presence of amino-acids consensus, poly(ADP-ribose)-binding motifs, that allows for non-covalent binding (Pleschke et al., 2000). Those consensuses are present in a wide variety of proteins with very divergent functions, ranging from structural proteins such as histones (Althaus et al., 1995), proteins involved in DNA repair to enzymes involved in regulation of DNA topology (Malanga & Althaus, 2005). Binding of negatively charged poly(ADP-ribose) polymers functionaly and structuraly modifies acceptor proteins (Panzeter et al., 1992).

The DNMT1 has two amino-acid consensus motifs for binding poly(ADP-ribose) polymers in its N-terminal domain. It was demonstrated that pARylated PARP-1 and DNMT1 form complex *in vivo*, and that either PARP-1-associated or free poly(ADP-ribose) polymers are able to inhibit catalytic activity of DNMT1 (Reale et al., 2005). The majority of PARP-1 molecules in normal cell is unmodified (D'Amours et al., 1999) and the mechanism directing the minority of, under physiological conditions, automodified PARP-1 molecules to CpG islands remains an open question. In an effort to elucidate the possible involvement of pARylation in the regulation of Dnmt1 gene promoter, Guastafierro et al. (2008) have examined transcription factors known to be subject to covalent poly(ADP-ribosyl)ation (Hassa et al., 2006). Highly conserved multifunctional transcription factor, CTCF (Ohlsson et al., 2001), attracted their attention based on its role in protection of DNA from methylation and functional dependence on pARylation. The key role of pARylation of CTCF in its insulator/ enhancer blocking function has been reported by Yu et al. in 2004. The role of CTCF in regulation of *IGF*2/*H19* imprinting has been abolished by treatment with PARP-1 inhibitor 3-AB. To establish whether the lack of CTCF pARylation is indeed responsible for the loss of *IGF*2 imprinting, the association of poly(ADP-ribose) polymers with H19 ICR was examined by ChIP on wild-type and mutant-type ICR containing CTCF-binding sites. The pARylation mark has been present only if the wild-type allele has been inherited maternally. In the lack of specific antibodies recognizing pARylated CTCF, however, it cannot be ruled out that CTCF binding to its target sites is necessary for activation of PARP-1 or other members of poly(ADP-ribose) polymerase family that would pARylate proteins other than CTCF in that region. Indeed, it has been demonstrated that, in addition to its previously recognized characteristic of being acceptor of poly(ADP-ribose) polymers, CTCF is interacting with PARP-1 and is able to activate PARP-1 in the absence of DNA nicks, whereby both proteins become pARylated and negatively affect DNA methylation machinery (Guastafierro et al., 2008). The epigenetic regulation of tumor suppressor *p16^INK4a* provided some insight into CTCF and PARP-1 DNA-binding and the influence of their pARylation relative to the expression of this gene and several other CTCF-regulated genes (Witcher & Emerson, 2009). Transcription of *p16^INK4a* depends on CTCF binding to a chromatin boundary ~ 2kb upstream of its transcription start site. In the absence of CTCF binding, *p16* is silenced in various types of tumor cells. When associated with the boundary element, in *p16*-expressing cells, CTCF is pARylated. However, no direct association of pARylated CTCF and PARP-1 could be detected in those *p16*-expressing cells. In *p16*-silenced cells, CTCF was not pARylated and bound to the methylated boundary element, but CTCF-PARP-1 complex could be readily detected by co-immunoprecipitation. Moreover, the authors found that pARylated CTCF dissociates from PARP-1, whereas pARylated CTCF remains associated with PARP-1 and loses its function, at this boundary element. Therefore, it is conceivable that deregulation of pARylation may impart the aberrant association of CTCF and PARP-1 and change the association of CTCF with its DNA binding sites. Relevant to the possible influence of the CTCFs binding to DNA and its protective role against DNA methylation is the recent report on the ability of CTCF to form an unusual DNA structure (MacPherson & Sadowski, 2010). Considering that, in addition to the classical view that PARP-1 is activated by DNA nicks, various non-linear DNA structures are able to activate this enzyme (Lonskaya et al., 2005), the property of CTCF to loop DNA may be yet another facet in connecting processes of DNA methylation and poly(ADP-ribosyl)ation.

7. Epigenomic therapy

The inhibition of DNMTs has been used in epigenetic cancer therapy, based on the idea that seems to be quite simple: what is hypermethylated, needs to be normomethylated. So, if we consider the act of removing the methylation marks from hypermethylated promoters of tumor suppressor genes whose protein products are involved in regulation of cell cycle, apoptosis and DNA repair as a therapeutic act, we may be well on a right way.

There are two kinds of DNA methylation inhibitors: nucleoside (Fig. 2) and non-nucleoside analogues. The consequences of nucleoside analogue incorporation into DNA (in this situation, in lieu of cytosines) is DNMT binding and blocking, causing depletion of overlay active enzyme molecules with DNA methyltransferase activity. Two DNA methylation inhibitors, cytidine analogs, were approved by the US Food and Administration (FDA), for the treatment of Myelodysplastic Syndrome and certain forms of leukemias: 5-azacytidine (azacytidine, Vidaza™), which was approved in May, 2004 and 5-aza-2'-deoxycytidine (5-azaCdR, decitabine, Dacogen™), which is a deoxyribose analog of 5-azacytidine, approved in May, 2006 (Figure 2).

The antineoplastic effects caused by these two drugs are related to targeted DNA demethylation (and consequential restoration of gene activity necessary for differentiation) and a direct cytotoxic effect on abnormal, rapidly dividing hematopoietic cells in the bone marrow. Non-proliferating cells are relatively insensitive to these two drugs, but their inherently toxic effects do produce certain side-effects.

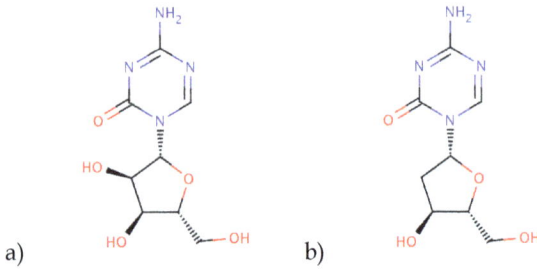

a)   b)

Fig. 2. 5-Azacytidine (a) and 5-Aza-2'-deoxycytidine (b)

7.1 5-azacytidine and decitabine

5-azacytidine was described as a DNMT inhibitor more than 30 years ago (Jones & Taylor, 1980). The first approved clinical targets for 5-azacytidine were Myelodysplastic Syndrome (MDS) and acute myeloid leukema. The drug can be applied subcutaneously or through IV infusion.

This nucleotide analogue becomes incorporated into DNA in place of cytosine after being modified by ribonucleotide reductase and subsequent phosphorylation. When DNMT1 „recognizes" it as a unmethylated substrate in a newly synthesized DNA strand and „approaches" it, it becomes trapped through covalent binding to the incorporated analogue. The reduced level of DNA methylation that follows represents the consequence of passive demethylation in consequent cell cycles due to the lack of functional enzyme. 5-azacytidine can also be phosphorylated by uridine-cytidine kinase and, as such, incorporated into RNA. When applied subcutaneously, it may cause: nausea, anemia, thrombocytopenia, vomiting, pyrexia, leukopenia, diarrhea, injection site erythema, constipation, neutropenia and ecchymosis. Most common adverse reactions through IV application, according to the FDA, included: petechiae, rigors, weakness and hypokalemia.

MDS is also the primary therapeutic indication for Dacogen™ , as in the case with Vidaza™. The most prominent side-effects associated with Dacogen™ treatment are neutropenia and thrombocytopenia.

The third analogue, zebularine, has several advantages over the two previously mentioned compounds. It is more stable, highly selective for cancer cells and, hence, far less toxic (Cheng et al., 2004). However, this potential drug seemed to fail after being very successful in a small pilot study, because a very high dosage of the drug were needed to obtain the desired antitumorous effect (Goffin & Eisenhauer, 2002). In all three cases, the facts which include non-selective DNA targeting resulting in side-effects, were the basis for approaching the problem in a different way. That is, developing compounds which target DNMTs directly, without prior DNA incorporation requirement.

7.1.1 New methods in exploring activity of methylation inhibitors

In 2008, Illumina Golden Gate arrays were used for direct characterization of the effects of azacytidine application in three different leukemia cell lines (HEL, HL-60, K562) and ten patients who fulfilled the WHO criteria for MDS. In the cell lines, the effect of the drug on

DNMT1 protein level differed, the lowest being in HL-60, while HEL cells appeared relatively resistant to DNMT1 depletion. Accordingly, HL-60 was considerably demethylated, while the HEL cell line did not exhibit significant change in global methylation level. After performing an array-based methylation profiling (1,505 CpGs representing 807 cancer-associated genes), the results were very interesting. In untreated cell lines, the number of methylated CpGs exceeded 80%. After treatment, more than 80% of spots became demethylated, but only in HL-60 and K562 cell lines. There was no consistent demethylation trend in HEL cells. Of importance, flow cytometry analysis showed similar overall cell cycle profiles in all three cell lines. The results obtained on patients' samples (6/10, as six patients completed at least one treatment course consisting of four cycles) differed significantly. In three patients, the methylation levels remained the same, while in three other persons the level of methylation decreased significantly, through a cyclic demethylation, following the cyclic administration of the drug (Stresemann et al., 2008; Stresmann & Lyko 2008).

In 2011., a genome-scale Infinium analysis (27,578 CG nucleotides; more than 14,475 associated genes) was performed on two human colon cancer cell lines (HCT116 and double knockout (DKO) HCT116, lacking DNMT1 and DNMT3B) and a HL-60 leukemia cell line treated with both azacytidine and decitabine. The bimodal peaks of methylation distribution was found in both treated and untreated cells, representing spots with low and high levels of methylation. These experiments not only showed more potent demethylation activity of decitabine when compared with azacytidine, but also preferential demethylation at specific loci and demethylation resistance of certain number of CGs, in HCT116. The results from these *in vitro* study shed new light on problems encountered in clinical work with these drugs: not only was the degree of demethylation of the whole genomic DNA higher than gene-specific demethylation (this is something that we do not want to happen, as these drugs were implemented in the clinic in order to demethylate the hypermethylated promoters of tumor suppressor genes), but also the spatial distribution of demethylated CpGs mimicked the distribution found in DKO HCT116. However, when the cluster of cancer-related genes associated CGs was analyzed separately, it turned out that 906 out of 2,125 were hypermethylated and both drugs were very efficient in removing methylation marks. It is hard to distinguish which gene (and joining CGs) represents the clean „cancer-related" gene. Many genes that were considered to be „inflammatory genes" or „metabolic genes", turned out to be "cancer genes", as well. One should be careful regarding this kind of clustering because we are currently far away from a complete understanding of how certain signaling proteins/pathways interact, regardless of to which cluster they were primarily asigned.

When performing computational modeling for the presence of transcription factors binding sites in 851 CpGs representing 644 genes, demethylation - sensitive and – resistant CGs showed different types of enrichment. For example, binding sites of Forkhead box (Fox) transcription factors were enriched in demethylation sensitive genes, while basic Helix-Loop-Helix transcription factor binding sites turned out to be enriched in demethylation resistant genes (Hagemann et al., 2011).

7.2 Non-nucleoside compounds

There are several more, potentially promising, non-nucleoside, candidates. Some of them are well known drugs/healing compounds, such as curcumin.

- Procaine (a well known local anesthetic) and its derivative, procainamide (a well known drug for treating cardiac arrhythmia) have shown demethylating activity in cancer cell lines of different origin. They were shown to be a specific inhibitors of DNMT1 (Lee et al, 2005). For example, in prostate cancer cells, procainamide restores *GSTP*1 expression through demethylation of *GSTP*1 promoter (Lin X et al., 2001). In lung cancer, these drugs induce demethylation of *WIF*-1 (Wnt Inhibitory Factor) promoter, a negative regulator of the Wnt-signaling pathway (Gao et al., 2009).
- Hydralazine: The methylation inhibitory role was shown to be specifically related to the inhibition of DNMT (Angeles et al., 2005). Its combination with magnesium valproate, seems to be promising in treating different types of malignant disease, including MDS (Candelaria et al., 2011).
- The inhibitory effect of (-)- epigallocatechin-3-gallate (EGCG), the healing compound from green tea, was shown in 2003 (Fang et al., 2003). It was then shown for the first time, that inhibition of DNA methylation can be inhibited by a commonly consumed dietary constituent. At the same time, these results suggested the potential use of EGCG for the prevention of cancer-related gene silencing. The authors measured the DNMT1 catalytic activity and performed molecular modeling of the interaction between EGCG and DNMT1. Finally, they proved reversal of hypermethylation through the reactivation of expression of several genes (*RAR*, *MGMT*, *p16INK*4a, and *hMLH*1).
- Genistein, the soy bean isoflavone, was also shown to influence DNMTs. Based on a literature search, there seems to be only one study exploring its efficacy as a DNA methyltransferase inhibitor (Li et al., 2009).

7.2.1 Curcumin

Curcumin (diferulolymethane) is the yellow pigment found in the cooking spice turmeric (*Curcuma longa linn*). Curcumin is a strong inhibitor of the NF-κB signaling pathway (Gupta et al., 2011). Having in mind the central role of NF-κB in many different signaling pathways, it is not surprising that this compound shows anti-inflammatory, anti-oxidant, antimicrobial and, finally, anticancer activity. Curcumin is currently being investigated for its chemopreventative efficacy in a variety of solid tumors. So far, most of the controlled clinical trials of curcumin are in phase I (Hatcher et al., 2008), suggesting that oral curcumin is more likely to be effective as a therapeutic agent in cancers of the gastrointestinal tract than in other tissues (Sharma et al., 2005). The results of one non-randomized, open-label, phase II clinical trial conducted in the U.S. were published recently, reporting on the first 25 patients with advanced pancreatic cancer. The patients did not receive any concomitant chemotherapy or radiotherapy. There was partial response in one patient and disease stabilization in other patient, for approximately 2.5 years (Dhillon et al., 2008). Another clinical phase I/II trial included 21 gemcitabine-resistant pancreatic cancer patients who received, like in Dhillon's study, 8 grams of curcumin daily, together with gemcitabine-based chemotherapy in this instance. This combination was shown to be „safe and feasible in patients with pancreatic cancer and warrants further investigation into its efficacy" (Kanai et al., 2011).

There are many efforts to improve curcumin's bioavailability. The most recent results reported on nanoparticle curcumin (Theracurmin), show that this form of curcumin can safely increase plasma curcumin levels in a dose-dependent manner at least up to 210 mg, without saturating the absorption system (Kanai et al., 2011).

We have shown that curcumin selectively inhibits the *H19* transcription in several different tumor cell lines, but not in non-tumorous cells. We do not think that protein-non-coding *H19* mRNA itself necessarily exerts any kind of vital oncogenic or tumor-suppressive function, but we do think that its mRNA presence indicates vivid, globally deregulated cellular transcription (Novak Kujundzic et al., 2008).

It has been confirmed that curcumin interacts directly with 33 proteins, one of them being DNMT1 (Liu et al., 2009). It is considered that this binding causes direct inhibition of the enzyme and represents the molecular basis for the DNA hypomethylating activity of curcumin (Liu et al., 2009).

8. 5-Hydroxymethylcytosine

As discussed in previous sub-chapters, methylated cytosine entered the spotlight of the international scientific community, primarily due to our understanding of how genes are, or need to be, regulated. We are just entering the era of full appreciation of the importance of one more cytidine modification, discovered in bacteriophage, in 1952. (Wyatt & Cohen, 1952). It was „rediscovered" in 1972. in rat tissue, but was neglected because the results did not seem to be reproducible (Penn et al., 1972). However, thanks to knowledge gained during these almost 60 years, combined with advances in technology, we are now learning about a sixth nucleotide in our genome (Münzel et al., 2011). Only two years ago, two papers in Science showed that mammalian DNA contains 5'-hydroxymethylcytosine (hmC; 5 hmC; 5-HOMEdC, Figure 3) (Kriaucionis & Heintz, 2009; Tahiliani et al., 2009).

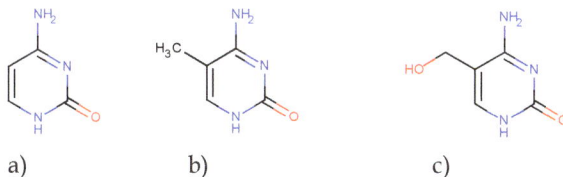

a)   b)   c)

Fig. 3. Cytosine and its modifications.

Unmodified cytosine (a), 5-methylcytosine (b), 5-hydroxymethylcytosine (c).

To date, based on a few *in vitro* experiments, it has been thought that hmC presents the major oxidative product of mC (Bienvenu et al. 1996; Wagner & Cadet, 2010).

Ideal for detecting the methylated cytosines, bisulfite fails on hmC. As a result, there is a problem with positioning the sixth nucleotide in the DNA molecule. There is hope from nanopore sequencing, because the first published results show the difference between mC and hmC in single-, and double-stranded DNA (Wanunu et al., 2011).

So far, most of the work on 5-hmC and TET group of proteins (TET1, TET2 and TET3) was performed on embryonic stem cells and there are only a few papers dealing with „the sixth nucleotide" in cancer. The TET proteins which can modify 5-methylcytosine in humans were initially discovered through a computational search showing these proteins as mammalian homologs of the trypanosome proteins JBP1 and JBP2, the enzymes proposed to oxidize the 5-methyl group of thymine (Tahiliani et al., 2009). Predictably, hmC levels decrease upon RNA interference-mediated depletion of TET1.

The role of hmC in cancer and its role in participating or in creating epigenomic networks, is almost entirely unknown. There are a few, recently published papers, showing its influence on the affinity (a strong decrease) with which the MBD proteins bind to DNA „occupied" by hmC. The 5-hmC immunoassay (developed by Liu at coworkers) was applied to various healthy human tissues and four cancer cell lines. The differences in hmC content among the tested cell lines were minor and insignificant. The tissue analyses showed that brain tissue has the highest content of hmC. Among all tissues tested, the lowest level was detected in lung, breast and placenta. When compared with mC distribution in various tissues (which published data showed a range of 1-2.5), the differences in hmC tissue-specific content were very strong. In cancer tissues, when compared to adjacent, non-tumorous tissue, the level of hmC was significantly lower; in one case 7.7-fold, in the other 28-fold (Li & Liu, 2011). So, the question was asked: is it possible that this increase occurs because the global level of methylated cytosines decreases in cancer cells?

Part of the answer was given in a paper published in September, 2011 (Haffner et al., 2011). The authors analyzed 78 carcinomas and 28 normal tissue samples (prostate, breast and colon). They have shown, by using immunohistochemical staining they developed, a significant decrease of hmC in tumorous tissues. There was also a significant difference in hmC tissue distribution: in normal tissue, the signals were strongest in the terminally differentiated luminal cells and far less strong in basal cells. In cancer tissues, the differences were very clear at the border between the tumor and non-tumorous tissue. However, although very prominent, these changes did not allow for any association with clinicopathological features, including the tumor grade (level of differentiation). Although there is a significant similarity between Haffner's and Li's result, the methods they used were quite different, which – to be sure, does not diminish the quality and importance of their results. In any event, the hmC story will need to be explored on many clinical samples before we allow ourselves to conclude anything about their prognostic significance in cancer patients.

Williams and colleagues published a very extensive study on TET protein family member function, showing the necessity of TET1 time-specific expression during development. In their experimental model, TET1 localized to both gene bodies and transcription start sites (TSS), being especially enriched at genes with high CpG content, while mC localized in regions with low CpG content. The results indicate „that TET1, by converting mC to hmC serves an important function in the regulation of DNA methylation fidelity" (Willimas et al., 2011). What, one should ask, is the consequence of the hmC presence in the gene promoters/bodies? It seems that, in a pure *in vitro* system, based on CMV promoter and HeLa cells extracts, the presence of hmCs in the gene promoter inhibits transcription, while their presence in the gene body does not directly inhibit transcription (Robertson et al., 2011).

9. Conclusion

In this chapter, we have covered aspects of deregulated DNA methylation in cancer, including a review of older data and introducing the most recent findings. By using this approach, we have tried to show maybe the most intriguing and certainly the most emerging aspects of molecular biology of a cancer cell, at the time of preparing this chapter. Certainly, the new exciting discoveries in the field of cancer epigenomics that we are presenting here are only part of emerging sets of data. The new papers with exciting findings are coming to scientific community almost on a daily base does and, for that reason, we did not allow to ourselves offering any hard conclusion, at this time period. We are aware that there are many more issues and mechanisms for discussion, such as the interactions of DNA and proteins, methylation related and unrelated, that we did not discuss. For that reason, we all look forward to future books and articles providing insight on these and like topics.

10. Acknowledgement

We are grateful to Dr. Višnja Stepanić and Mr. Zlatko Pandžić for their technical help in preparing this chapter. Our special thanks go to Mr. Aaron Etra who helped us with a careful revision of the text. Our research is supported by grant #098-0982464-2511 (PI:KGT), from the Ministry of Science, Education and Sport of the Republic of Croatia.

11. References

Aapola, U., Kawasaki, K., Scott, H.S. et al. (2000). Isolation and initial characterization of a novel zinc finger gene, DNMT3L, on 21q22.3, related to the cytosine-5-methyltransferase 3 gene family. *Genomics*, Vol.65, No.3, (May 2000), pp. 293-298, ISSN 0888-7543

Althaus, F.R., Bachmann, S., Hofferer, L. et al. (1995). Interactions of poly(ADP-ribose) with nuclear proteins. *Biochimie*, Vol.77, No.6, pp. 423-432, ISSN 0300-9084

Angeles, E., Vazquez-Valadez, V.H., Vazquez-Valadez, O. et al. (2005). Computational studies of 1-Hydrazinophtalazine (Hydralazine) as antineoplastic agent. Docking studies on methyltransferase. *Letters in Drug Design & Discovery*, Vol.2, No 4, (June 2005), pp. 282-286, ISSN: 1570-1808

Asada, K., Asada, R, Yoshiji, H. et al. (2006). DNA cytosine methylation profile in various cancer-related genes is altered in cultured rat hepatocyte cell lines as compared with primary hepatocytes. *Oncology Reports*, Vol.15, No.5, (May 2006), pp. 1241-1248, ISSN 1021-335X

Baylin, S.B. & Jones, P.A.(2011). A decade of exploring the cancer epigenome - biological and translational implications. *Nature Reviews. Cancer*. Vol.11, No.10, (September 2011), pp. 726-734, ISSN 1474-175X

Bell, A.C. & Felsenfeld, G. (2000). Methylation of a CTCF-dependent boundary controls imprinted expression of the Igf2 gene. *Nature*, Vol.405, No.6785, (May 2000), pp. 482-485, ISSN 0028-0836

Bienvenu, C., Wagner, J.R. & Cadet, J. (1996). Photosensitized oxidation of 5-Methyl-2'-deoxycytidine by 2-Methyl-1,4-naphthoquinone: Characterization of 5-(Hydroperoxymethyl)-2'-deoxycytidine and stable methyl group oxidation product. *Journal of the American Chemical Society*, Vol.118, No.46, (November 1996), pp. 11406-11411, ISSN 0002-7863

Biniszkiewicz, D., Gribnau, J., Ramsahoye, B. et al. (2002). Dnmt1 overexpression causes genomic hypermethylation, loss of imprinting, and embryonic lethality. *Molecular and Cellular Biology*, Vol.22, No.7, (April 2002), pp. 2124–2135, ISSN 0270-7306

Bock, C., Walter, J., Paulsen, M., Lengauer, T. (2007). CpG island mapping by epigenome prediction. *PloS Computational Biology*, 3(6):e110., doi: 10.1371/journal.pcbi.0030110, ISSN 1553-734X

Boehm, T.L. & Drahovsky, D. (1981). Hypomethylation of DNA in Raji cells after treatment with N-methyl-N-nitrosourea. *Carcinogenesis*, Vol.2, No.1, (January 1981), pp. 39–42, ISSN 0143-3334

Bostick, M., Kim, J., Estève, P-O. et al. (2007). UHRF1 plays a role in maintaining DNA methylation in mammalian cells. *Science*, Vol.317, No.5845, (September 2007), pp. 1760-1764, ISSN 0036-8075

Brandeis, M., Ariel, M. & Cedar, H. (1993). Dynamics of DNA methylation during development. *BioEssays: News and Reviews in Molecular, Cellular and Developmental Biology*, Vol.15, No.11, (November 1993), pp. 709-713, ISSN 0265-9247

Broët, P., Dalmasso, C., Tan, E.H. et al. (2011). Genomic profiles specific to patient ethnicity in lung adenocarcinoma. *Clinical* Cancer *Research*, Vol.17, No.11, (June 2011), pp. 3542-3550, ISSN 1078-0432

Caiafa, P., Reale, A., Allegra, P. et al. (1991). Histones and DNA methylation in mammalian chromatin. Differential inhibition by histone H1. *Biochimica et Biophysica Acta*, Vol.1090, No.1, (August 1991), pp. 38-42, ISSN 0006-3002

Candelaria, M., Herrera, A., Labardini, J. et al. (2011) Hydralazine and magnesium valproate as epigenetic treatment for myelodysplastic syndrome. Preliminary results of a phase-II trial. *Annals of Hematology*, Vol.90, No.4. (April 2011), pp 379-87, ISSN 0939-5555

Cheng, J.C., Yoo, C.B., Weisenberger, D.J. et al. (2004). Preferential response of cancer cells to zebularine. *Cancer Cell*, Vol.6, No.2, (August 2004), pp. 151-158, ISSN 1535-6108

Choi, S.H., Heo, K., Byun, H-M., An, W. Lu, W., Yang, A.S. Identification of preferential target sites for fuman DNA methyltransferases. (2010). *Nucleic Acids Research*, Vol.39, No.1, (September 2010), pp. 104-118, ISSN 0305-1048

Court, F., Baniol, M., Hagege, H. et al. (2011). Long-range chromatin interactions at the mouse Igf2/H19 locus reveal a novel paternally expressed long non-coding RNA. *Nucleic Acids Research*, Vol.39, No.14, (August 2011), pp. 5893-5906, ISSN 0305-1048

Cui, H., Onyango, P., Brandenburg, S. et al. (2002). Loss of imprinting in colorectal cancer linked to hypomethylation of H19 and IGF2. *Cancer Research*, Vol.62, No.22, (November 2002), pp. 6442-6446, ISSN 0008-547

Cui, H., Cruz-Correa, M., Giardiello, F.M. et al. (2003). Loss of IGF2 imprinting: a potential marker of colorectal cancer risk. *Science*, Vol.299, No.5613, (March 2003), pp. 1753-1755, ISSN 0036-8075

D'Erme, M., Zardo, G., Reale, A. et al (1996). Co-operative interactions of oligonucleosomal DNA with the H1e histone variant and its poly(ADP-ribosyl)ated isoform. *The Biochemical Journal*, Vol.316, No.2, (June 1996), pp. 475-480, ISSN 0264-6021

D'Amours, D., Desnoyers, S., D'Silva, I. et al. (1999). Poly(ADP-ribosyl)ation reactions in the regulation of nuclear functions. *The Biochemical Journal*, Vol.342, No.2, (September 1999), pp. 249-268, ISSN 0264-6021

Daniel, F.I., Cherubini, K., Yurgel, L.S. et al. (2011). The role of epigenetic transcription repression and DNA methyltransferases in cancer. *Cancer*, Vol.117, No.4, (February 2011), pp. 677-687, ISSN 0008-543X

D'Arcy, V., Pore, N., Docquier, F. et al. (2008). BORIS, a paralogue of the transcription factor, CTCF, is aberrantly expressed in breast tumours. *British Journal of Cancer*, Vol.98, No.3, (February 2008), pp. 571-579, ISSN 0007-0920

Dhillon, N., Aggarwal, B.B., Newman, R.A. et al. (2008). Phase II trial of curcumin in patients with advanced pancreatic cancer. *Clinical Cancer Research*, Vol.14, No.14, (July 2008), pp. 4491-4499, ISSN 1078-0432

Ehrlich, M. (2009). DNA hypomethylation in cancer cells. Epigenomics, Vol1, No.2, (December 2009), pp. 239-259, ISSN 1750-1911

Ehrlich, M., Jiang, G., Fiala, E. et al. (2002). Hypomethylation and hypermethylation of DNA in Wilms tumors. *Oncogene*, Vol.21, No.43, (September 2002), pp. 6694-6702, ISSN 0950-9232

Ehrlich, M., Woods, C.B., Yu, M.C. et al. (2006). Quantitative analysis of associations between DNA hypermethylation, hypomethylation, and DNMT RNA levels in ovarian tumors. *Oncogene*, Vol.25, No.18, (March 2006), pp. 2636–2645, ISSN 0950-9232

El-Naggar, A.K., Lai, S., Tucker, S.A. et al. (1999). Frequent loss of imprinting at the IGF2 and H19 genes in head and neck squamous carcinoma. *Oncogene*, Vol.18, No.50, (November 1999), pp. 7063-7069, ISSN 0950-9232

Espada, J. & Esteller, M. (2007). Epigenetic control of nuclear architecture. *Cellular and Molecular Life Sciences*, Vol.64, No.4, (February 2007), pp. 449-457, ISSN 1420-682X

Estève, P.O., Chang, Y., Samaranayake, M. et al. (2011). A methylation and phosphorylation switch between an adjacent lysine and serine determines human DNMT1 stability. *Nature Structural & Molecular Biology*, Vol.18, No.1, (January 2011), pp. 42-48, ISSN 1545-9985

Fan, H., Chen, L., Zhang, F. et al. (2011). MTSS1, a novel target of DNA methyltransferase 3B, functions as a tumor suppressor in hepatocellular carcinoma. *Oncogene*, in press, doi: 10.1038/onc.2011.411. ISSN 0950-9232

Fang, M.Z., Wang, Y. & Ai, N. (2003). Tea polyphenol (-)-epigallocatechin-3-gallate inhibits DNA methyltransferase and reactivates methylation-silenced genes in cancer cell lines. *Cancer Research*, Vol.63, No.22, (November 2003), pp. 7563-7570, ISSN 0008-5472

Feinberg, A.P. & Vogelstein, B. (1983). Hypomethylation distinguishes genes of some human cancers from their normal counterparts. *Nature*, Vol.301, No.5895, (January 1983), pp. 89–92, ISSN 0028-0836

Gall Trošelj, K. & Novak Kujundžić, R. (2010). Epigenomics – A Bird's Eye Perspective on the Genome. *Periodicum Biologorum*, Vol.112, No.4, (December 2010), pp. 411-418, ISSN 0031-5362

Gao, Z., Xu, Z., Hung, M.S. et al. (2009). Procaine and procainamide inhibit the Wnt canonical pathway by promoter demethylation of WIF-1 in lung cancer cells. *Oncology Reports*, Vol.22, No.6, (December 2009), pp. 1479-1484, ISSN 1021-335X

Gardiner-Garden, M. & Frommer, M. (1987). CpG islands in vertebrate genomes. *Journal of Molecular Biology*, Vol.196, No.2, (July 1987), pp. 261–282, ISSN 0022-2836

Goffin, J. & Eisenhauer, E. (2002). DNA methyltransferase inhibitors - state of the art. *Annals of Oncology*, Vol.13, No.11, (November 2002), pp. 1699-1716, ISSN 0923-7534

Goll, M.G. & Bestor, T.H. (2005). Eukaryotic cytosine methyltransferases. *Annual Review of Biochemistry*, Vol.74, (July 2005), pp. 481-514, ISSN 0066-4154

Goll, M.G., Kirpekar, F., Maggert, K.A. et al. (2006). Methylation of tRNA[Asp] by the DNA methyltransferase homolog Dnmt2. Science, Vol.311, No.5759, (January 2006), pp. 395-398, ISSN 0036-8075.

Gowher, H. & Jeltsch, A. (2002). Molecular enzymology of the catalytic domains of the Dnmt3a and Dnmt3b DNA methyltransferases. *Journal of Biological Chemistry*, Vol.277, No.23, (June 2002), pp. 20409-20414, ISSN 0021-9258

Grbesa, I., Marinkovic, M., Ivkic, M. et al. (2008). Loss of imprinting of IGF2 and H19, loss of heterozygosity of IGF2R and CTCF, and *Helicobacter pylori* infection in laryngeal squamous cell carcinoma. *Journal of Molecular Medicine (Berlin)*, Vol.86, No.9, (September 2008), pp. 1057-1066, ISSN 0946-2716

Guastafierro, T., Cecchinelli, B., Zampieri, M. et al. (2008). CCCTC-binding factor activates PARP-1 affecting DNA methylation machinery. *Journal of Biological Chemistry*, Vol.283, No.32, pp. 21873-21880, ISSN 0021-9258.

Gupta, S.C., Kim J.H., Kannappan, R. et al. (2011). Role of nuclear factor κB-mediated inflammatory pathways in cancer-related symptoms and their regulation by nutritional agents. *Experimental Biology and Medicine*. Vol.236, No.6, (June 2011), pp. 658-671, ISSN 1535-3702

Haffner, M.C., Chaux, A., Meeker, A.K. et al. (2011). Global 5-hydroxymethylcytosine content is significantly reduced in tissue stem/progenitor cell compartments and in human cancers. *Oncotarget*, Vol.2, No.8, (September 2011), pp. 627 - 637, Online ISSN: 1949-2553

Hagemann, S., Heil, O., Lyko, F. et al. (2011). Azacytidine and decitabine induce gene-specific and non-random DNA demethylation in human cancer cell lines. *PLoS ONE*, 6(3):e17388. doi:10.1371/journal.pone.0017388, ISSN 1932-6203

Hark, A.T., Schoenherr, C.J., Katz, D.J. et al. (2000). CTCF mediates methylation-sensitive enhancer-blocking activity at the H19/Igf2 locus. *Nature*, Vol.405, No.6785, (May 2000), pp. 486-489, ISSN 0028-0836

Hassa, P.O., Haenni, S.S., Elser, M. et al. (2006). Nuclear ADP-ribosylation reactions in mammalian cells: where are we today and where are we going? *Microbiology and Molecular Biology Reviews: MMBR*, Vol.70, No.3, (September 2006), pp. 789-829, ISSN 1092-2172

Hatcher, H., Planalp, R., Cho, J. et al. (2008). Curcumin: from ancient medicine to current clinical trials. *Cellular and Molecular Life Sciences*, Vol.65, No.11, (June 2008), pp. 1631-1652, ISSN 1420-682X

Hervouet, E., Lalier, L., Debien, E. et al. (2010). Disruption of Dnmt1/PCNA/UHRF1 interactions promotes tumorigenesis from human and mice glial cells. *PLoS ONE*, 5(6): e11333. doi: 10.1371/journal.pone.0011333, ISSN 1932-6203

Hervouet, E., Vallette, F.M. & Cartron, P-F. (2010). Impact of the DNA methyltransferases expression on the methylation status of apoptosis-associated genes in glioblastoma multiforme. *Cell Death & Disease*, 1(1):e8, doi: 10.1038/cddis.2009.7, ISSN 2041-4889

Hoffmann, M.J., Florl, A.R., Seifert, H.H. et al. (2005). Multiple mechanisms downregulate CDKN1C in human bladder cancer. *International Journal of Cancer*, Vol.114, No.3, (April 2005), pp. 406-413, ISSN 0020-7136

Hoffmann, M.J., Muller, M., Engers, R. et al. (2006). Epigenetic control of CTCFL/BORIS and OCT4 expression in urogenital malignancies. *Biochemical Pharmacology*, Vol.72, No.11, (November 2006), pp. 1577-1588, ISSN 0006-2952

Holliday, R. & Pugh, J.E. (1975). DNA modification mechanisms and gene activity during development. *Science*, Vol.187, No.4173, (January 1975), pp. 226-232, ISSN 0036-8075

Hong, J.A., Kang, Y., Abdullaev, Z. et al. (2005). Reciprocal binding of CTCF and BORIS to the NY-ESO-1 promoter coincides with derepression of this cancer-testis gene in lung cancer cells. *Cancer Research*, Vol.65, No.17, (September 2005), pp. 7763-7774, ISSN 0008-5472

Huang, Y.W., Kuo, C.T., Stoner, K., Huang, T.H., Wang, L.S. (2011). An overview of epigenetics and chemoprevention. *FEBS Letters*. Vol.585, No.13, (July 2011), pp. 2129-2136, ISSN 0014-5793

Jelinic, P. & Shaw, P. (2007). Loss of imprinting and cancer. *The Journal of Pathology*, Vol.211, No.3, (February 2007), pp.261-268, ISSN 0022-3417

Jelinic, P., Stehle, J.C. & Shaw, P. (2006). The testis-specific factor CTCFL cooperates with the protein methyltransferase PRMT7 in H19 imprinting control region methylation. *PLoS Biology*, 4(11): e355. doi:10.1371/journal.pbio.0040355, ISSN 1544-9173

Jones, P.A. & Taylor, S.M. (1980). Cellular differentiation, cytidine analogs and DNA methylation. Cell, Vol.20, No.1, (May 1980), pp. 85-93, ISSN 0092-8674

Jones, T.A., Ogunkolade, B.W., Szary, J. et al. (2011). Widespread expression of BORIS/CTCFL in normal and cancer cells. *PloS ONE*, 6(7): e22399. doi:10.1371/journal.pone.0022399, ISSN 1932-6203

Juarez-Salinas, H.L.V., Jacobson, E.L. & Jacobson, M.K. (1982). Poly(ADP-ribose) has a branched structure in vivo. *Journal of Biological Chemistry*, Vol.257, No.2, pp. 607-609, (January 1982), ISSN 0021-9258

Kaaks, R., Stattin, P., Villar, S. et al. (2009). Insulin-like growth factor-II methylation status in lymphocyte DNA and colon cancer risk in the Northern Sweden Health and Disease cohort. *Cancer Research*, Vol.69, No.13, (July 2009), pp. 5400-5405, ISSN 0008-5472

Kanai, M., Imaizumi, A., Otsuka, Y. et al. (2011) Dose-escalation and pharmacokinetic study of nanoparticle curcumin, a potential anticancer agent with improved bioavailability, in healthy human volunteers. *Cancer Chemotherapy and Pharmacology*, in press, ISSN 0344-5704

Kanai, M., Yoshimura, K., Asada, M. et al. (2011). A phase I/II study of gemcitabine-based chemotherapy plus curcumin for patients with gemcitabine-resistant pancreatic cancer. *Cancer Chemotherapy and Pharmacology*, Vol.68, No.1, (July 2011), pp. 157-164, ISSN 0344-5704

Kaneda, A., Tsukamoto, T., Takamura-Enya T. et al. (2004). Frequent hypomethylation in multiple promoter CpG islands is associated with global hypomethylation, but not with frequent promoter hypermethylation. *Cancer Science*, Vol.95, No.1, (January 2004), pp. 58-64, ISSN 1347-9032

Kholmanskikh, O., Loriot, A., Brasseur, F. et al. (2008). Expression of BORIS in melanoma: lack of association with MAGE-A1 activation. *International Journal of Cancer*, Vol.122, No.4, (February 2008), pp. 777-784, ISSN 0020-7136

Kim, J.Y., Siegmund, K.D., Tavaré, S. et al. (2005). Age-related human small intestine methylation: evidence for stem cell niches. *BMC Medicine*, 3:10, doi:10.1186/1741-7015-3-10, ISSN 1741-7015

Knudson, A.G., Jr. (1971). Mutation and cancer: statistical study of retinoblastoma. *Proceedings of the National Academy of Sciences of the United States of America*, Vol.68, No.4, (April 1971), pp. 820-823, ISSN 0027-8424

Kriaucionis, S. & Heintz, N. (2009). The nuclear DNA base 5-hydroxymethylcytosine is present in Purkinje neurons and the brain. *Science*, Vol.324, No.5929, (May 2009), pp. 929-930, ISSN 0036-8075

Lee, B.H., Yegnasubramanian, S., Lin, X. et al. (2005). Procainamide is a specific inhibitor of DNA methyltransferase 1. *Journal of Biological Chemistry*, Vol.280, No.49, (December 2005), pp. 40749-40756, ISSN 0021-9258

Leighton, P.A., Saam, J.R., Ingram, R.S. et al. (1995). An enhancer deletion affects both H19 and Igf2 expression. *Genes & Development*, Vol.9, No.17, (September 1995), pp. 2079-2089, ISSN 0890-9369

Li, W. & Liu, M. (2011). Distribution of 5-hydroxymethylcytosine in different human tissues. *Journal of Nucleic Acids*, Vol.286, No.28, (July 2011), pp. 24685-24693, ISSN 0021-9258

Li, Y., Liu, L., Andrews, L.G. et al. (2009). Genistein depletes telomerase activity through cross-talk between genetic and epigenetic mechanisms. *International Journal of Cancer*, Vol.125, No.2, (July 2009), pp. 286-296, ISSN 0020-7136

Lin, X., Asgari, K., Putzi, M.J. et al. (2001). Reversal of GSTP1 CpG island hypermethylation and reactivation of pi-class glutathione S-transferase (GSTP1) expression in human prostate cancer cells by treatment with procainamide. *Cancer Research*, Vol.61, No.24, (December 2001), pp. 8611-8616, ISSN 0008-5472

Lister, R., Pelizzola, M., Dowen, R.H. et al. (2009). Human DNA methylomes at base resolution show widespread epigenomic differences. *Nature*, Vol.462, No.7271, (November 2009), pp. 315-322, ISSN 0028-0836

Liu, Z., Xie, Z., Jones, W. et al. (2009). Curcumin is a potent DNA hypomethylation agent. *Bioorganic & Medicinal Chemistry Letters*, Vol.19, No.3, (February 2009), pp. 706-709, ISSN 0960-894X

Looijenga, L.H., Hersmus, R., Gillis, A.J. et al. (2006). Genomic and expression profiling of human spermatocytic seminomas: primary spermatocyte as tumorigenic precursor and DMRT1 as candidate chromosome 9 gene. *Cancer Research*, Vol.66, No.1, (January 2006), pp. 290-302, ISSN 0008-5472

Loukinov, D.I., Pugacheva, E., Vatolin, S. et al. (2002). BORIS, a novel male germ-line-specific protein associated with epigenetic reprogramming events, shares the same 11-zinc-finger domain with CTCF, the insulator protein involved in reading imprinting marks in the soma. *Proceedings of the National Academy of Sciences of the United States of America*, Vol.99, No.10, (May 2002), pp. 6806-6811, ISSN 0027-8424

Malanga, M. & Althaus, F.R. (2005). The role of poly(ADP-ribose) in the DNA damage signaling network. *Biochemistry and Cell Biology / Biochimie et Biologie Cellulaire*, Vol.83, No.3, pp. 354-364, (June 2005), ISSN 0829-8211

Mirisola, V., Mora, R., Esposito, A.I. et al. (2011). A prognostic multigene classifier for squamous cell carcinomas of the larynx. *Cancer Letters*, Vol.307, No.1, (August 2011), pp. 37-46, ISSN 0304-3835

Morison, I.M., Paton, C.J. & Cleverley, S.D. (2001). The imprinted gene and parent-of-origin effect database. *Nucleic Acids Research*, Vol.29, No.1, (January 2001), pp. 275-276, ISSN 0305-1048

Münzel, M., Globisch, D. & Carell, T. (2011). 5-Hydroxymethylcytosine, the sixth base of the genome. *Angewandte Chemie International Edition*, Vol.50, No.29, (July 2011), pp. 6460-6468, ISSN 1521-3773.

Murrell, A., Ito, Y., Verde, G. et al. (2008). Distinct methylation changes at the IGF2-H19 locus in congenital growth disorders and cancer. *PloS ONE*, 3(3): e1849. doi:10.1371/journal.pone.0001849, ISSN 1932-6203

Ndlovu, M.N., Denis, H., Fuks, F. (2011). Exposing the DNA methylation iceberg. *Trends in Biochemical Sciences*, Vol.36, No.7, (July 2011), pp. 381-387, ISSN 0968-0004

Novak Kujundžić, R., Grbeša, I., Ivkić, M. et al. (2008). Curcumin downregulates H19 gene transcription in tumor cells. Journal of Cellular Biochemistry, Vol.104, No.5, (August 2008), pp. 1781-1792, ISSN 0730-2312

Ohlsson, R., Renkawitz, R. & Lobanenkov, V. (2001). CTCF is a uniquely versatile transcription regulator linked to epigenetics and disease. *Trends in Genetics: TIG*, Vol.17, No.9, (September 2001), pp. 520-527, ISSN 0168-9525

Okano, M., Xie, S., Li, E. (1998). Dnmt2 is not required for de novo and maintenance methylation of viral DNA in embryonic stem cells. *Nucleic Acids Research*, Vol.26, No.11, (June 1998), pp. 2536–2540, ISSN 0305-1048

Okano, M., Bell, D. W., Haber, D. A. et al. (1999). DNA methyltransferases Dnmt3a and Dnmt3b are essential for de novo methylation and mammalian development. *Cell*, Vol.99, No.3, (October 1999), pp. 247-257, ISSN 0092-8674

Panzeter, P.L., Zweifel, B., Malanga, M. et al. (1993). Targeting of histone tails by poly(ADP-ribose). *Journal of Biological Chemistry*, Vol.268, No.24, (August 1993), pp. 17662-17664, ISSN 0021-9258

Penn, N.W., Suwalski, R., O'Riley, C. et al. (1972). The presence of 5-hydroxymethylcytosine in animal deoxyribonucleic acid. *Biochemical Journal*, Vol.126, No.4, (February 1972), pp. 781-790, ISSN 0264-6021

Pleschke, J.M., Kleczkowska, H.E., Strohm, M. et al. (2000). Poly(ADP-ribose) binds to specific domains in DNA damage checkpoint proteins. *Journal of Biological Chemistry*, Vol.275, No.52, (December 2000), pp. 40974-40980, ISSN 0021-9258

Ren, J., Singh, B.N., Huang Q., Li, Z., Gao, Y., Mishra, P. et al. (2011). DNA hypermethylation as a chemotherapy target. *Cellular Signalling*, Vol.23, No.27, (July 2011), pp. 1082-1093, ISSN 0898-6568

Renaud, S., Pugacheva, E.M., Delgado, M.D. et al. (2007). Expression of the CTCF-paralogous cancer-testis gene, brother of the regulator of imprinted sites (BORIS), is regulated by three alternative promoters modulated by CpG methylation and by CTCF and p53 transcription factors. *Nucleic Acids Research*, Vol.35, No.21, pp. 7372-7388, ISSN 0305-1048

Riggs, A.D. (1975). X inactivation, differentiation, and DNA methylation. *Cytogenetics and Cell Genetics*, Vol.14, No.1, (January 1975), pp. 9-25, ISSN 0301-0171

Risinger, J.I., Chandramouli, G.V., Maxwell, G.L. et al. (2007). Global expression analysis of cancer/testis genes in uterine cancers reveals a high incidence of BORIS expression. *Clinical Cancer Research,* Vol.13, No.6, (March 2007), pp. 1713-1719, ISSN 1078-0432

Robertson, J., Robertson, A.B. & Klungland A. (2011). The presence of 5-hydroxymethylcytosine at the gene promoter and not in the gene body negatively regulates gene expression. *Biochemical and Biophysical Research Communications,* Vol.411, No.1, (July 2011), pp. 40-43, ISSN 0006-291X

Rodríguez-Paredes, M. & Esteller M. (2011). Cancer epigenetics reaches mainstream oncology. *Nature Medicine,* Vol.17, No.3. (March 2011), pp. 330-339, ISSN 1078-8956

Sakatani, T., Wei, M., Katoh, M. et al. (2001). Epigenetic heterogeneity at imprinted loci in normal populations. *Biochemical and Biophysical Research Communications,* Vol.283, No.5, (May 2001), pp. 1124-1130, ISSN 0006-291X

Santoro, R., D'Erme, M., Mastrantonio, S. et al. (1995). Binding of histone H1e-c variants to CpG-rich DNA correlates with the inhibitory effect on enzymic DNA methylation. *The Biochemical Journal,* Vol.305, No.3, (February 1995), pp. 739-744, ISSN 0264-6021

Santoro, R., D'Erme, M., Reale, A. et al. (1993). Effect of H1 histone isoforms on the methylation of single- and double-stranded DNA. *Biochemical and Biophysical Research Communications,* Vol.190, No.1, (January 1993), pp. 86-91, ISSN 0006-291X

Santourlidis, S., Florl, A., Ackermann, R. et al. (1999). High frequency of alterations in DNA methylation in adenocarcinoma of the prostate. *Prostate,* Vol.39, No.3, (May 1999), pp. 166–174, ISSN 0270-4137

Schaefer, M., Hagemann S., Hanna, K., Lyko, F. (2009). Azacytidine inhibits RNA methylation at DNMT2 target sites in human cancer cell lines. Cancer Research, Vol.69, No.20, (October 2009), pp.8127-8132, ISSN 0008-547

Sharif, J., Muto, M., Takebayashi, S. et al. (2007). The SRA protein Np95 mediates epigenetic inheritance by recruiting Dnmt1 to methylated DNA. *Nature,* Vol.450, No.7171, (December 2007), pp.908-912, ISSN 0028-0836

Sharma, R.A., Euden, S.A., Platton, S.L. et al. (2004). Phase I clinical trial of oral curcumin: biomarkers of systemic activity and compliance. *Clinical Cancer Research,* Vol.10, No.20, (October 2004), pp. 6847-6854, ISSN 1078-0432

Smith, I.M., Glazer, C.A., Mithani, S.K. et al. (2009). Coordinated activation of candidate proto-oncogenes and cancer testes antigens via promoter demethylation in head and neck cancer and lung cancer. *PloS ONE* 4(3): e4961. doi:10.1371/journal.pone.0004961, ISSN 1932-6203

Spaventi, R., Pečur, L., Pavelic, K., et al. (1994). Human tumour bank in Croatia: a possible model for a small bank as part of the future European tumour bank network. *European Journal of Cancer,* Vol.30A, No.3, (September 2008), p 419, ISSN 0959-8049

Stresemann, C., Bokelmann, I., Mahlknecht, U. et al. (2008). Azacytidine causes complex DNA methylation responses in myeloid leukemia. *Molecular Cancer Therapeutics,* Vol.7, No.9, (September 2008), pp. 2998-3005, ISSN 1535-7163

Stresemann, C. &, Lyko, F. (2008). Modes of action of the DNA methyltransferase inhibitors azacytidine and decitabine. *International Journal of Cancer,* Vol.123, No.1, (July 2008), pp. 8-13, ISSN 0020-7136

Tahiliani, M., Koh, K.P., Shen, Y. et al. (2009). Conversion of 5-methylcytosine to 5-hydroxymethylcytosine in mammalian DNA by MLL partner TET1. *Science,* Vol.324, No.5929, (May 2009), pp. 930-935, ISSN 0036-8075

Takai, D., Gonzales, F.A., Tsai, Y.C. et al. (2001). Large scale mapping of methylcytosines in CTCF-binding sites in the human H19 promoter and aberrant hypomethylation in human bladder cancer. *Human Molecular Genetics*, Vol.10, No.23, (November 2001), pp. 2619-2626, ISSN 0964-6906

Takai, D. & Jones, P.A. (2002). Comprehensive analysis of CpG islands in human chromosomes 21 and 22. *Proceedings of the National Academy of Sciences of the United States of America*, Vol.99, No.6, (March 2002), pp. 3740–3745, ISSN 0027-8424

Tremblay, K.D., Duran, K.L. & Bartolomei, M.S. (1997). A 5' 2-kilobase-pair region of the imprinted mouse H19 gene exhibits exclusive paternal methylation throughout development. *Molecular and Cellular Biology*, Vol.17, No.8, (August 1997), pp. 4322-4329, ISSN 0270-7306

Ulaner, G.A., Vu, T.H., Li, T. et al. (2003a). Loss of imprinting of IGF2 and H19 in osteosarcoma is accompanied by reciprocal methylation changes of a CTCF-binding site. *Human Molecular Genetics*, Vol.12, No.5, (March 2003), pp. 535-549, ISSN 0964-6906

Ulaner, G.A., Yang, Y., Hu, J.F. et al. (2003b). CTCF binding at the insulin-like growth factor-II (IGF2)/H19 imprinting control region is insufficient to regulate IGF2/H19 expression in human tissues. *Endocrinology*, Vol.144, No.10, (October 2003), pp. 4420-4426, ISSN 0013-7227

Ushijima, T., Watanabe, N., Shimizu, K. et al. (2005). Decreased fidelity in replicating CpG methylation patterns in cancer cells. *Cancer Research*, Vol.65, No.1, (January 2005), pp. 11-17, ISSN 0008-5472

Ushijima, T., Watanabe, N., Okochi, E. et al. (2003). Fidelity of the methylation pattern and its variation in the genome. *Genome Research*, Vol.13, No.5, (May 2003), pp. 868-874, ISSN 1088-9051

Vu, T.H., Nguyen, A.H. & Hoffman, A.R. (2010). Loss of IGF2 imprinting is associated with abrogation of long-range intrachromosomal interactions in human cancer cells. *Human Molecular Genetics*, Vol.19, No.5, (March 2010), pp. 901-919, ISSN 0964-6906

Wagner, J.R. & Cadet, J. (2010). Oxidation reactions of cytosine DNA components by hydroxyl radical and one-electron oxidants in aerated aqueous solutions. *Accounts of Chemical Research*, Vol.43, No.4, (April 2010), pp. 564-571, ISSN 0001-4842

Wanunu, M., Cohen-Karni, D., Johnson, R.R. et al. (2011). Discrimination of methylcytosine from hydroxymethylcytosine in DNA molecules. *Journal of the American Chemical Society*, Vol.133, No.3, (January 2011), pp. 486–492, ISSN 0002-7863

Williams, K., Christensen, J., Pedersen, M.T. et al (2011). TET1 and hydroxymethylcytosine in transcription and DNA methylation fidelity. *Nature*, Vol.473, No.7347, (May 2011), pp. 343-348, ISSN 0028-0836

Witcher, M. & Emerson, B.M. (2009). Epigenetic silencing of the p16INK4a tumor suppressor is associated with loss of CTCF binding and a chromatin boundary. *Molecular Cell*, Vol.34, No.3, (May 2009), pp. 271-284, ISSN 097-2765

Woloszynska-Read, A., James, S.R., Link, P.A. et al. (2007). DNA methylation-dependent regulation of BORIS/CTCFL expression in ovarian cancer. *Cancer Immunity*, Vol.7, (December 2007), p. 21, ISSN 1424-9634

Wyatt, G.R. & Cohen, S.S. (1952). A new pyrimidine base from bacteriophage nucleic acids. *Nature*, Vol.170, No.4338, (December 1952), pp. 1072-1073, ISSN 0028-0836

Yu, W., Ginjala, V., Pant, V. et al. (2004). Poly(ADP-ribosyl)ation regulates CTCF-dependent chromatin insulation. *Nature Genetics*, Vol.36, No.10, pp. 1105-1110. ISSN 1061-4036.

Zardo, G. & Caiafa, P. (1998). The unmethylated state of CpG islands in mouse fibroblasts depends on the poly(ADP-ribosyl)ation process. *The Journal of Biological Chemistry*, Vol.273, No.26, (June 1998), pp. 16517-16520, ISSN 0021-9258

Zhang, H., Niu, B., Hu, J.F. et al. (2011). Interruption of intrachromosomal looping by CCCTC binding factor decoy proteins abrogates genomic imprinting of human insulin-like growth factor II. *The Journal of Cell Biology*, Vol.193, No.3, (May 2011), pp. 475-487, ISSN 0021-9525

Zhang, X., Gejman, R., Mahta, A. et al. (2010). Maternally expressed gene 3, an imprinted noncoding RNA gene, is associated with meningioma pathogenesis and progression. *Cancer Research*, Vol.70, No.6, pp. 2350-2358, (March 2010), ISSN 0008-5472

15

Effects of Dietary Nutrients on DNA Methylation and Imprinting

Ali A. Alshatwi and Gowhar Shafi
King Saud University Riyadh
Saudi Arabia

1. Introduction

DNA methylation is an enzymatic modification carried out by DNA methyltransferases. Alterations in DNA methylation patterns are the best understood epigenetic cause of disease and were first discovered in studies during the 1980s that focused on X chromosome inactivation (Avner and Heard 2001), genomic imprinting (Verona et al. 2006) and cancer (Feinberg and Tycko 2004). DNA methylation involves the addition of a methyl group to cytosines in CpG (cytosine/guanine) pairs (Ehrlich and Wang 1981, Laird and Jaenisch 1994). The added methyl group does not affect the base pairing itself, but the protrusion of methyl groups into the DNA major groove can affect DNA–protein interactions. Methylated CpGs are usually associated with silenced DNA, can block methylation sensitive proteins from binding to the DNA and are subject to high mutation rates. DNA methylation patterns are established and maintained by DNMTs, enzymes that are essential for proper gene expression patterns (Robertson 2002) (Figure 1). DNA methylation is an essential epigenetic

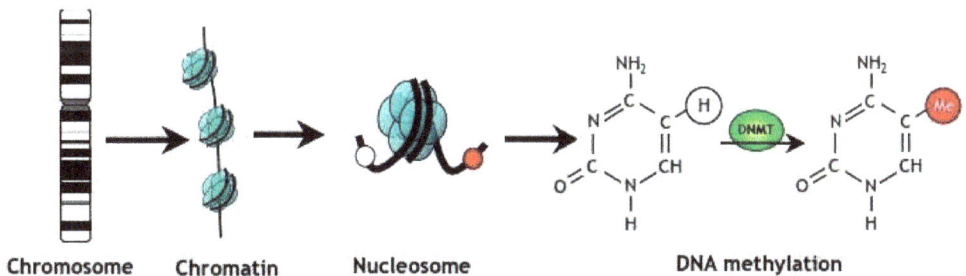

Fig. 1. Schematic representation of DNA methylation. Strands of DNA envelop histone octamers, forming nucleosomes. These nucleosomes are bundled together into chromatin, the building blocks of chromosomes. DNA methylation occurs at the 5′-position of cytosine residues in a reaction catalyzed by DNA methyltransferases (DNMTs). Together, these modifications create a unique epigenetic signature that regulates chromatin organization and gene expression.

mechanism of transcriptional control. This methylation plays a crucial role in maintaining cellular function, and alterations in methylation patterns may contribute to the development of cancer. Abnormal DNA methylation (global hypomethylation accompanied by region-specific hypermethylation) is frequently found in tumor cells. Global hypomethylation can result in chromosome instability, and hypermethylation has been linked with the silencing of tumor suppressor genes. DNA methylation predominantly involves the covalent addition of a methyl group (CH_3) to the 5' position of a cytosine that precedes a guanine in the DNA sequence. This is referred to as an epigenetic modification, as it does not alter the coding sequence of the DNA. The distribution of methylated CpG dinucleotides in the genome is asymmetric. In contrast to the relative paucity of CpGs in the genome as a whole, these dinucleotides can be found at high frequency in small stretches of DNA termed CpG islands.

2. DNA methylation in carcinogenesis

Differential patterns of DNA methylation in cancer have been recognized for more than two decades (Brown and Strathdee 2002). The situation has been confusing because virtually all types of cancer examined have both global hypomethylation and gene-specific hypermethylation in promoter regions (Baylin et al. 1998). Hypermethylation of promoter regions, which is associated with transcriptional silencing, is at least as common as actual DNA mutation as a mechanism for the inactivation of classical tumor suppressor genes in human cancers (Jones and Baylin 2002, Tsou et al. 2002). Additionally, a number of candidate tumor suppressor genes that are not commonly inactivated by mutation are transcriptionally silenced by this mechanism (Jones and Baylin 2002). The aberrant methylation of genes that suppress tumorigenesis appears to occur early in tumor development and increases gradually, ultimately leading to a malignant phenotype (Fearon and Vogelstein 1990, Kim and Mason 1995). Genes associated with tumorigenesis can be silenced by this epigenetic mechanism. In an excellent review on tumorigenesis, Hanahan and Weinberg (Hanahan and Weinberg 2000) have noted the major hallmarks of cancer. The crucial properties required to generate the characteristic malignant attributes associated with cancer are the ability to replicate without limitation, indifference to positive growth signals, disregard for growth inhibitory factors, evasion of programmed cell death, sustained angiogenesis, and the ability to invade and metastasize (Hanahan and Weinberg 2000). Each of these traits is influenced by a gene or set of genes. Failure to express the gene correctly and produce functional regulatory proteins leads to the uncontrolled pattern of cell behavior observed in a typical neoplasm. Hypermethylation is associated with the inactivation of virtually all pathways involved in the cancer process, including DNA repair, cell cycle regulation, apoptosis, carcinogen metabolism, hormonal response, and cell adherence (Momparler 2003, Esteller 2002, Costello and Plass 2001). Moreover, CpG island hypermethylation in human cancer is specific enough to enable the use of these aberrantly hypermethylated loci as biomarkers of malignant disease (Esteller 2003).

Although both global hypomethylation and regional DNA hypermethylation are well documented in cancer (Figure 2), the mechanisms behind these events remain unclear, particularly the paradox of why some DNA remains hypomethylated in the presence of increased DNA methyltransferase activity and expression. It has been suggested that the deregulation of DNA methyltransferases might lead to genome-wide hypomethylation in cancers (Pogribny et al. 2004). A significant correlation between overexpression of DNMT3b4, an inactive splice variant of DNMT3b, and DNA hypomethylation on peri-centromeric satellite regions of pre-neoplastic and neoplastic tissue provides support for this hypothesis (Saito et al. 2002). DNA methyl-transferases have also been found to bind with higher affinity to DNA strand breaks, abasic sites, and uracils than to similar hemimethylated CpG sites, consistent with their ancestral function as DNA repair enzymes (James et al. 2003). These same DNA lesions are often present in human pre-neoplastic cells, raising the possibility that DNA lesions may be a necessary prerequisite for the disruption of normal DNA methylation patterns in pre-neoplastic and neoplastic cells (James et al. 2003).

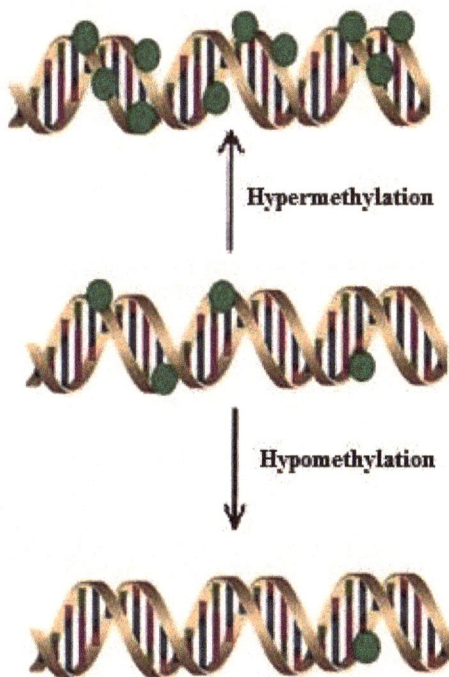

Fig. 2. Abnormal DNA methylation leads to the activation of oncogenes, genome instability and retrotransposon activation, and to the inactivation of genes such as tumor suppressor genes.

3. Diet and DNA methylation

Several bioactive food components can modulate DNA methylation and cancer susceptibility. Studies on dietary factors that are involved in one-carbon metabolism provide the most compelling data for an interaction between nutrients and DNA methylation. The one-carbon metabolism pathway influences the supply of donor methyl groups and consequently the biochemical pathways of methylation processes (Figure 3). These nutrients include vitamin B12, vitamin B6, folate, methionine, and choline. Folate plays the central role in one-carbon metabolism. In this pathway, a carbon unit from serine or glycine is transferred to tetrahydrofolate to form 5,10-methylenetetrahydrofolate (Scott and Weir 1998). Vitamin B6 is a necessary co-factor for glycine hydroxymethyltransferase in the synthesis of 5,10-methylenetetrahydrofolate (Ross 2003). This folate can be used for the synthesis of thymidine, can be oxidized to formyltetrahydrofolate for the synthesis of purines or can be reduced to 5-methyltetrahydrofolate and used to methylate homocysteine to form methionine (Ross 2003). The vitamin B_{12}–dependent enzyme methionine synthase (MS) catalyzes the synthesis of methionine from homocysteine. Methionine is subsequently converted to S-adenosylmethionine (SAM) by an ATP-dependent transfer of adenosine to methionine via methionine adenosyltransferase (Ross 2003). S-adenosylmethionine can then donate its labile methyl groups to more than 80 biological methylation reactions, including the methylation of DNA, RNA, and proteins (Choi and Mason 2001). When the folate source is inadequate, plasma and cellular levels of homocysteine increase. Although an alternative zinc-requiring enzyme in methionine synthesis, betaine-homocysteine methyltransferase, may partly compensate for the reduced MS activity, it is well known that dietary folate depletion alone is a perturbing force sufficient to diminish cellular SAM pools (Miller et al. 1994). This leads to a rise in cellular levels of S-adenosylhomocysteine (SAH), as the equilibrium of the SAH-homocysteine interconversion actually favors SAH synthesis. Hence, when homocysteine metabolism is inhibited (as in folate deficiency), cellular SAH will be increased. Increased SAH inhibits methyltransferase activity and, subsequently, DNA methylation reactions (De Cabo et al. 1995). The inhibition of DNA methylation resulting from insufficient dietary folate has also been associated with increased cancer susceptibility.

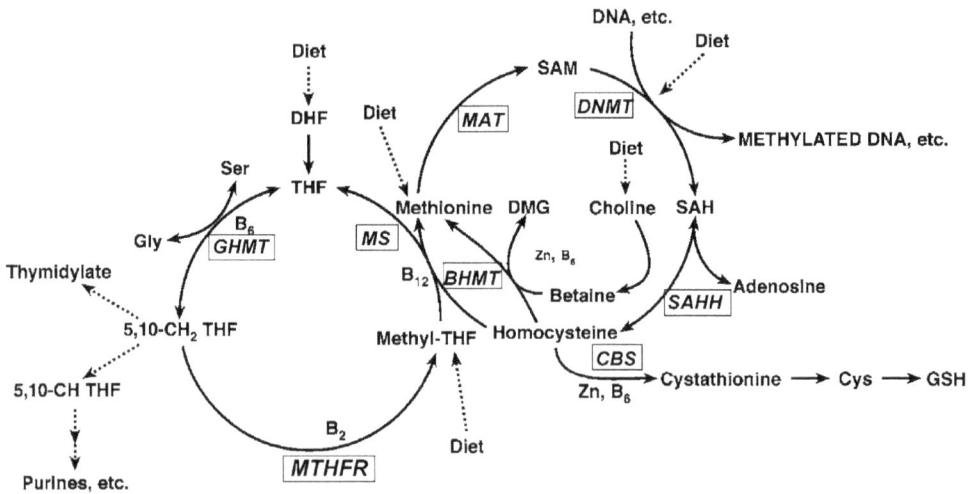

Fig. 3. Dietary factors, enzymes, and substrates involved in DNA methyl metabolism. Enzymes are shown in italics with a box around them.

A large number of epidemiologic and clinical studies suggest that dietary folate intake and blood folate concentrations are inversely associated with colorectal cancer risk (Kim 2004). Animal studies using chemical and genetically predisposed rodent models have provided considerable support for a causal relationship between folate depletion and colorectal carcinogenesis, as well as a dose-dependent protective effect of folate supplementation (Song et al. 2000). However, animal studies have also shown that the dose and timing of folate intervention are critical in providing safe and effective chemoprevention; exceptionally high supplemental folate concentrations as well as folate intervention after microscopic neoplastic foci are established in the colorectal mucosa promote rather than suppress carcinogenesis (Kim 2004). Animal studies have shown that folate deficiency causes DNA hypomethylation prior to the development of tumors (Jacob et al. 1998). DNA hypomethylation has also been found in the lymphocytes of humans on low dietary folate and can be reversed by folate repletion (Trasler et al. 2003). In contrast, folate deficiency with or without reductions in Dnmt1 did not affect either overall genomic DNA methylation levels or the methylation levels of two candidate genes, E-cadherin or p53, in normal and neoplastic intestinal tissue (Song et al. 2000). These studies suggest that the effects of folate deficiency on DNA methylation are highly complex, appear to depend on cell type, target organ, and stage of transformation and are gene- and site-specific.

Gene polymorphisms may modulate the effect of dietary folate on DNA methylation and cancer susceptibility. A single nucleotide substitution at position 677 of the methylenetetrahydrofolate reductase (MTHFR) gene (C677T) has been associated with a reduced risk of colon cancer (Curtin et al. 2004) but an increased risk of breast cancer (Alshatwi 2010, Beilby et al. 2004). When cellular folate is low, the presence of this TT genotype is associated with an increase in homocysteine concentration and DNA hypomethylation. These studies highlight the importance of taking into consideration interactions between folate status and key genes in the folate and one-carbon metabolic pathways when investigating the effect of folate on DNA methylation.

Several bioactive components in food have also been shown to affect DNA methylation. Many of these nutrients, including genistein, zinc, selenium and vitamin A have also been associated with cancer susceptibility. Although deficiencies of some food components induce global DNA hypomethylation, deficiencies of other food components induce global hypermethylation. For example, zinc deficiency (Wallwork and Duerre 1985) and retinoic acid excess (Rowling et al. 2002) has been shown to reduce the use of methyl groups from SAM in rat liver and to result in global DNA hypomethylation. Selenium deficiency decreased DNA methylation in Caco-2 cells and in rat liver and colon (Davis et al. 2000). In contrast, vitamin C deficiency has been linked with DNA hypermethylation in lung cancer (Haliwell 2001). There may be optimum amounts of certain dietary components that enable normal DNA methylation. For example, both absences and excesses of dietary arsenic have been shown to cause global hypomethylation in rat liver (Uthus 1993).

Phytoestrogens, such as genistein, are thought to be involved in preventing the development of certain prostate and mammary cancers by supporting the retention of a protective DNA methylation pattern (Beilby et al. 2004). The intake of genistein was positively correlated with changes in prostate DNA methylation at the CpG islands of specific mouse genes, as evaluated using mouse differential hybridization arrays (Day et al. 2002). Other phytoestrogens have also been shown to alter DNA methylation. Neonatal exposures to the phytoestrogens coumestrol and equol have been found to lead to specific hypermethylation of the c-Ha-ras proto-oncogene in pancreatic cancer (Lyn-Cook et al. 1995).

The effect of bioactive food components on gene-specific DNA methylation is less clearly understood than the effect of bioactive food components on global DNA methylation. In a rat model of hepatocellular carcinoma, a choline-deficient diet induced hypomethylation of CpG sites in the c-myc gene as well as the overexpression of this gene (Tsujiuchi et al. 1999). Methylation of the promoter region of p53 in Caco-2 cells decreased when cells were cultured in the absence of selenium (Davis et al. 2000). Arsenic has been shown to induce hypo-methylation of the 5' regulatory region of Ha-ras in animals (Okoji et al. 2002). Interestingly, dietary factors that modify global DNA methylation can simultaneously cause opposite effects on gene-specific methylation. For example, folate deficiency causes global DNA hypomethylation but hypermethylation of the 5' regulatory sequence of the H-cadherin gene (Jhaveri et al. 2001). Moreover, retinoic acid leads to global hypomethylation but region-specific hypermethylation. Thus, the pattern of gene-specific methylation may not be in concert with the track of overall alterations in genomic DNA methylation.

Nutrients have also been shown to affect gene transcription by altering exon-specific DNA methylation. For example, in animals fed methyl-deficient diets, increased levels of mRNA for c-fos, c-Ha-ras, and c-myc were correlated with hypomethylation at specific sites within the exons of these genes (Zapisek et al. 1992). In a hepatocarcinogenesis study with chronic dietary methyl deficiency, methylation in the p53 gene coding region decreased and then increased, which corresponded to high p53 mRNA levels in pre-neoplastic liver tissue and then lower p53 mRNA levels after tumor formation (Sohn et al. 2003). This suggests that methylation changes in the coding region of genes can affect gene transcription and that gene-specific methylation can vary during the carcinogenic process.

Bioactive food components can also modulate DNA methylation by interfering with DNMT activity. Green tea has been shown to inhibit carcinogenesis in many animal models (Fang et al. 2003). Recently, epigallocatechin-3-gallate, the major polyphenol from green tea, was found to inhibit DNMT activity by binding to the enzyme, which resulted in the reactivation of methylation-silenced genes in cancer cells (Fang et al. 2003). Rats fed selenium- or folate-deficient diets had significantly reduced liver and colon DNMT activity; however, the mechanism for this inhibitory effect is not known (Davis and Uthus 2003). Studies on cultured rat liver cells and in animals have shown that cadmium is another active inhibitor of DNMT (Takiguchi et al. 2003). Moreover, both cadmium and zinc inhibited DNMT activity in nuclear extracts from rats fed either a control or a methyl-deficient diet (Poirier and Vlasova 2002). The inhibitory effects of cadmium and zinc may be caused by the binding of these metals to a cysteine residue in the active site of DNMT (Poirier and Vlasova 2002).

4. Diet and gene interaction

Although many bioactive food components have been shown to modify epigenetic events and cancer susceptibility, looking at individual nutrients may be too simplistic.

Combinations of dietary deficiencies in methyl-related compounds such as folate, choline, and methionine cause decreased tissue SAM, global DNA hypomethylation and ultimately hepatic tumorigenesis in the absence of treatment with carcinogens (Steinmetz et al. 1998). The percentage of CpG sites that lose methyl groups on both DNA strands gradually increases in the liver after either folate or methyl deficiency, in spite of the presence of elevated DNA methyltransferase activity (Pogribny et al. 2004). Hence, it appears that DNA methyltransferase is incapable of methylating double-stranded unmethylated DNA present in the pre-neoplastic liver, and this may result in the establishment of a cancer-specific DNA methylation profile. Alterations in the DNA methylation profile may explain why these animals develop cancer in the absence of treatment with carcinogens. However, it is also possible that the increased mitogenesis leads to mutations, and it is a combination of mutagenesis and altered DNA methylation that leads to cancer. Moreover, other tissues, such as pancreas, spleen, kidney, and thymus, displayed no changes in the DNA methylation level or DNA methyltransferase activity after folate/methyl deficiency (Pogribny et al. 2004). These findings suggest that a folate/methyl-deficient diet causes specific DNA hypomethylation in the liver as well as hepatic tumorigenesis. These findings further support the hypothesis that alterations in the DNA methylation profile can cause cancer and that it is important to look at interactions between nutrients.

Diets that are deficient in choline or in both choline and methionine have also been shown to cause hepatocellular carcinoma in animals after 12–24 months of consumption (Brunaud et al. 2003). However, the carcinogenicity of a methyl-deficient diet is much higher when combined with vitamin B12 deficiency (Brunaud et al. 2003). Dietary arsenic has also been shown to interact with a methyl-deficient diet. The administration of arsenic alongside a methyl-deficient diet in mice resulted in genome-wide hypomethylation, as well as reduced methylation of the promoter region of the oncogene Ha-ras (Okoji et al. 2002). This process would be expected to induce expression of the oncogene and contribute to tumor development.

Another important dietary interaction that affects DNA methylation and cancer susceptibility occurs between alcohol and folate. Alcohol has been shown to cleave folate, impair folate absorption, increase folate excretion, and interfere with methionine synthase activity (Pufulete et al. 2003). A high alcohol intake may lead to localized folate deficiency and DNA hypomethylation, even when dietary folate intake and blood folate concentrations are normal (Pufulete et al. 2003).

Dietary selenium has been shown to modulate many of the adverse effects of folate deficiency, including alterations in one-carbon metabolism and aberrant crypt formation, a pre-neoplastic lesion (Davis and Uthus 2003). In contrast, SAM, SAH, and genomic methylation were not affected by any dietary interaction between selenium and folate. These results suggest that selenium alters some of the effects of folate deficiency, probably by shunting the buildup of homocysteine through the trans-sulfuration pathway.

5. Imprinting and DNA methylation

Several genes that are involved in mammalian developmental processes do not follow Mendelian genetics and are expressed in a mono-allelic fashion. DNA methylation is a critical element in this imprinting process. Methylation can mark specific alleles and establish a mono-allelic expression pattern of the imprinted genes. Much has been learned about the importance of DNA methylation in imprinting from the Dnmt1-deficient mouse. Loss of methylation on the imprinted Xist gene in Dnmt1-deficient cells does not activate Xist. This is surprising in light of the fact that the RNA product of the Xist gene is thought to spread an inactive state along one of the X-chromosomes in eutherian females. In contrast, the mono-allelic expression of the imprinted genes H19, Igf2 and Igf2r has been shown to be disrupted in Dnmt1 mutant cells. This observation established for the first time a causal link between DNA methylation and gene activity (Ramchandani et al. 1999). Differentially methylated regions (DMRs) in imprinted genes have been shown to serve as imprinting boxes that control the imprinting of the Igf2r gene (Cervoni et al. 1999) and the Prader Willi/Angelman syndrome domain (Bhattacharya et al. 1999). DMRs are also crucial elements in regulating reciprocal monoallelic expression of the maternal regulation of the imprinted I119 and Igf2 genes (Hendrich and Bird 1999). Experiments with Dnmt1 deficient mice have shown that the establishment of DMRs in imprinted genes requires methylation transmission through the germ line. Dnmt1 rescue in knockout ES cell lines did not restore imprinted methylation of the above-mentioned genes, even though Dnmt1 restoration was able to reestablish imprinting in the whole animal (Zhang et al. 1999). This is in accord with the fact that methylation of the Snrpn gene on the maternal allele is established during oogenesis and maintained thereafter (Ng et al. 1999).

6. MicroRNA and DNA methylation

MicroRNAs are a new class of noncoding, endogenous, small RNAs that regulate gene expression by translational repression, representing a new essential class of regulatory molecules. MicroRNAs can play essential roles in regulating DNA methylation and histone modifications, creating a highly organized feedback mechanism. Epigenetic mechanisms such as promoter methylation and histone acetylation can also alter microRNA expression. A connection between epigenetic phenomena and microRNA has been described in numerous physiological processes, and an altered balance between them represents one of the mechanisms leading to pathological conditions such as cancer. An abnormal expression of microRNA has been associated with the development or progression of human cancers through the alteration of cell proliferation and apoptosis processes (Iorio et al). A methyl-deficient diet, which induces tumors in rats, also induced prominent early changes in the expression of microRNA genes, including miR-34a, miR-127, miR-200b, and miR-16a, which are involved in the regulation of apoptosis, cell proliferation, cell-to-cell connection, and epithelial-mesenchymal transition in the rat (Tryndyak et al. 2009). Mice fed a methyl-deficient diet contracted nonalcoholic steatohepatitis, which was accompanied by alterations in the expression of several microRNAs including miR-29c, miR-34a, miR-155, and miR-200b. Interestingly, alterations in the expression of these microRNAs are paralleled by changes in the protein levels of their targets. These studies suggest that alterations in the expression of microRNAs are a prominent event during the development of cancer and nonalcoholic steatohepatitis caused by dietary methyl deficiency (Pogribny et al. 2010).

Similar to the methyl-deficient diet, folate deficiency induced a marked global increase in microRNA expression in human lymphoblastoid cells. miR-222 was significantly overexpressed under folate-deficient conditions in vitro. This finding was confirmed in vivo in human peripheral blood from individuals with low folate intake, suggesting that microRNA expression might be a potential biomarker of nutritional status in humans (Marsit et al. 2006). The Göttingen minipig, an animal model of obesity, was fed either a high cholesterol or a standard diet. Body weight, total cholesterol, and HDL were higher, and miR-122 was lower (1.4-fold; P < 0.0015) in pigs fed the high-cholesterol diet compared with those fed the standard diet, implicating this microRNA in obesity as well (Cirera et al. 2010).

A few reports have suggested that bioactive food components may reduce carcinogenesis through microRNA action [reviewed in (Saini et al. 2010)]. Genistein represses human uveal melanoma cells and murine chronic lymphocytic leukemia cells by altering miR-16 levels (Salerno et al. 2009). Curcumin represses human pancreatic cancer cells by upregulating miR-22 and downregulating miR-199a (Sun et al. 2008). Curcumin also upregulates miR-15a and miR-16 expression, which could inhibit the expression of B-cell lymphoma 2 (Bcl-2) and thereby induce apoptosis in MCF-7 breast cancer cells (Yang et al. 2009). Furthermore, miR-10a, a key mediator of metastatic behavior in pancreatic cancer, is a retinoic acid target. Retinoic acid receptor antagonists effectively repress miR-10a expression and block metastasis (Weiss et al. 2009). In neuroblastoma cells, miR-34a functions as a potential tumor suppressor, and retinoic acid-induced differentiation of the neuroblastoma cell line enhanced miR-34a expression and decreased expression of its target, E2F transcriptional factor 3 (Weiss et al. 2009).

7. Conclusion

Given that DNA methylation is at the heart of many phenotypic variations in health and disease, it seems likely that understanding and manipulating the epigenome holds enormous promise for preventing and treating common human illness. DNA methylation also offers an important window into understanding the role of interactions between the environment and the genome in causing disease, and in modulating these interactions to improve human health. Within the past two decades, scientists have discovered many details about the process of DNA methylation. Scientists now know that methylation plays a critical role in the regulation of gene expression, and they have also determined that this process tends to occur at certain locations within the genomes of different species. Moreover, DNA methylation has been shown to play a vital role in numerous cellular processes, and abnormal patterns of methylation have been implicated in several human diseases. However, as with other topics in the field of epigenetics, gaps remain in our knowledge of DNA methylation. As new laboratory techniques are developed and additional genomes are mapped, scientists will undoubtedly continue to uncover many of the unknowns of how, when, and where DNA is methylated, and for what purposes.

8. References

Avner P, Heard E. X-chromosome inactivation: counting, choice and initiation. Nat Rev Genet 2001;2:59-67.

Baylin SB, Herman JG, Graff JR, Vertino PM, Issa J-P. Alterations in DNA methylation: a fundamental aspect of neoplasia. Adv Cancer Res 72:141–196, 1998.

Beilby J, Ingram D, Hahnel R, Rossi E. Reduced breast cancer risk with increasing serum folate in a case-control study of the C677T genotype of the methylenetetrahydrofolate reductase gene. Eur J Cancer 40:1250–1254, 2004.

Bhattacharya SK, Ramchandani S, Cervoni N et al. A mammalian protein with specific demethylase activity for mCpG DNA. Nature 1999; 397:579-83.

Broday L, Peng W, Kuo MH, Salnikow K, Zoroddu M, Costa M. Nickel compounds are novel inhibitors of histone H4 acetylation. Cancer Res 60:238–241, 2000.

Brown R, Strathdee G. Epigenomics and epigenetic therapy of cancer. Trends Mol Med 8:S43–S48, 2002.

Brunaud L, Alberto JM, Ayav A, Gerard P, Namour F, Antunes L, Braun M, Bronowicki JP, Bresler L, Gueant JL. Effects of vitamin B and folate deficiencies on DNA methylation and carcinogenesis in rat liver. Clin Chem Lab Med 41:1012–1019, 2003.

Cervoni N, Bhattacharya S, Szyf M. DNA demethylase is a processive enzyme. J Biol Chem 1999; 274:8363-6.

Choi S.-W., Mason JB. Folate status: effects on pathways of colorectal carcinogenesis. J Nutr 132:2413S–2418S, 2002.

Cirera S, Birck M, Busk PK, Fredholm M. Expression profiles of miRNA-122 and its target CAT1 in minipigs (Sus scrofa) fed a highcholesterol diet. Comp Med. 2010;60:136–41.

Cooney CA, Dave AA, Wolff GL. Maternal methyl supplements in mice affect epigenetic variation and DNA methylation of offspring. J Nutr 132:2393S–2400S, 2002.

Costello JF, Plass C. Methylation matters. J Med Genet 38:285–303, 2001.

Curtin K, Bigler J, Slattery ML, Caan B, Potter JD, Ulrich CM. MTHFR C677T and A1298C polymorphisms: diet, estrogen, and risk of colon cancer. Cancer Epidemiol Biomarkers Prev 13:285–292, 2004.

Davis CD, Uthus EO, Finley JW. Dietary selenium and arsenic affect DNA methylation in vitro in Caco-2 cells and in vivo in rat liver and colon. J Nutr 130 2903–2909, 2000.

Davis CD, Uthus EO. Dietary folate and selenium affect dimethylhydrazine-induced aberrant crypt formation, global DNA methylation and one-carbon metabolism in rats. Nutr 133:2907–2914, 2003.

Day JK, Bauer AM, desBordes C, Zhuang Y, Kim B-E, Newton LG, Nehra V, Forsee KM, MacDonald RS, Besch-Williford C, Huang THM, Lubahn DB. Genistein alters methylation patterns in mice. J Nutr 132:2419S–2423S, 2002.

De Cabo SF, Santos J, Fernandez-Piqueras J. Molecular and cytological evidence of S-adenosyl-L-homocysteine as an innocuous undermethylating agent in vivo. Cytogenet Cell Genet 71:187–192, 1995.

Ehrlich M, Wang RY. 5-Methylcytosine in eukaryotic DNA. Science 1981;212:1350- Laird PW, Jaenisch R. DNA methylation and cancer. Hum Mol Genet 1994;3: 1487-95.

Esteller M. Cancer epigenetics: DNA methylation and chromatin alterations in human cancer. In: Llombart-Boschm F, Ed. New Trends in Cancer for the 21st Century. New York: Academic/Plenum Publishers, pp39–49, 2003.

Esteller M. CpG island hypermethylation and tumor suppressor genes: a booming present, a brighter future. Oncogene 21:5427–5440, 2002.

Fang MZ, Wang Y, Ai N, Hou Z, Sun Y, Lu H, Welsh W, Yang CS. Tea polyphenol (-)-epigallocatechin-3-gallate inhibits DNA methyltransferases and reactivates methylation-silenced genes in cancer cell lines. Cancer Res 63:7563–7570, 2003.

Fearon EA, Vogelstein B. A genetic model for colorectal tumorigenesis. Cell 61:759–767, 1990.

Feinberg AP, Tycko B. The history of cancer epigenetics. Nat Rev Cancer 2004;4: 143-53.

Furumai R, Matsuyama A, Kobashi N, Lee KH, Nishiyama M, Nakajima H, Tanaka A, Komatsu Y, Nishino N, Yoshida M, Horinouchi S. FK228 (depsipeptide) as a natural prodrug that inhibits class I histone deacetylases. Cancer Res 62:4916–4921, 2002.

Griseri P, Patrone G, Puppo F, Romeo G, Ravazzolo R, Ceccherini I. Rescue of human RET gene expression by sodium butyrate: a novel powerful tool for molecular studies in Hirschsprung disease. Gut 52:1154–1158, 2003.

Haliwell B. Vitamin C and genomic stability. Mutat Res 475:29–35, 2001.

Hanahan D, Weinberg RA. The hallmarks of cancer. Cell 100:57–70, 2000.

Hendrich B, Bird A. Identification and characterization of a family of mammalian methyl-CpG binding proteins. Mol Cell Biol 1998; 18:6538-47.

Iorio MV, Piovan C, Croce CM. Interplay between microRNAs and the epigenetic machinery: an intricate network. Biochim Biophys Acta. Epub May 20.

Jacob RA, Gretz DM, Taylor PC, James SJ, Pogribny IP, Miller BJ, Henning SM, Swendseid ME. Moderate folate depletion increases plasma homocysteine and decreases lymphocyte DNA methylation in postmenopausal women. J Nutr 128:1204–1212, 1998.

James SJ, Pogribny IP, Pogribna M, Miller BJ, Jernigan S, Melnyk S. Mechanisms of DNA damage, DNA hypomethylation, and tumor progression in the folate/methyl-deficient rat model of hepatocarcinogenesis. J Nutr 133:3740S-3747S, 2003.

Jhaveri MS, Wagner C, Trepel JB. Impact of extracellular folate levels on global gene expression. Mol Pharmacol 60:1288–1295, 2001.

Jones PA, Baylin SB. The fundamental role of epigenetic events in cancer. Nat Rev Genet 3:415–428. 2002.

Kim Y, Mason J. Folate, epithelial dysplasia and colon cancer. Proc Assoc Am Physicians 107:218–227, 1995.

Kim Y.-I. Folate and DNA methylation: a mechanistic link between folate deficiency and colorectal cancer? Cancer Epidemiol Biomarkers Prev 13:511–519, 2004.

Lee YW, Klein CB, Kargacin B, Salnikow K, Kitahara J, Dowjat K, Zhitkovich A, Christie NT, Costa M. Carcinogenic nickel silences gene expression by chromatin condensation and DNA methylation: a new model for epigenetic carcinogens. Mol Cell Biol 15:2547–2557, 1995.

Lu Z, Yang Y-T, Chen C-S, Davis M, Byrd JC, Eterton MR, Umar A, Chen C-S. Zn 2þ - chelating motif-tethered short-chain fatty acids as a novel class of histone deacetylase inhibitors. J Med Chem 47:467–474, 2004.

Lyn-Cook BD, Blann E, Payne PW, Bo J, Sheehan D, Medlock K. Methylation profile and amplification of proto-oncogenes in rat pancreas induced with phytoestrogens. Proc Soc Exp Biol Med 208:116–119, 1995.

Marsit CJ, Eddy K, Kelsey KT. MicroRNA responses to cellular stress. Cancer Res. 2006;66:10843–8.

Miller JW, Nadeau MR, Smith J, Smith D, Selhub J. Folate-deficiencyinduced homocysteinaemia in rats: disruption of S-adenosylmethionine=s co-ordinate regulation of homocysteine metabolism. Biochem J 298:415–419, 1994.

Momparler RL. Cancer epigenetics. Oncogene 22:6479–6483, 2003.

Ng HH, Zhang Y, Hendrich B et al. MBD2 is a transcriptional repressor belonging to the MeCP1 histone deacetylase complex . Nat Genet 1999; 23:58-61.

Okoji RS, Yu RC, Maronpot RR, Froines JR. Sodium arsenite administration via drinking water increases genome-wide and Ha-ras DNA hypomethylation in methyl-deficient C57BL/6J mice. Carcinogenesis 23:777–785, 2002.

Pogribny IP, James SJ, Jernigan S, Pogribna M. Genomic hypomethylation is specific for preneoplastic liver in folate/methyl deficient rats and does not occur in non-target tissues. Mutat Res 548:53–59, 2004.

Pogribny IP, Starlard-Davenport A, Tryndyak VP, Han T, Ross SA, Rusyn I, Beland FA. Difference in expression of hepatic microRNAs miR-29c, miR-34a, miR-155, and miR-200b is associated with strainspecific susceptibility to dietary nonalcoholic steatohepatitis in mice. Lab Invest. Epub 2010 Jun 14.

Poirier LA, Vlasova TI. The prospective role of abnormal methyl metabolism in cadmium toxicity. Environ Health Perspect 110:793–795, 2002.

Pufulete M, Al-Ghnaniem R, Leather AJ, Appleby P, Gout S, Terry C, Emery PW, Sanders TA. Folate status, genomic DNA hypomethylation, and risk of colorectal adenoma and cancer: a case control study. Gastroenterology 124:1240–1248, 2003.

Ramchandani S, Bhattacharya SK, Cervoni N et al. DNA methylation is a reversible biological signal. Proc Natl Acad Sci USA 1999; 96:6107-12.

Robertson KD. DNA methylation and chromatin — unraveling the tangled web. Oncogene 2002;21:5361-79.

Ross SA. Diet and DNA methylation interactions in cancer prevention. Ann NY Acad Sci 983:197–207, 2003.

Rowling MJ, McMullen MH, Schalinske KL. Vitamin A and its derivatives induce hepatic glycine N-methyltransferase and hypomethylation of DNA in rats. J Nutr 132:365–369, 2002.

Saini S, Majid S, Dahiya R. Diet, microRNAs and prostate cancer. Pharm Res. 2010;27:1014–26. 16 Choi and Friso

Saito Y, Kanai Y, Sakamoto M, Saito H, Ishii H, Hirohashi S. Overexpression of a splice variant of DNA methyltransferase 3b, DNMT3b4, associated with DNA hypomethylation on pericentromeric satellite regions during human hepatocarcinogenesis. Proc Natl Acad Sci U S A 99:10060-10065, 2002.

Salerno E, Scaglione BJ, Coffman FD, Brown BD, Baccarini A, Fernandes H, Marti G, Raveche ES. Correcting miR-15a/16 genetic defect in New Zealand Black mouse model of CLL enhances drug sensitivity. Mol Cancer Ther. 2009;8:2684–92.

Scott JM, Weir DG. Folic acid, homocysteine and one-carbon metabolism: a review of the essential biochemistry. J Cardiovas Risk 5:223–227, 1998.

Sohn KJ, Stempak JM, Reid S, Shirwadkar S, Mason JB, Kim YI. The effect of dietary folate on genomic and p53-specific DNA methylation in rat colon. Carcinogenesis 24:81–90, 2003.

Song J, Medline A, Mason JB, Gallinger S, Kim YI. Effects of dietary folate on intestinal tumorigenesis in the apcMin mouse. Cancer Res 60:5434–5440, 2000.

Steinmetz KL, Pogribny IP, James SJ, Pitot HC. Hypomethylation of the rat glutathione S-transferase pi (GSTP) promoter region isolated from methyl-deficient livers and GSTP-positive liver neoplasm. Carcinogenesis 19:1487–1494, 1998

Sun M, Estrov Z, Ji Y, Coombes KR, Harris DH, Kurzrock R. Curcumin (diferuloylmethane) alters the expression profiles of microRNAs in human pancreatic cancer cells. Mol Cancer Ther. 2008;7:464–73.

Takiguchi M, Achanzar WE, Qu W, Li G, Waalkes MP Effects of cadmium on DNA-(cytosine-5) methyltransferase activity and DNA methylation status during cadmium-induced cellular transformation Exp Cell Res 286:355–365, 2003.

Trasler J, Deng L, Melnyk S, Pogribny I, Hiou-Tim F, Sibani S, Oakes C, Li E, James SJ, Rozen R. Impact of Dnmt1 deficiency, with and without low folate diets, on tumor numbers and DNA methylation in Min mice. Carcinogenesis 24:39–45, 2003.

Tryndyak VP, Ross SA, Beland FA, Pogribny IP. Down-regulation of the microRNAs miR-34a, miR-127, and miR-200b in rat liver during hepatocarcinogenesis induced by a methyl-deficient diet. Mol Carcinog. 2009;48:479–87.

Tsou JA, Hagen JA, Carpenter CL, Laird-Offringa IA. DNA methylation analysis: a powerful new tool for lung cancer diagnosis. Oncogene 21:5450–5461, 2002.

Tsujiuchi T, Tsutsumi M, Sasaki Y, Takahama M, Konishi Y. Hypomethylation of CpG sites and c-myc gene overexpression in hepatocellular carcinomas, but not hyperplastic nodules, induced by a choline-deficient L-amino acid-defined diet in rats. Jpn J Cancer Res 90:909–913, 1999.

Uthus EO. Estimation of safe and adequate daily intake for arsenic. In: Mertz W, Abernathy CO, Olin SS, Eds. Risk Assessment of Essential Elements. Washington, DC: ILSI Press, pp273–282, 1994.

Verona RI, Mann MR, Bartolomei MS. Genomic imprinting: intricacies of epigenetic regulation in clusters. Annu Rev Cell Dev Biol 2003;19:237-59.

Wallwork JC, Duerre JA. Effect of zinc deficiency on methionine metabolism, methylation reactions and protein synthesis in isolated perfused rat liver. J Nutr 115:252–262, 1985.

Waterland RA, Jirtle RL. Early nutrition, epigenetic changes at transposons and imprinted genes, and enhanced susceptibility to adult chronic diseases. Nutrition 20:63–68, 2004.

Weiss FU, Marques IJ, Woltering JM, Vlecken DH, Aghdassi A, Partecke LI, Heidecke CD, Lerch MM, Bagowski CP. Retinoic acid receptor antagonists inhibit miR-10a expression and block metastatic behavior of pancreatic cancer. Gastroenterology. 2009;137:2136–45.e1-7.

Yang J, Cao Y, Sun J, Zhang Y. Curcumin reduces the expression of Bcl2 by upregulating miR-15a and miR-16 in MCF-7 cells. Med Oncol. Epub 2009 Nov 12.

Zapisek WF, Cronin GM, Lyn-Cook BD, Poirier LA. The onset of oncogene hypomethylation in the livers of rats fed methyl-deficient, amino acid-defined diets. Carcinogenesis 13:1869–1872, 1992.

Zhang Y, Ng HH, Erdjument-Bromage H et al. Analysis of the NuRD subunits reveals a histone deacetylase core complex and a connection with DNA methylation. Genes Dev 1999; 13:1924-35.

16

Epigenetic Alteration of Receptor Tyrosine Kinases in Cancer

Anica Dricu et al.,[*]
Department of Biochemistry,
University of Medicine and Pharmacy of Craiova,
Romania

1. Introduction

In the process of cellular carcinogenesis, genetic and epigenetic mechanisms contribute to abnormal expression of genes. Conrad Waddington introduced the term "epigenetics" in 1942 (Goldberg, Allis et al. 2007) to describe heritable changes in gene expression that have no connection with changes in DNA sequence (Yoo and Jones 2006; Goldberg, Allis et al. 2007). The molecular basis of epigenetics is complex and involves DNA methylation, histone modifications, chromatin remodeling and microRNAs (Esteller 2006).

The most studied epigenetic mechanism is DNA methylation (Urdinguio, Sanchez-Mut et al. 2009) defined as the covalent addition of a methyl group at the 5′-position of cytosines within CpG dinucleotides, a process usually related to gene silencing in eukaryotes (Fig. 1A). In the human genome, the CpG dinucleotides are generally gathered in regions called CpG islands, preferentially located in promoter regions and usually not containing 5-methylcytosines (Bird 2002; Esteller 2008). Between 60% and 90% of all CpGs are methylated in mammals, to prevent chromosomal instability (Ehrlich, Gama-Sosa et al. 1982).

While DNA methylation of CpG islands is required by many physiological processes during normal development, aberrant DNA methylation can cause several pathologies, including tumor formation. (Bestor 2000; Reik, Dean et al. 2001). In the process of cellular carcinogenesis, several epigenetic alterations contribute to the abnormal expression of genes. For example, aberrant DNA methylation patterns (hypermethylation and hypomethylation) have been described in a large number of human malignancies (Fig. 1B). Hypermethylation is one of the major epigenetic modifications and typically occurs at CpG islands in the promoter region, leading to gene inactivation (Fig. 1B). Gene promoters that are aberrantly methylated during tumor development offer valuable insight on the biological pathways

[*] Stefana Oana Purcaru[1], Raluca Budiu[2], Roxana Ola[3], Daniela Elise Tache[1], Anda Vlad[2,4]
[2]*Department of Obstetrics, Gynecology and Reproductive Sciences, University of Pittsburgh School of Medicine and Magee Women's Research Institute, Pittsburgh, PA*
[3]*Department of Biochemistry and Developmental Biology, Institute of Biomedicine, University of Helsinki,*
[4]*Department of Immunology, University of Pittsburgh School of Medicine, Pittsburgh, PA,*
[2,4]*USA,*
[3]*Finland*

that are commonly interrupted during tumorigenesis (Klarmann, Decker et al. 2008; Suzuki, Toyota et al. 2008) . Gene promoter hypermethylation is related to the inhibition of cancer-related genes such as tumor suppressor genes and DNA mismatch repair genes (Feinberg and Tycko 2004; Yoo and Jones 2006) (Fig. 1B). Several key tumor suppressor genes have been found to exhibit promoter hypermethylation more often than genetic disruption (Chan, Glockner et al. 2008).

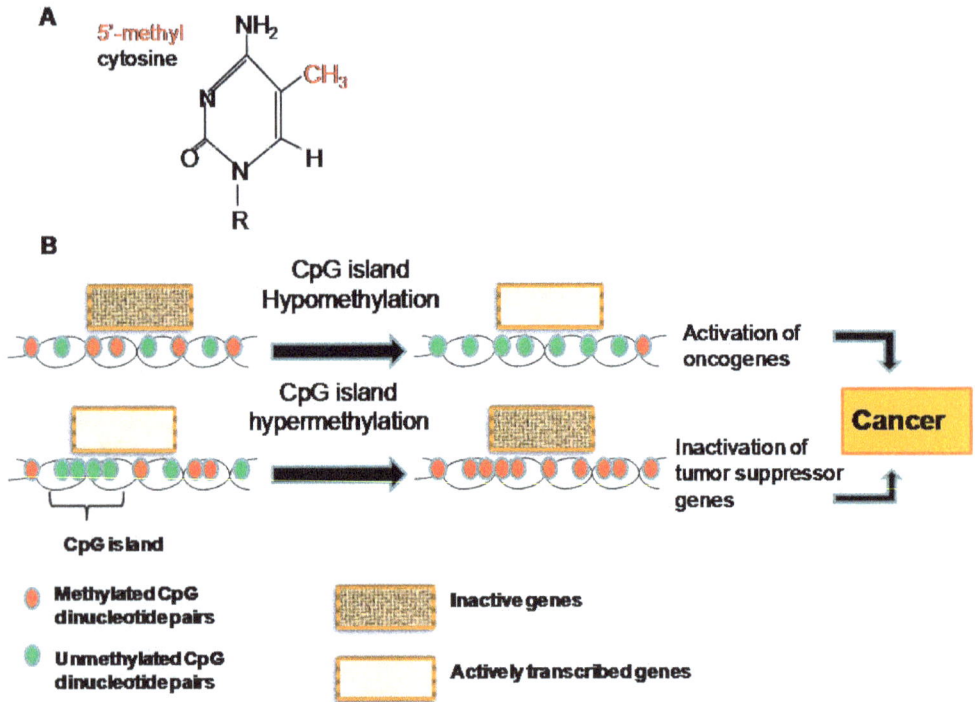

Fig. 1. Aberrant DNA methylation is involved in in the development of cancer.

DNA methylation is a process of covalent addition of a methyl group at the 5'-position of cytosines (A) within CpG dinucleotides. DNA hypermethylation is commonly interrupted during tumorigenesis and is related to the inhibition of tumor suppressor genes and DNA mismatch repair genes while DNA hypometylation appears to be involved in tumor cell development by activation of oncogenes (B).

Global hypomethylation has also been implicated in the development and progression of cancer, through different mechanisms (Yoo and Jones 2006). DNA hypomethylation, which is the first epigenetic event identified in cancer cell (Feinberg and Tycko 2004), appears to be involved in tumor cell development by activation of oncogenes, generation of chromosomal instability and loss of imprinting (Feinberg and Tycko 2004) (Fig. 1B). The low level of DNA methylation in tumors can increase the expression of several oncoproteins involved in cell growth and survival, apoptosis and cell cycle regulation (Jun, Woolfenden et al. 2009).

Among all oncogenes, receptor tyrosine kinases (RTKs) play crucial roles in the control of cancer cell growth and differentiation. RTKs are transmembrane proteins with intrinsic intracellular tyrosine kinase activity (Fig. 2).

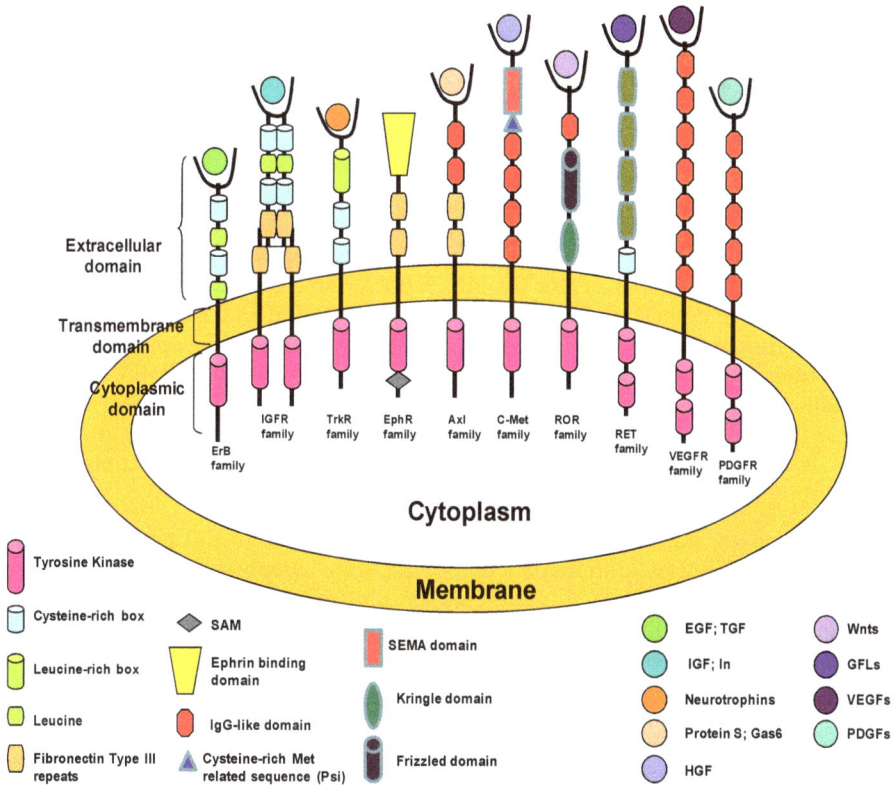

Fig. 2. Human Receptor Tyrosine Kinase (RTK) Families.

RTKs are membrane proteins, composed by an extracellular ligand-binding domain and an intracellular catalytic domain. Based on the structure of their intracellular domain and the nature of their specific ligand, RTKs are grouped in distinct families: ErbB (epidermal growth factor receptor) family, IGF-1R (Insulin-like Growth Factor 1 Receptor) family; TRK (tropomyosin receptor kinase) family, EphR (erythroprotein-producing hepatoma amplified sequence receptor) family, AXL (AXL receptor) family, c-Met (hepatocyte growth factor receptor or Scatter factor receptor) family, ROR (retinoid-related orphan receptors) family, RET (rearranged during transformation) family, VEGFR, (vascular endothelial growth factor receptor) family and PDGFR (platelet-derived growth factor receptor).

RTKs overexpression or overactivation has been described in almost all tumor types. The human genome is reported to contain 58 genes encoding receptor protein kinase, grouped into 20 classes, or subfamilies, based upon their kinase domain sequence (Robinson, Wu et al. 2000; Lemmon and Schlessinger 2010). Many RTK genes contain a typical CpG island and alterations in RTK promoter methylation have been linked to cancer development and progression (Datta, Kutay et al. 2008). Furthermore, emerging data suggests that acquired resistance to conventional cancer therapy results from progressive accumulation of RTK epigenetic modifications.

We review herein the epigenetic alterations of ten RTK families (Fig. 1), discuss their role in tumor development and implications for the response of cancer cells to conventional therapy.

2. ErbB family

ErbB protein family, or epidermal growth factor receptor (EGFR) family, consists of four structurally related receptors, with a well described tyrosine kinase activity: ErbB-1, also called epidermal growth factor receptor (EGFR), ErbB-2, named HER2 in humans and neu in rodents, ErbB-3, also named HER3 and ErbB-4 or HER4 (Bublil and Yarden 2007). ErbB-1 and ErbB-4 bind different polypeptide extracellular ligands including the epidermal growth factor (EGF), transforming growth factor α (TGFα), amphiregulin, betacellulin, epiregulin, heparin binding EGF, epigen, and neuregulins 1-4, which share a conserved epidermal growth factor (EGF) domain, giving rise to a diverse signaling network (Citri and Yarden 2006). ErbB2 is an "orphan" receptor and ErbB3 lacks the tyrosine kinase activity, hence, both signal through ErbB family heterodimers (Alimandi, Romano et al. 1995).

ErbB receptors/ligands are involved in many fundamental processes during organogenesis and adulthood: cell growth, differentiation, proliferation, apoptosis, motility, invasion, repair, survival and cell-cell interaction (Yarden and Sliwkowski 2001). In tumors, there is abnormal signalling via the ErbB pathway, mostly as a result of receptor overexpression or constitutive activation (Yarden and Sliwkowski 2001).

The *egfr* proto-oncogene localizes on chromosome 7p12, and the promoter region was described as a CG rich sequence, which lacks a TATA box and displays a large CpG island that extends into exon 1 (Kageyama, Merlino et al. 1988). EGFR expression is primarily regulated at the mRNA levels (Xu, Richert et al. 1984; Merlino, Ishii et al. 1985) and many human malignancies with epithelial origin including brain, head and neck squamous cell carcinoma (HNSCC), breast, esophagus, gastric, colon and lung cancers are characterized by EGFR overexpression. In glioma, lung, ovarian and breast cancers, mutations in the *egfr* gene are responsible for the receptor overactivity (Moscatello, Holgado-Madruga et al. 1995). Amplifications at *egfr* locus were also described in gliomas (Li, Chang et al. 2003), colorectal cancers, and less frequently in breast cancer (Al-Kuraya, Schraml et al. 2004; Ooi, Takehana et al. 2004). Genetic polymorphism of *egfr* in the intron 1 region, involving CA single sequence repeats (SSR), constitutes another mechanism that may influence *egfr* gene transcription (Gebhardt, Zanker et al. 1999). More recently it was shown that DNA methylation may also be responsible for the aberrant transcription of *egfr* in neoplastic cells (Kulis and Esteller 2010). Petrangeli et al., attempted for the first time to reveal the implication of DNA methylation in the control of EGFR function in breast cancer patients.

However, a methylation profile comparison of *egfr* promoter in tumoral versus perineoplastic tissues could not identify any differences in *egfr* gene methylation (Petrangeli, Lubrano et al. 1995). In colorectal cancer, Montero et al., identified *egfr* methylation only in the adjacent normal colon tissue but not in the corresponding tumor tissue analyzed (Montero, Diaz-Montero et al. 2006). In contrast with these results, Scartozzi et al., identified *egfr* promoter methylation in 39% of the colorectal tumors samples (Scartozzi, Bearzi et al. 2011). Furthermore, in a subsequent study of a larger cohort, the same authors showed that 58% of patients treated with monoclonal antibodies targeting EGFR had *egfr*, monoallelic or biallelic methylated (Scartozzi, Bearzi et al. 2011) and revealed a direct correlation between *egfr* methylation status and clinical outcome. However, no correlation between the methylated *egfr* and absence of protein expression was evident, as EGFR protein was detected in samples displaying a methylated *egfr*.

Different methylation density in the *egfr* promoter was also identified in three breast cancer cell lines (MB435, CAMA1, and MB453), in SAOS-2 human osteosarcoma cell line, SF-539 glioma cell line, and in two hematopoietic cell lines Raji and Raji Dac where a clear association between the methylated *egfr* and the loss of protein expression was evident (Montero, Diaz-Montero et al. 2006). Furthermore, a low density of methylated cytosines within the *egfr* promoter was also identified in 20% of primary breast, 11% of lung and 35% of head and neck tumors. It is therefore apparent that *egfr* promoter methylation is one potential mechanism responsible for the transcriptional control of *egfr* oncogene in cancer.

The second member of the EGF receptor family, Her-2/neu proto-oncogene, located on human chromosome 17q21 contains, similarly to EGFR, a large CpG island within the gene promoter region. Even though the regulatory region of the gene displays more than 10 potential sites of methylation, no aberrant methylation of *her2* could be detected in different tumor samples, including breast, lung, colon, head and neck and cell lines (Montero, Diaz-Montero et al. 2006). Nonetheless, in epithelial ovarian cancer, the expression of DNA demethylase (dMTase) correlates with poor methylation *of her2*, suggesting that dMTase enzyme is involved in Her2 promoter demethylation (Hattori, Sakamoto et al. 2001). In addition, overexpression of the rat neu oncogene in mammary tumors of MMTV/c-neu transgenic mice induced demethylation of the MMTV promoter (Zhou, Chen et al. 2001). In the same mouse model Kmieciak et al., later showed that Interferon γ (IFN-γ) mediates epigenetic changes in neu oncogene, resulting in neu antigen loss and tumor escape (Kmieciak, Knutson et al. 2007).

Only a paucity of data describing the transcriptional control by any epigenetic mechanisms of Her 3 and Her 4 currently exists. DNA methylation of a CpG island in the promoter region seems to be involved in *her4* transcriptional suppression in breast cancer cell lines and primary breast carcinomas. Moreover, Das, P.M., et al., demonstrated a direct correlation between the *her4* methylation status and HER4 protein expression and proposed *her4* promoter methylation as a negative prognostic indicator at least in breast cancer (Das, Thor et al. 2010).

Identification of DNA methylation as a regulatory mechanism and a prognostic marker for three out of four EGF receptors provides promising new alternatives to targeting ERB signaling with better selectivity, safety and efficacy.

3. IGF receptor family

The insulin-like growth factor (IGF) family is composed of two ligands – (IGF-1 and IGF-2), two receptors – (IGF-1R and IGF-2R), six high-affinity IGF-binding proteins – (IGFBP 1-6) and several associated IGFBP degrading enzymes (proteases). The IGF family has a critical role in the development and maintenance of normal tissue homeostasis and it appears to be involved a number of pathological states, including cancer.

Insulin- like growth factor type-1 receptor (IGF-1R), the main receptor of IGF family, is a transmembrane heterotetramer, linked to the PI3K/Akt and MAPK signal transduction pathways. Approximately 75% of the IGF-1R promoter region consists of cytosine and guanine. The IGF-1R promoter is also TATA-less and CCAAT-less and contains several potential binding sites for Sp1, ETF and AP-2 nuclear transcription factors (Werner, Stannard et al. 1990). IGF-1R is overexpressed in several types of cancer and their involvement in cancer cells' response to treatment has also been reported (Kanter-Lewensohn, Dricu et al. 2000; Cosaceanu, Budiu et al. 2007). In the human IGF-1R promoter region, bioinformatic analysis revealed the presence of multiple CpG dinucleotides (Schayek, Bentov et al. 2010). The presence of CpG islands in the promoter region of IGF-1R gene supports the theory that an altered methylation status of the IGF-1R promoter may be responsible for the IGF-1R oncogene overexpression in various cancers. In 2010, we described for the first time the partial methylation of IGF-1R promoter in three subtypes of non-small cell lung cancer (NSCLC): large cell lung cancer, squamous cell carcinoma and adenocarcinoma. We found the same level of IGF-1R promoter methylation in all NSCLC subtypes and no correlation between the gene methylation level and receptor protein expression (Ola 2010).

While several studies suggest IGF-1R overexpresion as a key in prostate cancer initiation and progression (Hellawell, Turner et al. 2002; Turney, Turner et al. 2011), Chott at al. demonstrate that loss of IGF-1R may contribute to prostate cancer progression (Chott, Sun et al. 1999). In line with these findings, Schayek et al., 2010, showed that progression towards metastatic stages in prostate cancer is correlated with a dramatic reduction in both total IGF-1R protein levels and basal phospho-IGF-1R values, which reflects a decrease in IGF-1R activation (Schayek, Bentov et al. 2010). Furthermore, studies on the IGF-1R methylation of six prostate cancer cell lines showed that the IGF-1R promoter is unmethylated in all examined cell lines, suggesting that the gene silencing in metastatic prostate cancer is probably not caused by direct gene promoter methylation (Schayek, Bentov et al. 2010). However, our studies from two glioblastoma lines sugget that IGF-1R is, at least in this setting, partially methylated (Ola 2010).

Many promising agents that inhibit the key enzymes involved in establishing and maintaining the epigenetic changes have been identified and tested in various types of pathologies. One of those agents is S-Adenosylmethionine (SAM), a methyl donor agent. Several studies have reported DNA hyper-methylation, after SAM exposure (Fuso, Cavallaro et al. 2001; Detich, Hamm et al. 2003). In contrast, evidence from our studies on S-Adenosylmethionine-induced cytotoxicity in both glioblastoma and non-small cell lung cancer cell lines demonstrates that SAM does not affects the IGF-1R methylation status (Ola 2010; Ola 2010). While IGF1R epigenetics seems important, the exact mechanisms underlying IGF-1R overexpression in cancer are far from being understood.

Insulin- like growth factor type-2 receptor (IGF-2R, also known as the cation-independent mannose 6-phosphate receptor), the second cell-surface receptor of the IGF family, is a 275 ± 300 kDa glycoprotein, comprised of an extracytoplasmic domain made up of 15 contiguous cysteine-rich repeats, a single transmembrane-spanning domain and a carboxyl -terminal cytoplasmic domain (Ghosh, Dahms et al. 2003). The receptor does not appear to have any protein kinase activity but the carboxyl-terminal cytoplasmic domain has been shown to facilitate endocytosis (Probst, Puxbaum et al. 2009) and intracellular sorting of lysosomal enzymes (Munier-Lehmann, Mauxion et al. 1996). IGF-2R is a maternally expressed gene regulated by epigenetic modifications and one of the classical examples of tissue-specific and species-dependent imprinted genes. There are several studies demonstrating that IGF2 ligand is also an imprinted gene and it has been proved that loss of imprinting causes proliferation of transformed cells in Wilm's tumors, by elevating the level of available IGF2 ligand (Steenman, Rainier et al. 1994; Vu 1996). Both IGF2 and IGF2R genes possess a CG-rich and TATA-less promoter (Szentirmay, Yang et al. 2003). In mice, methylation of region 2, a region rich in cytosine-guanine doublets in the second intron of IGF-2R, represents the imprinting signal that maintains expression of the maternal allele (Stoger, Kubicka et al. 1993). Wutz and his colleagues tested the role of region 2 and the influence of chromosome location on IGF-2R imprinting, using mouse YAC-T1/P and YAC-T1/PΔR2 transgenes (Wutz, Smrzka et al. 1997). Their results show that deletion of region 2 of the IGF-2R gene is associated with loss of imprinting and restoration of biallelic IGF2R expression, demonstrating the primary role for the region 2 and the negligible role for chromosomal location in IGF2R imprinting (Wutz, Smrzka et al. 1997).

Increasing evidence from genetic studies currently implicate IGF-2R as a tumour suppressor in a variety of malignancies. Loss of heterozygosity (LOH) at the gene locus on 6q26 ± 27 has been reported for a number of tumour types, including breast (Chappell, Walsh et al. 1997), liver (Oka, Waterland et al. 2002), lung (Kong, Anscher et al. 2000) and head and neck (Jamieson 2003) cancers. Further investigations of the involvement of both IGF-1R and IGF-2R and theirs interaction with the ligands are required to fully understand the role of these complex molecules in cancer.

Unlike genetic changes, epigenetic events do not alter the DNA code and are potentially reversible. Therefore, reactivation of epigenetically silenced genes like IGF2-R could provide attractive therapeutic opportunities.

4. C-MET receptor family

The *c-met* oncogene is a tyrosine kinase receptor binding to the hepatocyte growth factor (HGF) also known as the scatter factor (SF). Signals through c-Met are required for normal mammalian development and fundamental processes like cell migration, differentiation, proliferation, cell growth, branching and angiogenesis (Birchmeier, Birchmeier et al. 2003). Although HGF is the only known ligand for C-Met, glial cell derived neurotrophic factor (GDNF) also seems to be able to activate C-Met, albeit indirectly (Popsueva, Poteryaev et al. 2003).

C-met gene has been mapped to chromosome 7q31 and its regulatory region is GC-rich. It contains no TATA box, and displays a CpG island with an increased frequency of CpG, suggesting that an aberrant transcriptional regulation may play a critical role in oncogene activation (Seol and Zarnegar 1998).

Many studies to date provide evidence that dysregulation of c-Met activity is a key event in the initiation and progression of carcinogenesis (Maulik, Shrikhande et al. 2002). Mutations in *c-met* were identified in papillary renal carcinoma, gastric and liver cancer, small and non–small cell lung cancers, and head and neck squamous cell carcinomas (Peruzzi and Bottaro 2006). Furthermore, amplification of *c-met* in gastric, colorectal and lung carcinoma seems to inversely correlate with survival (Kuniyasu, Yasui et al. 1992; Zeng, Weiser et al. 2008; Lee, Seo et al. 2011).

The possibility that an altered methylation of *c-met* promoter induces overexpression of C-MET protein was, for the first time, addressed in papillary carcinoma although no changes in c-Met promoter methylation could be detected in tumour tissue compared to normal tissue (Scarpino, Di Napoli et al. 2004). Evidence from Morozov et al demonstrated that the death-domain associated protein (Daxx) (a component of a multiprotein repression complex), preferentially binds to the *c-met* promoter together with Histone deacetylase 2 (HDAC2), leading to *c-met* transcriptional repression (Morozov, Massoll et al. 2008). However, DNA methylation seems not to be involved in the Daxx-mediated repression of the *c-met* promoter (Morozov, Massoll et al. 2008). Further studies are needed in order to elucidate the possible implication of c-met promoter methylation in carcinogenesis.

Despite the transcriptional repression of *c-met* proto-oncogene by chromatin modifications, direct evidence of DNA methylation as a responsible mechanism for *c-met* promoter function is missing and needs to be further investigated.

5. Trk receptor family

Tropomyosin-receptor-kinase (Trk) receptors belong to a family of tyrosine kinases that control synaptic strength and plasticity in the mammalian nervous system (Huang and Reichardt 2003). Trk receptors bind neurotrophins and trigger downstream activation of several signaling cascades that affect normal physiological processes, like neuronal survival and differentiation (Segal 2003).

TrkA, TrkB, and TrkC are the three most common types of trk receptors. Each of them has different binding affinity to certain types of neurotrophins. The receptors act through different intracellular pathways and the differences in their signaling can generate a diversity of biological responses (Segal 2003). The Trk oncogene was initially identified in a colon carcinoma and its identification led to the discovery of the first member, TrkA (Huang and Reichardt 2003). The oncogene was produced by a mutation in chromosome 1 that resulted in the fusion of the first seven exons of tropomyosin to the transmembrane and cytoplasmic domains of the then-unknown TrkA receptor (Martin-Zanca, Hughes et al. 1986).

Accumulating evidence now suggests that TrkA, TrkB and TrkC play an important role in the malignant behavior of cancer cells, like increased metastasis, proliferation and survival (Sclabas, Fujioka et al. 2005; Jin, Kim et al. 2010). Studies on different tumors show that TrkC seems to be highly expressed in neuroblastoma, medulloblastoma, (Segal, Goumnerova et al. 1994; Yamashiro, Liu et al. 1997; Grotzer, Janss et al. 2000) and breast cancer (Bardelli, Parsons et al. 2003; Wood, Calhoun et al. 2006). Furthermore, overexpression of TrkA was reported in papillary thyroid and colon carcinoma (Nakagawara 2001), and of TrkB in malignant keratinocytes (Slominski and Wortsman 2000) and pancreatic cancer (Sclabas,

Fujioka et al. 2005). Recent evidece from Jiin et al show that while Trks might be important for liver cancer metastasis and are highly expressed during the course of tumor progression, no aberrant promoter methylation could be detected. In contrast, low or undetectable expression level of Trk receptors in normal liver cell lines demonstrated a dramatically increased pattern of methylation (Jin, Lee et al. 2011).

Given their high level of expression in cancer and their roles in metastasis, current studies now attempt to decipher their functions as oncogenic tyrosine kinases during malignant transformation, and to define their potential as attractive targets for therapeutic intervention (Shawver, Slamon et al. 2002).

6. Eph receptor family

The erythroprotein-producing hepatoma amplified sequence (Eph) receptor tyrosine kinase family is the largest family of tyrosine kinases, comprising at least 14 Eph receptors and 8 ligands (Manning, Whyte et al. 2002). Based upon sequence similarities in their extracellular domains and their ability to bind to ligands, Eph receptors are divided into two groups, EphA and EphB. The EphA members (EphA1-A8 and A10) bind to the glycosyl-phospahtidyl-inositol (GPI) - linked ligands EFNA1-5, whereas EphB1-B4 and B6 belonging to the EphB receptor family binds the transmembrane ligands EFNB1-3 (Pasquale 2004). Upon ephrin binding, Eph receptors are clustered, phosphorylated and kinase activated (Bruckner and Klein 1998). Different pathways have been shown to be implicated in Eph signaling, including activation of the MAPK/ERK pathways by EphB2 receptor (Huusko, Ponciano-Jackson et al. 2004), inhibition of the Ras/ERK1/2 signaling cascade by EphA2, and phosphorylation of Src kinases and Akt by EphA2 and EphA4 (Guo, Miao et al. 2006).

Various human tissues differentially express the Eph/ephrin family (Hafner, Schmitz et al. 2004; Fox, Tabone et al. 2006; Hafner, Becker et al. 2006). Eph involvement in developmental processes, especially in embryonic development, vasculature and nervous system formation has been reported (Kullander and Klein 2002; Himanen and Nikolov 2003).

A growing body of evidence currently indicates that *Eph* gene family also plays an important role in carcinogenesis and tumor progression in many types of cancer (Andres, Reid et al. 1994; Walker-Daniels, Coffman et al. 1999; Liu, Ahmad et al. 2002; Kinch, Moore et al. 2003; Fang, Brantley-Sieders et al. 2005; Noren, Foos et al. 2006; Merlos-Suarez and Batlle 2008). Eph receptors were reported to have both tumor promoting activity, by being highly expressed in different human cancers, (i.e. breast, colon, melanomas, lung and prostate cancer) (Zelinski, Zantek et al. 2001; Surawska, Ma et al. 2004; Nakada, Drake et al. 2006; Foubert, Silvestre et al. 2007), and suppressing activity by acting as tumor suppressors (Batlle, Bacani et al. 2005; Noren, Foos et al. 2006). Some tumors are characterized by loss of expression of Eph receptors (EphB2 and B4 receptors in colorectal cancer and EphB6 receptor in breast cancer) (Alazzouzi, Davalos et al. 2005; Davalos, Dopeso et al. 2006; Fox and Kandpal 2006). Gene promotor methylation of *epha2, epha3, epha7, ephb2, ephb4* and *ephb6* was found in several human solid tumors, including breast, colorectal and prostate cancer, suggesting that an aberrant CpG island methylation, located in the promoter region of tumor related genes, may determine inactivation (Dottori, Down et al. 1999; Alazzouzi, Davalos et al. 2005; Wang, Kataoka et al. 2005; Davalos, Dopeso et al. 2006; Fox and Kandpal 2006; Nosho, Yamamoto et al. 2007). In addition, epigenetic silencing by hypermethylation of *EPH/EPRIN* genes seems to contribute to the pathogenesis of acute lymphoblastic leukemia, where *ephb4* acts as a tumor suppressor (Kuang, Bai et al. 2010).

In breast cancer, *EphA2* appears to be important for the progression of the tumor from noninvasive to an invasive phenotype (Fox and Kandpal 2004), potentially via ErbB2 signaling (Brantley-Sieders, Zhuang et al. 2008). Fu et al. demosntrated that EphA5 is frequently down regulated in breast cancer cell lines and tumor tissues *via* aberrant methylation of its promoter, a finding significantly associated with the clinico-pathologic tumor grade and metastasis to the lymph nodes (Fu, Wang et al. 2010). The authors also reported the presence of a dense GpG island with transcriptional and translational start sites at the 5' end of the *EphA5* gene. Such molecular architecture is entirely consistent with that described for other genes, known as targets for epigenetic silencing during tumorigenesis (Feltus, Lee et al. 2003), an early and frequent event in the development of breast cancer (Lehmann, Langer et al. 2002).

Altogether, these reports suggest that methylation of Eph genes might be used as a potential marker for cancer diagnosis and prognosis and underscore the need for further studies that better define their translational potential.

7. AXL receptor family

The Axl receptors belong to the TAM (Tyro-Axl-Mer) receptor tyrosine kinases family (Lai and Lemke 1991; O'Bryan, Frye et al. 1991). Axl is a 140-kDa protein, with an extracellular region comprised of two immunoglobulin-like domains and fibronectin type III repeats, a transmembrane region and a cytoplasmic domain with kinase activity (O'Bryan, Frye et al. 1991). There are two ligands for Axl: protein S and growth-arrest specific 6 (Gas6), the latter binding with higher affinity to Axl (Stitt, Conn et al. 1995; Varnum, Young et al. 1995). The main downstream Axl signaling is triggered through the phosphatidylinositol 3 kinase (PI3K) pathway (Collett, Sage et al. 2007), although in some circumstances the Janus kinase-signal transducers and activator of transcription (STAT) (Rothlin, Ghosh et al. 2007) or the p38 mitogen-activated protein kinase pathway (Allen, Linseman et al. 2002) may also be induced. The Axl receptor also cooperates with the cytokine receptor signaling network in order to regulate many biologic functions (Budagian, Bulanova et al. 2005; Gallicchio, Mitola et al. 2005) like cell survival, proliferation, adhesion and migration (Hafizi and Dahlback 2006). It contributes to vascular smooth muscle homeostasis and regulates endothelial cell migration and vascular network formation (Korshunov, Mohan et al. 2006; Collett, Sage et al. 2007).

Axl is overexpressed in several human cancers (Craven, Xu et al. 1995; Ito, Ito et al. 1999; Berclaz, Altermatt et al. 2001; Sun, Fujimoto et al. 2004) where it plays important roles in tumor angiogenesis (Holland, Powell et al. 2005; Li, Ye et al. 2009) and metastasis as demonstrated by studies in lung (Shieh, Lai et al. 2005), prostate (Sainaghi, Castello et al. 2005), breast (Meric, Lee et al. 2002), gastric (Wu, Li et al. 2002) andrenal cell carcinomas (Chung, Malkowicz et al. 2003), as well as glioblastomas (Hutterer, Knyazev et al. 2008). Interestingly, Axl overexpression was reported to induce Imatinib resistance in gastrointestinal stromal tumors (Mahadevan, Cooke et al. 2007), Herceptin resistance in breast cancer (Liu, Greger et al. 2009) or chemotherapy resistance in acute myeloid leukemia (Hong, Lay et al. 2008), supporting its applicability as a potential therapeutic biomarker.

The invasive capacity of cancer cells was shown to be reduced by downregulation of Axl expression, as evidenced in breast and lung cancer (Holland, Powell et al. 2005; Shieh, Lai et al. 2005). Liu R et al. recently showed that Axl is induced in Kaposi sarcoma and Kaposi sarcoma herpesvirus (KSHV) transformed endothelial cells, but also in Kaposi sarcoma cell lines lacking KSHV(Liu, Gong et al. 2010).

Axl is transcriptionally regulated by the Sp1/Sp3 transcription factors and further controlled by CpG island methylation (Mudduluru and Allgayer 2008). Mudduluru et al. showed that transactivation of Axl gene is controled by MZF1, which induces migration, invasion and in vivo metastasis formation (Mudduluru, Vajkoczy et al. 2010). In a subsequent study, focused on the the epigenetic regulation of Axl at post-transcriptional level by micro-RNAs, the same authors showed that three micro-RNAs, miR-34a, miR-199a and miR-199b were frequently methylated and that their expression levels inversely correlates with Axl expression in three types of cancer: non-small cell lung cancer, colorectal cancer and breast cancer (Mudduluru, Ceppi et al. 2011).

TAM receptor inhibitions in animal xenograft tumor models of glioblastoma and breast cancer have provided preliminary validation of this receptor family as a cancer therapy target and further translational studies are needed in order to fully establish its potential.

8. ROR receptor family

Two human Ror RTK-encoding genes, *hRor1* and *hRor2*, the first Ror family members, and two rat partial complementary DNAs (cDNAs), *rRor1* and *rRor2* were identified nearly 10 years ago (Masiakowski and Carroll 1992). Ror family members have a conserved domain structure; the extracellular regions contain immunoglobulin (Ig), cysteine-rich (CRD) and kringle domains, all of which are thought to mediate protein-protein interactions. The intracellular domain of Rors contains a tyrosine kinase domain, two regions rich in serine and threonine separated by a region rich in prolines. There is a high similarity between the two Ror proteins Ror1 and Ror2 from each species. For instance, hRor1 and hRor2 have a 58% overall amino acid identity and 68% amino acid identity within the kinase domains (Forrester 2002).

ROR1 belongs to the RTK family of orphan receptors connected with muscle specific kinase and neurotrophin receptors (Masiakowski and Carroll 1992; Valenzuela, Stitt et al. 1995; Glass, Bowen et al. 1996). ROR has a predicted 937 amino acids sequence including an Ig-like domain, cysteine-rich domain, kringle domain, tyrosine kinase domain and proline-rich domain (Yoda, Oishi et al. 2003) .

An alternative spliced form of *hRor1*, called truncated *hRor1* (Reddy, Phatak et al. 1996) has been reported in fetal and adult central nervous system, and was identified in human leukemia and lymphoma cell lines, as well as in a variety of neuroectodermal cancers. However, the function and significance of truncated hRor1 remain unclear (Reddy, Phatak et al. 1996; Forrester 2002).

Although ROR1 is situated on chromosomal region 1p31.3, which is a region where chromosomal aberrations are not frequently found in hematological malignancies, results from gene expression profiling studies show a 43.8-fold increase of ROR1 in chronic lymphocytic leukemia (CLL) cells (Klein, Tu et al. 2001). Furthermore, activation of NF-kB in

CLL cells give rise to a functionally active ROR1 protein via Wnt5a and nonclassical Wnt-signaling pathway, supporting a role for ROR1 in the pathogenesis of CLL (Fukuda, Lu et al. 2004). Due to the high level of Ror1 surface expression in leukemic cells and its absence in normal (nonactivated) blood leukocytes (Daneshmanesh, Mikaelsson et al. 2008), ROR1 is now considered a candidate structure for targeted therapy, including monoclonal antibody-based therapies (Reddy, Phatak et al. 1996; Daneshmanesh, Mikaelsson et al. 2008). However, the exact mechanisms behind ROR1 gene regulation are yet to be fully delineated. Despite the recent evidence from fluorescence *in situ* hybridization (FISH) studies (Daneshmanesh, Mikaelsson et al. 2008) showing that the expression of RORr1 may not be related to genomic aberrations but rather epigenetic regulation, further evidence is needed.

Receptor tyrosine kinase-like orphan receptor 2 (ROR2), a transmembrane protein, is a member of a conserved family of tyrosine kinase receptors implicated in many developmental processes, including chondrogenesis (Sammar, Stricker et al. 2004), osteoblastogenesis (Liu, Bhat et al. 2007) and neural differentiation (Matsuda, Nomi et al. 2001). ROR2 mutations in humans result in dominant brachydactyly type B and Robinow syndrome (Ali, Jeffery et al. 2007; Lara, Calvanese et al. 2010).

ROR2 exerts its role in cell differentiation primarily via the Wnt signalling pathway (Angers and Moon 2009), employed by several extracellular effectors, membrane proteins, intracellular signal transducers and nuclear gene regulators that transmit extracellular signals to the nucleus as precise instructions for regulating specific genes (Aguilera, Munoz et al. 2007). Canonical Wnt is the signalling pathway involving β-catenin. Beta catenin-independent signals can also be induced by Wnt effectors via the non-canonical Wnt signalling pathway. The manner in which ROR2 realises its primary role, to mediate WNT5A signals within the Wnt signalling pathway, is still unclear (Lara, Calvanese et al. 2010). It was initially demonstrated in 293 cells that ROR2 mediates WNT5A-dependent inhibition of canonical Wnt signalling downstream of β-catenin, at the level of TCF-mediated transcription (Mikels and Nusse 2006). Subsequently it was shown that ROR2 mediates WNT5A dependent JNK (c-JUN NH_2-terminal protein kinase) activation and regulates convergent extension movements in *Xenopus* gastrulation (Schambony and Wedlich 2007) while in osteoblastic cells it enhances WNT1 and antagonise WNT3 activities (Billiard, Way et al. 2005). The inhibition of the β-catenin-dependent Wnt signalling pathway is mediated by ROR2 (Billiard, Way et al. 2005; Mikels and Nusse 2006; MacLeod, Hayes et al. 2007). Furthermore, by promoting constitutive Wnt signaling, the aberrant epigenetic repression of other Wnt inhibitors such as WIF-1, DKK1, SFRP1 and SFRP2 directly promotes tumourigenesis in colon cancer cells (Suzuki, Gabrielson et al. 2002; Mazieres, He et al. 2004; Aguilera, Fraga et al. 2006; Aguilera, Munoz et al. 2007). Inhibitor of the canonical Wnt signalling pathway in certain molecular contexts (Mikels and Nusse 2006), the ROR2 extracellular ligand WNT5A, is also aberrantly repressed by promoter hypermethylation in acute lymphoblastic leukaemia (Roman-Gomez, Jimenez-Velasco et al. 2007) and in colon cancer (Ying, Li et al. 2008), and its absence is tumourigenic in these tumour types.

ROR2 positively modulates Wnt3a-activated canonical signaling in the H441 lung carcinoma cell line (Li 2008). The Wnt signalling pathway is essential to cell differentiation and cancer. A primary mechanism of colon cancer development is the genetic and epigenetic changes of components of the canonical Wnt signalling pathway (Aguilera, Munoz et al. 2007). ROR2 is overexpressed in oral (Kobayashi, Shibuya et al. 2009) and renal cancer (Wright, Brannon et al. 2009) and in osteosarcoma (Enomoto, Hayakawa et al. 2009). In osteosarcoma cells, the same suppressed expression of ROR2 (or its extracellular effector, WNT5A) diminishes invadopodium formation and inhibits cell invasiveness (Enomoto, Hayakawa et al. 2009). These studies underline the complex role of ROR2 in cancer and its function both as a promoter and suppressor of tumor formation, depending on the tumour type and molecular context.

Furthermore, ROR2 can mediate WNT5A-dependent activation of JNK, a member of the non-canonical Wnt pathway in mice (Oishi, Suzuki et al. 2003; Schambony and Wedlich 2007) and it can govern the WNT5A-dependent inhibition of canonical Wnt signalling downstream of β-catenin stabilization (Mikels and Nusse 2006). Considering the tumour type, ROR2 signals can therefore reveal a preference for β-catenin/TCF-dependent genes or for non-canonical Wnt pathways. There are two possibilities which support the evidences: firstly, in renal cancer and osteosarcoma activation of the non-canonical Wnt signalling kinase JNK mediates the pro-tumourigenic role of ROR2 (Enomoto, Hayakawa et al. 2009; Wright, Brannon et al. 2009) and secondly, restoration of ROR2 activity increased the inhibition of β-catenin reporter genes in colon cancer cells with constitutive Wnt signalling activity (MacLeod, Hayes et al. 2007; Lara, Calvanese et al. 2010).

As suggested for ROR2, WNT5A might have both a tumour-promoting and suppressing role. A tumour-suppressing effect has been reported in many studies, and it is downregulated in a number of cancers such as colorectal and ductal breast cancer, neuroblastoma and leukaemia (McDonald and Silver 2009). It was shown that WNT5A repression in colon and haematopoietic tumours is intervened by aberrant promoter hypermethylation (Roman-Gomez, Jimenez-Velasco et al. 2007; Ying, Li et al. 2008). WNT5A also reveals a tumour-promoting role in diseases such as non-small-cell lung cancer, melanoma, breast, gastric, pancreatic and prostate cancers (McDonald and Silver 2009). ROR2 knockout mice phenocopy most of the alterations seen in the WNT5A knockout mice, further supporting the intimate association between ROR2 and WNT5A and their complementary roles in cancer. While ROR2/WNT5A epigenetic downregulation would benefit tumours, such as colon and haematopoietic cancers typically driven by canonical Wnt signaling, their upregulation could be advantageous to cancers which are driven by non-canonical Wnt signaling (Lara, Calvanese et al. 2010).

Ror2 repression by aberrant promoter hypermethylation was reported in human colon cancer where epigenetic-dependent loss of ROR2 can promote tumour growth in colon cancer cells. Furthermore, ROR2 was reported to be overexpressed and have oncogenic properties in other tumour types such as oral cancer (Kobayashi, Shibuya et al. 2009), renal cancer (Wright, Brannon et al. 2009) and osteosarcoma (Enomoto, Hayakawa et al. 2009).

9. RET receptor family

Ret gene encodes for a receptor tyrosine kinase shared by four ligands, all members of the glial cell-derived neurotrophic factor (GDNF) family with important roles in neuronal survival: GDNF, artemin (ARTN), neurturin (NRTN) and persephin (PSPN) (Sariola and Saarma 2003). Signaling through RET is required for the development of the enteric nervous system, metanephric kidney and for the process of spermatogenesis (Schuchardt, D'Agati et al. 1994; Meng, Lindahl et al. 2000). The human *RET* gene is localized on chromosome 10 (10q11.2) and contains 21 exons (Ceccherini, Bocciardi et al. 1993). Nine hundred base pairs within the promoter region of the *Ret* protooncogene contains 95 5'-CG-3' dinucleotide pairs, suggesting that *Ret* transcriptional activity might be regulated by DNA methylation (Munnes, Patrone et al. 1998).

Mutations of *Ret* are frequently reported in thyroid carcinoma and in Hirschsprung's (HSCR) disease (Pelet, Attie et al. 1994). Germline mutations in *Ret* induce constitutive RET protein activation and are the cause of multiple endocrine neoplasia type 2 (MEN 2). Also, *Ret* rearrangements were identified as a frequent pathogenic event that occurs in papillary thyroid carcinoma (PTC) (Jhiang, Sagartz et al. 1996).

The DNA methylation profile of *Ret* gene promoter was for the first time addressed in human HSCR disease (Munnes, Patrone et al. 1998). By bisulfite sequencing, Munnes, M., et al., reported that the *Ret* promoter is completely lacking 5-mC in both, malignant and normal samples (Munnes, Patrone et al. 1998). However, the methylation profile of the genomic region upstream of the *Ret* promoter indicated methylation in some CG sequences, with consequences on promoter activity (Munnes, Patrone et al. 1998). Recent investigations on the DNA methylation profiles at the *Ret* locus in medullary thyroid carcinoma identified a significantly lower degree of methylation in tumor cells compared with normal thyroid tissues (Angrisano, Sacchetti et al. 2011), providing supporting evidence for Ret epigenetics in thyroid cancer.

Furthermore, both histone and DNA methylation modifications seem to contribute to retinoic acid (RA)-mediated RET activation in neuroblastoma cells (Angrisano, Sacchetti et al. 2011) where several changes in methylation and acetylation profile of the core histones H3K4, H3K9, H3K9 and H3K27 within the *Ret* promoter region trigger modifications of Ret transcription. In addition, *in vitro* treatment of neuroblastoma cells (SK-N-BE) with a demethylating agent, seams to trigger an increase of RET expression even in the absence of RA (Angrisano, Sacchetti et al. 2011), indicating the possibility of using the methylation state of *Ret* proto-oncogene as a tumoral prognostic marker in cancer and raising the prospect of employing demethylating drugs in cancer therapy.

10. VEGF receptor family

Vascular endothelial growth factor receptors – 1 (VEGFR-1) and – 2 (VEGFR-2) play a critical role in physiologic and pathologic angiogenesis, including that associated with cancer (Carmeliet and Jain 2000; Ferrara 2005). Moreover, the expression of VEGF, VEGFR-1 and VEGFR-2 was reported in many solid tumors (Smith, Baker et al. 2010) of the colon (Kobayashi, Sugihara et al. 2008), ovary (Boocock, Charnock-Jones et al. 1995), breast (de Jong, van Diest et al. 2001), lung (Takahama, Tsutsumi et al. 1999) and prostate (Ferrer, Miller et al. 1999).

The 5'-flanking region of the VEGFR-1 gene contains a CpG island, where several putative transcription factor binding sites such as a TATA box and a binding protein/activating transcription factor (CREB/ATF) element have been described (Morishita, Johnson et al. 1995). In 2003, Yamada et al. reported an aberrant promoter methylation of the 5' region of the VEGFR1 detected by cDNA microarray screening analysis in prostate cancer cell lines compared to primary prostate samples (Yamada, Watanabe et al. 2003). The VEGFR-1 promoter was methylated in 38.1% of primary prostate cancer samples, in contrast to results from benign prostate samples where no promoter methylation could be detected. These results demonstrate that promoter methylation of VEGFR-1 plays a key role in silencing of this gene in prostate cancer cells (Yamada, Watanabe et al. 2003). However, the VEGFR-1 promoter methylation status was not elucidated in other cancer types. In 2009, Kim et al. evaluated the VEGF and VEGFR genes promoter methylation in different cancer cell lines and primary solid cancers (Kim, Hwang et al. 2009). While the VEGF promoter was not methylated in any of the cancer cell lines tested (colon, stomach, lung, melanoma, breast or thyroid cancers) VEGFR-1 as well as VEGFR-2 showed variable hypermethylation in the tested cancer cell lines, with an increased frequency in the stomach and colon cancer (Kim, Hwang et al. 2009). The study also showed a negative correlation between promoter methylation and expression of VEGFR genes, providing new insight on the epigenetic mechanisms underlying VEGFR expression in tumor tissues.

11. PDGFR family

Plateled-derived growth factor family includes four ligands: PDGFA, PDGFB PDGFC and PDGFD that regulate cell proliferation, cellular differentiation, cell growth, development and many diseases including cancer. The PDGFs bind to the protein tyrosine kinase receptors PDGF receptor-α and -β. These two receptor isoforms dimerize upon binding the ligand dimer, leading to three possible receptor combinations: -αα, -ββ and -αβ.

PDGFRα plays an important role in the normal development of neurons as well as in the pathogenesis of different disorders of the central nervous system. PDGFs and PDGFRα are coexpressed in human gliomas and glioma cell lines (Hermanson, Funa et al. 1992; Lokker, Sullivan et al. 2002). On the other hand, PGFRa was reported to be overexpressed in malignant gliomas (Fleming, Saxena et al. 1992; Joensuu, Puputti et al. 2005; Carapancea, Cosaceanu et al. 2007) which leads to cell proliferation, invasion, and resistance to apoptosis. PDGFR activation initiates the signaling cascades that comprise the Ras/Raf/MAPK and PI3K/AKT pathways (Dong, Jia et al. 2011).

Several other neoplasia have also been associated with disregulated expression of PGFRα, in particular ovarian cancer (Lassus, Sihto et al. 2004), osteosarcoma (Sulzbacher, Birner et al. 2003), breast cancer (Carvalho, Milanezi et al. 2005) and laryngeal small cell carcinoma (Kanazawa, Nokubi et al. 2011). Following their intial studies showing that the human PDGFRα contains 10 polymorphic sites that give rise to 5 haplotypes, H1, H2a, H2b, H2g and H2d (Joosten, Toepoel et al. 2001), Joosten and colleagues later reported data on PDGFRa haplotype promoter's modifications and its consequences for the genetic predisposition of individuals to develop glioblastoma multiforme (Toepoel, Joosten et al. 2008). These studies show that H1 allele has a low activity in glioblastoma cell lines and is associated with allele-specific DNA methylation and histone deacetylation. Furthermore, epigenetic repression causes a low-activity of PDGFRa H1 allele during glial development and provides a reduce risk of glioblastoma development (Toepoel, Joosten et al. 2008).

12. Conclusion

Tumors are characterized by acquired somatic mutations and epigenetic alterations in genes that are crucial for differentiation, proliferation and survival pathways. Receptor tyrosine kinases and their downstream signalling pathways play key roles in cancer development and are widely studied for their potential as therapeutic targets. It is increasingly evident that deciphering the mechanisms of RTK gene regulation in cancer is essential for the future development of new and improved therapies. Cancer epigenetics is one of the most rapidly expanding fields and current comprehensive epigenomic approaches will likely lead to a better understanding of the epigenetic regulations of RTK genes and their roles in proliferation, differentiation and cell growth and will open the door for the development of new treatment strategies based on these mechanisms. The advent of sensitive technology and the increasing evidence from recent studies provide a solid rationale for future exploration of epigenetics in mainstream oncology. Furthermore, the integration of epigenetics with data from genomics and transcriptomics will dramatically increase our understanding of tumorigenesis and will potentially yield better biomarkers for early detection, prognosis and therapy responses. The heterogeneity and genetic complexity of tumors is daunting, but the improvement in our knowledge of the pathogenetic mechanisms underlying RTK-induced transformation, coupled with the increasing availability of agents that target these pathways, offer unique opportunities for cancer research.

13. Acknowledgments

Grant support: 134/2011 UEFISCDI Romania to AD and DOD Ovarian Cancer Academy grant W8IXWH-10-1-0525 to AMV.

14. References

Aguilera, O., M. F. Fraga, et al. (2006). "Epigenetic inactivation of the Wnt antagonist DICKKOPF-1 (DKK-1) gene in human colorectal cancer." *Oncogene* 25(29): 4116-4121.

Aguilera, O., A. Munoz, et al. (2007). "Epigenetic alterations of the Wnt/beta-catenin pathway in human disease." *Endocr Metab Immune Disord Drug Targets* 7(1): 13-21.

Al-Kuraya, K., P. Schraml, et al. (2004). "Prognostic relevance of gene amplifications and coamplifications in breast cancer." *Cancer Res* 64(23): 8534-8540.

Alazzouzi, H., V. Davalos, et al. (2005). "Mechanisms of inactivation of the receptor tyrosine kinase EPHB2 in colorectal tumors." *Cancer Res* 65(22): 10170-10173.

Ali, B. R., S. Jeffery, et al. (2007). "Novel Robinow syndrome causing mutations in the proximal region of the frizzled-like domain of ROR2 are retained in the endoplasmic reticulum." *Hum Genet* 122(3-4): 389-395.

Alimandi, M., A. Romano, et al. (1995). "Cooperative signaling of ErbB3 and ErbB2 in neoplastic transformation and human mammary carcinomas." *Oncogene* 10(9): 1813-1821.

Allen, M. P., D. A. Linseman, et al. (2002). "Novel mechanism for gonadotropin-releasing hormone neuronal migration involving Gas6/Ark signaling to p38 mitogen-activated protein kinase." *Mol Cell Biol* 22(2): 599-613.

Andres, A. C., H. H. Reid, et al. (1994). "Expression of two novel eph-related receptor protein tyrosine kinases in mammary gland development and carcinogenesis." *Oncogene* 9(5): 1461-1467.

Angers, S. and R. T. Moon (2009). "Proximal events in Wnt signal transduction." *Nat Rev Mol Cell Biol* 10(7): 468-477.

Angrisano, T., S. Sacchetti, et al. (2011). "Chromatin and DNA methylation dynamics during retinoic acid-induced RET gene transcriptional activation in neuroblastoma cells." *Nucleic Acids Res* 39(6): 1993-2006.

Bardelli, A., D. W. Parsons, et al. (2003). "Mutational analysis of the tyrosine kinome in colorectal cancers." *Science* 300(5621): 949.

Batlle, E., J. Bacani, et al. (2005). "EphB receptor activity suppresses colorectal cancer progression." *Nature* 435(7045): 1126-1130.

Berclaz, G., H. J. Altermatt, et al. (2001). "Estrogen dependent expression of the receptor tyrosine kinase axl in normal and malignant human breast." *Ann Oncol* 12(6): 819-824.

Bestor, T. H. (2000). "The DNA methyltransferases of mammals." *Hum Mol Genet* 9(16): 2395-2402.

Billiard, J., D. S. Way, et al. (2005). "The orphan receptor tyrosine kinase Ror2 modulates canonical Wnt signaling in osteoblastic cells." *Mol Endocrinol* 19(1): 90-101.

Birchmeier, C., W. Birchmeier, et al. (2003). "Met, metastasis, motility and more." *Nat Rev Mol Cell Biol* 4(12): 915-925.

Bird, A. (2002). "DNA methylation patterns and epigenetic memory." *Genes Dev* 16(1): 6-21.

Boocock, C. A., D. S. Charnock-Jones, et al. (1995). "Expression of vascular endothelial growth factor and its receptors flt and KDR in ovarian carcinoma." *J Natl Cancer Inst* 87(7): 506-516.

Brantley-Sieders, D. M., G. Zhuang, et al. (2008). "The receptor tyrosine kinase EphA2 promotes mammary adenocarcinoma tumorigenesis and metastatic progression in mice by amplifying ErbB2 signaling." *J Clin Invest* 118(1): 64-78.

Bruckner, K. and R. Klein (1998). "Signaling by Eph receptors and their ephrin ligands." *Curr Opin Neurobiol* 8(3): 375-382.

Bublil, E. M. and Y. Yarden (2007). "The EGF receptor family: spearheading a merger of signaling and therapeutics." *Curr Opin Cell Biol* 19(2): 124-134.

Budagian, V., E. Bulanova, et al. (2005). "A promiscuous liaison between IL-15 receptor and Axl receptor tyrosine kinase in cell death control." *EMBO J* 24(24): 4260-4270.

Carapancea, M., D. Cosaceanu, et al. (2007). "Dual targeting of IGF-1R and PDGFR inhibits proliferation in high-grade gliomas cells and induces radiosensitivity in JNK-1 expressing cells." *J Neurooncol* 85(3): 245-254.

Carmeliet, P. and R. K. Jain (2000). "Angiogenesis in cancer and other diseases." *Nature* 407(6801): 249-257.

Carvalho, I., F. Milanezi, et al. (2005). "Overexpression of platelet-derived growth factor receptor alpha in breast cancer is associated with tumour progression." *Breast Cancer Res* 7(5): R788-795.

Ceccherini, I., R. Bocciardi, et al. (1993). "Exon structure and flanking intronic sequences of the human RET proto-oncogene." *Biochem Biophys Res Commun* 196(3): 1288-1295.

Chan, T. A., S. Glockner, et al. (2008). "Convergence of mutation and epigenetic alterations identifies common genes in cancer that predict for poor prognosis." *PLoS Med* 5(5): e114.

Chappell, S. A., T. Walsh, et al. (1997). "Loss of heterozygosity at the mannose 6-phosphate insulin-like growth factor 2 receptor gene correlates with poor differentiation in early breast carcinomas." *Br J Cancer* 76(12): 1558-1561.

Chott, A., Z. Sun, et al. (1999). "Tyrosine kinases expressed in vivo by human prostate cancer bone marrow metastases and loss of the type 1 insulin-like growth factor receptor." *Am J Pathol* 155(4): 1271-1279.

Chung, B. I., S. B. Malkowicz, et al. (2003). "Expression of the proto-oncogene Axl in renal cell carcinoma." *DNA Cell Biol* 22(8): 533-540.

Citri, A. and Y. Yarden (2006). "EGF-ERBB signalling: towards the systems level." *Nat Rev Mol Cell Biol* 7(7): 505-516.

Collett, G. D., A. P. Sage, et al. (2007). "Axl/phosphatidylinositol 3-kinase signaling inhibits mineral deposition by vascular smooth muscle cells." *Circ Res* 100(4): 502-509.

Cosaceanu, D., R. A. Budiu, et al. (2007). "Ionizing radiation activates IGF-1R triggering a cytoprotective signaling by interfering with Ku-DNA binding and by modulating Ku86 expression via a p38 kinase-dependent mechanism." *Oncogene* 26(17): 2423-2434.

Craven, R. J., L. H. Xu, et al. (1995). "Receptor tyrosine kinases expressed in metastatic colon cancer." *Int J Cancer* 60(6): 791-797.

Daneshmanesh, A. H., E. Mikaelsson, et al. (2008). "Ror1, a cell surface receptor tyrosine kinase is expressed in chronic lymphocytic leukemia and may serve as a putative target for therapy." *Int J Cancer* 123(5): 1190-1195.

Das, P. M., A. D. Thor, et al. (2010). "Reactivation of epigenetically silenced HER4/ERBB4 results in apoptosis of breast tumor cells." *Oncogene* 29(37): 5214-5219.

Datta, J., H. Kutay, et al. (2008). "Methylation mediated silencing of MicroRNA-1 gene and its role in hepatocellular carcinogenesis." *Cancer Res* 68(13): 5049-5058.

Davalos, V., H. Dopeso, et al. (2006). "EPHB4 and survival of colorectal cancer patients." *Cancer Res* 66(18): 8943-8948.

de Jong, J. S., P. J. van Diest, et al. (2001). "Expression of growth factors, growth factor receptors and apoptosis related proteins in invasive breast cancer: relation to apoptotic rate." *Breast Cancer Res Treat* 66(3): 201-208.

Detich, N., S. Hamm, et al. (2003). "The methyl donor S-Adenosylmethionine inhibits active demethylation of DNA: a candidate novel mechanism for the pharmacological effects of S-Adenosylmethionine." *J Biol Chem* 278(23): 20812-20820.

Dong, Y., L. Jia, et al. (2011). "Selective inhibition of PDGFR by imatinib elicits the sustained activation of ERK and downstream receptor signaling in malignant glioma cells." *Int J Oncol* 38(2): 555-569.

Dottori, M., M. Down, et al. (1999). "Cloning and characterization of EphA3 (Hek) gene promoter: DNA methylation regulates expression in hematopoietic tumor cells." *Blood* 94(7): 2477-2486.

Ehrlich, M., M. A. Gama-Sosa, et al. (1982). "Amount and distribution of 5-methylcytosine in human DNA from different types of tissues of cells." *Nucleic Acids Res* 10(8): 2709-2721.

Enomoto, M., S. Hayakawa, et al. (2009). "Autonomous regulation of osteosarcoma cell invasiveness by Wnt5a/Ror2 signaling." *Oncogene* 28(36): 3197-3208.

Esteller, M. (2006). "The necessity of a human epigenome project." *Carcinogenesis* 27(6): 1121-1125.

Esteller, M. (2008). "Epigenetics in evolution and disease." *Lancet*

Fang, W. B., D. M. Brantley-Sieders, et al. (2005). "A kinase-dependent role for EphA2 receptor in promoting tumor growth and metastasis." *Oncogene* 24(53): 7859-7868.

Feinberg, A. P. and B. Tycko (2004). "The history of cancer epigenetics." *Nat Rev Cancer* 4(2): 143-153.

Feltus, F. A., E. K. Lee, et al. (2003). "Predicting aberrant CpG island methylation." *Proc Natl Acad Sci U S A* 100(21): 12253-12258.

Ferrara, N. (2005). "VEGF as a therapeutic target in cancer." *Oncology* 69 Suppl 3: 11-16.

Ferrer, F. A., L. J. Miller, et al. (1999). "Expression of vascular endothelial growth factor receptors in human prostate cancer." *Urology* 54(3): 567-572.

Fleming, T. P., A. Saxena, et al. (1992). "Amplification and/or overexpression of platelet-derived growth factor receptors and epidermal growth factor receptor in human glial tumors." *Cancer Res* 52(16): 4550-4553.

Forrester, W. C. (2002). "The Ror receptor tyrosine kinase family." *Cell Mol Life Sci* 59(1): 83-96.

Foubert, P., J. S. Silvestre, et al. (2007). "PSGL-1-mediated activation of EphB4 increases the proangiogenic potential of endothelial progenitor cells." *J Clin Invest* 117(6): 1527-1537.

Fox, B. P. and R. P. Kandpal (2004). "Invasiveness of breast carcinoma cells and transcript profile: Eph receptors and ephrin ligands as molecular markers of potential diagnostic and prognostic application." *Biochem Biophys Res Commun* 318(4): 882-892.

Fox, B. P. and R. P. Kandpal (2006). "Transcriptional silencing of EphB6 receptor tyrosine kinase in invasive breast carcinoma cells and detection of methylated promoter by methylation specific PCR." *Biochem Biophys Res Commun* 340(1): 268-276.

Fox, B. P., C. J. Tabone, et al. (2006). "Potential clinical relevance of Eph receptors and ephrin ligands expressed in prostate carcinoma cell lines." *Biochem Biophys Res Commun* 342(4): 1263-1272.

Fu, D. Y., Z. M. Wang, et al. (2010). "Frequent epigenetic inactivation of the receptor tyrosine kinase EphA5 by promoter methylation in human breast cancer." *Hum Pathol* 41(1): 48-58.

Fukuda, T., D. Lu, et al. (2004). "Restricted Expression of the Orphan Tyrosine Kinase Receptor ROR1 in Chronic Lymphocytic Leukemia." *ASH Annual Meeting Abstracts* 104(11): 772-.

Fuso, A., R. A. Cavallaro, et al. (2001). "Gene silencing by S-adenosylmethionine in muscle differentiation." *FEBS Lett* 508(3): 337-340.

Gallicchio, M., S. Mitola, et al. (2005). "Inhibition of vascular endothelial growth factor receptor 2-mediated endothelial cell activation by Axl tyrosine kinase receptor." *Blood* 105(5): 1970-1976.

Gebhardt, F., K. S. Zanker, et al. (1999). "Modulation of epidermal growth factor receptor gene transcription by a polymorphic dinucleotide repeat in intron 1." *J Biol Chem* 274(19): 13176-13180.

Ghosh, P., N. M. Dahms, et al. (2003). "Mannose 6-phosphate receptors: new twists in the tale." *Nat Rev Mol Cell Biol* 4(3): 202-212.

Glass, D. J., D. C. Bowen, et al. (1996). "Agrin acts via a MuSK receptor complex." *Cell* 85(4): 513-523.

Goldberg, A. D., C. D. Allis, et al. (2007). "Epigenetics: a landscape takes shape." *Cell* 128(4): 635-638.

Grotzer, M. A., A. J. Janss, et al. (2000). "TrkC expression predicts good clinical outcome in primitive neuroectodermal brain tumors." *J Clin Oncol* 18(5): 1027-1035.

Guo, H., H. Miao, et al. (2006). "Disruption of EphA2 receptor tyrosine kinase leads to increased susceptibility to carcinogenesis in mouse skin." *Cancer Res* 66(14): 7050-7058.

Hafizi, S. and B. Dahlback (2006). "Signalling and functional diversity within the Axl subfamily of receptor tyrosine kinases." *Cytokine Growth Factor Rev* 17(4): 295-304.

Hafner, C., B. Becker, et al. (2006). "Expression profile of Eph receptors and ephrin ligands in human skin and downregulation of EphA1 in nonmelanoma skin cancer." *Mod Pathol* 19(10): 1369-1377.

Hafner, C., G. Schmitz, et al. (2004). "Differential gene expression of Eph receptors and ephrins in benign human tissues and cancers." *Clin Chem* 50(3): 490-499.

Hattori, M., H. Sakamoto, et al. (2001). "DNA demethylase is expressed in ovarian cancers and the expression correlates with demethylation of CpG sites in the promoter region of c-erbB-2 and survivin genes." *Cancer Lett* 169(2): 155-164.

Hellawell, G. O., G. D. Turner, et al. (2002). "Expression of the type 1 insulin-like growth factor receptor is up-regulated in primary prostate cancer and commonly persists in metastatic disease." *Cancer Res* 62(10): 2942-2950.

Hermanson, M., K. Funa, et al. (1992). "Platelet-derived growth factor and its receptors in human glioma tissue: expression of messenger RNA and protein suggests the presence of autocrine and paracrine loops." *Cancer Res* 52(11): 3213-3219.

Himanen, J. P. and D. B. Nikolov (2003). "Eph receptors and ephrins." *Int J Biochem Cell Biol* 35(2): 130-134.

Holland, S. J., M. J. Powell, et al. (2005). "Multiple roles for the receptor tyrosine kinase axl in tumor formation." *Cancer Res* 65(20): 9294-9303.

Hong, C. C., J. D. Lay, et al. (2008). "Receptor tyrosine kinase AXL is induced by chemotherapy drugs and overexpression of AXL confers drug resistance in acute myeloid leukemia." *Cancer Lett* 268(2): 314-324.

Huang, E. J. and L. F. Reichardt (2003). "Trk receptors: roles in neuronal signal transduction." *Annu Rev Biochem* 72: 609-642.

Hutterer, M., P. Knyazev, et al. (2008). "Axl and growth arrest-specific gene 6 are frequently overexpressed in human gliomas and predict poor prognosis in patients with glioblastoma multiforme." *Clin Cancer Res* 14(1): 130-138.

Huusko, P., D. Ponciano-Jackson, et al. (2004). "Nonsense-mediated decay microarray analysis identifies mutations of EPHB2 in human prostate cancer." *Nat Genet* 36(9): 979-983.

Ito, T., M. Ito, et al. (1999). "Expression of the Axl receptor tyrosine kinase in human thyroid carcinoma." *Thyroid* 9(6): 563-567.

Jamieson, T. A. (2003). "M6P/IGF2R loss of heterozygosity in head and neck cancer associated with poor patient prognosis." *BMC Cancer* 3:4.

Jhiang, S. M., J. E. Sagartz, et al. (1996). "Targeted expression of the ret/PTC1 oncogene induces papillary thyroid carcinomas." *Endocrinology* 137(1): 375-378.

Jin, W., G. M. Kim, et al. (2010). "TrkC plays an essential role in breast tumor growth and metastasis." *Carcinogenesis* 31(11): 1939-1947.

Jin, W., J. J. Lee, et al. (2011). "DNA methylation-dependent regulation of TrkA, TrkB, and TrkC genes in human hepatocellular carcinoma." *Biochem Biophys Res Commun* 406(1): 89-95.

Joensuu, H., M. Puputti, et al. (2005). "Amplification of genes encoding KIT, PDGFRalpha and VEGFR2 receptor tyrosine kinases is frequent in glioblastoma multiforme." *J Pathol* 207(2): 224-231.

Joosten, P. H., M. Toepoel, et al. (2001). "Promoter haplotype combinations of the platelet-derived growth factor alpha-receptor gene predispose to human neural tube defects." *Nat Genet* 27(2): 215-217.

Jun, H. J., S. Woolfenden, et al. (2009). "Epigenetic regulation of c-ROS receptor tyrosine kinase expression in malignant gliomas." *Cancer Res* 69(6): 2180-2184.

Kageyama, R., G. T. Merlino, et al. (1988). "A transcription factor active on the epidermal growth factor receptor gene." *Proc Natl Acad Sci U S A* 85(14): 5016-5020.

Kanazawa, T., M. Nokubi, et al. (2011). "Atypical carcinoid of the larynx and expressions of proteins associated with molecular targeted therapy." *Auris Nasus Larynx* 38(1): 123-126.

Kanter-Lewensohn, L., A. Dricu, et al. (2000). "Expression of insulin-like growth factor-1 receptor (IGF-1R) and p27Kip1 in melanocytic tumors: a potential regulatory role of IGF-1 pathway in distribution of p27Kip1 between different cyclins." *Growth Factors* 17(3): 193-202.

Kim, J. Y., J. H. Hwang, et al. (2009). "The expression of VEGF receptor genes is concurrently influenced by epigenetic gene silencing of the genes and VEGF activation." *Epigenetics* 4(5): 313-321.

Kinch, M. S., M. B. Moore, et al. (2003). "Predictive value of the EphA2 receptor tyrosine kinase in lung cancer recurrence and survival." *Clin Cancer Res* 9(2): 613-618.

Klarmann, G. J., A. Decker, et al. (2008). "Epigenetic gene silencing in the Wnt pathway in breast cancer." *Epigenetics* 3(2): 59-63.

Klein, U., Y. Tu, et al. (2001). "Gene expression profiling of B cell chronic lymphocytic leukemia reveals a homogeneous phenotype related to memory B cells." *J Exp Med* 194(11): 1625-1638.

Kmieciak, M., K. L. Knutson, et al. (2007). "HER-2/neu antigen loss and relapse of mammary carcinoma are actively induced by T cell-mediated anti-tumor immune responses." *Eur J Immunol* 37(3): 675-685.

Kobayashi, H., K. Sugihara, et al. (2008). "Messenger RNA expression of vascular endothelial growth factor and its receptors in primary colorectal cancer and corresponding liver metastasis." *Ann Surg Oncol* 15(4): 1232-1238.

Kobayashi, M., Y. Shibuya, et al. (2009). "Ror2 expression in squamous cell carcinoma and epithelial dysplasia of the oral cavity." *Oral Surg Oral Med Oral Pathol Oral Radiol Endod* 107(3): 398-406.

Kong, F. M., M. S. Anscher, et al. (2000). "M6P/IGF2R is mutated in squamous cell carcinoma of the lung." *Oncogene* 19(12): 1572-1578.

Korshunov, V. A., A. M. Mohan, et al. (2006). "Axl, a receptor tyrosine kinase, mediates flow-induced vascular remodeling." *Circ Res* 98(11): 1446-1452.

Kuang, S. Q., H. Bai, et al. (2010). "Aberrant DNA methylation and epigenetic inactivation of Eph receptor tyrosine kinases and ephrin ligands in acute lymphoblastic leukemia." *Blood* 115(12): 2412-2419.

Kulis, M. and M. Esteller (2010). "DNA methylation and cancer." *Adv Genet* 70: 27-56.

Kullander, K. and R. Klein (2002). "Mechanisms and functions of Eph and ephrin signalling." *Nat Rev Mol Cell Biol* 3(7): 475-486.

Kuniyasu, H., W. Yasui, et al. (1992). "Frequent amplification of the c-met gene in scirrhous type stomach cancer." *Biochem Biophys Res Commun* 189(1): 227-232.

Lai, C. and G. Lemke (1991). "An extended family of protein-tyrosine kinase genes differentially expressed in the vertebrate nervous system." *Neuron* 6(5): 691-704.

Lara, E., V. Calvanese, et al. (2010). "Epigenetic repression of ROR2 has a Wnt-mediated, pro-tumourigenic role in colon cancer." *Mol Cancer* 9: 170.

Lassus, H., H. Sihto, et al. (2004). "Genetic alterations and protein expression of KIT and PDGFRA in serous ovarian carcinoma." *Br J Cancer* 91(12): 2048-2055.

Lee, J., J. W. Seo, et al. (2011). "Impact of MET amplification on gastric cancer: possible roles as a novel prognostic marker and a potential therapeutic target." *Oncol Rep* 25(6): 1517-1524.

Lehmann, U., F. Langer, et al. (2002). "Quantitative assessment of promoter hypermethylation during breast cancer development." *Am J Pathol* 160(2): 605-612.

Lemmon, M. A. and J. Schlessinger (2010). "Cell signaling by receptor tyrosine kinases." *Cell* 141(7): 1117-1134.

Li, B., C. M. Chang, et al. (2003). "Resistance to small molecule inhibitors of epidermal growth factor receptor in malignant gliomas." *Cancer Res* 63(21): 7443-7450.

Li, C. (2008). "Ror2 modulates the canonical Wnt signaling in lung epithelial cells through cooperation with Fzd2." *BMC Mol Biol.* 23: 9-11.

Li, Y., X. Ye, et al. (2009). "Axl as a potential therapeutic target in cancer: role of Axl in tumor growth, metastasis and angiogenesis." *Oncogene* 28(39): 3442-3455.

Liu, L., J. Greger, et al. (2009). "Novel mechanism of lapatinib resistance in HER2-positive breast tumor cells: activation of AXL." *Cancer Res* 69(17): 6871-6878.

Liu, R., M. Gong, et al. (2010). "Induction, regulation, and biologic function of Axl receptor tyrosine kinase in Kaposi sarcoma." *Blood* 116(2): 297-305.

Liu, W., S. A. Ahmad, et al. (2002). "Coexpression of ephrin-Bs and their receptors in colon carcinoma." *Cancer* 94(4): 934-939.

Liu, Y., R. A. Bhat, et al. (2007). "The orphan receptor tyrosine kinase Ror2 promotes osteoblast differentiation and enhances ex vivo bone formation." *Mol Endocrinol* 21(2): 376-387.

Lokker, N. A., C. M. Sullivan, et al. (2002). "Platelet-derived growth factor (PDGF) autocrine signaling regulates survival and mitogenic pathways in glioblastoma cells: evidence that the novel PDGF-C and PDGF-D ligands may play a role in the development of brain tumors." *Cancer Res* 62(13): 3729-3735.

MacLeod, R. J., M. Hayes, et al. (2007). "Wnt5a secretion stimulated by the extracellular calcium-sensing receptor inhibits defective Wnt signaling in colon cancer cells." *Am J Physiol Gastrointest Liver Physiol* 293(1): G403-411.

Mahadevan, D., L. Cooke, et al. (2007). "A novel tyrosine kinase switch is a mechanism of imatinib resistance in gastrointestinal stromal tumors." *Oncogene* 26(27): 3909-3919.

Manning, G., D. B. Whyte, et al. (2002). "The protein kinase complement of the human genome." *Science* 298(5600): 1912-1934.

Martin-Zanca, D., S. H. Hughes, et al. (1986). "A human oncogene formed by the fusion of truncated tropomyosin and protein tyrosine kinase sequences." *Nature* 319(6056): 743-748.

Masiakowski, P. and R. D. Carroll (1992). "A novel family of cell surface receptors with tyrosine kinase-like domain." *J Biol Chem* 267(36): 26181-26190.

Matsuda, T., M. Nomi, et al. (2001). "Expression of the receptor tyrosine kinase genes, Ror1 and Ror2, during mouse development." *Mech Dev* 105(1-2): 153-156.

Maulik, G., A. Shrikhande, et al. (2002). "Role of the hepatocyte growth factor receptor, c-Met, in oncogenesis and potential for therapeutic inhibition." *Cytokine Growth Factor Rev* 13(1): 41-59.

Mazieres, J., B. He, et al. (2004). "Wnt inhibitory factor-1 is silenced by promoter hypermethylation in human lung cancer." *Cancer Res* 64(14): 4717-4720.

McDonald, S. L. and A. Silver (2009). "The opposing roles of Wnt-5a in cancer." *Br J Cancer* 101(2): 209-214.

Meng, X., M. Lindahl, et al. (2000). "Regulation of cell fate decision of undifferentiated spermatogonia by GDNF." *Science* 287(5457): 1489-1493.

Meric, F., W. P. Lee, et al. (2002). "Expression profile of tyrosine kinases in breast cancer." *Clin Cancer Res* 8(2): 361-367.

Merlino, G. T., S. Ishii, et al. (1985). "Structure and localization of genes encoding aberrant and normal epidermal growth factor receptor RNAs from A431 human carcinoma cells." *Mol Cell Biol* 5(7): 1722-1734.

Merlos-Suarez, A. and E. Batlle (2008). "Eph-ephrin signalling in adult tissues and cancer." *Curr Opin Cell Biol* 20(2): 194-200.

Mikels, A. J. and R. Nusse (2006). "Purified Wnt5a protein activates or inhibits beta-catenin-TCF signaling depending on receptor context." *PLoS Biol* 4(4): e115.

Montero, A. J., C. M. Diaz-Montero, et al. (2006). "Epigenetic inactivation of EGFR by CpG island hypermethylation in cancer." *Cancer Biol Ther* 5(11): 1494-1501.

Morishita, K., D. E. Johnson, et al. (1995). "A novel promoter for vascular endothelial growth factor receptor (flt-1) that confers endothelial-specific gene expression." *J Biol Chem* 270(46): 27948-27953.

Morozov, V. M., N. A. Massoll, et al. (2008). "Regulation of c-met expression by transcription repressor Daxx." *Oncogene* 27(15): 2177-2186.

Moscatello, D. K., M. Holgado-Madruga, et al. (1995). "Frequent expression of a mutant epidermal growth factor receptor in multiple human tumors." *Cancer Res* 55(23): 5536-5539.

Mudduluru, G. and H. Allgayer (2008). "The human receptor tyrosine kinase Axl gene--promoter characterization and regulation of constitutive expression by Sp1, Sp3 and CpG methylation." *Biosci Rep* 28(3): 161-176.

Mudduluru, G., P. Ceppi, et al. (2011). "Regulation of Axl receptor tyrosine kinase expression by miR-34a and miR-199a/b in solid cancer." *Oncogene* 30(25): 2888-2899.

Mudduluru, G., P. Vajkoczy, et al. (2010). "Myeloid zinc finger 1 induces migration, invasion, and in vivo metastasis through Axl gene expression in solid cancer." *Mol Cancer Res* 8(2): 159-169.

Munier-Lehmann, H., F. Mauxion, et al. (1996). "Re-expression of the mannose 6-phosphate receptors in receptor-deficient fibroblasts. Complementary function of the two mannose 6-phosphate receptors in lysosomal enzyme targeting." *J Biol Chem* 271(25): 15166-15174.

Munnes, M., G. Patrone, et al. (1998). "A 5'-CG-3'-rich region in the promoter of the transcriptionally frequently silenced RET protooncogene lacks methylated cytidine residues." *Oncogene* 17(20): 2573-2583.

Nakada, M., K. L. Drake, et al. (2006). "Ephrin-B3 ligand promotes glioma invasion through activation of Rac1." *Cancer Res* 66(17): 8492-8500.

Nakagawara, A. (2001). "Trk receptor tyrosine kinases: a bridge between cancer and neural development." *Cancer Lett* 169(2): 107-114.

Noren, N. K., G. Foos, et al. (2006). "The EphB4 receptor suppresses breast cancer cell tumorigenicity through an Abl-Crk pathway." *Nat Cell Biol* 8(8): 815-825.

Nosho, K., H. Yamamoto, et al. (2007). "Genetic and epigenetic profiling in early colorectal tumors and prediction of invasive potential in pT1 (early invasive) colorectal cancers." *Carcinogenesis* 28(6): 1364-1370.

O'Bryan, J. P., R. A. Frye, et al. (1991). "axl, a transforming gene isolated from primary human myeloid leukemia cells, encodes a novel receptor tyrosine kinase." *Mol Cell Biol* 11(10): 5016-5031.

Oishi, I., H. Suzuki, et al. (2003). "The receptor tyrosine kinase Ror2 is involved in non-canonical Wnt5a/JNK signalling pathway." *Genes Cells* 8(7): 645-654.

Oka, Y., R. A. Waterland, et al. (2002). "M6P/IGF2R tumor suppressor gene mutated in hepatocellular carcinomas in Japan." *Hepatology* 35(5): 1153-1163.

Ola, R. (2010). "CpG Island Promotor of Igf-1r is Partial Methylated in Non Small Cell Lung Cancer." *Journal of Radiotherapy and Medical Oncology* XVI(3): 140-147.

Ola, R. (2010). "S-Adenosylmethionine - induced cytotoxicity in glioblastoma cells does not affect the Igf-1r methylation." *Revista Romana de Medicina de Laborator* 18(3/4): 51-59.

Ooi, A., T. Takehana, et al. (2004). "Protein overexpression and gene amplification of HER-2 and EGFR in colorectal cancers: an immunohistochemical and fluorescent in situ hybridization study." *Mod Pathol* 17(8): 895-904.

Pasquale, E. B. (2004). "Eph-ephrin promiscuity is now crystal clear." *Nat Neurosci* 7(5): 417-418.

Pelet, A., T. Attie, et al. (1994). "De-novo mutations of the RET proto-oncogene in Hirschsprung's disease." *Lancet* 344(8939-8940): 1769-1770.

Peruzzi, B. and D. P. Bottaro (2006). "Targeting the c-Met signaling pathway in cancer." *Clin Cancer Res* 12(12): 3657-3660.

Petrangeli, E., C. Lubrano, et al. (1995). "Gene methylation of oestrogen and epidermal growth factor receptors in neoplastic and perineoplastic breast tissues." *Br J Cancer* 72(4): 973-975.

Popsueva, A., D. Poteryaev, et al. (2003). "GDNF promotes tubulogenesis of GFRalpha1-expressing MDCK cells by Src-mediated phosphorylation of Met receptor tyrosine kinase." *J Cell Biol* 161(1): 119-129.

Probst, O. C., V. Puxbaum, et al. (2009). "The mannose 6-phosphate/insulin-like growth factor II receptor restricts the tumourigenicity and invasiveness of squamous cell carcinoma cells." *Int J Cancer* 124(11): 2559-2567.

Reddy, U. R., S. Phatak, et al. (1996). "Human neural tissues express a truncated Ror1 receptor tyrosine kinase, lacking both extracellular and transmembrane domains." *Oncogene* 13(7): 1555-1559.

Reik, W., W. Dean, et al. (2001). "Epigenetic reprogramming in mammalian development." *Science* 293(5532): 1089-1093.

Robinson, D. R., Y. M. Wu, et al. (2000). "The protein tyrosine kinase family of the human genome." *Oncogene* 19(49): 5548-5557.

Roman-Gomez, J., A. Jimenez-Velasco, et al. (2007). "WNT5A, a putative tumour suppressor of lymphoid malignancies, is inactivated by aberrant methylation in acute lymphoblastic leukaemia." *Eur J Cancer* 43(18): 2736-2746.

Rothlin, C. V., S. Ghosh, et al. (2007). "TAM receptors are pleiotropic inhibitors of the innate immune response." *Cell* 131(6): 1124-1136.

Sainaghi, P. P., L. Castello, et al. (2005). "Gas6 induces proliferation in prostate carcinoma cell lines expressing the Axl receptor." *J Cell Physiol* 204(1): 36-44.

Sammar, M., S. Stricker, et al. (2004). "Modulation of GDF5/BRI-b signalling through interaction with the tyrosine kinase receptor Ror2." *Genes Cells* 9(12): 1227-1238.

Sariola, H. and M. Saarma (2003). "Novel functions and signalling pathways for GDNF." *J Cell Sci* 116(Pt 19): 3855-3862.

Scarpino, S., A. Di Napoli, et al. (2004). "Papillary carcinoma of the thyroid: methylation is not involved in the regulation of MET expression." *Br J Cancer* 91(4): 703-706.

Scartozzi, M., I. Bearzi, et al. (2011). "Epidermal growth factor receptor (EGFR) gene promoter methylation and cetuximab treatment in colorectal cancer patients." *Br J Cancer* 104(11): 1786-1790.

Schambony, A. and D. Wedlich (2007). "Wnt-5A/Ror2 regulate expression of XPAPC through an alternative noncanonical signaling pathway." *Dev Cell* 12(5): 779-792.

Schayek, H., I. Bentov, et al. (2010). "Progression to metastatic stage in a cellular model of prostate cancer is associated with methylation of the androgen receptor gene and transcriptional suppression of the insulin-like growth factor-I receptor gene." *Exp Cell Res* 316(9): 1479-1488.

Schuchardt, A., V. D'Agati, et al. (1994). "Defects in the kidney and enteric nervous system of mice lacking the tyrosine kinase receptor Ret." *Nature* 367(6461): 380-383.

Sclabas, G. M., S. Fujioka, et al. (2005). "Overexpression of tropomysin-related kinase B in metastatic human pancreatic cancer cells." *Clin Cancer Res* 11(2 Pt 1): 440-449.

Segal, R. A. (2003). "Selectivity in neurotrophin signaling: theme and variations." *Annu Rev Neurosci* 26: 299-330.

Segal, R. A., L. C. Goumnerova, et al. (1994). "Expression of the neurotrophin receptor TrkC is linked to a favorable outcome in medulloblastoma." *Proc Natl Acad Sci U S A* 91(26): 12867-12871.

Seol, D. W. and R. Zarnegar (1998). "Structural and functional characterization of the mouse c-met proto-oncogene (hepatocyte growth factor receptor) promoter." *Biochim Biophys Acta* 1395(3): 252-258.

Shawver, L. K., D. Slamon, et al. (2002). "Smart drugs: tyrosine kinase inhibitors in cancer therapy." *Cancer Cell* 1(2): 117-123.

Shieh, Y. S., C. Y. Lai, et al. (2005). "Expression of axl in lung adenocarcinoma and correlation with tumor progression." *Neoplasia* 7(12): 1058-1064.

Slominski, A. and J. Wortsman (2000). "Neuroendocrinology of the skin." *Endocr Rev* 21(5): 457-487.

Smith, N. R., D. Baker, et al. (2010). "Vascular endothelial growth factor receptors VEGFR-2 and VEGFR-3 are localized primarily to the vasculature in human primary solid cancers." *Clin Cancer Res* 16(14): 3548-3561.

Steenman, M. J., S. Rainier, et al. (1994). "Loss of imprinting of IGF2 is linked to reduced expression and abnormal methylation of H19 in Wilms' tumour." *Nat Genet* 7(3): 433-439.

Stitt, T. N., G. Conn, et al. (1995). "The anticoagulation factor protein S and its relative, Gas6, are ligands for the Tyro 3/Axl family of receptor tyrosine kinases." *Cell* 80(4): 661-670.

Stoger, R., P. Kubicka, et al. (1993). "Maternal-specific methylation of the imprinted mouse Igf2r locus identifies the expressed locus as carrying the imprinting signal." *Cell* 73(1): 61-71.

Sulzbacher, I., P. Birner, et al. (2003). "Expression of platelet-derived growth factor-AA is associated with tumor progression in osteosarcoma." *Mod Pathol* 16(1): 66-71.

Sun, W., J. Fujimoto, et al. (2004). "Coexpression of Gas6/Axl in human ovarian cancers." *Oncology* 66(6): 450-457.

Surawska, H., P. C. Ma, et al. (2004). "The role of ephrins and Eph receptors in cancer." *Cytokine Growth Factor Rev* 15(6): 419-433.

Suzuki, H., E. Gabrielson, et al. (2002). "A genomic screen for genes upregulated by demethylation and histone deacetylase inhibition in human colorectal cancer." *Nat Genet* 31(2): 141-149.

Suzuki, H., M. Toyota, et al. (2008). "Frequent epigenetic inactivation of Wnt antagonist genes in breast cancer." *Br J Cancer* 98(6): 1147-1156.

Szentirmay, M. N., H. X. Yang, et al. (2003). "The IGF2 receptor is a USF2-specific target in nontumorigenic mammary epithelial cells but not in breast cancer cells." *J Biol Chem* 278(39): 37231-37240.

Takahama, M., M. Tsutsumi, et al. (1999). "Expression of vascular endothelial growth factor and its receptors during lung carcinogenesis by N-nitrosobis(2-hydroxypropyl)amine in rats." *Mol Carcinog* 24(4): 287-293.

Toepoel, M., P. H. Joosten, et al. (2008). "Haplotype-specific expression of the human PDGFRA gene correlates with the risk of glioblastomas." *Int J Cancer* 123(2): 322-329.

Turney, B. W., G. D. Turner, et al. (2011). "Serial analysis of resected prostate cancer suggests up-regulation of type 1 IGF receptor with disease progression." *BJU Int* 107(9): 1488-1499.

Urdinguio, R. G., J. V. Sanchez-Mut, et al. (2009). "Epigenetic mechanisms in neurological diseases: genes, syndromes, and therapies." *Lancet Neurol* 8(11): 1056-1072.

Valenzuela, D. M., T. N. Stitt, et al. (1995). "Receptor tyrosine kinase specific for the skeletal muscle lineage: expression in embryonic muscle, at the neuromuscular junction, and after injury." *Neuron* 15(3): 573-584.

Varnum, B. C., C. Young, et al. (1995). "Axl receptor tyrosine kinase stimulated by the vitamin K-dependent protein encoded by growth-arrest-specific gene 6." *Nature* 373(6515): 623-626.

Vu, T. H. (1996). "Alterations in the promoter-specific imprinting of the insulin-like growth factor-II gene in Wilms' tumor." *J Biol Chem.*

Walker-Daniels, J., K. Coffman, et al. (1999). "Overexpression of the EphA2 tyrosine kinase in prostate cancer." *Prostate* 41(4): 275-280.

Wang, J., H. Kataoka, et al. (2005). "Downregulation of EphA7 by hypermethylation in colorectal cancer." *Oncogene* 24(36): 5637-5647.

Werner, H., B. Stannard, et al. (1990). "Cloning and characterization of the proximal promoter region of the rat insulin-like growth factor I (IGF-I) receptor gene." *Biochem Biophys Res Commun* 169(3): 1021-1027.

Wood, L. D., E. S. Calhoun, et al. (2006). "Somatic mutations of GUCY2F, EPHA3, and NTRK3 in human cancers." *Hum Mutat* 27(10): 1060-1061.

Wright, T. M., A. R. Brannon, et al. (2009). "Ror2, a developmentally regulated kinase, promotes tumor growth potential in renal cell carcinoma." *Oncogene* 28(27): 2513-2523.

Wu, C. W., A. F. Li, et al. (2002). "Clinical significance of AXL kinase family in gastric cancer." *Anticancer Res* 22(2B): 1071-1078.

Wutz, A., O. W. Smrzka, et al. (1997). "Imprinted expression of the Igf2r gene depends on an intronic CpG island." *Nature* 389(6652): 745-749.

Xu, Y. H., N. Richert, et al. (1984). "Characterization of epidermal growth factor receptor gene expression in malignant and normal human cell lines." *Proc Natl Acad Sci U S A* 81(23): 7308-7312.

Yamada, Y., M. Watanabe, et al. (2003). "Aberrant methylation of the vascular endothelial growth factor receptor-1 gene in prostate cancer." *Cancer Sci* 94(6): 536-539.

Yamashiro, D. J., X. G. Liu, et al. (1997). "Expression and function of Trk-C in favourable human neuroblastomas." *Eur J Cancer* 33(12): 2054-2057.

Yarden, Y. and M. X. Sliwkowski (2001). "Untangling the ErbB signalling network." *Nat Rev Mol Cell Biol* 2(2): 127-137.

Ying, J., H. Li, et al. (2008). "WNT5A exhibits tumor-suppressive activity through antagonizing the Wnt/beta-catenin signaling, and is frequently methylated in colorectal cancer." *Clin Cancer Res* 14(1): 55-61.

Yoda, A., I. Oishi, et al. (2003). "Expression and function of the Ror-family receptor tyrosine kinases during development: lessons from genetic analyses of nematodes, mice, and humans." *J Recept Signal Transduct Res* 23(1): 1-15.

Yoo, C. B. and P. A. Jones (2006). "Epigenetic therapy of cancer: past, present and future." *Nat Rev Drug Discov* 5(1): 37-50.

Zelinski, D. P., N. D. Zantek, et al. (2001). "EphA2 overexpression causes tumorigenesis of mammary epithelial cells." *Cancer Res* 61(5): 2301-2306.

Zeng, Z. S., M. R. Weiser, et al. (2008). "c-Met gene amplification is associated with advanced stage colorectal cancer and liver metastases." *Cancer Lett* 265(2): 258-269.

Zhou, H., W. D. Chen, et al. (2001). "MMTV promoter hypomethylation is linked to spontaneous and MNU associated c-neu expression and mammary carcinogenesis in MMTV c-neu transgenic mice." *Oncogene* 20(42): 6009-6017.

Permissions

The contributors of this book come from diverse backgrounds, making this book a truly international effort. This book will bring forth new frontiers with its revolutionizing research information and detailed analysis of the nascent developments around the world.

We would like to thank Dr. Tatiana Tatarinova and Dr. Owain Kerton, for lending their expertise to make the book truly unique. They have played a crucial role in the development of this book. Without their invaluable contribution this book wouldn't have been possible. They have made vital efforts to compile up to date information on the varied aspects of this subject to make this book a valuable addition to the collection of many professionals and students.

This book was conceptualized with the vision of imparting up-to-date information and advanced data in this field. To ensure the same, a matchless editorial board was set up. Every individual on the board went through rigorous rounds of assessment to prove their worth. After which they invested a large part of their time researching and compiling the most relevant data for our readers. Conferences and sessions were held from time to time between the editorial board and the contributing authors to present the data in the most comprehensible form. The editorial team has worked tirelessly to provide valuable and valid information to help people across the globe.

Every chapter published in this book has been scrutinized by our experts. Their significance has been extensively debated. The topics covered herein carry significant findings which will fuel the growth of the discipline. They may even be implemented as practical applications or may be referred to as a beginning point for another development. Chapters in this book were first published by InTech; hereby published with permission under the Creative Commons Attribution License or equivalent.

The editorial board has been involved in producing this book since its inception. They have spent rigorous hours researching and exploring the diverse topics which have resulted in the successful publishing of this book. They have passed on their knowledge of decades through this book. To expedite this challenging task, the publisher supported the team at every step. A small team of assistant editors was also appointed to further simplify the editing procedure and attain best results for the readers.

Our editorial team has been hand-picked from every corner of the world. Their multi-ethnicity adds dynamic inputs to the discussions which result in innovative outcomes. These outcomes are then further discussed with the researchers and contributors who give their valuable feedback and opinion regarding the same. The feedback is then collaborated with the researches and they are edited in a comprehensive manner to aid the understanding of the subject.

Apart from the editorial board, the designing team has also invested a significant amount of their time in understanding the subject and creating the most relevant covers. They scrutinized every image to scout for the most suitable representation of the subject and create an appropriate cover for the book.

The publishing team has been involved in this book since its early stages. They were actively engaged in every process, be it collecting the data, connecting with the contributors or procuring relevant information. The team has been an ardent support to the editorial, designing and production team. Their endless efforts to recruit the best for this project, has resulted in the accomplishment of this book. They are a veteran in the field of academics and their pool of knowledge is as vast as their experience in printing. Their expertise and guidance has proved useful at every step. Their uncompromising quality standards have made this book an exceptional effort. Their encouragement from time to time has been an inspiration for everyone.

The publisher and the editorial board hope that this book will prove to be a valuable piece of knowledge for researchers, students, practitioners and scholars across the globe.

List of Contributors

Michael Hackenberg, Guillermo Barturen and José L. Oliver
Dpto. de Genética, Facultad de Ciencias, Universidad de Granada,Granada, Lab. de Bioinformática, Inst. de Biotecnología, Centro de Investigación Biomédica, Granada, Spain

Karthika Raghavan and Heather J. Ruskin
Centre for Scientific Computing and Complex Systems Modeling (SCI SYM), School of Computing, Dublin City University, Ireland

Tomoko Takamiya
School of Pharmacy, Nihon University, Japan

Saeko Hosobuchi and Yasufumi Murakami
Tokyo University of Sciences, Japan

Hisato Okuizumi
Genetic Resources Center, National Institute of Agrobiological Sciences (NIAS), Japan

Kaliyamoorthy Seetharam
Tamil Nadu Agriculture University, India

Eran Elhaik
McKusick - Nathans Institute of Genetic Medicine, Johns Hopkins University School of Medicine, Baltimore, USA
Department of Mental Health, Johns Hopkins University Bloomberg School of Public Health, USA

Tatiana Tatarinova
Faculty of Advanced Technology, University of Glamorga, Wales

David R. McCarthy and Michelle M. Hanna
RiboMed Biotechnologies, Inc., Carlsbad, CA, USA

Philip D. Cotter
Department of Pathology, Children's Hospital and Research Center, Oakland, CA, USA

Michelle M. Hanna
Arizona Cancer Center, University of Arizona, Tucson, AZ, USA

Mark A. Pook
Division of Biosciences, School of Health Sciences and Social Care, Brunel University, Uxbridge, UK

Michael Moffat, James P. Reddington and Richard R. Meehan
MRC Human Genetics Unit, IGMM, Western General Hospital, Edinburgh, UK

Sari Pennings
Queen's Medical Research Institute, University of Edinburgh, Edinburgh, UK

Takahiro Arima, Hiroaki Okae, Hitoshi Hiura, Naoko Miyauchi and Fumi Sato
Department of Informative Genetics, Environment and Genome Research Center, Tohoku University Graduate School of Medicine, 2-1 Seiryo-cho, Aoba-ku, Sendai, Japan

Akiko Sato and Chika Hayashi
Departments of Obstetrics and Gynecology, Tohoku University Graduate School of Medicine, Sendai, Japan

Alicia Postiglioni, Rody Artigas, Andrés Iriarte, Wanda Iriarte and Nicolás Grasso
Depto. de Genética y Mejora Animal, Facultad de Veterinaria, Universidad de la República, Montevideo, Uruguay

Gonzalo Rincón
Department of Animal Science, University of California, Davis, USA

Makoto Nagai
Ishikawa Prefectural Livestock Research Center, Japan

Makiko Meguro-Horike and Shin-ichi Horike
Frontier Science Organization, Kanazawa University, Japan

Makiko Meguro-Horike
JSPS Research Fellow, Japan

Johannes F. Wentzel and Pieter J. Pretorius
Division for Biochemistry, North-West University, South Africa

Yi Qiu, Daniel Shabashvili, Xuehui Li, Priya K. Gopalan, Min Chen and Maria Zajac-Kaye
Department of Anatomy and Cell Biology and Department of Medicine, College of Medicine, University of Florida, Gainesville, Florida, USA

Kristen H. Taylor and Michael X. Wang
University of Missouri-Columbia, USA

Koraljka Gall Trošelj, Renata Novak Kujundžić and Ivana Grbeša
Rudjer Boskovic Institute, Division of Molecular Medicine, Zagreb, Croatia

Ali A. Alshatwi and Gowhar Shafi
King Saud University Riyadh, Saudi Arabia